NORMAN MYERS
SCOTT E. SPOOLMAN

ENVIRONMENTAL
Issues & Solutions

A Modular Approach

BROOKS/COLE
CENGAGE Learning™

Australia • Brazil • Japan • Korea • Mexico • Singapore • Spain • United Kingdom • United States

BROOKS/COLE
CENGAGE Learning™

**Environmental Issues and Solutions:
A Modular Approach, International Edition**
Norman Myers, Scott E. Spoolman

Publisher: Yolanda Cossio

Development Editor: Jake Warde

Assistant Editor: Alexis Glubka

Editorial Assistant: Lauren Crosby

Media Editor: Alexandria Brady

Marketing Manager: Tom Ziolkowski

Marketing Communications Manager:
Darlene Macanan

Content Project Manager: Hal Humphrey

Design Director: Rob Hugel

Art Director: John Walker

Manufacturing Planner: Karen Hunt

Rights Acquisitions Specialist: Don Scholtman

Production Service: Thompson Steele, Inc.

Photo Researcher: Abigail Reip

Text Researcher: Pablo d'Stair

Copy Editor: Deborah Thompson

Illustrator: Patrick Lane, ScEYEnce Studios, Inc.

Interior Design: Riezebos Holzbaur/Tim Heraldo

Cover Image: © Cameron
Davidson/Corbis

Compositor: Thompson Steele, Inc.

International Edition:

ISBN-13: 978-1-4354-6232-8

ISBN-10: 1-4354-6232-7

Cengage Learning International Offices

Asia
www.cengageasia.com
tel: (65) 6410 1200

Australia/New Zealand
www.cengage.com/au
tel: (61) 3 9685 4111

Brazil
www.cengage.com.br
tel: (55) 11 3665 9900

India
www.cengage.co.in
tel: (91) 11 4364 1111

Latin America
www.cengage.com.mx
tel: (52) 55 1500 6000

UK/Europe/Middle East/Africa
www.cengage.co.uk
tel: (44) 0 1264 332 424

**Represented in Canada by
Nelson Education, Ltd.**
www.nelson.com
tel: (416) 752 9100 / (800) 668 0671

Cengage Learning is a leading provider of customized learning solutions with office locations around the globe, including Singapore, the United Kingdom, Australia, Mexico, Brazil, and Japan. Locate your local office at **www.cengage.com/global**.

For product information and free companion resources:
www.cengage.com/international

Visit your local office: **www.cengage.com/global**

Visit our corporate website: **www.cengage.com**

FSC
www.fsc.org
MIX
Paper from
responsible sources
FSC® C011825

Printed in Canada
1 2 3 4 5 6 7 15 14 13 12

Environmental Issues and Solutions

Brief Contents

Contents

3 Urbanization 55

4 Food Resources 83

5 Energy Efficiency and Renewable Energy   115

6 Nonrenewable Energy 145

7 Mineral Resources 175

8 Species Extinction   201

Cigdem Sean Cooper/Shutterstock.com

9 Land Degradation   225

10 Water Resources 261

11 Water Pollution 289

12 Air Pollution 319

13 Climate Change 349

14 Wastes 381

15 Environmental Health Hazards   411

Preface

This book is focused on and organized around critical environmental issues—one in each of 14 modules. Each module describes an environmental problem or a set of related problems and gives equal weight to possible solutions to those problems. We believe that, in order to think about and apply such solutions, students need to believe that something can be done and have reasons to hope for a bright future. With that in mind, along with full descriptions of each of the major environmental problems, we have provided real stories about what people are doing to tackle the resulting challenges.

What Makes This Book Different?

Many environmental science textbooks start with several chapters introducing the basics of science, followed by another set of chapters devoted to various environmental problems. We heard from numerous instructors who would prefer a book that gets more quickly into specific environmental issues while providing enough background in science to enable students to understand each issue. We also found that flexibility in a textbook is an important priority because of the many different ways in which introductory environmental science courses are organized.

To help you meet these needs, we have taken a new and different approach in writing this book. First, we have focused sharply on *issues and solutions*, and second, we have written a *modular textbook*, with 14 of the total 15 modules each devoted entirely to a specific environmental issue. The first module is an introduction to science and sustainability concepts that form a foundation for the course. We strongly recommend that each course include this module as an introduction. Then, each of the other 14 issues-oriented modules is designed to stand on its own, so that students can read and understand it without having to read any other modules, except for the first

one. Thus, with this book, you can use any number of the 14 issues modules in any order desired without missing important content, or you can use them all.

This textbook also provides *frequent reviews* of the material. Each module has several brief question lists entitled "Let's Review," placed at regular intervals throughout the module. These lists give students a chance to review and reinforce material immediately after they have studied it. Also, taken together, these questions provide a quick review of an entire module, which makes them useful for preparing for quizzes and tests.

While this book can help your students review and reinforce reading material, it will also allow them to apply the material they have studied through *critical thinking activities* that appear at the end of each module. These activities are designed to prompt students to re-read parts of the module, to think about what they have read, and to apply it in some way to their own daily lives.

How Is This Book Organized?

The first module, "Environmental Science and Sustainability," introduces and explains the fundamentals of environmental science that are necessary for understanding the material presented in the rest of the modules. It also introduces students to *sustainability*, an essential concept underlying the content of all the modules. It is our view that environmental problems arise from unsustainable use of the earth's natural resources and life-support systems. By focusing on how to use these resources and systems more sustainably, we believe that we can find and apply solutions to our environmental problems.

Modules 2–14 address the issues of human population growth, urbanization, food production, energy efficiency and renewable energy resources, nonrenewable energy resources, mineral resources, species extinction, land degradation, water resources, water pollution, air pollution, climate change, wastes, and environmental health hazards. Each of these modules is designed to stand on its own and to be used independently of all other issues modules. From the first module, the basic science concepts that apply are raised again and applied in each issues module.

Important note: because each of the 14 modules stands on its own, various sections of necessary background material appear in more than one module. Students who run across such duplication can review and reinforce materials they have learned. They will also see how this basic background science material, including the principles of sustainability, applies to many diverse environmental issues.

How Is Each Module Organized?

Why Should You Care about Sustainability?	**2**
What Do You Need to Know?	**2**
What Are the Problems?	**21**
What Can Be Done?	**24**
A Look to the Future	**28**
What Would You Do?	**29**

Each module begins with the question "Why Should You Care?" We apply this question to the issue covered in the module and then suggest some reasons why readers might care about the issue (see pp. 56, 176, and 320). We then turn to a section titled "What Do You Need to Know?" This is where we include the additional science background that the reader will need in order to understand the rest of the material in the module and to begin thinking about solutions to the problems presented (see pp. 84, 262, and 382).

The next section is entitled "What Are the Problems?" Here we explore the particular environmental issue addressed in the module (see pp. 59, 155, and 239). This section is based on research and data available from scientific journals, government agencies, news outlets, businesses, and other sources of pertinent information. We strive to be up-to-date in reporting data, and have inserted the most recent information into the modules as late as possible in the book production process.

The problems section is followed by a section called "What Can Be Done?" Here, we examine proposed solutions to environmental problems as well as solutions that have been tried and shown to be useful (see pp. 44, 103, and 278). There is plenty of good news in the area of environmental issues—stories of people who are designing and trying new products, technologies, and ways of doing business that are helping us to be more sustainable in how we live and work.

The next section, "A Look to the Future," is intended to get students to think about what lies down the road with regard to the issue they are studying. In some cases, this section is more focused on a particular problem that will grow more troublesome in the future (see pp. 51, 316, and 378). In other cases, this section deals more with a promising new practice or technology that could help us to live more sustainably, especially relating to the issue at hand in the module (see pp. 80, 198, and 285).

The last text section is called "What Would You Do?" This section is intended to help students relate the material in the module to their own lives. We report here what people around the world are doing, on personal and local levels, to address the issue covered in the module (see pp. 113, 143, and 287), and we hope that this will prompt students to think about what they can do to deal with the issues.

Each module contains several features designed to reinforce or illustrate the concepts presented. One of these features, called "For Instance," is designed to showcase real-world stories about how environmental issues have affected particular areas or groups of people in the world (see pp. 68, 126, and 269). Some of these stories involve examples of how people have found a solution to some environmental problem. Every module contains at least one of these illustrative case studies.

Each module also contains at least one "Making a Difference" feature about a particular individual or group of people who have succeeded in some effort to find more sustainable ways of living or of doing business (see pp. 76, 257, and 282). These, too, are illustrative cases intended to inspire students or to spur them to think of ways in which they might make a positive difference in the world.

Another feature occurring at least once in each module is the "Key Idea," a brief summary statement of a major concept that we hope students will take away from their reading of the module (see pp. 85, 119, and 230). Similarly, we have included a feature called "Consider This," each of which is a brief statement about one or more impressive facts or statistics that we hope will stay with students and stimulate their thinking as they read (see pp. 58, 72, and 241). Another similar feature is the "Numbers" box that appears in a few places within each module, emphasizing a single, dramatic statistic that illustrates the material being covered (see pp. 64, 236, and 276).

As noted above, we provide several "Let's Review" boxes in every module. Each of these is a set of one to six questions asking students to review and recall important facts or data that they have just read. These short lists are placed at the ends of many subsections and major sections, at regular intervals throughout each module (see pp. 90, 148, and 204).

In some of the modules, we have included a two-page photo essay, entitled "The Big Picture," that dramatically illustrates an important concept or set of facts. For example, in Module 4 "Food Resources," you will find a photo essay showing the three major industrialized food production systems that the world now relies on (pp. 88–89). For other examples, see pp. 22–23; 136–137; and 168–169.

In all of the modules, we have included informative photos, illustrations, and diagrams to help make the textual information real and relevant to students in their lives (see Figure 3.4, p. 58; Figure 4.9, p. 92; Figure 8.24, p. 220; and Figure 10.8, p. 267). In many of the modules, we also include graphs to give students quantitative information wherever it will make the material clearer (see Figure 3.22, p. 71; Figure 5.15, p. 127; and Figure 10.11, p. 270).

Finally, we close each module with three important features. First is the "Key Terms" list—those terms that have appeared in boldface earlier in the module, along with their definitions, and again in italics in the regularly appearing review features. This list includes page number references for all of the key terms' definitions (see pp. 30, 114, and 200).

The second feature at the end of each module is the list of "Thinking Critically" questions and exercises. While the "Let's Review" questions throughout each module serve as a factual review, these module-ending questions ask students to think about what they have read and to reach some conclusions about what the material means in the real world. Some of these exercises ask students to undertake a research project in their community or to chronicle their own habits that might be affecting the environment (see pp. 82, 200, and 348). Similar to other features described here, these exercises are intended to help students see the relevance of the modules' material in their lives.

Finally, each module closes with instructions for "Learning Online" (see p. 54), showing students how to access supplemental online materials, including flashcards, practice quizzes, videos, and more.

Supplements for Students

A multitude of electronic supplements available to students take the learning experience beyond the textbook:

- *Aplia™* is an online interactive learning tool that improves comprehension and outcomes by increasing student effort and engagement. It includes innovative teaching materials and automatically graded assignments with detailed, immediate explanations on every question. This system has been used by more than 1 million students at more than 1,800 institutions.
- *WebTutor* on WebCT or Blackboard provides qualified adopters of this textbook with access to a full array of study tools, including flashcards, practice quizzes, animations, exercises, Weblinks, and more.
- *CourseMate* 🔵 is another helpful supplement that includes:
  - An interactive eBook, with highlighting, note taking, and search capabilities
  - Interactive learning tools including:
    - Quizzes
    - Flashcards
    - Videos
    - Animations
    - and more!

Go to **http://login.cengagebrain.com** to access these resources.

- *Global Environment Watch*. Updated several times a day, the Global Environment Watch is a focused portal into GREENR—the Global Reference on the Environment, Energy, and Natural Resources—an ideal one-stop site for classroom discussion and research projects. This resource center keeps courses up-to-date with the most current news on the environment. Users get access to information from trusted academic journals, news outlets, and magazines, as well as statistics, an interactive world map, videos, primary sources, case studies, podcasts, and much more. Login or purchase access at **www.cengagebrain.com/shop/ISBN/9780538735605**.
- *Virtual Field Trips in Environmental Issues*. This supplement brings the field to you, with dynamic panoramas,

videos, photographs, maps, and quizzes covering important topics within environmental science. A case study approach covers the issues of keystone species, the role of climate change in extinctions, invasive species, the evolution of a species in relation to its environment, and an ecosystem approach to sustaining biodiversity. Students are engaged, interacting with real issues to help them think critically about the world around them.

Finally, the following materials for this textbook are available on the companion website at **www.cengagebrain .com/shop/ISBN/9780538735605**.

- *Module Summaries* guide students in reading and studying each module.
- *Flashcards* and *Glossary* allow students to test their mastery of each module's Key Terms.
- *Module Tests* provide multiple-choice practice quizzes.
- Information on a variety of *Green Careers* and tips for sustainable living.
- *What Can You Do?* offers students resources for effecting change at the personal level on key environmental issues.
- *Weblinks* offer an extensive list of websites with news and research related to each module.

Supplements for Instructors

- *PowerLecture*. This DVD, available to adopters, allows you to create custom lectures in Microsoft Power-Point® using lecture outlines, all of the figures and photos from the text, bonus photos, and animations. PowerPoint's editing tools allow use of slides from other lectures, modification or removal of figure labels and leaders, insertion of your own slides, saving slides as JPEG images, and preparation of lectures for use on the web.
- *Instructor's Manual*. Available to adopters, the Instructor's Manual has been thoughtfully built to make creating your lectures even easier. Its features include an introduction that has a section on how to approach using the modular text, a sample syllabus that includes activities for the course, Global Environment Watch assignment templates, answers to the end-of-module questions and more. Also available on PowerLecture.
- *Test Bank*. Available to adopters, the Test Bank contains questions and answers in a variety of formats, including multiple choice, true/false, completion, short answer,

and essay questions. All learning objectives from the text are covered. Also available on PowerLecture.

- *CourseMate* ☻. How do you assess your students' engagement in your course? How do you know if your students have read the material or viewed the resources you've assigned? How can you tell if your students are struggling with a concept? With Course-Mate ☻, you can use the included Engagement Tracker to assess student preparation and engagement. Use the tracking tools to see progress for the class as a whole or for individual students. Identify students at risk early in the course. Find out which concepts are the most difficult for your class. Monitor time on task. Keep your students engaged.
- *BBC Videos for Environmental Science*. This large library includes short, informative video clips that cover current news stories on environmental issues from around the world. These clips are a great way to start a lecture or spark a discussion. Available on DVD with a workbook, on the PowerLecture DVD, and within Course-Mate ☻ and *WebTutor* with additional Internet activities.
- *ExamView*. This full-featured program helps you create and deliver customized tests (both print and online) in minutes, using its complete word processing capabilities. Available on the PowerLecture DVD.

Help Us Improve This Book and Its Supplements

Let us know how you think we can improve this book. If you find any errors or incomplete or confusing explanations, please e-mail us about them at:

spoolman@tds.net

Most errors can be corrected in subsequent printings of the modules, as well as in future editions of this book.

Acknowledgments

We wish to thank everyone on our top-notch production team, listed on the copyright page, all of whom have put their creative talents to work on this project. Our special thanks go to development editor Jake Warde, production editors Hal Humphrey and Nicole Barone, copy editor Deborah Thompson, layout expert Judy Maenle, photo researcher Abigail Reip, artist Patrick Lane, media editor

Alexandria Brady, assistant editor Alexis Glubka, editorial assistant Lauren Crosby, and all of Cengage Learning's hard-working sales staff. Finally, we have benefited in many ways from the guidance and inspiration of life sciences publisher Yolanda Cossio and her energetic and highly talented team.

Many people come together to create a complete learning package. We are grateful to the following people for helping to create the diverse set of learning tools accompanying this text:

Robert Harrison, University of Washington—PowerPoint lecture outlines

Jason Hlebakos, Mt. San Jacinto College—Instructor's manual

Scott Brame, Clemson University—Test bank

Edward Guy, Lakeland Community College—Online quizzing

Dan L. McNally, Bryant University—Global vs. Local activities

Brian Mooney, Johnson & Wales University—*What Would You Do?* activities

John Soares, Higher Education Freelance Writer—Additional *What Can Be Done?* information for online use

Lu Anne Clark, Lansing Community College—JoinIn student response questions

Heidi Marcum, Baylor University—Animation storyboarding and advising

Kara Kuvakas, Brandman University—Media correlation and accuracy review of the test bank

Mary Puglia, Central Arizona College—Updates to the Green Careers and weblinks

Finally, we wish to give special thanks to our reviewers for their carefully considered comments on the new approach we are taking with this book, as well as for their diligence in providing detailed comments on the specific content areas of each module. We list them below.

Norman Myers
Scott Spoolman

Cumulative List of Reviewers

Matthew Abbott, *Des Moines Area Community College*

David Aborn, *University of Tennessee, Chattanooga*

Suzanne M. Albach, *Washtenaw Community College*

Walter Arenstein, *San Jose State University*

Allison Beauregard, *Northwest Florida State College*

Aaron Binns, *Florida State University*

Michael Bourne, *Wright State University*

Judy Bramble, *DePaul University*

Scott Brame, *Clemson University*

Robert Bruck, *North Carolina State University*

Dale Burnside, *Lenoir-Rhyne College*

Catherine Carter, *Georgia Perimeter College - Decatur*

Lu Anne Clark, *Lansing Community College*

John Dunning, *Purdue University*

Joshua Eckenrode, *Vance-Granville Community College*

Brian Fath, *Towson University*

Brad Fiero, *Pima Community College, West Campus*

Caitie Finlayson, *Florida State University*

Ginger Fisher, *University of Northern Colorado*

Marcia Gillette, *Indiana University, Kokomo*

Edward Guy, *Lakeland Community College*

Rob Harrison, *University of Washington*

Keith Hench, *Kirkwood Community College*

Jason Hlebakos, *Mount San Jacinto Community College*

Chasidy Hobbs, *University of West Florida*

LeRoy F. Humphries III, *Fayetteville Technical Community College*

Catherine Hurlbut, *Florida State College at Jacksonville*

Heather Jezorek, *University of South Florida*

David Johnson, *Michigan State University*

Richard Jurin, *University of Northern Colorado*

John Keller, *College of Southern Nevada*

Cindy Klevickis, *James Madison University*

David Knowles, *East Carolina University*

James Kubicki, *Pennsylvania State University*

Kara Kuvakas, *Brandman University*

Katherine LaCommare, *Lansing Community College*

Elizabeth Larson, *Arizona State University*

Jennifer Latimer, *Indiana State University*

Kurt M. Leuschner, *College of the Desert*

Heidi Marcum, *Baylor University*

Dan McNally, *Bryant University*

Kiran Misra, *Edinboro University of Pennsylvania*

Brian Mooney, *Johnson & Wales University*

Rennee Moore, *Solano Community College*

Eric Myers, *South Suburban College*

Jason Neff, *University of Colorado, Boulder*

Barry Perlmutter, *College of Southern Nevada*

Alain Plante, *University of Pennsylvania*

Mary Puglia, *Central Arizona College*

Virginia Rivers, *Truckee Meadows Community College*

Susan Rolke, *Franklin Pierce University*

Amanda Senft, *Bellevue College*

W. Aaron Shoults-Wilson, *Roosevelt University*

John Sulik, *Florida State University*

Nathan Thomas, *Shippensburg University*

Jamey Thompson, *Hudson Valley Community College*

Kip Thompson, *Ozarks Technical College*

Jill Trepanier, *Florida State University*

Mike Tveten, *Pima Community College, Northwest Campus*

Katherine Van de Wal, *Community College of Baltimore County*

Richard Waldren, *University of Nebraska, Lincoln*

David Wyatt, *Sacramento City College*

About the Authors

Norman Myers

As an interdisciplinary environmental scientist, Norman Myers has been a pioneer in identifying and looking for solutions to some of the world's major environmental problems. He has BA and MA degrees from the University of Oxford, and a PhD from the University of California, Berkeley. His main areas of expertise are systems ecology, ecosystem management, biodiversity, mass extinction, tropical forests, population growth, resource consumption and poverty, resource economics, environmental security, and sustainable development.

Myers has carried out research and worked as an environmental consultant in more than 50 countries. He has also served as a visiting professor at more than 20 colleges and universities, including Oxford University, Duke University, the University of Vermont, Cornell University, Stanford University, Harvard University, and the University of California, Berkeley.

Myers' pioneering research led him to estimate that we were losing one species every day, on average. He also helped to alert the world to the destruction of tropical forests, explained the economic and ecological reasons for preserving biodiversity, and co-authored a groundbreaking interdisciplinary book describing the state of the planet (*The New Atlas of Planet Management*, published in 1984 and updated in 2005). In addition, Myers originated the idea of preserving biodiversity by identifying and protecting biodiversity hotspots. He has also written about the need for environmental security as a basis for political stability; perverse government subsidies and how they harm the environment; and institutional roadblocks that have thwarted many solutions to environmental problems.

Myers has written 20 books that have seen total sales of more than 1 million copies in 11 languages. He has published more than 300 professional papers in scientific journals and more than 300 popular articles in newspapers and magazines. He has also served as an adviser to many academic bodies, governments, and international agencies, including the U.S. National Academy of Sciences, the White House, NASA, the World Bank, the United Nations, and the U.S. Departments of State, Defense, and Energy. Myers has received more than 20 major environmental awards and honors, including the Blue Planet Prize, the Volvo Environment Prize, and the UNEP Environment Prize.

Scott Spoolman

Scott Spoolman is a writer with over 25 years of experience in educational publishing. He has had a lifelong interest in the environmental sciences and believes that education provides the best hope for our dealing with environmental problems and finding more sustainable ways to work, play, and live on our planet.

Spoolman holds a master's degree in science journalism from the University of Minnesota. He has authored numerous articles in the fields of science, environmental engineering, politics, and business. He worked as an acquisitions editor on a series of college forestry textbooks. He has also worked as a consulting editor in the development of more than 70 college and high school textbooks in the natural and social sciences fields. With G. Tyler Miller, Jr., Spoolman has coauthored several editions of *Living in the Environment*, *Environmental Science*, and *Sustaining the Earth*.

In his free time, he enjoys exploring the forests and waters of his native Wisconsin along with his family—his wife, environmental educator Gail Martinelli, and his children, Will and Katie.

CENGAGE LEARNING'S COMMITMENT TO SUSTAINABLE PRACTICES

We, the authors of this textbook, and Cengage Learning, the publisher, are committed to making the publishing process as sustainable as possible. This involves four basic strategies:

- *Using sustainably produced paper*. The book-publishing industry is committed to increasing the use of recycled fibers, and Cengage Learning is always looking for ways to increase this content. Cengage Learning works with paper suppliers to maximize the use of paper that contains only wood fibers that are certified as sustainably produced, from the growing and cutting of trees all the way through paper production.

- *Reducing resources used per book*. The publisher has an ongoing program to reduce the amount of wood pulp, virgin fibers, and other materials that go into each sheet of paper used. New, specially designed printing presses also reduce the amount of scrap paper produced per book.

- *Recycling*. Printers recycle the scrap paper that is produced as part of the printing process. Cengage Learning also recycles waste cardboard from shipping cartons, along with other materials used in the publishing process.

- *Process improvements*. In years past, publishing has involved using a great deal of paper and ink for writing and editing of manuscripts, copyediting, reviewing page proofs, and creating illustrations. Almost all of these materials are now saved through use of electronic files. Except for our review of page proofs, very little paper and ink were used in the preparation of this textbook.

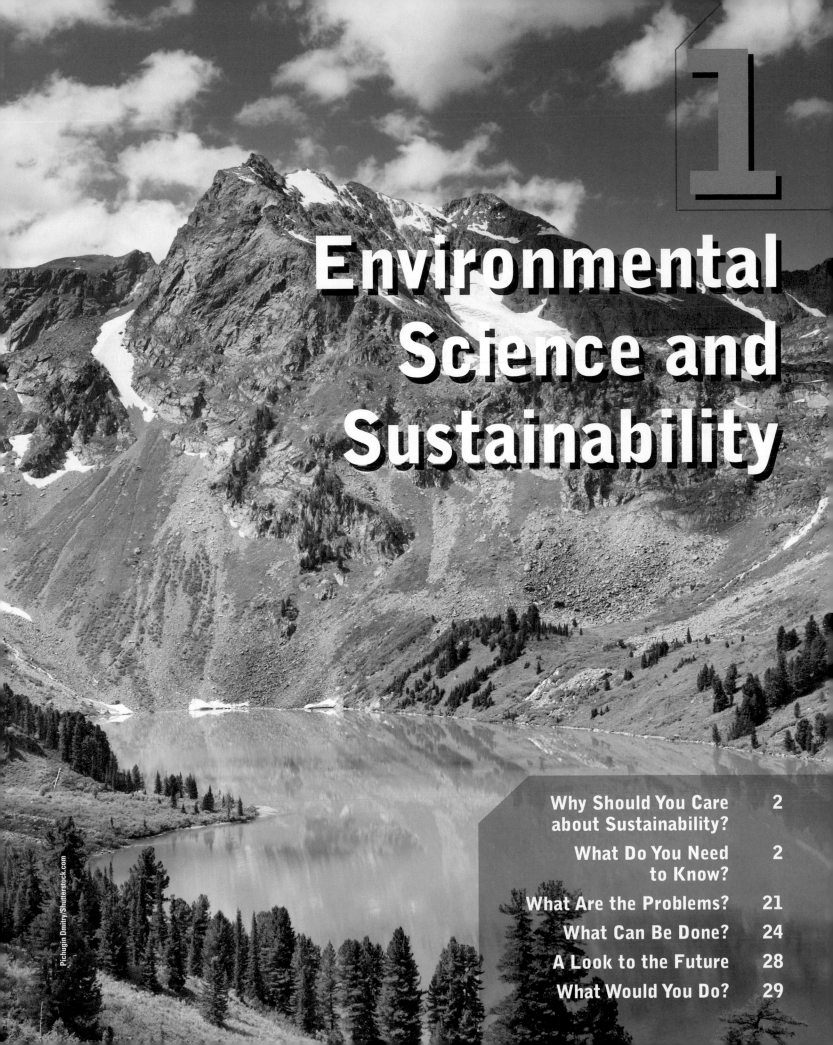

1

Environmental Science and Sustainability

Why Should You Care about Sustainability?

Sustainability is the capacity of the earth's natural systems and our human cultural systems to survive, flourish, and adapt to changing environmental conditions into the long-term future. The earth's life-support system consists of four major components: the *geosphere* (crust, core, and mantle), the *hydrosphere* (water), the *atmosphere* (air), and the *biosphere* (living things) (Figure 1.1).

The photo of the mountain lake area on the opening page of this module shows the air, water, and plants that, along with an endless supply of energy from the sun, sustain the earth's life and human economies. Figure 1.2 shows two of the almost 2 million identified forms of life that help to keep us alive and support our lifestyles.

When we don't overload or degrade the earth's natural systems, they can sustain themselves over time. For example, rivers and lakes can provide us with freshwater as long as we don't withdraw or pollute the water faster

than natural processes can replace or purify it. Likewise, forests can renew themselves as long as we don't remove their trees faster than new trees can grow back.

However, we have been destroying, overusing, degrading, and polluting these and many other components of the earth's life-support system. In other words, we have been living unsustainably as more and more people use and waste more and more of the earth's resources and pollute its air, water, and soil. No one knows how long we can keep doing this before the earth's natural processes limit our actions in unpleasant ways, but there are some alarming signs that we are approaching such limits.

For example, one result of our living unsustainably is that other *species*, or forms of life, are disappearing—going *extinct*—at an alarming rate, mostly because humans have taken over or eliminated the places where they live. Scientists estimate that during this century, if we continue on this path, we are likely to drive one-fourth to one-half of the world's species to extinction.

There are two key reasons why we should care deeply about sustainability. One is self-interest: the long-term survival of our own species and cultures depends on our learning how to live more sustainably. The other is the moral or ethical view that we do not have the right to crowd out or eliminate large numbers of the planet's other species and the places where they live. In this and other modules in this book, we examine our unsustainable practices and suggest some ways in which we can try to live more sustainably.

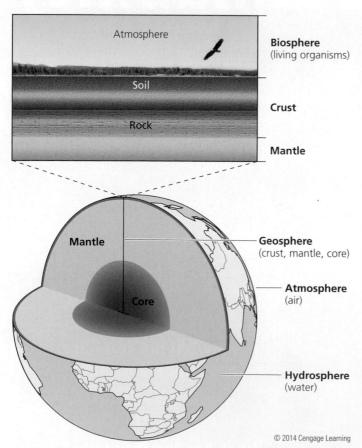

FIGURE 1.1 This diagram, like the opening photo of this module, shows the major components of the earth's life-support system.

© 2014 Cengage Learning

What Do You Need to Know?

Environmental Science

The **environment** is everything around us. It includes energy from the sun and all living and nonliving things in the air, water, and soil with which we can interact (module-opening photo and Figure 1.2). It is a complex web of relationships that connects us with one another and with the world in which we live.

Many scientific and technological advances have made our lives better. However, in the end we are still dependent on the environment for air, water, food, shelter, energy, and everything else we need in order to stay alive and be healthy. In other words, we are part of, and not apart from, the rest of nature.

This textbook consists of a set of modules that introduce you to **environmental science**, a study of how humans interact with the environment of living things

FIGURE 1.2 The plant and peacock shown here are two living components of our life-support system.

(plants, animals, and bacteria) and nonliving things (air, water, and energy). The three goals of environmental science are to:

- Learn how the natural world works,
- Understand the major environmental issues we face, and
- Find ways to deal with these environmental issues.

Ecology—the biological science that focuses on how living things interact with one another and with their environment—is an important component of environmental science. Scientists refer to living things as **organisms** and they classify organisms into **species**, or groups of organisms with distinctive traits.

Ecologists often focus on ecosystems. An **ecosystem** consists of one or more groups of organisms interacting among themselves and with the nonliving matter and energy in their physical environment within a specified place. For example, a forest ecosystem consists of plants (especially trees), animals, and tiny organisms (mostly bacteria) all interacting with one another, with solar energy, and with the chemicals in the forest's air, water, and soil.

Environmental science and ecology are sciences, unlike *environmentalism*, which is a political and philosophical movement devoted to promoting environmental awareness and the protection of the earth's life-support systems for all forms of life.

This is not your typical textbook, which consists of a series of chapters that are often related to and dependent on one another. This initial module provides an introduction to science, matter, energy, life, and sustainability and

serves as necessary background for the 14 other modules in this book. Each of those modules explores a particular environmental issue and possible ways to deal with that issue. The topics of the other modules are population growth, cities, food production, energy, minerals, species extinction, land degradation, water resources, water pollution, air pollution, climate change, wastes, and human health.

Each of these additional 14 modules stands alone and can be used in any order depending on the design of the course you are taking. Each module also includes detailed scientific background material not covered in this module that is necessary for understanding problems related to the issue and their possible solutions. To allow for flexibility of use and to enable each module to stand alone, we have duplicated some information among various modules.

KEY idea We should learn about environmental science because it is a study of earth's life-support system that keeps us alive and supports our economies and lifestyles.

Let's **REVIEW**

- What is *sustainability*? Give two reasons for caring about sustainability.
- Define *environment*. What is *environmental science* and what are its three main goals?
- What is *ecology*? Define and distinguish among *organisms*, *species*, and *ecosystems*.
- Distinguish between environmentalism and environmental science.

The Scientific Process

Science is the human effort to understand how the natural world works by making observations and measurements and carrying out experiments. Science is based on the assumption that events in the natural world occur in orderly cause-and-effect patterns.

We often hear about the scientific method, but there is no single method that all scientists use consistently for conducting science. Instead, scientists use a variety of methods that tend to fall into a general process for uncovering secrets of the natural world.

Here is a rough outline of what scientists typically do, although not necessarily in this order, with an example from a real scientific experiment:

- *Ask a question to be investigated.* For example, why are the fish in a particular lake dying (Figure 1.3)?
- *Find out what is already known* by searching the scientific literature to find out what scientists have learned about fish deaths and lakes.
- *Gather* **scientific data**, or factual information pertaining to the investigation, to help answer the original question. Make observations and measurements about factors such as chemicals in the lake and in the bodies of the dead fish.
- *Come up with a* **scientific hypothesis**—an attempt to explain the data. For example, perhaps the fish are dying because a chemical pesticide used on nearby croplands is running off the fields and into the lake.
- *Test the hypothesis.* Carry out a *controlled experiment* by choosing two small, very similar lakes located near

each other in a remote, undeveloped area and making one of them the *control site*—the lake that is left to natural processes. The other would be the *experimental site*—the lake to be studied. In the latter lake, scientists would add small amounts of the chemical pesticide being used by local farmers. They would measure the concentrations of the pesticide in the water, record any observed fish deaths, and measure the concentrations of the pesticide in dead fish from both lakes for a year.

- *Evaluate the results.* Let's assume that the results do not support the hypothesis that a particular pesticide is the culprit. The two lakes show no major difference in fish deaths.
- *Come up with new hypotheses.* Some other possible causes of the fish deaths might be runoff of a different pesticide, runoff of fertilizer, a disease, or a fungus.
- *Test each new hypothesis.* Let's assume that the idea that the fish deaths were due to a particular fungus (which we will call fungus X) is the only hypothesis supported by further observations and measurements.
- *Repeat the experiment* to see if its results are *reproducible*—whether the same or other scientists can perform the experiment in the same manner with the same results. In other words, scientists *retest the hypothesis*. Let's assume that the results are reproduced.
- *Publish the results.* Scientists, once they have reproducible results, communicate the data and hypotheses to others in their field for evaluation, further testing, and feedback. Scientists typically publish their results in professional journals, on professional websites, in government reports, or as presentations at professional meetings.
- *Test and retest until there is general agreement.* Let's assume that there is general agreement among the experts in this field that the fungus X hypothesis is the most useful explanation of the fish deaths.
- *Upgrade the hypothesis to a* **scientific theory**—a thoroughly tested and widely accepted scientific hypothesis or a group of related hypotheses.

In our everyday lives, each of us uses this general approach to help us make decisions and solve problems (Figure 1.4).

<figure>**FIGURE 1.3** A scientist might ask, "What killed these fish?"</figure>

Let's **REVIEW**

- What is *science*?
- Outline the general scientific process.
- Define and distinguish among *scientific data*, a *scientific hypothesis*, and a *scientific theory*.

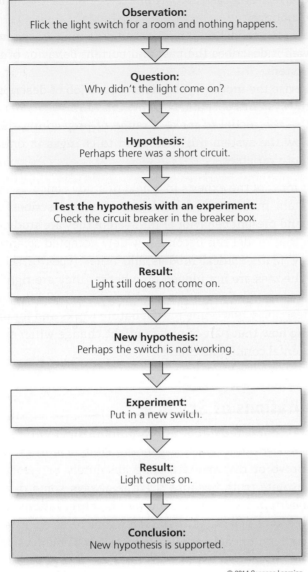

```
┌─────────────────────────────────────┐
│          Observation:                │
│ Flick the light switch for a room    │
│ and nothing happens.                 │
└─────────────────────────────────────┘
                  ↓
┌─────────────────────────────────────┐
│          Question:                   │
│ Why didn't the light come on?        │
└─────────────────────────────────────┘
                  ↓
┌─────────────────────────────────────┐
│          Hypothesis:                 │
│ Perhaps there was a short circuit.   │
└─────────────────────────────────────┘
                  ↓
┌─────────────────────────────────────┐
│ Test the hypothesis with an          │
│ experiment:                          │
│ Check the circuit breaker in the     │
│ breaker box.                         │
└─────────────────────────────────────┘
                  ↓
┌─────────────────────────────────────┐
│          Result:                     │
│ Light still does not come on.        │
└─────────────────────────────────────┘
                  ↓
┌─────────────────────────────────────┐
│          New hypothesis:             │
│ Perhaps the switch is not working.   │
└─────────────────────────────────────┘
                  ↓
┌─────────────────────────────────────┐
│          Experiment:                 │
│ Put in a new switch.                 │
└─────────────────────────────────────┘
                  ↓
┌─────────────────────────────────────┐
│          Result:                     │
│ Light comes on.                      │
└─────────────────────────────────────┘
                  ↓
┌─────────────────────────────────────┐
│          Conclusion:                 │
│ New hypothesis is supported.         │
└─────────────────────────────────────┘
```

© 2014 Cengage Learning

FIGURE 1.4 These are the key steps for using the scientific process to solve a problem.

Curiosity, Skepticism, Reproducibility, and Peer Review

The success of the scientific process depends on four factors: *curiosity, skepticism, reproducibility,* and *peer review*. Good scientists are very curious about how the world works. They are also highly skeptical about new data and hypotheses until they can test and verify the information and findings with lots of evidence. Scientists require their colleagues to show their evidence and explain the reasoning behind their proposed scientific ideas or hypotheses. They also require that any experimental results that support a hypothesis must be reproducible.

A crucial part of the scientific process is **peer review** in which scientists openly report details of their experimental design, the results, and the reasoning behind their hypotheses for other scientists working in the same field (their *peers*) to look at, evaluate, and criticize. When a scientist submits an article for publication in a professional journal, several experts typically review the article, and this can lead to further testing.

Through such peer review and retesting, some of the preliminary results and hypotheses produced by scientists eventually are classified as reliable while other results and hypotheses might be viewed as unreliable. Through this usually lengthy weeding-out process, scientists openly debate and test the accuracy of data and the validity of hypotheses. Such skepticism and debate among peers in the scientific community is essential to the scientific process.

Critical Thinking and Creativity

Scientists use logical reasoning and critical thinking skills in learning about the natural world. Such skills help scientists and the rest of us to evaluate evidence and arguments, to distinguish between facts and opinions, and to develop and defend informed positions on issues.

Thinking critically involves three important steps:

1. Be skeptical about everything you read and hear.
2. Look at the evidence and evaluate it and any related information and opinions with careful consideration of the sources. Validating information and separating evidence from opinions is especially important in the Internet age where one can be exposed to often unreliable data and opinion from uninformed individuals posing as experts.
3. Identify and evaluate your personal assumptions, biases, and beliefs, and separate them from the facts.

Logic and critical thinking are very important tools in science, but imagination, creativity, and intuition are just as vital. Most major scientific advances are the results of creative people coming up with new and better ways to help us understand how various parts of the natural world work.

Let's **REVIEW**

- List four factors on which the scientific process depends.
- What is *peer review* and why is it important?
- List three steps you can use in thinking critically.
- Describe the roles of critical thinking and creativity in science.

Scientific Theories and Laws

Facts or data are the raw materials of the scientific process. But the main goal of science is to come up with reliable and useful theories and laws that explain how the natural world works.

When a large body of observations and measurements supports a scientific hypothesis or group of related hypotheses, it becomes a scientific theory. You might think of a theory as something that need not be taken seriously, but this is far from the truth. Any scientific theory has been thoroughly tested, debated, supported by extensive evidence, and accepted as being useful by most scientists in a particular field of study. Because of this rigorous process, it is difficult for any scientific idea to reach this level of acceptance and, thus, scientific theories are rarely overturned.

Another reliable and very important result of the scientific process is a **scientific law**: a well-tested and widely accepted description of something that happens in nature over and over again in the same way. The *law of gravity* is an example. After making thousands of observations and measurements of objects falling from different heights, scientists came up with the following scientific law: All objects fall towards the earth's surface at predictable speeds.

We can break a speed limit or some other law of society. But we cannot break a scientific law, assuming the data used to come up with it were accurate. For example, you could throw a penny up in the air a hundred or a thousand times, but the law of gravity says that it will come back down every time at a certain speed.

> ## KEY idea
> Scientific theories and laws, based on extensive data collection, debate, and testing, are the most important and reliable results of science.

Scientific Models

Modeling is another tool that scientists use to help them understand the behavior of complex systems such as lakes, forests, and the earth's climate. A **scientific model** is any simplified version of a more complex system that we can more easily observe and understand. Scientists often use mathematical models. Here are the key steps they typically use to develop and evaluate such a model:

- Identify the key *variables*, or factors that can affect a system, and determine how they interact.
- Convert this information into a set of mathematical equations.
- Run the model on a high-speed computer to see how well it describes the past and current behavior of the system.
- Revise the model until it does a good job of describing the system's past and current behavior.
- Use the model to make a range of projections about how the system might respond to changes in one or more variables.

If most of the experts in a field of study conclude that a certain mathematical model accurately describes the past and present behavior of the system they are studying, then that model can become a widely accepted scientific theory. Mathematical models and other types of scientific theories are not judged on whether they are right or wrong. Instead, they are evaluated on how useful they are for describing how some part of nature works and for projecting how that part of nature might change when environmental conditions change.

Limitations of Science

Let's look at three important limitations of environmental science and science in general. First, scientists cannot prove or disprove anything absolutely or establish an absolute truth because there is always some degree of uncertainty in the measurements, observations, and models that scientists make. The goal of science is to reduce such uncertainty to the lowest possible level. For this reason, some scientists steer clear of the word *proof* because the popular understanding of the word is *absolute proof*.

Instead of focusing on absolute proof, scientists generally work to establish varying *degrees of certainty* about how well scientific hypotheses, models, theories, and laws explain things. For example, you would rarely find a scientist saying that "Cigarettes cause lung cancer." Instead, a scientist might say something like: "Evidence from thousands of studies strongly indicates that long-term smokers have a much higher chance of developing lung cancer than do nonsmokers."

A second limitation of science is that scientists are human and thus cannot be totally unbiased about their experimental results and hypotheses. However, science is based on skepticism, debate, peer review, and thorough testing for reproducibility—a process that helps to uncover or greatly reduce personal bias and discourage the falsifying of scientific results.

A third limitation, which is especially important in environmental science, is that most systems in nature are extremely complex. Widely tested and accepted scientific models and other types of scientific theories can improve our knowledge of natural systems. But nature is so complex that our understanding of how it works will always be incomplete and will always include some degree of uncertainty. This explains why, when we change a natural system, for example, by cutting down a forest or emitting chemicals into the atmosphere, there will always be some unexpected side effects.

Despite these limitations, science is the most useful tool we have for increasing our understanding of how nature works and for projecting how it might behave in the future. Science is a dynamic, open-ended, creative, and exciting human endeavor.

KEY idea

Science is devoted to using experimentation, critical thinking, and creativity to find useful and widely accepted theories and laws that explain, with a high degree of certainty, how some part of nature works.

Let's REVIEW

- Define *scientific law* and *scientific model*. Distinguish among a scientific theory, a scientific law, and a scientific model.
- Explain why scientific theories and laws are the most important results of science.
- What are three limitations of science?
- If scientists can never prove anything absolutely, what is their basic goal?

FIGURE 1.5 Carbon in the form of coal (a) and diamond (b), mercury (c), and copper (d) are chemical elements.

Scientific research reveals that matter and energy are the basic components of the physical world. Let's look more closely at the nature of these vital components of the world within and around us.

Components of Matter

Matter is anything that has mass and takes up space. Matter can exist in three *physical states*—solid, liquid, and gas. For example, water exists as solid ice, liquid water, or water vapor, depending mostly on its temperature.

Matter also exists in two major *chemical forms*—elements and compounds. A **chemical element** is a fundamental type of matter that has a unique set of properties and cannot be broken down into simpler substances by chemical means. For example, the elements carbon (found in coal and diamonds), mercury, and copper (Figure 1.5) cannot be broken down chemically into any other substance.

Chemists use a one- or two-letter symbol to represent each element. Examples are carbon (C), hydrogen (H), oxygen (O), nitrogen (N), phosphorus (P), sulfur (S), sodium (Na), calcium (Ca), chlorine (Cl), gold (Au), copper (Cu), lead (Pb), and mercury (Hg).

Some matter is composed of one element, such as carbon, mercury, or gold. But most matter consists of **compounds**: combinations of two or more different chemical elements held together in fixed proportions. For example, water is a compound made of the elements hydrogen and oxygen, chemically combined with one another in a particular way.

Atoms, Ions, and Molecules

The three basic building blocks of elements and compounds are atoms, ions, and molecules. An **atom** is the most fundamental building block. It is the smallest unit of matter into which an element can be broken down and still have its characteristic chemical properties. According

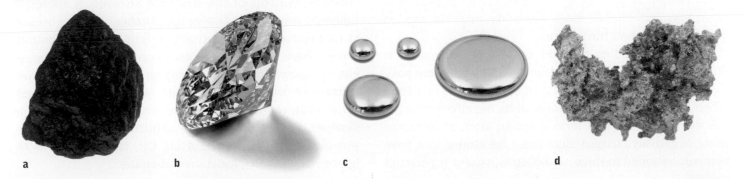

a   b   c   d

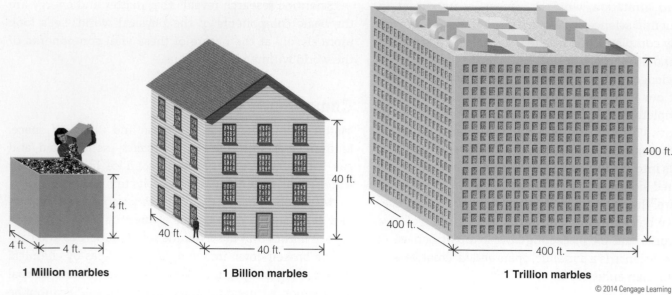

1 Million marbles   **1 Billion marbles**   **1 Trillion marbles**

© 2014 Cengage Learning

FIGURE 1.6 One million marbles will fit in this box, one billion in this house, and one trillion in this office building.

to the *atomic theory*—the most widely accepted scientific theory in chemistry—all elements are made up of atoms.

The period at the end of this sentence is wide enough to hold a row of more than 3 million hydrogen atoms sitting side by side.

The universe is made up of unimaginably small things, such as atoms, and unimaginably large things, such as galaxies. To describe the sizes of such things, scientists often use numbers such as millions, billions, and trillions. But such quantities are very hard to grasp. Figure 1.6 might give you an idea of what such large numbers represent.

Atoms are made of three types of *subatomic particles*: *neutrons* with no electrical charge, *protons* with one positive electrical charge (+) each, and *electrons* with one negative electrical charge (−) each. Every atom has an extremely small center, or *nucleus*, which contains its protons and neutrons, and one or more electrons in rapid motion somewhere around the nucleus.

A second building block of some types of matter is an **ion**: an atom or a group of atoms with one or more positive (+) or negative (−) electrical charges. Positive ions form when an atom loses one or more of its negatively charged electrons. Negative ions form when an atom gains one or more negatively charged electrons. Like atoms, ions have symbols assigned to them, and chemists use a superscript

after the symbol of an ion to indicate how many positive or negative electrical charges it has. Examples of some common ions used in this book are the hydrogen (H^+), sodium (Na^+), calcium (Ca^{2+}), chloride (Cl^-), nitrate (NO_3^-), carbonate (CO_3^{2-}), sulfate (SO_4^{2-}), and phosphate (PO_4^{3-}) ions.

Many compounds are made of oppositely charged ions. For example, the basic unit of sodium chloride ($NaCl$), or table salt (Figure 1.7a), is a pairing of the ions Na^+ and Cl^- whose attractive forces hold the structural units together. Another example is calcium carbonate ($CaCO_3$), which is the major component of seashells (Figure 1.7b), coral, limestone, and marble. It consists of pairs of calcium (Ca^{2+}) and carbonate (CO_3^{2-}) ions.

Other compounds, including water and carbon dioxide, are made of molecules, the third building block of matter. A **molecule** is a combination of two or more atoms of the same or different elements held together by chemical bonds.

Chemists use a **chemical formula** as a shorthand way to show the number of each type of atom or ion in each structural unit of a compound. Each formula includes the chemical symbols of the elements involved. A subscript following a symbol indicates the number of atoms or ions of that element in each structural unit of the compound. Where there is only one atom or ion of an element present, no subscript is used. Examples of chemical formulas commonly used in this book are H_2O (water), O_3 (ozone), CO (carbon monoxide), CO_2 (carbon dioxide), SO_2 (sulfur dioxide), H_2SO_4 (sulfuric acid), NO (nitric oxide), NO_2 (nitrogen dioxide), HNO_3 (nitric acid), CH_4 (methane), $C_6H_{12}O_6$ (glucose) and $CaCO_3$ (calcium carbonate).

FIGURE 1.7 These two chemical compounds—table salt **(a)** and calcium carbonate **(b)**—are each composed of oppositely charged ions.

Let's **REVIEW**

- What is *matter*?
- Define *chemical element* and *compound* and give two examples of each.
- Define *atom*, *ion*, and *molecule* and give an example of each. Define *chemical formula* and give two examples.

> **KEY idea** All matter consists of elements and compounds, which in turn are composed of atoms, ions, and molecules.

Physical and Chemical Changes

Matter can undergo physical and chemical changes. When a sample of matter undergoes a **physical change** its chemical composition does not change. Pulverize a seashell (Figure 1.7b), which is mostly calcium carbonate ($CaCO_3$), and it is still $CaCO_3$. When liquid water (H_2O) boils and becomes water vapor, it still consists of H_2O molecules.

In a **chemical change**, also called a **chemical reaction**, there is a change in the chemical composition of the substances involved. Chemists use a *chemical equation* to represent how chemicals are rearranged in a chemical reaction. For example, coal (Figure 1.5a) consists mostly of carbon atoms. When completely burned, its carbon (C) atoms react with oxygen gas (O_2) in the atmosphere to form the gaseous compound carbon dioxide (CO_2). Chemists use the following shorthand equation to represent this chemical reaction:

$$\text{carbon} + \text{oxygen} = \text{carbon dioxide}$$
$$C + O_2 = CO_2$$

This equation reveals that whenever we completely burn a carbon-containing fossil fuel such as coal, oil, or natural gas, the carbon in the fuel reacts with oxygen in the air to produce carbon dioxide, a chemical that helps to warm the atmosphere. Indeed, without carbon dioxide and water vapor in the atmosphere, the earth would be a frigid and essentially lifeless planet.

The Law of Conservation of Matter

Whenever matter undergoes a change, the atoms, ions, or molecules involved are rearranged into different spatial patterns (physical changes) or chemical combinations (chemical changes). However, many thousands of experiments have revealed that regardless of how we change matter, we can never create or destroy any of the atoms involved. This is the scientific law known as the **law of conservation of matter**: *whenever matter undergoes a physical or chemical change, no atoms are created or destroyed.*

This law is important in environmental science because it means we cannot destroy or even really get rid of our wastes. It tells us that the atoms of everything that we have ever "thrown away" are still around somewhere in some physical or chemical form. For example, when we bury wastes in a sanitary landfill (Figure 1.8, p. 10) or burn them in an incinerator, the atoms in the waste materials will always be around in one form or another.

We can convert many waste chemicals to less harmful forms. An exception would be a toxic metal such as mercury (Figure 1.5c) because it is a chemical element that cannot be broken down or destroyed. Thus, this indestructible element circulates in the atmosphere and returns to the land and to bodies of water. This means that it can

FIGURE 1.8 We can never really throw anything away because of the law of conservation of matter.

build up in water and soil and in the bodies of fish, polar bears in the Arctic, and people. Mercury can also react with other elements and form various mercury compounds that are even more toxic than elemental mercury.

Let's **REVIEW**

- Distinguish between a *physical change* and a *chemical change*, and give an example of each.
- What is the *law of conservation of matter* and why is it important?

KEY idea   Whenever matter undergoes a physical or chemical change, no atoms are created or destroyed (the law of conservation of matter).

Types of Energy

Scientists define **energy** as the capacity to do work or to transfer heat. When you pick this book up and put it on your desk, for example, you must use a certain amount of force to move the book a certain distance. In scientific terms, you are doing *work* (work = force × distance). Another example of energy is the heat that flows from a hot stove to a pan of boiling water.

Scientists distinguish between two major types of energy, called kinetic energy and potential energy. **Kinetic**

energy is any energy associated with movement. Moving matter, such as water flowing in a stream, has kinetic energy. The blades of wind turbines (Figure 1.9) are turned by the kinetic energy of the gaseous molecules in a mass of moving air, or *wind*. The turbine then converts this kinetic energy to electrical energy (flowing electrons), which is another form of kinetic energy. *Heat* is another type of kinetic energy. Whenever two objects at different temperatures come in contact, heat flows from the warmer object to the cooler object.

Another form of kinetic energy is **electromagnetic radiation**, in which energy travels as a *wave* generated by changes in an electrical or magnetic field. Electromagnetic radiation occurs in many different forms, each having a different *wavelength* (the distance between two successive peaks or troughs in the wave) and *energy content*, the intensity level of the energy. Examples are X-rays and

FIGURE 1.9 Kinetic energy in moving masses of air that we call *wind* can turn the blades of wind turbines and produce electricity.

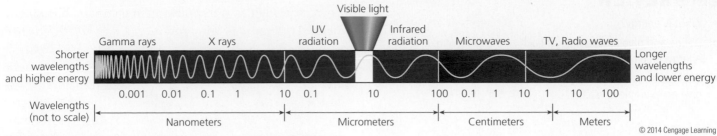

Shorter wavelengths and higher energy

Gamma rays   X rays   UV radiation   Visible light   Infrared radiation   Microwaves   TV, Radio waves

Longer wavelengths and lower energy

Wavelengths (not to scale)

0.001 0.01 0.1 1 10 0.1 10 100 0.1 1 10 1 10 100

Nanometers   Micrometers   Centimeters   Meters

© 2014 Cengage Learning

FIGURE 1.10 The sun releases a variety of different forms of kinetic energy called electromagnetic radiation.

ultraviolet (UV) radiation, both with short wavelengths and high energy content, and visible light and infrared (IR) radiation, both having longer wavelengths and lower energy content (Figure 1.10).

The other major type of energy is **potential energy**, or energy that is stored and potentially available for use. Examples include the water in a reservoir behind a dam (Figure 1.11) and the chemical energy stored in the carbon atoms of a lump of coal.

We can convert potential energy to kinetic energy. If you hold a rock in front of you at a certain height, it has potential energy. Drop the rock and its potential energy changes into kinetic energy as it falls through the air. As water in a reservoir flows through pipes in a dam to the river below, the potential energy it had in the reservoir changes into kinetic energy that we can use to spin turbines in the dam and produce electricity—another form of

kinetic energy. When you burn a lump of coal, its potential energy becomes a combination of heat, or kinetic energy, and light.

Energy Quality

Different types of energy vary in their usefulness, referred to as **energy quality**, a measure of the capacity of any type of energy to do useful work. **High-quality energy** has a high capacity to do useful work because it is concentrated. Examples are very-high-temperature heat, high-speed wind, concentrated sunlight, and energy released when we burn coal or gasoline.

Energy that is more dispersed and has little capacity to do useful work is called **low-quality energy**. For example, the heat stored in the moving molecules that are widely dispersed in the atmosphere or in an ocean is low-quality energy, because of its low average temperature and low capacity for useful work.

Stas Volik/Shutterstock.com

FIGURE 1.11 The water stored in this reservoir has potential energy.

Let's **REVIEW**

- What is *energy*?
- Define and give an example of each of the following: *kinetic energy, electromagnetic radiation,* and *potential energy.*
- Distinguish between *high-quality* and *low-quality energy* and give an example of each.

Two Laws of Thermodynamics

Thermodynamics is the study of what happens when energy is converted from one form to another in physical and chemical changes. After studying millions of such changes, scientists have formulated the **first law of thermodynamics**, also known as the **law of conservation of energy**. According to this scientific law, *whenever energy is converted from one form to another in a physical or chemical change, no energy is created or destroyed.* In other words, we cannot get more energy out of a physical or chemical change than we put in. This is a basic rule that we have never been able to violate.

You might think that, because energy cannot be destroyed, we will never run out of it. But scientists have learned that whenever we use any form of energy, there is a drop in its energy quality. For example, when we burn a tank of gasoline in a car or run down the battery in a cell phone, what we lose is energy quality, the amount of energy available for performing useful work.

Thousands of experiments have also shown that *whenever energy is converted from one form to another in a physical or chemical change, it is always degraded to a lower-quality, less useful form of energy.* This is a statement of the **second law of thermodynamics**. Usually, the resulting lower-quality energy ends up as waste heat (infrared radiation, Figure 1.10) that flows into the environment. Once in the environment, the random motion of air or water molecules further disperses this heat and makes it even less useful as an energy resource. The second law of thermodynamics is another basic rule of nature that we have never been able to violate.

We can recycle many forms of matter such as aluminum and paper. But the second law of thermodynamics tells us that once the concentrated, high-quality chemical energy in gasoline or in a morsel of food is released, it is automatically degraded to low-quality heat that we cannot recapture and recycle or reuse as high-quality energy.

A good example of the loss of energy quality takes place in the incandescent lightbulb (Figure 1.12, left). Only about 5–10% of the electrical energy flowing into

FIGURE 1.12 Incandescent bulbs (left) are being replaced with more energy-efficient and much longer-lasting fluorescent bulbs (middle), and LED bulbs (right).

it produces light. The other 90–95% enters the environment as low-quality heat. This explains why some energy experts say we should call it a heat bulb and why consumers are increasingly preferring more efficient types of lightbulbs (Figure 1.12, middle and right). Another example is the internal combustion engine used in most vehicles. It converts about 87% of the energy stored in gasoline to low-quality heat that does not help to move the car.

NUMB3RS

5–10%
Percentage of the electricity used by an incandescent lightbulb that is converted to light

CONSIDER this When we burn gasoline in the internal combustion engine of a car, only about 13% of the high-quality chemical energy in the gasoline, and 13% of the money we spend for the gasoline, is used to move the car. The other 87% of the energy released from the gasoline eventually flows into the atmosphere as heat.

Let's **REVIEW**

- State the *first law of thermodynamics* (or *law of conservation of energy*).
- State the *second law of thermodynamics.*
- Use the second law of thermodynamics to explain why we cannot recycle high-quality energy.
- Give an example of how energy can be degraded during a physical or chemical change.

- Whenever energy is converted from one form to another in a physical or chemical change, no energy is created or destroyed (first law of thermodynamics).

- Whenever energy is converted from one form to another in a physical or chemical change, we end up with lower-quality or less usable energy than we started with (second law of thermodynamics).

Natural Capital

One of the major components of sustainability is **natural capital**—the resources and ecological services provided by nature that keep us and other species alive and support human economies (Figure 1.13). A **resource** is anything that we can take from the environment to meet our needs and wants. Matter and energy resources that are provided by nature and are essential or useful to humans are called **natural resources**. Important functions provided by nature that help humans and other life forms to survive are called **natural** or **ecosystem services**. Examples are pollination, topsoil renewal, and air and water purification. This natural capital helps to sustain the **biosphere**—the parts of the earth's life-support system where the genes, species, and ecosystems that make up its biodiversity are found.

Some of the resources that support us are **renewable resources**—those that natural processes can replenish, whether in days or in hundreds of years. They are inexhaustible resources as long as we do not use them up faster than nature can renew them. Examples are forests, grasslands, wildlife, freshwater, fresh air, and fertile topsoil. Another inexhaustible resource is direct solar energy, which we can convert to electricity through the use of

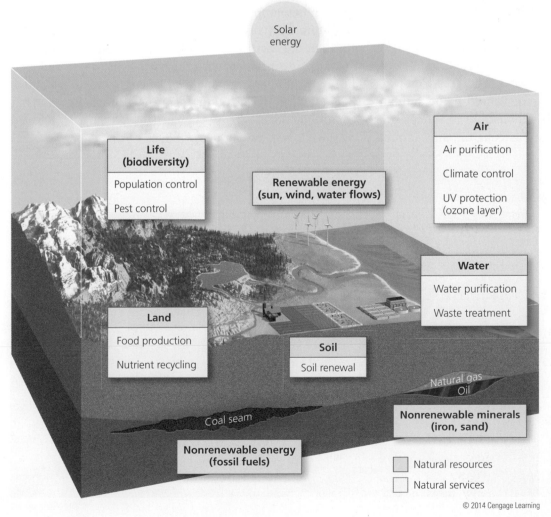

FIGURE 1.13 Natural capital consists of *natural resources* (blue boxes) and *natural services* (yellow boxes) that support and sustain the earth's life and our economies.

© 2014 Cengage Learning

FIGURE 1.14 Using solar cells, we can produce electricity by capturing direct solar energy.

solar cells (Figure 1.14). Direct solar energy also helps to generate wind, an indirect form of solar energy that we can use to produce electricity (Figure 1.9). Another renewable form of energy is heat from the earth's interior (geothermal energy).

Nonrenewable resources are resources that exist in a fixed quantity, or *stock*, in the earth's crust. Because it takes millions of years for geological processes to renew these resources, they are nonrenewable on the human time scale. Examples of such exhaustible resources are *energy resources* (fossil fuels such as coal, oil, and natural

gas), *metallic mineral resources* (such as copper and aluminum), and *nonmetallic mineral resources* (such as salt and sand).

KEY idea Our lives and economies depend on energy from the sun and on the *natural capital*, including *natural resources* and *natural services*, provided by the earth.

Let's **REVIEW**

- What is *natural capital*? What is a *resource*?
- Distinguish between *natural resources* and *natural services* (or *ecosystem services*), and give two examples of each. Define *biosphere*.
- Distinguish between *renewable resources* and *nonrenewable resources*, and give two examples of each.

Life-Sustaining Solar Energy

Energy from the sun helps to sustain the planet's natural capital that supports life and our economies (Figure 1.13). High-quality (very useful) energy flows from the sun through the earth's life-support system. As the incoming solar energy interacts with the earth's air, water, soil, and life, and changes from one form of energy to another, it is degraded into lower-quality (less useful) waste heat that eventually flows back into space (Figure 1.15) as explained by the second law of thermodynamics.

FIGURE 1.15 Life depends on the continuous flow of high-quality energy from the sun to the earth. *See an animation based on this figure at* **www.cengagebrain.com.**

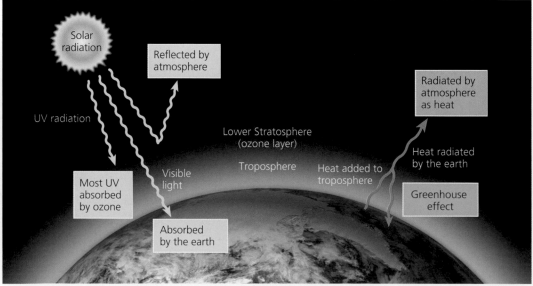

© 2014 Cengage Learning

During this process, certain chemicals such as carbon dioxide and water vapor in the atmosphere's innermost layer (the *troposphere*) absorb and release some of the out-going flow of energy, and this heats the atmosphere and the earth's surface in what is called the **greenhouse effect** (Figure 1.15). The gases that have this warming effect, including water vapor, carbon dioxide, and methane, are called **greenhouse gases**. When atmospheric levels of these gases increase over a prolonged period of time, the atmosphere tends to warm up, and when their levels drop, the atmosphere cools. Without energy from the sun and this natural atmospheric warming, life as we know it would not exist on the earth.

Along with life-sustaining energy, sunlight also contains harmful *ultraviolet (UV) radiation* (Figure 1.10) that can damage plant tissues and cause skin cancers and cataracts of the eye. Fortunately for us and most other forms of life, molecules of oxygen gas (O_2) in a layer of the atmosphere called the *stratosphere* (lying just above the troposphere) filter out some of the sun's harmful UV radiation by inter-acting with it to form molecules of ozone (O_3). This process degrades some of the UV radiation to lower-quality heat that flows back into space. Life forms on the land and near the surface of the water in oceans, lakes, and rivers sur-vive partly because of ozone in the stratosphere.

Incoming solar energy also supports **photosynthesis**—the chemical process used by green plants to make the chemicals necessary for their survival, by converting solar energy to chemical energy that plants store in their tissues. These life-sustaining chemicals are called **nutrients**. In turn, these plants provide food and energy for humans and other animals that eat the plants.

FIGURE 1.16 Two centers of the earth's biodiversity are (a) tropical rain forests and (b) coral reefs.

Life-Sustaining Biodiversity

Biodiversity is the earth's astonishing variety of species, the genes they contain, and the natural ecosystems in which they live. It is a major component of the earth's nat-ural capital (Figure 1.13).

Biodiversity plays a vital role in sustaining life in three major ways. First, its key components interact with one another and with the earth's soil, water, and air, and pro-vide us with natural resources such as food, wood, fibers, and various forms of energy (Figure 1.13, blue boxes). Sec-ond, biodiversity plays critical roles in providing free natu-ral services (Figure 1.13, yellow boxes) that help to preserve the quality of the air and water, maintain the fertility of soils, decompose and recycle wastes, and control popula-tions of pests. Third, biodiversity provides countless ways for life to adapt to changing environmental conditions.

Two ecosystems that serve as centers of biodiversity are tropical rain forests (Figure 1.16a) and coral reefs (Figure 1.16b). Because biodiversity is critical for sustaining life on the earth, scientists argue that it would be wise to protect and preserve these and all other centers of biodiversity.

Let's **REVIEW**

- How does energy from the sun sustain life? What is the *green-house effect* and how does it help to sustain life? What are *greenhouse gases*?
- How do chemicals in the stratosphere protect us from the sun's harmful ultraviolet radiation?
- What is *photosynthesis*? Define *nutrients*.
- Define *biodiversity* and explain the three-part role it plays in sustaining life on the earth.
- What are two important centers of biodiversity?

S. Borisov/Shutterstock.com

Paweł Borówka/Shutterstock.com

Energy Flow in Ecosystems

Ecologists classify the organisms living in ecosystems into different *feeding levels*, or *trophic levels*, based on how the organisms get the chemicals (nutrients) and energy that sustain them. Similarly, scientists also classify organisms as either producers or consumers.

Producers are those species that make their own nutrients from compounds and energy that they get from their environment. For example, plants use photosynthesis to capture some of the solar energy that reaches their leaves, and they use this energy to combine carbon dioxide with water to form compounds that serve as their nutrients.

All other organisms in an ecosystem are called **consumers**. They cannot produce the nutrients they need and must get them by feeding on other organisms or on their remains. Most producers that live on land are *green plants* (Figure 1.2a). In aquatic ecosystems, the major producers are *phytoplankton*—tiny organisms, mostly algae, floating in open water—and aquatic plants growing near the shore.

Decomposers are consumers that break down the remains of dead organisms and the wastes of living organisms by feeding on them. Decomposers, most of which are bacteria or fungi, act as nature's nutrient recyclers. They return most of the nutrients contained in the wastes and remains that they feed on to the soils and water in the environment for use by producers.

Almost all of the earth's life is sustained by the *one-way flow of high-quality energy* from the sun through organisms (where some of it is temporarily stored in chemical form) and into the environment as lower-quality energy in the form of heat (Figure 1.15). This energy flows through ecosystems from one trophic level to another in a **food chain**—a sequence of organisms in which energy is transferred through feeding interactions among producers and consumers (Figure 1.17).

Each time energy flows from one trophic level to the next in a food chain, some of it is lost to the environment as low-quality energy, usually in the form of waste heat, as explained by the second law of thermodynamics. Typically, only about 10% of the chemical energy available at one trophic level gets transferred to the next trophic level. This means that there is much less chemical energy available to organisms at higher trophic levels than there is at lower levels. It explains why there are many more grasshoppers that feed on plants than there are tigers that feed on other animals.

Most organisms feed on a variety of other organisms and, in turn, are eaten or decomposed by more than one

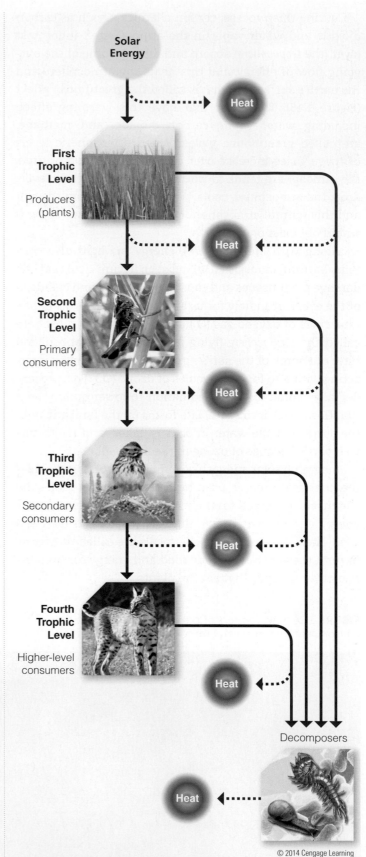

FIGURE 1.17 This diagram illustrates a food chain. *See an animation based on this figure at* **www.cengagebrain.com**.

Photo credits (top to bottom): Paolo Neo/Public-Domain-Photos.com; Peter Häger/Public Domain Pictures.net; Boyd Amanda, U.S. Fish and Wildlife Service; artwork by Patrick Lane; artwork by Patrick Lane.

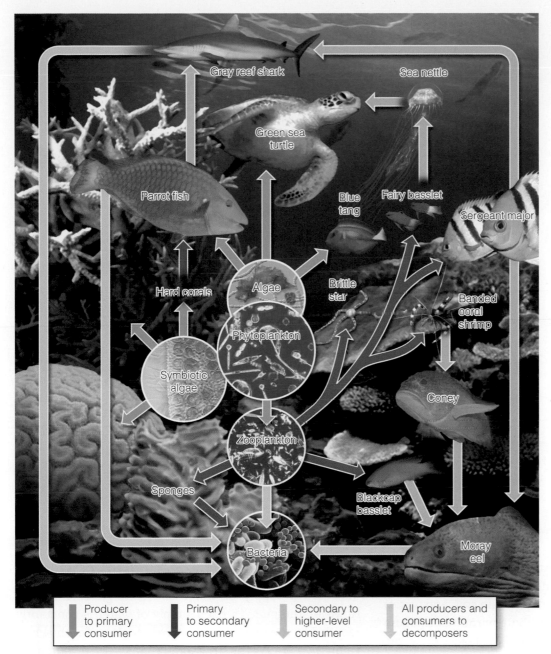

FIGURE 1.18 This food web is composed of complex feeding interactions among organisms in a coral reef ecosystem. Organisms are not drawn to scale.

Gray reef shark

Sea nettle

Green sea turtle

Parrot fish

Blue tang

Fairy basslet

Sergeant major

Hard corals

Algae

Brittle star

Banded coral shrimp

Phytoplankton

Symbiotic algae

Coney

Zooplankton

Sponges

Blackcap basslet

Bacteria

Moray eel

| Producer to primary consumer | Primary to secondary consumer | Secondary to higher-level consumer | All producers and consumers to decomposers |

© 2014 Cengage Learning

type of consumer. So each organism is part of a number of different food chains that interact in complex ways to form a feeding network called a **food web** (Figure 1.18).

Let's **REVIEW**

- Define *producer*, *consumer*, and *decomposer*, and give an example of each.
- Distinguish between a *food chain* and a *food web*.
- What happens to energy as it flows through a food chain or food web?

Chemical Cycling

Although the earth continually gets energy from the sun, it does not get any new inputs of matter (except for a sprinkling of cosmic dust and a limited number of strikes by meteorites and asteroids). Over billions of years, life on planet Earth has developed ways to recycle this fixed supply of matter and especially the nutrients necessary for life processes. These nutrients that cycle through the earth's life-support systems (Figure 1.2) include water (H_2O) and various compounds that contain carbon (C), nitrogen (N), phosphorus (P), and sulfur (S).

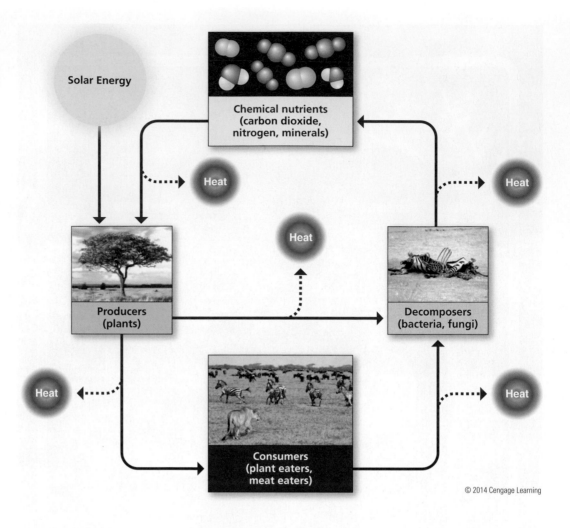

FIGURE 1.19 Life is sustained by the one-way flow of energy from the sun and by the cycling of key chemicals. *See an animation based on this figure at* www.cengagebrain.com.

Solar Energy

Chemical nutrients (carbon dioxide, nitrogen, minerals)

Heat

Heat

Heat

Producers (plants)

Decomposers (bacteria, fungi)

Heat

Heat

Consumers (plant eaters, meat eaters)

© 2014 Cengage Learning

Thus, one of nature's key natural services is **chemical cycling** (or *nutrient cycling*)—the continual circulation of key chemicals from the environment (air, soil, and water) through organisms and back to the environment (Figure 1.19). Important examples are the hydrologic, or water, cycle and the carbon, nitrogen, phosphorus, and sulfur cycles, all driven directly or indirectly by incoming solar energy and gravity. Because these and other nutrients move about in continuous and unending cycles through producers, consumers, and decomposers, *there is very little waste of such chemicals in nature.*

KEY idea   Life on the earth is sustained by a combination of the one-way flow of energy from the sun and by the recycling of nutrients.

The Water Cycle and the Carbon Cycle

The earth's supply of water does not grow or shrink. However, the planet's fixed supply of water is constantly recycled by the **water cycle**, or **hydrologic cycle**, shown in Figure 1.20. We get the freshwater that we use by removing it from *surface water* stored in the world's streams, rivers, and lakes, and from *groundwater* stored in *aquifers*, underground layers of sand, gravel, and permeable rock.

The earth's water changes from liquid to vapor or to ice and back again as it moves through the four parts of earth's life-support system (Figure 1.1). The water cycle, driven by solar energy, involves three major processes: **evaporation**, the conversion of liquid water in the oceans, rivers, lakes, and soils into water vapor in the atmosphere; **transpiration**, the evaporation of water from plant leaves into the atmosphere; and **precipitation**, the return of water to the planet's surface in the forms of dew, rain, sleet, snow, and hail. These processes help to purify water, thus providing an essential ecosystem service.

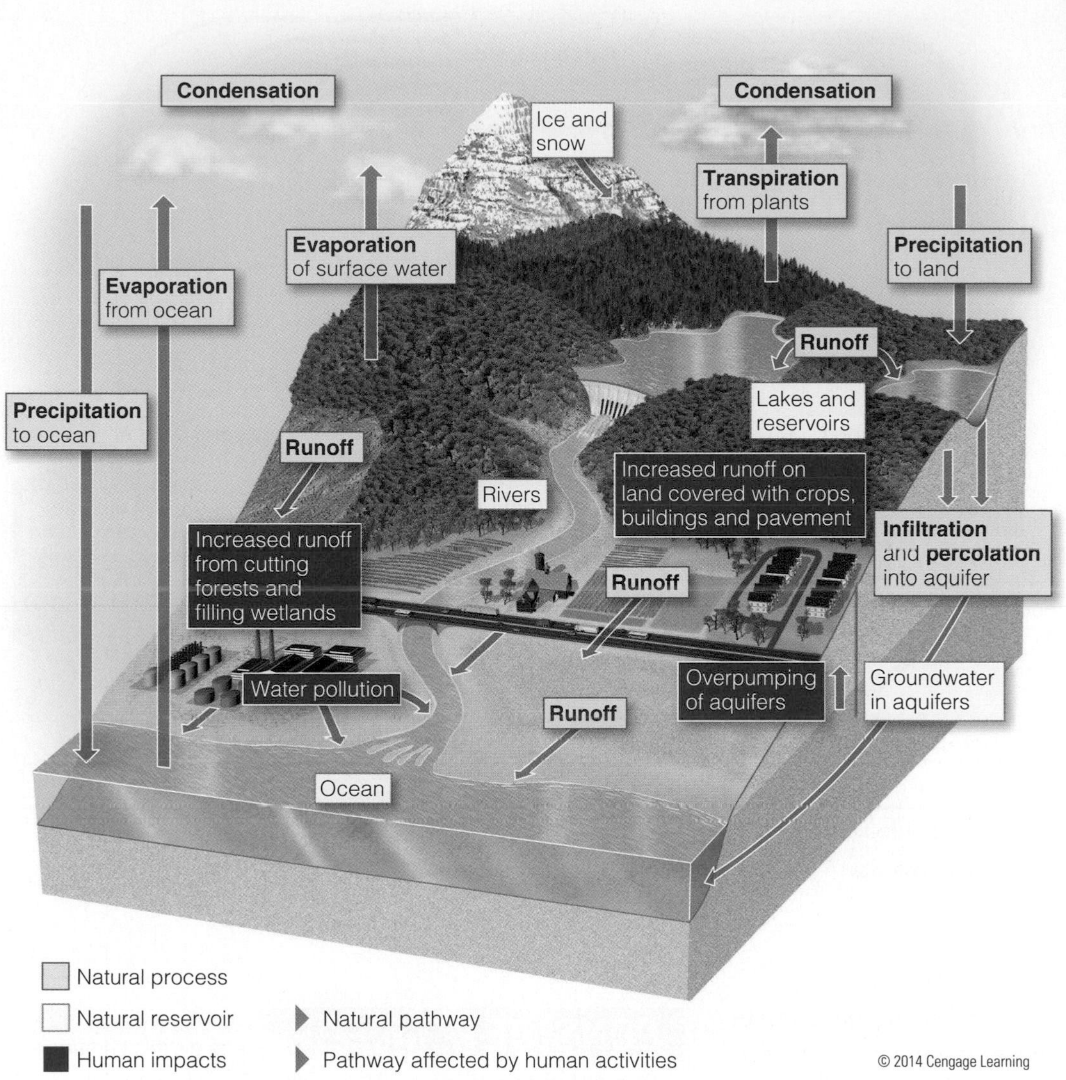

FIGURE 1.20 This diagram illustrates the water cycle (blue arrows) and human impacts on this cycle (red arrows). *See an animation based on this figure at* www.cengagebrain.com.

Condensation

Ice and snow

Condensation

Transpiration from plants

Evaporation of surface water

Precipitation to land

Evaporation from ocean

Runoff

Precipitation to ocean

Lakes and reservoirs

Runoff

Increased runoff on land covered with crops, buildings and pavement

Rivers

Increased runoff from cutting forests and filling wetlands

Infiltration and **percolation** into aquifer

Runoff

Water pollution

Runoff

Overpumping of aquifers

Groundwater in aquifers

Runoff

Ocean

☐ Natural process
☐ Natural reservoir   ▶ Natural pathway
■ Human impacts   ▶ Pathway affected by human activities

© 2014 Cengage Learning

The water cycle collects, purifies, recycles, and distributes massive amounts of water. However, humans can disrupt and even deplete these vital natural services when we remove surface water from rivers, lakes, and streams, as well as groundwater from aquifers faster than the water cycle can renew these water sources.

Another vital chemical cycle is the **carbon cycle**, which circulates life-sustaining carbon (C) endlessly through parts of the earth's life-support system (Figure 1.21, p. 20). A key part of this cycle is the compound carbon dioxide (CO_2). Plants remove CO_2 from the atmosphere through photosynthesis and convert it into complex carbon compounds (such as glucose) that plants and animals use as sources of energy. As plants and animals survive by breaking down these compounds, they return CO_2 to the air and water, completing the cycle.

Carbon dioxide is also a key greenhouse gas. When we add CO_2 to the atmosphere faster than the carbon cycle can remove it, its concentration in the atmosphere increases, as has been happening now for more than 120 years. Scientists have demonstrated that rising CO_2 concentrations have led to a warmer atmosphere. Many scientists are concerned that this atmospheric warming over the last several decades is leading to long-term changes in the planet's **climate**: the earth's overall weather conditions such as precipitation and temperature averaged over a period of at least three decades.

It is important to distinguish between climate and *weather*, which is determined mostly by changes in atmospheric temperatures, precipitation, cloud cover, and other factors, over periods of hours to years. It is easy to observe changes in the weather, and weather trends—such as general warming or cooling, and rainy or dry periods—commonly taking place over several years or even a couple of decades.

FIGURE 1.21 This is the carbon cycle, shown by the blue arrows; the red arrows show the human impacts on this cycle. *See an animation based on this figure at* www.cengagebrain.com.

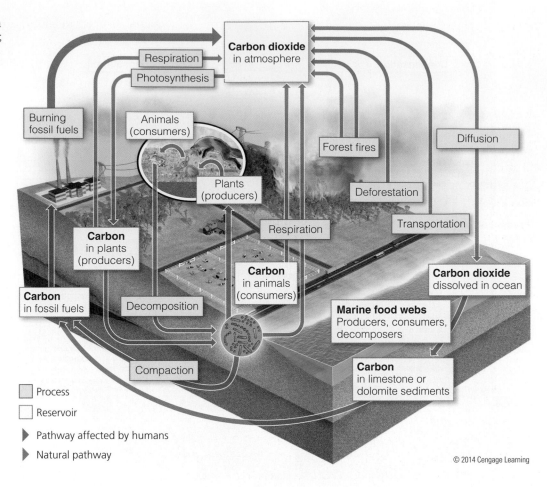

Respiration

Photosynthesis

Carbon dioxide in atmosphere

Burning fossil fuels

Animals (consumers)

Forest fires

Diffusion

Plants (producers)

Deforestation

Respiration

Transportation

Carbon in plants (producers)

Carbon in animals (consumers)

Carbon dioxide dissolved in ocean

Carbon in fossil fuels

Decomposition

Marine food webs Producers, consumers, decomposers

Compaction

Carbon in limestone or dolomite sediments

Process

Reservoir

▶ Pathway affected by humans

▶ Natural pathway

© 2014 Cengage Learning

However, conclusions about changes in the climate depend on much longer-term observations. Climate scientists require data on average annual temperatures, precipitation levels, and other factors from periods of at least 30 years in order to make observations or conclusions about climate change in any particular area of the planet. A colder or warmer year or even a decade of colder or warmer weather does not indicate climate change. However, 30 or more years of overall increases or decreases in average annual temperatures or precipitation in a given area likely would indicate a changing climate.

Since the beginning of agriculture about 10,000 years ago, we have, for the most part, been living in a fairly mild climate. But during that time, and especially since 1950, we have been adding CO_2 to the atmosphere faster than the carbon cycle could remove it, and at increasing rates, mostly by burning carbon-containing fossil fuels (coal, oil, and natural gas). We have also cut down vast areas of forests, which had been removing CO_2 as part of the carbon cycle, faster than they could grow back. These changes to the water and carbon cycles have led to problems that we explore in other modules of this textbook.

CONSIDER this
The earth's endless chemical cycling connects all past, present, and future forms of life to each other and to their environment. Some of the carbon atoms found in your skin may once have been part of a dinosaur's claw, a maple leaf, or a piece of limestone. Some of the nitrogen atoms in the breath you just took may have been inhaled by a cave dweller who lived 30,000 years ago or by Julius Caesar of ancient Rome. And your great-great-grandchildren might someday inhale some of those same nitrogen atoms.

Let's REVIEW

- What is *chemical cycling* (or *nutrient cycling*)?
- What two major natural processes sustain life on the earth?
- Describe the earth's *water cycle* (or *hydrologic cycle*) and explain the processes of *evaporation*, *transpiration*, and *precipitation*.
- Explain the nature and importance of the *carbon cycle* and the role of carbon dioxide in this cycle.
- What is *climate* and how does the carbon cycle affect it?

What Are the Problems?

High-Consumption Economic Growth, Pollution, and Wastes

Most of today's advanced industrialized countries have **high-consumption**, **high-waste economies**—economic systems that attempt to stimulate economic growth by using more matter and energy resources every year to produce more goods and services for more people (Figure 1.22). Such economies convert much of the high-quality matter and energy resources they use into waste, pollution, and low-quality heat that end up in the environment.

A group of economists called *ecological economists* contend that in the long run, such high-consumption, high-waste economic systems are not sustainable for two reasons. First, they tend to degrade and deplete irreplaceable natural resources and natural services that make up the earth's life-sustaining natural capital (Figure 1.13). Second, they can eventually exceed the capacity of the environment to handle the growing amounts of wastes and pollutants that they produce. These economists argue that we must shift from the types of economic growth that harm our life-support systems to types of economic growth that help to sustain these systems. We explore such possibilities in other modules of this textbook.

According to a number of these ecological economists, one of the root causes of such environmental degradation and waste is the fact that market prices usually do not account for the harmful environmental and health effects of providing goods and services. This means that consumers do not have accurate information about the harmful impacts of producing and using the goods and services they buy, and this encourages environmental harm. To such economists, until this underpricing is corrected, we will continue our unsustainable use of topsoil, forests, oceans, the atmosphere, and many other irreplaceable forms of natural capital.

Let's REVIEW

- Describe the key components of a *high-consumption, high-waste economy* and explain why it depends on the earth's natural capital.
- Explain why ecological economists contend that high-consumption, high-waste economies are not sustainable in the long run.
- Why do some economists push for finding ways to include the estimated harmful environmental and health costs in the market prices of goods and services?

Environmental Degradation and Ecological Footprints

There is growing scientific evidence that we are damaging our own life-support system as more and more people consume more and more resources to sustain economic growth. When we use or pollute a renewable resource faster than natural processes can replace it or clean it up, we say it has become *degraded*, or less useful. When we use up a nonrenewable resource, we say it has been *depleted*. Both cases illustrate a process called **environmental degradation** (see *The Big Picture*, pp. 22–23). When we degrade or deplete the earth's natural resources, we also reduce or disrupt the ecosystem services they provide, such as natural water purification and cycling of nutrients.

The effects of environmental degradation and depletion by human activities can be described as an **ecological footprint**—a rough measure of our environmental impact on the earth's renewable resources and life-support systems. It is an estimate of the area of ecologically productive land and water needed to provide the population of a country or area with renewable resources indefinitely, and to absorb and recycle the resulting pollution and wastes. A **per capita ecological footprint** is the average ecological footprint of an individual in a country or area.

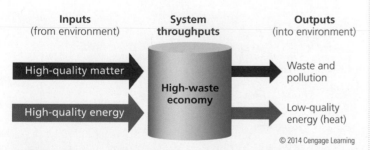

FIGURE 1.22 Industrialized countries have *high-consumption, high-waste economies. See an animation based on this figure at* www.cengagebrain.com.

Our Growing Environmental Impact

Examples of environmental degradation include large-scale cutting of tropical forests, drought, or long-term, excessively dry weather, topsoil erosion, species extinction, and air and water pollution.

Some forms of degradation such as drought are natural, but can be worsened by human activities. For example, large-scale cutting of a tropical forest can disrupt the forest region's water cycle, leading to less precipitation and making a drought worse. Similarly, while species naturally go extinct every year, human activities such as the clearing of forests and the expansion of cities can eliminate habitat for endangered species such as the Sumatran tiger.

S. Chamnanrith-UNEP/Peter Arnold, Inc.

Deforestation

viki2vin/Shutterstock.com

Drought

22

Severe Erosion from Overgrazing of Livestock

Endangered Tiger

Air Pollution

Water Pollution

23

In 2008, scientists at the World Wildlife Fund (WWF) and the Global Footprint Network estimated that humanity's global ecological footprint is at least 30% larger than the earth's estimated long-term ecological capacity to support us and other forms of life (Figure 1.23).

CONSIDER this

- We would need five or six earths to sustain a worldwide level of renewable resource use per person equal to the current U.S. level.

- There is only one earth.

FIGURE 1.23 The upper graph shows the total ecological footprint of the human population, along with projections. The lower graphs show total and per capita ecological footprints for selected countries. (Data from World Wildlife Fund and Global Footprint Network, *Living Planet Report*, 2008)

Let's REVIEW

- Define and give six examples of *environmental degradation*.
- Define *ecological footprint* and *per capita ecological footprint*.
- How many earths would we need to sustain current levels of nonrenewable resource use indefinitely?

What Can Be Done?

Three Scientific Principles of Sustainability

If we are living unsustainably, then our goal should be to learn how to live more sustainably. The only real model for sustainability is the earth, which has sustained an enormous variety of species during the roughly 3.5-billion-year history of life on the planet, despite many catastrophic changes in environmental conditions.

So how has life on planet Earth sustained itself for so long? Analysis by environmental scientist and educator Tyler Miller (see *Making a Difference*, at right) suggests that the long-term sustainability of the earth's diverse forms

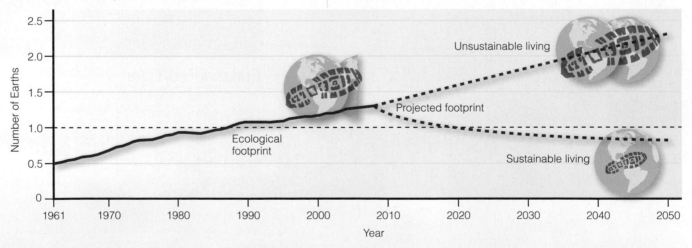

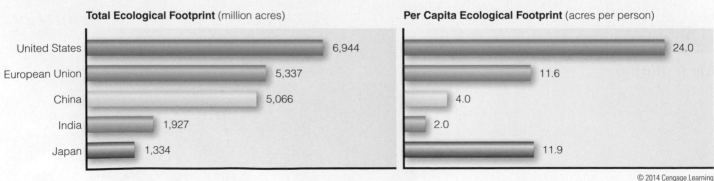

© 2014 Cengage Learning

MAKING A difference

G. Tyler Miller, Jr.—Champion of Sustainability

Professor and author Tyler Miller made an important career change. He had earned a PhD in physical chemistry, but his background also included physics, energy, ecology, and environmental economics. He decided to switch his professional interests from chemistry to environmental science, and in 1970, he wrote his first book exploring relationships between energy (thermodynamics) and the environment. In 1971, he started one of the first undergraduate environmental studies programs in the United States.

Miller believes that the most important thing any of us can do is to understand and apply three very simple scientific principles of sustainability (Figure 1.24, p. 26). He arrived at these principles by asking how the earth—the only truly sustainable system—has sustained an amazing variety of life for at least 3.5 billion years despite several highly disruptive changes in environmental conditions.

Including multiple editions of several books, Miller has written more than 60 volumes for introductory courses in environmental science, basic ecology, energy studies, and environmental chemistry. His first introductory environmental science textbook, published in 1975, has served as the model for many similar textbooks published since then. Miller has received two honorary doctorate degrees for his contributions to environmental education.

Beginning in 1985, Miller spent 10 years living in a forest in North Carolina in order to deepen his understanding of environmental science and work on his books. He converted an old school bus into living quarters and environmental science laboratory. To do this, he thoroughly insulated the bus, attached a small sunroom to help heat the structure, and buried tubes to bring in air cooled by the earth (geothermal cooling) at a cost of about $1 per summer. He also experimented with solar hot water systems, energy-efficient windows, and waterless composting toilets. He studied organic gardening and biological pest control, composted his food wastes, used natural vegetation instead of grass around his home and laboratory, and studied the plants, animals, and natural systems around him. In 1995 he left the woods and has since applied what he learned to urban living while he continues to study and write.

Miller's books have had an important impact on several million students throughout the world. He has received many letters and e-mails from students who have said that studying from one of his environmental science books changed their lives and motivated them to pursue further studies in the environmental sciences.

---

of life is explained by three easily understood **scientific principles of sustainability** (Figure 1.24, p. 26). Without solar energy flow, chemical cycling, and biodiversity, life as we know it would not exist. Miller suggests that we can use these three scientific principles to guide our efforts in making a cultural shift toward living more sustainably over the next few decades.

These principles of sustainability are:

1. **Solar energy:** The one-way flow of energy from the sun through living organisms and back to the environment (Figures 1.15) warms the earth, helps plants to provide food through photosynthesis, and powers indirect forms of renewable solar energy such as wind.
2. **Chemical cycling:** Natural processes continually cycle the chemicals necessary for life from the environment's soil, water, and air through organisms and back to the environment (Figures 1.19, 1.20, and 1.21).
3. **Biodiversity:** The earth's great variety of genes, species, and ecosystems are a part of the life-sustaining processes of energy flow and chemical cycling (Figure 1.19).

Biodiversity also plays a key role in ecosystem services (Figure 1.13, blue boxes) such as the renewal of topsoil, pest control, and air and water purification. Interactions among species for resources such as food, water, and a place to live also provide population control by limiting the population sizes of all species. In addition, the earth's storehouse of different genes provides countless ways for life to survive and adapt to major changes in environmental conditions. Such changes have wiped out much of the earth's life at least five times in the past. Within 5–10 million years after each of these *mass extinctions*, natural processes led to a new diverse mixture of species and ecosystems.

KEY idea Our lives and economies depend on three scientific principles of sustainability: the one-way flow of energy from the sun, the continual recycling of the chemicals needed by all forms of life, and earth's biodiversity.

FIGURE 1.24 This diagram shows the three inter-connected scientific principles of sustainability, or lessons from nature, that can help us learn to live more sustainably.

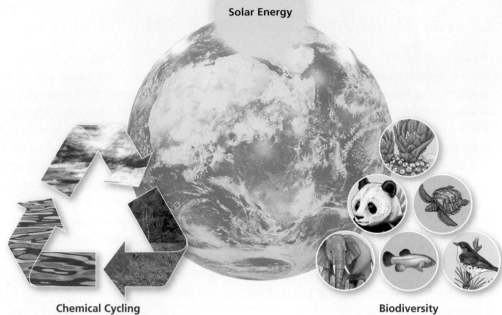

Solar Energy

Chemical Cycling

Biodiversity

© 2014 Cengage Learning

Let's **REVIEW**

Summarize the three *scientific principles of sustainability*.

We use these and other principles to help us evaluate various proposed solutions for each of the major environmental issues discussed in this book's 14 other modules. There are thousands of success stories about efforts to solve environmental problems, and we also include many of these in those modules. Many of these environmental successes are taking place today on college and university campuses around the world (see the following *For Instance*).

FOR **INSTANCE...**

Campus Sustainability Efforts

Sustainability is alive and thriving on many college and university campuses. Such efforts often begin when students, faculty, and administrators work together to make a comprehensive *environmental audit* of their campus—a process in which they gather information and ideas about how to make their institutions more environmentally sustainable. Such research also helps students to learn how to make measurements of environmental quality and practices on their campuses.

Through such sustainability efforts, campuses often cut costs and save money, using a variety of strategies. For example, many of them focus on reducing energy waste by insulating campus buildings and plugging air leaks, installing energy-efficient windows, and using energy-efficient appliances, lighting, and heating and cooling systems. Some schools have installed highly visible meters to monitor energy use in their dormitories, and some have posted this information online. This measure has led to competitions among students to see whose dormitories can be the most energy efficient.

Another campus strategy is to decrease climate-changing carbon dioxide emissions, mostly by reducing energy waste. Many campuses are also getting more energy from low-carbon sources such as the sun, wind, and biomass heating plants. Some are reducing car use by encouraging walking, bicycling, car-sharing, and the use of buses. Conserving water is another way to promote sustainability. Some campuses are reducing water waste by fixing leaks, using natural vegetation to reduce the need for watering plants, and installing water-efficient showerheads and sink faucets. Some use waterless urinals and waterless toilets designed to convert human wastes to compost.

Some schools focus on cutting their outputs of wastes by reducing resource use, keeping track of the amounts of trash being thrown away, emphasizing recycling and

FIGURE 1.25 These students at Connecticut College are dumping food scraps into a large composter that will convert this waste to a useful fertilizer for use in growing crops.

composting (Figure 1.25), and promoting reuse of resources by collecting unwanted furniture and other items for sale or for donation to charity organizations. Since 2000, several hundred colleges have participated in an annual 10-week *Recycle Mania* competition—started by students—to see which schools can recycle or reuse the largest percentage of their solid wastes.

Other strategies include:

- emphasizing sustainable design for new buildings and for renovating existing buildings;
- sustaining biodiversity by protecting natural areas on campus and restoring degraded areas;
- using natural landscaping with areas of native vegetation in place of grass lawns, thereby reducing the need for watering as well as the use of fertilizers, herbicides, and pesticides;
- requiring the use of recycled paper, certifiably sustainable wood products, and biodegradable and nontoxic cleaning products in campus operations;
- promoting sustainable food production by growing organic food or buying it from nearby organic farms for use in campus food systems; and integrating sustainability concepts and practices throughout the entire curriculum as Arizona State University did when it opened its School of Sustainability, the first in the United States, in 2008.

All of these strategies involve the application of one or more of nature's three scientific principles of sustainability

(Figure 1.24). Here are three more of the thousands of examples of sustainability efforts being carried out in colleges and universities:

Students at Wisconsin's Northland College played a major role in designing a dormitory and learning center that gets much of its heat and electricity from passive solar design, solar cells, and a wind turbine (Figure 1.26). The building uses waterless toilets that convert human waste to a compost that can be used to fertilize campus vegetation. The building also contains furniture, carpeting, and window shades made from recycled materials.

At Ohio's Oberlin College, viewed as one of the greenest U.S. colleges, environmental science faculty and students worked together to design a nationally recognized environmental studies building. It has an ecological system for purifying wastewater in open tanks containing a diversity of algae and other organisms. This wastewater treatment process mimics how nature converts wastes into resources and is powered mostly by sunlight collected in the building's passive solar greenhouse.

In Cupertino, California, students and faculty at De Anza College won the highest award ever granted to a U.S. community college for sustainable building design. Their design applies all three scientific principles of sustainability. First, it emphasizes use of the sun for natural lighting and heating, and for producing electricity with the use of solar cells. Displays near the building's entrances show how much electricity its rooftop solar-cell panels are generating. Secondly, the designers applied the chemical cycling principle by using structural steel, furniture, carpeting, toilet seats, and other items made out of recycled

FIGURE 1.26 Students helped to plan and build the Environmental Living and Learning Center at Northland College in Ashland, Wisconsin.

materials and by stressing reduction of solid and hazardous waste outputs. Finally, the designers applied the biodiversity principle by focusing on preservation of natural areas near the building.

What type of sustainability programs does your school have?

A Look to the Future

What will be our environmental future? No one knows, of course, but two things seem assured. First, new and surprising events and changes will unfold, and along with them will come new problems and new possibilities. Second, we are all deeply involved in influencing the possible futures we face, whether we appreciate it or not.

We live in what might be called the *environmental century*, and we face a vital and fascinating challenge. It is likely that no future generation will have the same chance that we have to influence the world of the future. We are at a fork in the road with a choice between two paths. One choice is to continue degrading our life-support system. The other is to learn how to live more sustainably.

The biggest challenge we face is to reduce our ecological footprints (Figure 1.23) as a way to improve the quality of life for many people today without reducing the prospects for future generations. *This is the challenge of sustainability.* The Industrial Revolution, which began about 250 years ago, was a remarkable global transformation. Many environmental leaders say that during this century, we need to bring about another global transformation in the form of a *sustainability revolution.* Figure 1.27 shows some of the major shifts that could be involved in such a cultural change.

In 2010, environmental scientist and educator G. Tyler Miller (see *Making a Difference*, p. 25) proposed three **social science principles of sustainability** (Figure 1.28), which, along with his other three science-based principles of sustainability (Figure 1.24), could help us to make this cultural shift. They are **(1)** *full-cost pricing* (from economics), in which we find ways to include in the market prices of the goods and services we use the harmful environmental and health costs of producing and using those goods and services; **(2)** *win-win solutions* (from political science), in which we learn to work together to deal with environmental problems and find solutions that will benefit the largest possible number of people, as well as the earth itself, now and in the long-term future; and **(3)** *a responsibility to future*

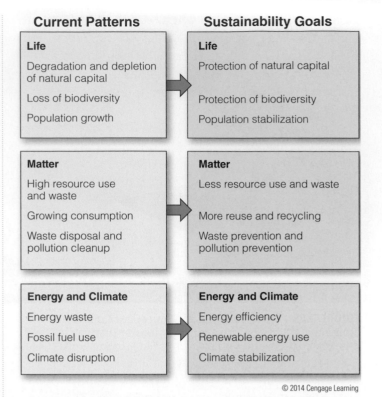

Current Patterns	Sustainability Goals
Life	**Life**
Degradation and depletion of natural capital	Protection of natural capital
Loss of biodiversity	Protection of biodiversity
Population growth	Population stabilization
Matter	**Matter**
High resource use and waste	Less resource use and waste
Growing consumption	More reuse and recycling
Waste disposal and pollution cleanup	Waste prevention and pollution prevention
Energy and Climate	**Energy and Climate**
Energy waste	Energy efficiency
Fossil fuel use	Renewable energy use
Climate disruption	Climate stabilization

© 2014 Cengage Learning

FIGURE 1.27 Environmental leaders call for us to make a number of shifts in emphasis in order to bring about an *environmental* or *sustainability revolution* within this century.

© 2014 Cengage Learning

FIGURE 1.28 These three *social science principles of sustainability* could help us make a transition to a more environmentally and economically sustainable future.

generations (from ethics) in which we accept our responsibility to pass on to them the earth's life-support systems in as healthy a condition as that which we have enjoyed.

Based on our history of cultural change, it is likely that a sustainability revolution can happen much faster than we might think. Social scientists point out that major social changes usually occur when just 5–10% of the people become convinced that they are necessary and take the lead to convince others to bring about such change.

You have the opportunity to take part in meeting the environmental challenges of the future. You can start by thinking about and making environmentally beneficial changes in your own life and in your community. As G. Tyler Miller has stated:

> *Some say that making the shift to sustainable societies is idealistic and unrealistic. Others say that it is idealistic, unrealistic, and dangerous to continue assuming that our present course is sustainable, and they warn that we have precious little time to change. If certain individuals had not had the courage to forge ahead with ideas that others called idealistic and unrealistic, very few of the human achievements that we now celebrate would have come to pass.*

Let's **REVIEW**

- List nine major cultural shifts that environmental leaders say we need to make in order to bring about a sustainability revolution.
- List three *social science principles of sustainability* that could help us to make a transition to a more sustainable future.

What Would You Do?

The four major types of human activities that have the greatest harmful impacts on the environment are *food production and consumption, transportation, home energy use,* and *overall resource use.* Most environmental scientists have focused on one or more of these areas as they study environmental problems and search for solutions. At the personal level, informed individuals around the world are taking the same approach, looking for ways in which they can shrink their ecological footprints. Here are some of the ideas and practices they are adopting.

Food Production and Consumption

- Reducing meat consumption
- Buying or growing organic food
- Buying locally grown food

Transportation

- Driving energy-efficient vehicles that get at least 40 miles per gallon
- Reducing car use by walking, bicycling, car-sharing, and using mass transit
- Working at home if possible

Home Energy Use

- Heavily insulating houses and work spaces, plugging air leaks, and installing energy-efficient windows
- Using energy-efficient heating and cooling systems, lights, and appliances
- Using the sun, geothermal energy, and other renewable energy sources to provide as much space heating and water heating as possible

Resource Use

- Reducing, reusing, recycling, composting, and sharing
- Focusing on prevention of pollution and waste
- Using electricity generated by renewable energy resources such as the sun, wind, flowing water, and geothermal energy

Many of the more effective and sustainable solutions to the environmental problems we face will involve changing the way we think about, and act in, the world. This will involve each of us examining our *environmental worldview*—our view about how the earth works, how we interact with the environment that sustains us, and what our role in the world should be.

It starts with this question: How can I live more sustainably?

KEYterms

atom, p. 7
biodiversity, p. 15
biosphere, p. 13
carbon cycle, p. 19
chemical change, p. 9
chemical cycling, p. 18
chemical element, p. 7
chemical formula, p. 8
chemical reaction, p. 9
climate, p. 19
compound, p. 7
consumers, p. 16
decomposers, p. 16
ecological footprint, p. 21
ecology, p. 3
ecosystem, p. 3
ecosystem services, p. 13
electromagnetic radiation,
    p. 10
energy, p. 10
energy quality, p. 11

environment, p. 2
environmental
    degradation, p. 21
environmental science,
    p. 2
evaporation, p. 18
first law of
    thermodynamics (law
    of conservation of
    energy), p. 12
food chain, p. 16
food web, p. 17
greenhouse effect, p. 15
greenhouse gases, p. 15
high-consumption,
    high-waste economy,
    p. 21
high-quality energy, p. 11
hydrologic cycle, p. 18
ion, p. 8
kinetic energy, p. 10

law of conservation of
    matter, p. 9
low-quality energy,
    p. 11
matter, p. 7
molecule, p. 8
natural capital, p. 13
natural resources, p. 13
natural services, p. 13
nonrenewable resources,
    p. 14
nutrients, p. 15
organism, p. 3
peer review, p. 5
per capita ecological
    footprint, p. 21
photosynthesis, p. 15
physical change, p. 9
potential energy, p. 11
precipitation, p. 18
producers, p. 16

renewable resources, p. 13
resource, p. 13
science, p. 4
scientific data, p. 4
scientific hypothesis, p. 4
scientific law, p. 6
scientific model, p. 6
scientific principles of
    sustainability, p. 25
scientific theory, p. 4
second law of
    thermodynamics, p. 12
social science principles
    of sustainability, p. 28
species, p. 3
sustainability, p. 2
transpiration, p. 18
water cycle, p. 18

THINKINGcritically

1. Explain how you used the scientific process at some point to help you solve a problem.

2. How would you respond if someone told you that we need not sharply reduce mercury emissions from coal-burning industrial or power plants because scientists have not absolutely proven that inhaling mercury from such emissions has ever killed anyone?

3. What are three things you would do to cut waste in human societies by copying the earth's chemical cycling principle of sustainability?

4. Use the second law of thermodynamics to explain why we cannot recycle the high-quality chemical energy stored in gasoline.

5. Describe two ways in which you could apply each of the three scientific principles of sustainability (Figure 1.24) to your daily activities.

6. For each of the three social science principles of sustainability (Figure 1.28), give an example of how a violation of that principle might have resulted in harm to someone or something in your community or in the world.

LEARNINGonline

Access an interactive eBook and module-specific interactive learning tools, including flashcards, quizzes, videos and more in your Environmental Science CourseMate, accessed through **CengageBrain.com**.

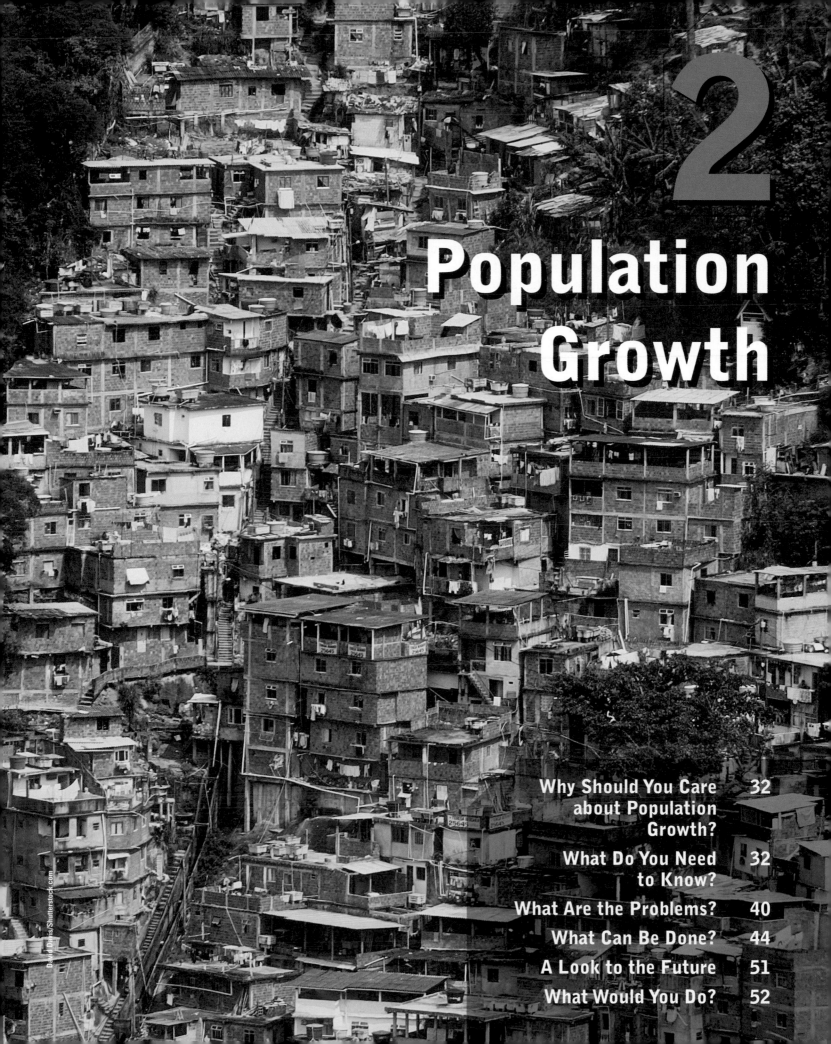

2

Population Growth

Why Should You Care about Population Growth?

The human population on our planetary home now stands at 7 billion and every year it grows by about 83 million. By 2050, there may be 9.3 billion of us. This means that in less than four decades, we could be adding almost twice the number of people now living in China to the earth's population. Are there too many of us? Are there too few of us? These are important questions.

Why should we care about the growth of the human population? There are two major reasons. First, each of us depends on the earth's life-support systems for food, shelter, clean water and air, energy, and other vital resources such as wood, iron, and aluminum. Adding more people to the population increases the need for these natural resources as well as natural services such as chemical cycling and renewal of topsoil.

Also, average income per person is rising in many countries, and most of the world's more affluent consumers tend to use more of the earth's natural resources and services. As a result, average per-person, or *per capita*, natural resource use is very high in wealthy countries such as the United States (Figure 2.1a) and much lower in poorer countries such as India (Figure 2.1b).

The second major reason for caring about population growth is the strong and growing scientific evidence that we are degrading our life-support system with our rapidly growing *ecological footprints* (see Module 1, Figure 1.23, p. 24). Each newcomer adds to this planetary stress, especially in areas of the world where people are using a lot of resources.

While our ecological footprints are growing larger, we are not meeting the basic needs for many of the people alive today. In 2008, the World Bank estimated that 1.4 billion people—one of every six persons on the planet—were living in extreme poverty and trying to survive on the equivalent of less than $1.25 a day. This number equals China's entire population and amounts to about 4.5 times the U.S. population. This raises a crucial question: If we have failed to meet the basic needs for 1.4 billion people today, what will happen in 2050 when there may be 2.3 billion more of us?

In this module, we look at the basics of *population dynamics*, or how populations change, and we apply these principles to the human population. We explore some of the problems arising from population growth and consider some ways to deal with those problems.

What Do You Need to Know?

Population Growth and Limits to Growth

An important concept in population dynamics is **exponential growth**, the growth of any quantity at a fixed percentage per unit of time. For example, a population can grow at 2% per year, a rate that will double its size every 35 years. Exponential population growth starts out slowly, but eventually, the population's size increases rapidly. This is because, even if the percentage of growth remains constant, at 2%, the number of people added every year gets

FIGURE 2.1 Compare these photos: (a) an American family of four from Pearland, Texas, with their major possessions, and (b) a family of five from the village of Shingkhey, Bhutan with all their possessions.

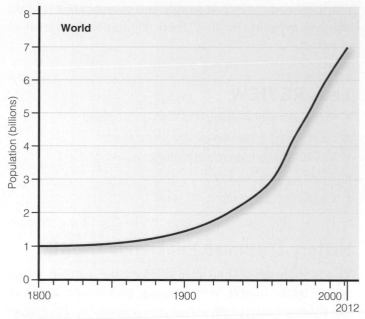

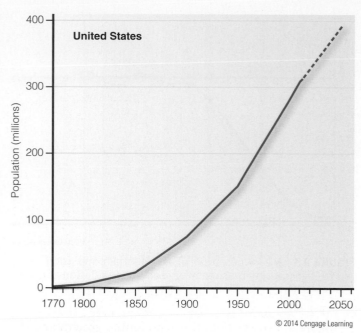

© 2014 Cengage Learning

FIGURE 2.2 Plotting exponential world population growth on a graph (left) shows that it started out slowly but accelerated rapidly. The graph on the right (with a projection to 2050) shows how the U.S. population has also grown exponentially.

larger because it is 2% of a larger and larger total. So the population grows by a much larger amount in each unit of time, and eventually it explodes (Figure 2.2).

A striking example of exponential growth is the human population. It took from the time when humans emerged, nearly 200,000 years ago, until around 1800 to add the first billion people. The second billion were added during the next 130 years (by 1930), the third billion in the next 30 years (by 1960), and the fourth billion within the next 14 years (by 1974). Since then, another billion have been added about every 12–13 years, and experts project that we will add another 1 billion people during the 12 years between 2011 and 2023. Figure 2.2 (right) shows the similar exponential growth of the U.S. population, with a projection to 2050.

Some life forms, including many species of bacteria and insects, have an even more astonishing ability to increase their numbers. Individuals in these populations tend to reproduce many times during their life spans and to have many offspring each time they reproduce. For example, some species of bacteria can reproduce a generation of millions every 20 minutes. Without any controls on the population growth of these species, they could produce enough offspring to form a 1-foot deep layer of microscopic organisms over the earth's entire surface in less than 2 days.

In reality, this kind of exponential growth will not happen because it is limited by certain natural factors, called **limiting factors**. These include the availability of resources necessary for survival, such as water, food, and space, as well as environmental threats to survival, such as diseases and predators. Extensive scientific research has shown that in the real world, every growing population eventually reaches some size limit imposed by one or more limiting factors.

The combination of limiting factors that act together to control the growth of a population helps to determine the **carrying capacity**, or the maximum population of a given species that an area of land or a volume of water can sustain indefinitely. As the size of a population nears the carrying capacity of its environment, its rate of growth typically slows because of decreasing supplies of key resources such as food, water, and space.

In other words, there are always limits to population growth in nature because there are not enough resources to sustain an ever-growing population of any species. This is the result of the scientific sustainability principle of biodiversity (Module 1, Figure 1.24, p. 26), which has to do with how the earth's variety of species interact through feeding and other relationships to limit the population sizes of all species.

KEY idea

Every growing population encounters natural limits to its growth.

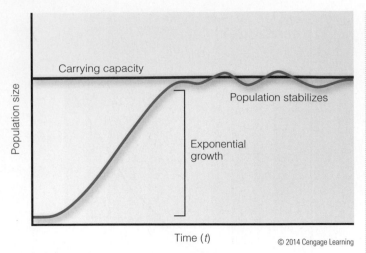

FIGURE 2.3 A J-shaped curve of exponential population growth becomes an S-shaped curve as the population encounters limits to its growth.

Figure 2.3 shows both the exponential growth and the slowdown in the growth of a population that reaches its environment's carrying capacity. The left half of the curve shows a J-shaped curve of exponential growth. The right half shows how the growth rate slows and eventually levels off, becoming an S-shaped curve as limiting factors begin kicking in.

Sometimes, populations grow so rapidly that they temporarily exceed their environment's carrying capacity. In such a case, the population dies back in a *population crash* because its environment cannot support it. This happened to a herd of 26 reindeer placed on a small island in the Bering Sea in 1910 (Figure 2.4). At first, they had plenty of food, but by 1935, their population had exploded to 2,000 as

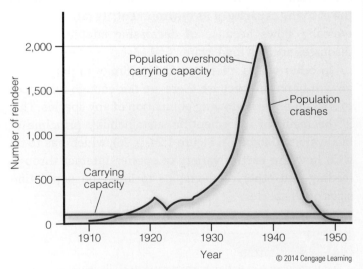

FIGURE 2.4 This graph shows the exponential growth and population crash of a herd of reindeer on a small island in the Bering Sea.

they depleted their food supply and overshot the island's carrying capacity. By 1950, their population had crashed to only 8.

Let's REVIEW

- Define and give an example of *exponential growth*.
- Define and give an example of a *limiting factor*.
- Why is it that no population can keep growing forever?
- Define *carrying capacity*.
- Explain how a population can crash.

History of Human Population Growth

The human population has followed the pattern represented by the J-curve of exponential growth and is now in the sharply rising part of the curve (Figure 2.5). This exponential growth is mostly due to three factors:

1. We learned to use energy from coal, oil, and other fuels and developed a variety of machines and other technologies. These advances have allowed us to expand into almost all of earth's climate zones.
2. Modern agriculture has allowed us to produce larger and longer-lasting food supplies by using technologies such as farm machinery (Figure 2.6), irrigation, chemical fertilizers and pesticides, and factory fishing boats.
3. Human death rates dropped sharply below birth rates as we improved sanitation and health care and developed medicines to help control infectious diseases.

CONSIDER this Every day, the earth's population grows by more than 227,000 people. In about 43 days, this increase becomes larger than the population of Los Angeles County, California; in 9 months, it exceeds the population of France; and in about 4 years, it equals the population of the United States.

Although the world's population is growing by 83 million per year, the annual exponential *rate* of population growth has been declining and is projected to continue dropping. Note on the graph in Figure 2.7 that in 2012, the average annual rate of growth of the human population was just over 1%, and that in 1965, the rate was about 2%. Even though the rate of growth has been cut in half, the population is still growing exponentially (Figure 2.5). For example, 1% of 7 billion people (the population in 2012) is

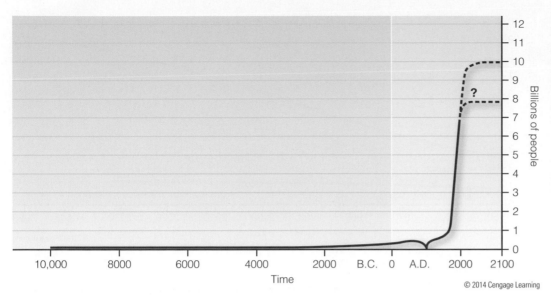

FIGURE 2.5 The human population has grown exponentially as represented by a J-shaped curve. In 2011, United Nations population experts projected that the population will reach about 9.3 billion by 2050 and may level off at about 10 billion by 2100 with the J-shaped curve changing to an S-shaped curve. (Data from the World Bank and United Nations, 2011; this figure is not to scale)

© 2014 Cengage Learning

70 million, which is more than 2% of 3 billion (the population in 1965), or 60 million.

The world's annual population growth is unevenly distributed between poorer and richer countries, and the UN projects that this gap will widen (Figure 2.8, p. 36). During 2012, only 1 of every 80 infants was born in one of the world's **more-developed countries**—highly industrialized countries where the average level of income per person and rate of consumption are relatively high. The other

79 babies were added to the world's **less-developed countries**—those countries with lower average levels of income and low rates of consumption. In 2012, about 82% of the world's people lived in less-developed countries. Many of these people live in encampments or crowded slums in cities such as Rio de Janeiro.

In 2012, China had the world's largest population, with almost 1.4 billion people, or one of every 5 people on the planet. India ranked second with almost 1.3 billion people. Together, China and India were home for 38% of the earth's people. The United States, with 313 million people, was the third most populous country with 4.5% of the world's people.

FIGURE 2.6 High-tech farming equipment and other technologies have enabled humans to greatly expand their ability to raise large quantities of crops such as the wheat in this field.

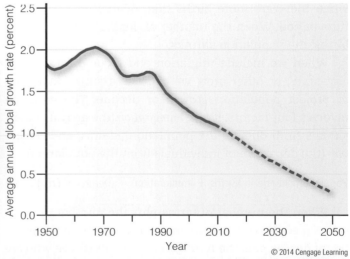

© 2014 Cengage Learning

FIGURE 2.7 The annual rate of population growth has been declining since 1960 and is projected to continue dropping.

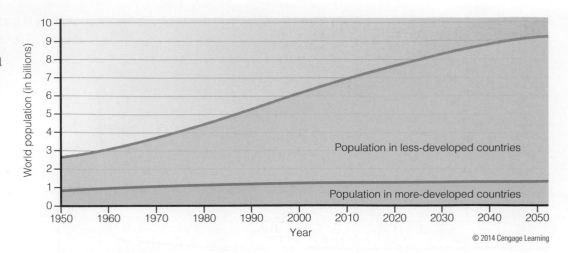

FIGURE 2.8 The United Nations projects that the difference in population growth rates between more-developed and less-developed countries will grow. (Data from Population Reference Bureau and United Nations Population Division, 2012)

© 2014 Cengage Learning

Let's **REVIEW**

- What is the current size of the human population?
- What three factors caused the world's population to grow exponentially?
- Define and distinguish between *more-developed countries* and *less-developed countries*.
- How were the people who were added to the population in 2012 distributed between more-developed and less-developed countries?
- What are the world's three most populous countries?

Dynamics of Human Population Growth

The populations of countries, cities, states, and other defined areas grow or shrink as a result of three factors acting together: births, deaths, and **migration**, or the movement of people from one population to another. Putting migration aside for a moment, a population grows when there are more births than deaths in a particular time period. When the number of deaths exceeds that of births, the population shrinks.

When we include migration and focus on a country, city, state, or other area, we can use a simple equation to project population growth or decline. The equation involves four factors: *births*, *immigration* (the arrival of individuals from outside the population), *deaths*, and *emigration* (the departure of individuals from the population).

Population change = (Births + Immigration) − (Deaths + Emigration)

When the number of births plus the number of *immigrants* (people who are moving in) is larger than the number of deaths plus the number of *emigrants* (those who are leaving), an area's population grows; when the reverse is true, its population declines.

The birth factor in human population growth is commonly measured by **total fertility rate (TFR)**: the average number of children born to the women in a population during their reproductive years. This number is important because it helps to determine the rate of growth of a population. If a population has a TFR of well over 2.0, it will grow, because each couple is having more than enough children to replace themselves when they die. If the TFR is lower than 2.0, the population will eventually start to decline, because not enough people will be born to replace each couple.

TFRs have generally been dropping in most countries of the world. Figure 2.9 shows the changes in average TFRs between 1950 and 2012 for the world, the more-developed countries, and the less-developed countries, with projections to 2050.

TFRs vary widely and tend to be higher in less-developed countries. One reason for this is that in poorer cultures, large families are common because many poor couples rely on their children to care for them in their old age. In countries with old-age pension programs, there is less of a tendency for people to have large families.

Another reason for higher TFRs in less-developed countries is that infants are more likely to die in poorer countries, so having several children might insure the survival of at least a few. Also, in these cultures, larger numbers of children can provide more help by hauling daily drinking water (Figure 2.10), gathering wood for heating and cooking, or tending crops and livestock.

NUMB3RS

2.5

The world's total fertility rate, which needs to reach 2.0 to eventually halt global population growth

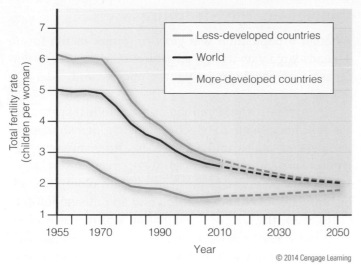

FIGURE 2.9 This graph shows key total fertility rates from 1950 to 2012 with projections to 2050. (Data from UN Population Division)

Researchers are also finding that TFRs can be affected by the availability or lack of education and employment opportunities for women. In countries where women are free to go to school and to get jobs outside of the home, TFRs tend to be lower than they are in countries where women are expected to stay at home. In less-developed countries, on average, women who have no formal education typically have two more children than do women who have a high school education, according to UN studies. Similarly, TFRs tend to be lower in countries where family planning services are available, which is more likely the case in urban areas than in rural areas.

KEY idea

The average number of children born to fertile women in a population (*total fertility rate or TFR*) is a major factor that largely determines the population's size. The average TFR for the world has been dropping but is not yet low enough to stop the world's population growth in the near future.

Two other important factors in human population growth are related to the death factor of the equation. One is **life expectancy**, the average number of years a person in a population can expect to live. It is calculated by statisticians who consider a host of factors such as diseases present in the population, occupations, accident rates, and other health hazards. The higher the life expectancy, the more likely a population is to remain stable or to grow. Between 1955 and 2012, global life expectancy increased from 48 years to 70 years (78 years in more-developed countries and 68 years in less-developed countries) and is projected to reach 74 by 2050.

FIGURE 2.10 These girls in India are carrying water to their homes.

The other important factor is the **infant mortality rate**, or the number of babies, out of every 1,000 born, who die in their first year of life. Because infant mortality is related to a country's general level of health care and nutrition, it is viewed as one of the best measures of a society's quality of life. If this number is high, it can slow a population's rate of growth. The lower this rate is, the more likely the population is to remain stable or to grow. Between 1965 and 2012, the infant mortality rates dropped from 20 to 6 in more-developed countries and from 118 to 48 in less-developed countries. Even with these sharp declines, an average of 11,000 infants (most of them in less-developed countries) die every day from preventable health causes.

Let's REVIEW

- Define *migration,* and write a simple equation to show how four key factors affect human population change.
- What is the *total fertility rate (TFR)*? How does it affect population growth?
- Give three reasons why it can make sense for couples living in poverty to have a large number of children.
- Define *life expectancy* and *infant mortality rate* and explain how they can affect population growth.

Population Growth in the United States

Between 1900 and 2012, the population of the United States grew about fourfold, from 76 million to 313 million (Figure 2.2, right). The country added 79 million people to its population between 1946 and 1964—a period of high birth rates called the *baby boom*. At the peak of the baby boom in 1957, the TFR reached 3.7 children per woman. Since then, it has declined, and in most years since 1972, has been at or below 2.1 children per woman.

The rate of population growth in the United States slowed because of this drop in the TFR. But the U.S. population is still growing faster than the populations of all other more-developed countries and of China, and it shows no sign of leveling off. According to the U.S. Census Bureau, the U.S. population grew in 2012 by about 3.2 million, an average of 1 additional person every 10 seconds. Most of these newcomers were newborn babies, but around 1 million were immigrants, mostly from Latin America, Asia, and Europe.

The U.S. Census Bureau projects that between 2012 and 2050, the U.S. population will grow from 313 million to 423 million. Immigration is projected to account for at least a third of this projected growth.

Population Age Structures

Whether a population will grow or shrink in the future depends largely on its **age structure**: the numbers or percentages of people in its young, middle, and older age groups.

To represent these numbers, population experts use *age-structure diagrams*, which show the percentages or numbers of males and females in three different age groups: *pre-reproductive* (ages 0–14), *reproductive* (ages 15–44), and *post-reproductive* (ages 45 and older). Excluding immigration and emigration and an unusual jump in death rates, the size of a population will increase when most individuals are in, or will soon enter, the reproductive age group. It will drop when most individuals are older than the reproductive age group, and it will remain about the same when there are roughly equal numbers of people in the three age groups.

Figure 2.11 shows age-structure diagrams for countries with rapid, slow, zero, and negative population growth rates.

KEY idea   The numbers or percentages of people in a population's young, middle, and older age groups determine how fast the population will grow or decline.

The populations of countries like Guatemala (Figure 2.11, far left), having large percentages of people younger than age 15, are growing rapidly. In fact, their birth numbers will grow for several more decades, even if these countries can cut their TFRs to 2.0. This is because the numbers of girls entering their prime reproductive years in these countries are very large. In other words, these countries are in the J-curve stage of their exponential population growth (Figure 2.5).

Because many less-developed countries like Guatemala have large pre-reproductive populations, there are dramatic differences between the age structures of a typical more-developed country (Figure 2.11, middle and right) and a typical less-developed country (Figure 2.11, left). These differences help to explain why all but a small percentage of the projected population growth in this century will occur in less-developed countries (Figure 2.8). The people there who are younger than age 15, being future mothers and fathers, will play a key role in their countries' future population growth rates. About 29% of the people in less-developed countries are younger than age 15, compared to 16% in the more-developed countries.

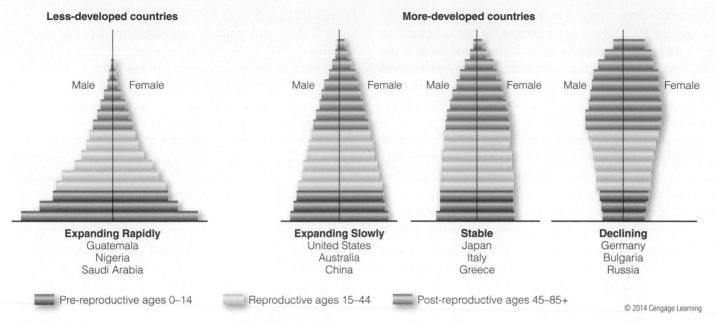

Less-developed countries / More-developed countries

Male | Female   Male | Female   Male | Female   Male | Female

Expanding Rapidly
Guatemala
Nigeria
Saudi Arabia

Expanding Slowly
United States
Australia
China

Stable
Japan
Italy
Greece

Declining
Germany
Bulgaria
Russia

Pre-reproductive ages 0–14   Reproductive ages 15–44   Post-reproductive ages 45–85+

© 2014 Cengage Learning

FIGURE 2.11 Typical population age-structure diagrams for countries with rapid (1.5–3%), slow (0.3–1.4%), zero (0–0.2%), and negative (declining) population growth rates. (Data from Population Reference Bureau)

A change in the distribution of people among a country's three major age groups can have long-lasting economic and social impacts. For example, in the United States, as the baby-boom generation has aged, the U.S. age structure has changed (Figure 2.12).

FIGURE 2.12 The baby-boom generation created a bulge that has moved upward through the U.S. age structure as baby boomers have aged. The two diagrams on the right reflect population projections. (Data from U.S. Census Bureau)

Baby boomers, born between 1946 and 1964, now make up over a third of the U.S. population and thus have considerable political and economic power. Between 2011 and 2019, this large group will be moving into retirement, and analysts fear that this will put a strain on the country's Social Security and Medicare systems, to be funded largely by the generations to follow. This will create challenges for the U.S. government and economy.

This is an example of how aging can affect a population. Another effect is that the population of a country can decline as the percentage of its people aged 65 or older

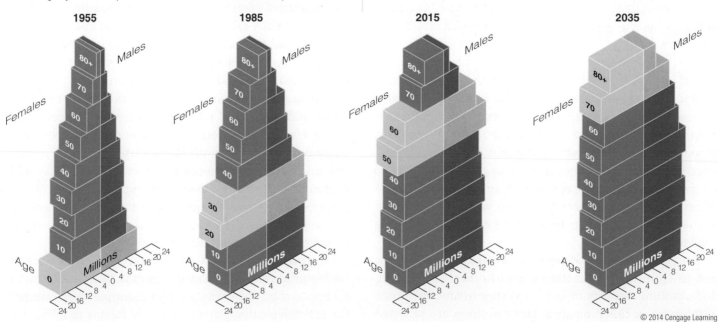

© 2014 Cengage Learning

rises and the percentage of its people aged 15 and younger falls (Figure 2.11, right). For example, in 2012, Japan had the world's highest percentage of elderly people and the world's lowest percentage of young people, and Japan severely restricts immigration. As a result, its population is projected to drop by 24% from 128 million to 95 million people between 2012 and 2050.

A slow population decline can usually be managed. But rapid population decline can cause serious economic and social problems. Government budgets are strained as expenses for public services such as health care rise. At the same time, there are fewer workers paying the taxes needed to support the growing number of older people. A country with a rapidly declining population can also experience a labor shortage unless it encourages immigration of young workers from other countries.

Let's REVIEW

- Summarize the projections for U.S. population growth. How does U.S. population growth compare with that of other more-developed countries?

- What is the *age structure* of a population and how does it influence the future size of a population?

- Summarize the differences between the age structures of a typical less-developed country and a typical more-developed country.

- Explain why a population with a large percentage of young people will keep growing for many decades even if its women have only one or two children, on average.

- What are two possible effects of a rapid decline in the population of a country due to aging?

What Are the Problems?

Growing Ecological Footprints

In Module 1, we discussed renewable and nonrenewable natural resources (pp. 13–14). Because their supplies are fixed, we can deplete nonrenewable resources such as copper and iron. However, renewable resources such as forests and the fresh water in streams can be replenished as long as we do not use or pollute them faster than natural processes can renew them. When we do misuse them in these ways, they can become degraded and possibly depleted (see Module 1, p. 21).

For example, much of the water we use to irrigate crops and to supply cities, factories, and homes comes from underground deposits of water, called *groundwater*, stored in formations called *aquifers*. Most aquifers are renewed

by rain that percolates down through the soil. However, if we drill wells and remove water from this type of aquifer faster than it is replenished, the available supply of water can shrink. In this way, we can deplete the aquifer.

Another resource that is being depleted in many parts of the world where populations are growing is topsoil. Wherever farmers grow crops year after year to feed more mouths without adding nutrients to replace those that are taken up by their crops, those topsoil nutrients provided by nature become exhausted. Then, to replace nature's free soil nutrients, farmers are forced to use costly fertilizers, which can cause water pollution.

We are also depleting some mineral resources. As populations grow and as resource use per person increases, several countries have exhausted their stocks of important minerals such as copper and gold, and must rely on other countries to supply these resources at great cost. Today, the United States imports all or nearly all of its supplies of 25 key nonrenewable mineral resources.

Still another important example of resource depletion is the overexploitation of many wild species. Numerous ocean fish species have been overfished to the point where their populations are depleted. Rare and endangered plants and animals are being collected, captured, or killed for use as food or as part of the global trade (illegal as well as legal) in ornamental plants, exotic pets (Figure 2.13), and animal hides, horns, and other body parts. This is a growing factor in the sharply rising rate of biodiversity loss.

Resource depletion is an example of what biologist Garrett Hardin (1915–2003) called the *tragedy of the commons*. Historically, most users of widely available renewable resources such as ocean fish have assumed that the supplies of such resources are unlimited. They have also assumed that their own small use of any resource will not affect the resource's supply, especially since it is renewable anyway. This kind of reasoning works when there are a small number of users; however, as the number of users grows and their total rate of resource use outpaces the resource's rate of replenishment, the resource can eventually be depleted. Then it is no longer available to anyone, and that is the tragedy of the commons. Today with 7 billion people using growing amounts of resources and producing more wastes and pollution, humanity faces the ultimate tragedy of the commons—the degradation of the planet's life-support system.

A growing population and growing per capita ecological footprints can also lead to other forms of environmental degradation and depletion. One example is *deforestation*, the extensive cutting and burning of forests to make way

FIGURE 2.13 This scarlet macaw is found in several of the tropical rain forests of Central and South America. These birds are endangered due to loss of habitat and because they are captured and sold as pets, often illegally.

for crops, grazing land, settlement, and the expansion of cities. Often, the result is erosion of a forest's topsoil to the point where the forest cannot grow back (Figure 2.14).

Clearing a forest or grassland and replacing it with roads, parking lots, houses, and other buildings have the same effect. The processes of mining and producing mineral resources also degrade vast areas of land, pollute air and water, and destroy wildlife habitat.

Another important example of environmental degradation resulting in a tragedy of the commons is the rising global average level of carbon dioxide (CO_2) in the atmosphere. This has resulted primarily from the large-scale burning of fossil fuels and the removal of forests that absorb CO_2 from the atmosphere, as both the human population and resource use per person have grown. Higher atmospheric levels of CO_2 and other greenhouse gases are adding to the problem of atmospheric warming, which is projected to disrupt the earth's climate during this century.

In some cases, we can deplete a renewable resource to the point where it cannot recover. For example, some diverse tropical forests (Figure 2.15, p. 42) that have been cleared cannot grow back, largely because of a severe loss of topsoil (Figure 2.14). In such cases, where environmental degradation has caused an irreversible ecological change in a natural system, that system has reached an **ecological tipping point**.

KEY idea Environmental degradation from a combination of population growth and rising rates of resource use per person can lead to irreversible ecological tipping points that drastically change natural systems.

FIGURE 2.14 This is the result of extreme tropical deforestation in an area of Thailand. The clearing of this forest caused such severe soil erosion that the forest will not be able to grow back here.

S. Borisov/Shutterstock.com

FIGURE 2.15 This highly diverse tropical rain forest is located in Australia. If it were extensively and repeatedly cleared to the point where its soil became severely eroded as in the area shown in Figure 2.14, it would likely reach an irreversible ecological tipping point that would prevent the regrowth of the forest in the cleared area.

Let's REVIEW

- Give an example of how a resource can be depleted. What is the tragedy of the commons? Give two examples of this effect.
- How does population growth affect resource depletion and other forms of environmental degradation? Give four examples.
- Define and give an example of an *ecological tipping point*.

The IPAT Model

The United States is a more-developed nation with a population of about 313 million. India is a less-developed nation with a population that is about 4 times as large as that of the United States. However, the U.S. population has a much higher overall environmental impact. How can this be? One of the major reasons is that the average American consumes about 30 times more resources (Figure 2.1a) than the average citizen of India consumes (Figure 2.1b). In general, in more-developed countries, each person, on average, consumes much more in the way of resources and generates much more pollution and waste than people in less-developed countries do.

To compare the environmental impacts of various countries, scientists John Holdren and Paul Ehrlich developed a simple model, called the **IPAT model**, which brings together three major factors that impact the environment: population size (P), resource consumption per person, or

affluence (A), and the harmful and beneficial environmental effects of technologies (T). The IPAT model shows that the environmental impact (I) of human activities depends primarily on how these three factors interact, as represented by this simple equation:

$$\text{Impact (I)} = \text{Population (P)} \times \text{Affluence (A)} \times \text{Technology (T)}$$

While the ecological footprint model (see Module 1, Figure 1.23, p. 24) focuses on the environmental impact of the human use of renewable resources, the IPAT model shows the environmental impact of the per capita use of renewable and nonrenewable resources. Figure 2.16 shows how the three IPAT factors interact.

Most countries have high environmental impacts for a variety of reasons. The IPAT model reveals that there are two major types of these impacts. **Overpopulation impacts** occur when population size (P) is the biggest factor in a country's total environmental impact (Figure 2.16, top). This sort of environmental impact is common in less-developed countries where environmental degradation, such as deforestation and depletion of topsoil, often results from a growing number of poor people trying to survive by using these resources. Even though average resource use per person in these countries is low, the total resource use is high because of the large and growing population.

In more-developed countries, the environmental degradation typically comes from **overconsumption impacts** that occur when affluence (A) is the biggest factor in a country's total environmental impact (Figure 2.16, bottom). In these countries, the high rate of resource use per person leads to high levels of waste, pollution, and resource depletion and degradation.

Some forms of technology have a high environmental impact because they increase the T factor. Examples are gas-guzzling motor vehicles and farm machinery (Figure 2.6), polluting factories (Figure 2.17a) and power plants that burn coal. Other technologies can lower an environmental impact by decreasing the T factor. Examples of technologies that prevent or reduce pollution are air pollution control systems, solar cells, and wind turbines (Figure 2.17b).

What the IPAT model also tells us is that in countries with large populations, like China and India, when affluence and use of harmful technologies grow, as they are doing in both countries, the environmental impacts will also grow sharply. Further, in a country such as the United States with high affluence and technology factors, if the population continues to grow rapidly, the overall environmental impact will continue to grow unless widespread

Less-Developed Countries

 ×  ×  =

More-Developed Countries

 ×  ×  =

| Population (**P**) | × | Consumption per person (affluence, **A**) | × | Technological impact per unit of consumption (**T**) | = | Environmental impact of population (**I**) |

© 2014 Cengage Learning

FIGURE 2.16 This simplified representation of the IPAT model compares the environmental impacts of populations in less-developed and more-developed countries. Red arrows of different lengths show varying harmful effects of all three factors; green arrows of different lengths show varying beneficial effects of technology. The arrows are not intended to show precise measurements.

Photo credits: Top row (left to right). www.asia-insider-photos.com; Simon/pixabay.com; Simon/pixabay.com; Vladimir Melnik/Shutterstock.com. Bottom row (left to right). Flashon Studio/Shutterstock.com; James Armitage/wikipedia.org; Bobby Mikul/www.publicdomainpictures.net; Ropable/wikipedia.org.

use of less harmful technologies (Figure 2.17b) can lessen that impact.

Some environmental scientists and economists argue that the increasingly rapid development and use of environmentally beneficial technologies could offset the harmful environmental impacts of a growing population and rising rates of resource use per person. Others argue that beneficial technologies probably will not be able to keep up with affluence, harmful technologies, and population growth, and that these factors will continue to increase the overall harmful human impact on the environment. This is an important debate.

KEY idea A country or area can have a significant environmental impact either because of a large population or because of a high level of resource use per person.

FIGURE 2.17 This coal-burning industrial plant in India (a) adds large quantities of carbon dioxide, soot, and other pollutants to the atmosphere. By contrast, the use of solar cells and wind turbines to produce electricity (b) has a very low harmful impact.

The Earth's Carrying Capacity

There is growing concern that as our ecological footprints grow and spread across the earth's surface, we are likely to reach several ecological tipping points (Figure 2.14) and overwhelm the carrying capacity of more and more of the planet's natural systems. In 2008, scientists at the World Wildlife Fund (WWF) and the Global Footprint Network estimated that humanity's global ecological footprint was at least 30% higher than the earth's estimated long-term ecological capacity to support us and other forms of life (see Module 1, Figure 1.23, top, p. 24).

The exact size of the earth's carrying capacity for the human population is a subject of debate. The estimates of experts range from about 2 billion to as many as 50 billion people. The high-end estimates carry the assumption that humans will develop technologies that will allow more and more people to live on the planet at higher levels of resource use without increasing our overall environmental impact.

However, many environmental scientists warn that such technological optimism could be dangerous if we continue to expand the population and its resource use per person, and fail to develop environmentally beneficial technologies to offset our resulting harmful environmental impacts. They suggest that, given the environmental problems we face today with a population of 7 billion, we may be nearing the earth's carrying capacity. Some argue that we already have exceeded it and will eventually face a dieback of the human population imposed by natural processes that have always controlled the populations of other species that exceeded the carrying capacities of their environment. Other experts, including some economists, believe that continued economic growth can provide enough resources for tens of billions of people without serious environmental harm.

Most environmental experts argue that sooner or later, we will face two serious consequences if we fail to slow and eventually halt human population growth by sharply lowering birth rates, and fail to find ways to reduce the harmful environmental impacts of resource consumption. First, in some areas, health and environmental conditions will deteriorate, and death rates will rise. This is already happening in parts of Africa and southern Asia, partly because of a combination of eroding soils, lower food production, water shortages, severe poverty, and conflicts over control of vital resources.

Second, a growing population of consumers will expand the already large ecological footprints in more-developed countries such as the United States and in rapidly developing countries, such as China and India. This will likely lead to spreading resource depletion and environmental degradation.

NUMB3RS

5

Number of earths needed to sustain today's global population at current U.S. levels of consumption

Let's REVIEW

- What is the *IPAT model,* and how can we use it to estimate environmental impacts?
- Define and distinguish between *overpopulation impacts* and *overconsumption impacts.* Which of these is likely to be larger in a more-developed country?
- Summarize the debate about the earth's carrying capacity for humans. What are two possible consequences of failing to slow population growth and lessen the harmful impacts of consumption?

What Can Be Done?

Promoting Economic Development

Researchers have examined the birth and death rates of western European countries that became industrialized during the 19th century. Using these data, they proposed a hypothesis of population change called the **demographic transition**: a four-phase process in which a country's birth and death rates change as the country becomes industrialized, and its population growth rate declines. Figure 2.18 shows the four stages of such a transition.

In recent years, several nations have climbed out of poverty in a short time through economic development. For example, within two decades, South Korea transformed itself from a poor nation with a rapidly growing population to a modern, industrialized nation (Figure 2.19) with a slowly growing population.

Some experts believe that, over the next several decades, most of the world's developing countries will undergo a demographic transition, mostly through a combination of economic development and family planning. But other analysts warn that some extremely poor countries with fast-growing populations could become mired in the transitional phase (Figure 2.18, Stage 2) of the demographic transition. This can occur if environmental degradation becomes too devastating to allow for

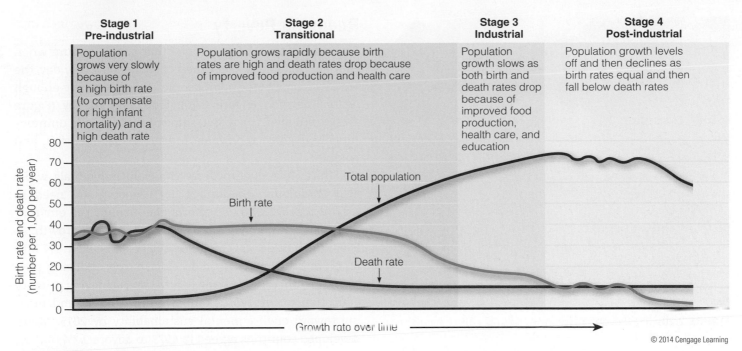

Stage 1 Pre-industrial	Stage 2 Transitional	Stage 3 Industrial	Stage 4 Post-industrial
Population grows very slowly because of a high birth rate (to compensate for high infant mortality) and a high death rate	Population grows rapidly because birth rates are high and death rates drop because of improved food production and health care	Population growth slows as both birth and death rates drop because of improved food production, health care, and education	Population growth levels off and then declines as birth rates equal and then fall below death rates

Total population

Birth rate

Death rate

Growth rate over time

© 2014 Cengage Learning

FIGURE 2.18 The *demographic transition*, experienced by many countries that have become industrialized, can take place in four stages.

improvements in health care and food production that are part of Stage 2.

Let's **REVIEW**

- State the *demographic transition* hypothesis and describe the four phases of the demographic transition.
- Explain why some countries may have difficulty in making such a transition.

FIGURE 2.19 These are high-speed express trains in Seoul, South Korea.

Steve Vidler/SuperStock

Empowering Women

In most countries, women have fewer rights and educational and economic opportunities than men have. Yet within their families, women provide more unpaid health care, globally, than that provided by all of the world's organized health services. In rural areas of Asia, Africa, and Latin America, women and girls do well over half of the daily work devoted to growing food, finding and hauling water, and gathering and carrying firewood (Figure 2.20, p. 46). Women account for about two-thirds of all the hours of work performed every day throughout the world but receive only a tenth of the world's income and own less than 2% of its land.

In many societies, sons are valued more than daughters, and parents often make their school-age daughters work at home instead of sending them to school. As a result, about 900 million girls—almost 3 times the entire U.S. population—do not attend elementary school. And almost 2 of every 3 illiterate adults and 7 of every 10 people in the world who suffer from poverty are women.

CONSIDER this

On a global basis, women:

- Account for two-thirds of all the hours of work performed but receive only one-tenth of the world's income.
- Own less than 2% of the world's land and make up 64% of illiterate adults and 70% of the poor.

FIGURE 2.20 These young girls from a rural village near the Kalahari Desert in the African country of Botswana typically spend two hours a day, two or three times a week, searching for and carrying firewood.

Many experts argue that societies have a much better chance of making a demographic transition and slowing their population growth when they promote measures to improve conditions for women. Helping women to become educated is one way to do this.

Also, a number of studies have shown that women tend to have fewer children when they have the ability to control their fertility and when they have paying jobs outside the home. In many of the world's less-developed countries, women are learning how to gain these advantages through family planning, including birth control, and through employment outside their homes. Increasingly, women in these countries are getting small loans to start businesses and work their way out of poverty. Many experts believe these trends will help to slow population growth and reduce poverty and environmental degradation.

> **KEY idea** Women tend to have fewer children if they can receive an education, have the ability to control their own fertility, and hold a paying job outside the home.

Let's REVIEW

- Describe some of the inequalities that women, especially poor women, face throughout most of the world.
- How can empowering women help to slow population growth?

Reducing Poverty

Poverty is the set of conditions that people endure when they are not able to meet their basic needs. Every day, the world's desperately poor people struggle to get enough water, firewood (Figure 2.20), food, and money (Figure 2.21) to survive. As a result, some of these people unintentionally degrade renewable forests, grasslands, soil, and wildlife.

Efforts to reduce poverty can help individuals while also stimulating economies. By reducing the need to exploit resources, these efforts also benefit the environment, and they tend to help slow population growth. In 2000, the world's governments set goals—called UN Millennium Development Goals—for sharply reducing poverty and hunger, improving health care, achieving primary education for everyone, empowering women, and moving toward environmental sustainability by 2015. More-developed countries agreed in 1979 to devote 0.7% of their annual national income toward achieving these goals. But the average amount donated in most years has been 0.25% of national income. The United States, the world's richest nation, has been donating only 0.16% of its national income.

FIGURE 2.21 Every day, this child searches through an open dump in Manila, the Philippines, to collect items to sell. He and his family live in or near the dump.

Businesses and governments can help to cut poverty and reduce population growth by providing funds and other assistance toward achieving the UN Millennium Development Goals. For example, nearly 1.6 billion people who live in rural villages around the world are not connected to an electrical grid. Governments and businesses could fund solar energy technology to provide electricity for many of these villages (Figure 2.22), and they could also help poor people to work themselves out of poverty by providing small loans (see *Making a Difference* that follows).

Let's **REVIEW**

- Define *poverty* and list three ways in which businesses and governments can help to reduce it.

- Explain the importance of microloans for the poor in reducing poverty and slowing population growth.

FIGURE 2.22 This system of solar cells provides electricity for a remote village in Niger, Africa

MAKING A difference

Muhammad Yunus and Microloans for the Poor

One very promising way to help poor people who want to work their way out of poverty is to provide them with small loans for purposes such as starting a small business or buying seeds and fertilizer for growing food crops. Most of the world's poor people do not have a credit record or the means to qualify for conventional loans.

In 1983, economist Muhammad Yunus (Figure 2.A) started the Grameen (Village) Bank in Bangladesh, a country with a high poverty rate and a rapidly growing population. Since then, the bank has provided a total of $7.4 billion in *microloans* of $100 to $1,000 at very low interest rates to 7.6 million impoverished people (97% of them women) who cannot qualify for loans at traditional banks. Almost all of these loans have been used by women to plant crops, to start small businesses, or to buy livestock, bicycles for transportation, or small irrigation pumps.

To promote loan repayment, the bank puts borrowers into groups of five. If a group member fails to make a weekly payment, other members have to make it instead. As a result, the bank has made a solid profit and the average repayment rate on its microloans has been 95% or higher—much higher than the average repayment rate for loans by conventional banks.

About half of the bank's microborrowers have moved above the poverty line and improved their lives within 5 years. Between 1975 and 2005, this innovative approach, along with the hard work of the people receiving the microloans, helped to reduce the poverty rate in Bangladesh from 74% to 40%. In addition, birth rates are generally lower among most of the borrowers.

Muhammad Yunus and his Grameen Bank developed a new business model that combined the power of market capitalism with the desire for a more humane world. In 2006, Yunus and his colleagues at the bank jointly won the Nobel Peace Prize for their pioneering use of microcredit. Microcredit banks based on the Grameen model have spread to 58 countries, including the United States. This approach has helped at least 100 million people worldwide to work their way out of poverty.

Unfortunately, some people have taken advantage of Grameen Bank's successes by making small loans at very high interest rates, and this is casting a shadow over microlending in some areas. However, Yunus and his supporters have not given up on their efforts to help millions more to escape poverty and improve their lives.

FIGURE 2.A Economist Muhammad Yunus's efforts have helped 7.6 million Bangladeshis to better their lives, which has in turn helped Bangladesh to slow its population growth.

Promoting Family Planning

In many countries, an important step in making the demographic transition has been to employ **family planning**, a set of programs designed to provide information and clinical services that can help couples to decide how many children to have and when to have them. These services usually take the form of information on birth control and birth spacing, and health care for pregnant women and infants.

Family planning has played a major role in lowering birth rates in many countries (see *Making a Difference*, below), although it has met with certain challenges in other areas of the world (see the following *For Instance*). It has also played a role in reducing the number of abortions performed each year, as well as the number of deaths of mothers and fetuses during pregnancy. The UN Population Division and other population agencies have estimated that family planning has accounted for at least 55% of the drop in TFRs in the world's less-developed countries between 1960 and 2010, from 6.0 to 2.7.

According to the UN population experts, in spite of the effectiveness of family planning programs in slowing population growth, two problems remain. One is that about 42% of all pregnancies in less-developed countries are unplanned, and more than half of them end with abortion. The other is that in less-developed countries, family planning programs are not available to an estimated 160 million couples that want them.

KEY idea Family planning has helped to slow population growth, reduce the number of abortions, and decrease the number of deaths of mothers and fetuses during pregnancy.

MAKING A difference

Slowing Population Growth in Thailand

Can a country reduce its population growth within a small number of years? Thailand did, mostly through government-supported family planning and with the leadership of Mechai Viravaidya (Figure 2.B)—a public relations genius and former government economist.

In 1971, Thailand's population was growing at a very rapid rate of 3.2% per year, and the average Thai family had 6.4 children. The Thai government adopted a policy to sharply reduce its population growth rate. Fifteen years later, the country's population growth rate had been cut in half to 1.6%. By 2012, the rate had fallen to 0.5% and the average number of children per family was 1.6.

A combination of strong government support for family planning, a high literacy rate among women (90%), and support of family planning by the country's influential Buddhist leaders has played a key role in this achievement. However, Mechai Viravaidya is credited by many for leading the way by establishing the nonprofit Population and Community Development Association (PCDA), a grassroots organization that has helped to implement family planning as a national goal.

Viravaidya has used an array of creative methods to publicize his campaign. For example, volunteer PCDA workers hand out condoms at festivals, movie theaters, and even traffic jams, and develop ads and witty songs about contraceptive use. The PCDA promotes condoms as a contraceptive, but this has the added benefit of providing protection from the threat of HIV/AIDS. Another strategy was to establish a loan program to help people install running water and toilets in their homes in return for agreeing to participate in family planning programs. Low-rate loans have also been offered to rural farming couples who agree to practice family planning.

The PCDA and its 12,000 volunteers have reached more than 10 million Thais in 18,000 villages. Between 1971 and 2012, these efforts have helped to raise the percentage of married Thai women (ages 15–49) using modern birth control methods from 15% to 77%. This rate is much higher than that found in most other countries.

Courtesy of Population and Community Development Association

FIGURE 2.B Mechai Viravaidya has played a major role in Thailand's successful efforts to reduce both its population growth and its rate of HIV/AIDS infection.

According to recent estimates by the UN Population Fund:

- Without family planning, there would be 8.5 billion people on the planet instead of 7 billion.

- Fulfilling the world's unmet needs for family planning could, every year, prevent 52 million unwanted pregnancies, 22 million induced abortions, 1.4 million infant deaths, and 142,000 pregnancy-related deaths.

Let's REVIEW

- What is *family planning* and how has it affected population growth? How has it benefited the health of many people?

- Describe two problems that hinder family planning efforts.

Promoting economic development, combatting poverty, empowering women, and encouraging family planning are all major challenges that analysts argue should be tackled without delay, on both national and international levels. These analysts warn that, if we cannot successfully implement these measures, the world will see an exploding number of *environmental refugees*—people who are homeless and starving, due to the effects of environmental degradation. (See *A Look to the Future*, p. 51.)

FIGURE 2.23 This is home for this impoverished family in Bihar, India.

Jeremyrich/Dreamstime.com

FOR INSTANCE...

Let's Compare Population Control Programs

India's Moderate Success in Reducing Population Growth

India established the world's first national family planning program in 1952 when its population was nearly 400 million. By 2012, after 6 decades of trying to control its population growth, India had 1.2 billion people.

Between 1952 and 2012, the number of people added to India's population each year rose from 5 million to 18 million—the world's highest increase in population growth rate. The United Nations projects that India will become the most populous country by 2015, and by 2050, it will have a population of about 1.7 billion.

India faces a number of serious poverty and environmental problems that could worsen as its population continues to grow rapidly. Although India has 17% of the world's population, it occupies only 2.4% of the world's land area. About 70% of its people live in the country's 550,000 rural villages. In 2012, estimates of the size of India's rapidly growing middle class ranged from 100 million to 170 million people—between a third and half the size of the U.S. population. On the other hand, about 76% of India's people are poor or very poor (Figure 2.23), struggling to live on around $2 or less per day. And almost half of the country's labor force is unemployed or underemployed.

Despite the Indian government's family planning efforts, the average TFR in India in 2012 was 2.6 children per woman. Only 47% of all Indian couples were using modern methods for controlling births, even though 9 of every 10 couples had access to these methods.

Some of India's family planning program supporters argue that the results of the programs have been disappointing for several reasons, including poor planning, bureaucratic inefficiency, the low status of women (despite constitutional guarantees of

NUMB3RS

1.7

The population of India is projected to reach 1.7 billion in the year 2050

FIGURE 2.24 These computer software engineers are members of the rapidly growing middle class in India.

equality), extreme poverty, and lack of administrative and financial support. This is in sharp contrast to Thailand's more successful program (see *Making a Difference*, p. 48).

Two factors help to account for larger families in India. Because of extensive poverty, most poor couples decide to have several children to help them survive and to take care of them in old age. In addition, a strong cultural preference for male children in parts of India encourages couples to keep having children until they have one or more boys.

India has undergone rapid economic growth. As members of its growing middle class (Figure 2.24) use more resources per person, the country's total and per capita ecological footprints (see Module 1, Figure 1.23, p. 24) will grow. On the other hand, further economic growth could help the country to slow its population growth by making a demographic transition (Figure 2.18).

China's One-Child Policy

In the 1960s, China's large population was growing so rapidly that there was a serious threat of mass starvation. To avoid this, government officials decided to establish a family planning and birth control program. It became the world's most comprehensive, strict, and intrusive of such efforts.

China's goal is to sharply reduce fertility by promoting one-child families (Figure 2.25). The government provides contraceptives, sterilizations, and abortions for married couples. In addition, married couples pledging to have no more than one child receive a number of benefits including better housing, more food, free health care, salary bonuses, and preferential job opportunities for their child. Many (but not all) couples who break their pledge lose these benefits.

Since this program began, China has made great strides in slowing its population growth. Between 1972 and 2012, China slashed its population growth rate from 2.3% to 0.5% and reduced its TFR from 5.7 to 1.5 children per woman. (By comparison, in 2012, the population growth rate for the United States was 0.6% and its TFR was 2.0.)

Despite this success, China has the world's largest population and, in 2012, added about 7 million people to its population. The United Nations projects that China's population will peak at about 1.46 billion in 2033 and then will begin to decrease slowly. In addition, since 1980, China has undergone rapid economic growth. This has helped an estimated 400 million Chinese to work their way out of poverty and become middle-class consumers (Figure 2.26). This will add to the growth of the country's overall ecological footprint.

FIGURE 2.25 This mural in Guangzhou, China, promotes the government's one-child policy.

FIGURE 2.26 This modern shopping center in Shanghai, China, caters to the country's large and growing number of middle-class consumers.

The one-child policy has presented other problems. Because of a strong preference for male children, China has a rapidly growing bride shortage. Young girls in some rural areas of China are being kidnapped and sold in other parts of the country as brides for single men. In some cases, pregnant Chinese women get abortions if their ultrasound scans show that their fetus is female.

Also, because there are fewer children, the average age of China's population is increasing rapidly. This shift in age structure means that there will be fewer children and grandchildren to care for the growing number of older people, and there will be fewer workers to support the economy. These factors may lead to some relaxation of the government's one-child population policy.

Let's REVIEW

- Summarize the story of India's efforts to slow its population growth.
- How successful have these efforts been?
- Describe China's one-child policy and how it is implemented.
- What are the major benefits of China's policy and what are some problems resulting from it?

A Look to the Future

Environmental Refugees

In some parts of the world, population growth and the resulting environmental degradation are creating **environmental refugees**—people who have been forced to leave their homes because of environmental degradation that has left them without the resources, such as soil, food, and water, that they need in order to survive and live well. Other refugees have migrated because of wars within and between their countries for control over these resources.

Of the roughly 830 million people added to the global population during the past decade, a good proportion of them have struggled to survive on what amounts to $1.25 per day or less. Many of these people are forced to live in environments that are too wet or too dry, or where the land is too steep for sustainable agriculture. Many have had to flee their homes because of severe drought, excessive soil erosion, floods (Figure 2.27, p. 52), and other environmental hazards.

The growing throngs of environmental refugees, estimated to have totaled 25 million in the mid-1990s, rose by an average of 1 million a year since then to a total of around 40 million in 2011. This total is greater than that for all other types of refugees—including people who have

FIGURE 2.27 Refugees from the Democratic Republic of Congo, Africa, arriving at a refugee camp in 2008.

Croplands and pastures in many parts of this region have dried out to the point where they can no longer support agriculture or grazing. Severe droughts are projected to affect two-fifths of the planet's land surface during the last half of this century.

Agencies such as the UN High Commissioner for Refugees are reluctant to officially recognize this new category of refugees, referring to them as "displaced persons" or "destitute migrants," even though they are in fact environmental refugees, having been driven to seek refuge as a result of deteriorating environmental conditions. These agencies point out that they have their hands full with other types of refugees and that their budgets are already stretched to the limit.

Growing numbers of environmental refugees could soon start to pose new threats to international stability, especially if many of them try to find shelter in the more-developed nations. According to many experts, the refusal to recognize the existing number of environmental refugees, as well as the potential for much larger numbers to come, is a recipe for disaster. The U.S. Central Intelligence Agency and several U.S. military experts see this problem as a major threat to global peace and social and economic stability. To these and many other experts, it is a humanitarian problem that deserves immediate attention from the international community.

> **KEY idea** By the end of this century, there could be 200 million to 500 million environmental refugees displaced from their homes because of degraded land, water shortages, drought, and flooding.

fled civil wars, political oppression, and religious and ethnic persecution—put together.

Scattered throughout the developing world are 250 million people threatened by land degradation resulting from severe drought. Another 745 million people suffer from acute water shortages. If the earth's atmosphere continues to get warmer, climate scientists project that, because of the resulting climate change, the number of environmental refugees could eventually soar to at least 250 million and conceivably twice that—10 to 20 times today's number for all other types of refugees.

For example, the 3-foot rise in sea level that is likely to result from projected climate change would threaten at least 100 million people who live in low-lying coastal areas that would be flooded before the end of this century. Some scientists project a much larger rise in sea levels. Two-thirds of the world's large cities are located at least partially in such areas, and the populations of these cities are continually growing. Many of these coastal city dwellers could become environmental refugees. In addition, there could easily be another 50 million refugees driven from their land by drought in areas such as sub-Saharan Africa.

Let's REVIEW

- What are *environmental refugees*? Summarize the threats they face.
- Why is the issue of environmental refugees an urgent issue that the international community should deal with now, even though it is largely a future problem?

What Would You Do?

This module began with a discussion of why we should care about population growth. We have now examined some of the problems that stem from population growth and some of the possible ways of dealing with those problems. However, as with so many issues that involve the environment, dealing with overpopulation and overconsumption problems ultimately depends on what individuals do—on the daily choices and decisions that each of us makes.

There is a growing awareness of population issues and many people are thinking about these issues and making choices that will help the world to deal with them. Here are some of the steps they are taking:

Thinking carefully about how many children to have.
- Many people are considering the environmental impact of having a child. Some have opted to have no children or just one or two. Others are choosing to adopt children, rather than to bring new ones into the world. And many have chosen a combination of the options above.
- Family planning professionals often advise people, before having children of their own, to find out what raising a child is like by caring for infants or young children of friends or family members over a period of time.
- Another part of family planning is to consider carefully the financial implications of raising a child. According to a 2010 report by the U.S. Department of Agriculture, the typical cost of raising a child through age 17 in the United States is about $220,000 for a middle-income family, with college costs adding another $100,000 to $200,000.

Supporting efforts to slow population growth.
- Individuals can support the UN Millennium Development Goals by contacting elected officials at the local, state, and federal levels and urging them to support these goals and by asking friends to do the same.
- People can also encourage lending agencies and governments to make microloans to poor people who want to work their way out of poverty (see *Making a Difference*, p. 47).

- Individuals can learn about and support government and private family planning programs that are compatible with their beliefs.

Reducing personal ecological footprints.
Four major factors determine the ecological footprint and overall environmental impact of one's lifestyle: transportation, food, home energy use, and resource consumption and waste. People are using interesting strategies to reduce their environmental impacts, especially those related to these four factors. In most instances, they find that when they reduce their impacts, they also save money. Here are some examples:
- People can calculate their own ecological footprints by visiting several websites that can be found easily by using "ecological footprint" as a key term in searches.
- More people are driving energy-efficient vehicles that get at least 40 miles per gallon; in addition, they are driving less and are consolidating trips, walking, bicycling, carpooling, and using mass transit.
- Many people are reducing meat consumption by one or more meals each week, and buying locally grown and seasonal food, or growing their own food.
- An increasing number of homeowners are heavily insulating their houses, plugging air leaks, and using energy-efficient heating and cooling systems, lights, and appliances.
- Many consumers are examining their buying habits and reducing their level of consumption. They are also reusing many of the items they buy, and many try to buy only recyclable products and then to recycle them.

Given these ideas and other information that you gained from reading this module, what are some steps that you would take to help deal with the environmental issue of population growth?

KEYterms

age structure, p. 38
carrying capacity, p. 33
demographic transition,
   p. 44
ecological tipping point,
   p. 41
environmental refugees,
   p. 51

exponential growth,
   p. 32
family planning, p. 48
infant mortality rate,
   p. 38
IPAT model, p. 42
less-developed country,
   p. 35

life expectancy, p. 37
limiting factors, p. 33
migration, p. 36
more-developed country,
   p. 35
overconsumption impacts,
   p. 42

overpopulation impacts,
   p. 42
poverty, p. 46
total fertility rate (TFR),
   p. 36

THINKINGcritically

1. What would you say if someone asked you why he or she should care about population growth in the world and in his or her own country? What about you? Do you really care about how many people there are on the earth, in your country, or in the area where you live? Explain.

2. On average, how many people are added to the world's human population every second? Using this current rate, calculate how many people will likely be added to the population by the time you reach the age of 70. Do you believe that this number of people will be within the earth's carrying capacity for humans? Explain.

3. Should everyone have the right to have as many children as they want? Explain. Would you support this right even if it degrades the earth's life support systems and makes it harder for future generations to live comfortably? Explain.

4. Why do you think reducing poverty tends to slow population growth? Think of at least three reasons and explain.

5. Why do you think there has been little effort to reduce population growth in the United States? Would you support or oppose efforts to slow population growth in the United States? Explain. If you support slowing this growth, what are three strategies that you think might be useful for accomplishing this?

6. Considering the major requirements in your life (such as food, transportation, and housing), in what ways do you think you could or should reduce your ecological footprint? List the steps you could take to begin making these changes. If you do not believe that you should do this, explain your thinking.

LEARNINGonline

Access an interactive eBook and module-specific interactive learning tools, including flashcards, quizzes, videos and more in your Environmental Science CourseMate, accessed through **CengageBrain.com**.

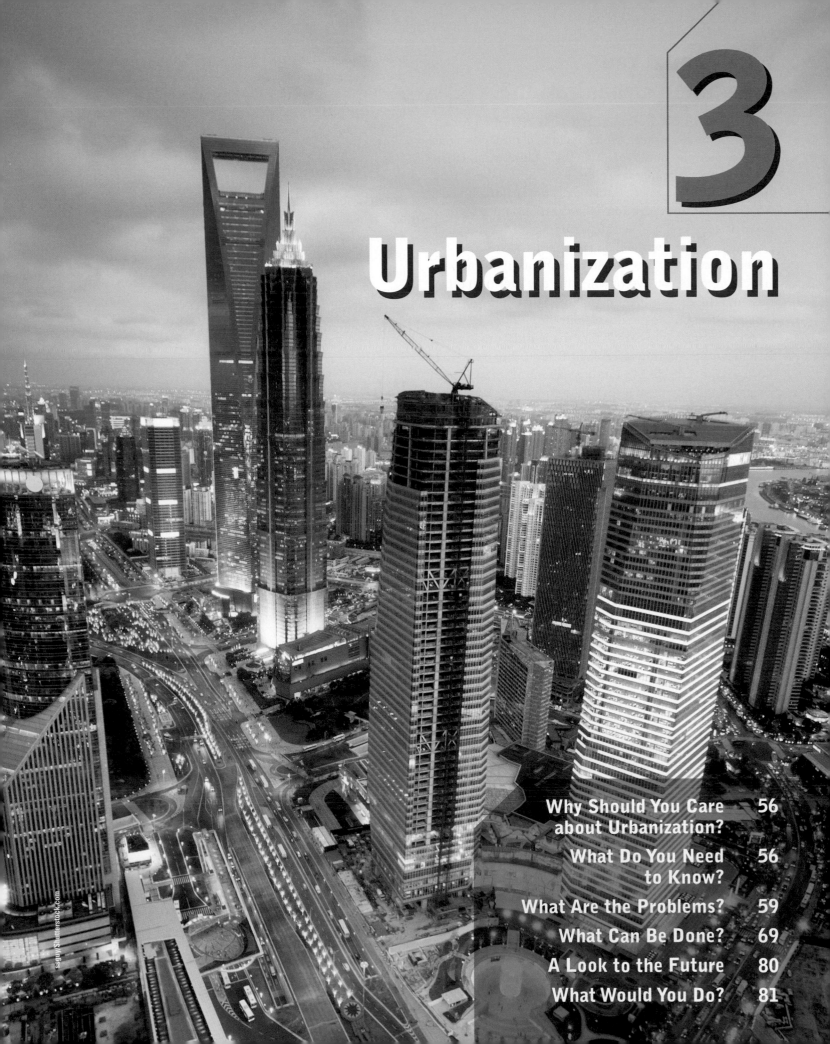

3

Urbanization

Why Should You Care about Urbanization?

In 2012, 51% of the world's people, and 79% of all Americans, lived in *urban areas*, which include central cities (see module-opening photo of Shanghai, China and Figure 3.1a) and suburban areas (Figure 3.1b). Since about 1950, urban areas have grown dramatically as part of a broader process called **urbanization**—the creation and growth of urban areas, measured as the percentage of the people in a country or in the world living in urban areas.

One reason why you should care about urbanization is that chances are you live, or will some day live, in an urban area, and the environmental quality of that area will have a major effect on the quality of your life.

Another reason to care is that the world's urban areas each have an enormous ecological footprint (see Module 1, Figure 1.23, p. 24), and just about all urban areas are growing. This means that these areas are large contributors to the global environmental problems we face, including air and water pollution, projected climate disruption, and depletion of resources such as topsoil, minerals, freshwater supplies, and wetlands.

Even if you don't live in an urban area, you have reasons to care about the environmental quality of these areas. Many of the air pollutants, water pollutants, and solid wastes coming from cities end up in the rural areas around them. In addition, rural areas are exploited through industrialized agriculture, the clear-cutting of forests, and mining operations in order to provide resources for urban dwellers. Such operations often disturb soils, pollute air and water, degrade or destroy forests, and eliminate wildlife habitats.

On the other hand, urbanization can also have beneficial effects on the environment. Concentrating people in urban areas can lead to more efficient use of resources, preservation of land and important ecosystems services, and conservation of the vital topsoil that we need for growing food. In this module, we consider these and other benefits, and we explore the major environmental problems that result from urbanization. We look at how we can improve the livability and sustainability of both existing and new urban areas, and we consider how we can help make the transition to more sustainable urban living.

What Do You Need to Know?

Some Key Terms

An **urban area**, or **city** (Figure 3.1), is normally defined as an area having a total population of 2,500 people or more with a relatively high density of people and of human-built structures in comparison to the area surrounding it. (This defined lower population limit varies greatly among different countries, but we will use the number most commonly used in the United States.) A **rural area** is any populated area that is not classified as an urban area and in some cases includes small towns and villages (Figure 3.2).

FIGURE 3.1   Mexico City, Mexico (a), is one of the world's largest cities, with a population of about 21 million people. Many of the world's cities are surrounded by suburbs such as this one (b) in Southern California.

FIGURE 3.2 Many of the world's rural populations live in villages such as this one in the southern African country of Malawi.

Magdalena Bujak/Shutterstock.com

In this module, we refer to the world's countries as belonging to one of two broad categories. The first is **more-developed countries**—highly industrialized countries in which the average levels of income and consumption per person are relatively high compared to all other countries. The other category is **less-developed countries**, those countries that are not industrialized or are less industrialized, with lower average levels of income and consumption per person. About 82% of the world's people now live in less-developed countries.

All countries fall within a range between the two extremes. The United States is a more-developed country with very high average income. China and India are less-developed countries that are, fairly quickly, becoming more developed with average incomes rising. Haiti and Somalia—the least developed of all countries—are examples of less-developed countries with very low average incomes.

In using these terms, we don't mean to say that a less-developed country is inferior to a more-developed country. The terms simply point to differences in the degree of economic development, or industrialization, that a country has undergone. In fact, some people argue that countries can lose ancient and highly valued traditions as they become more developed, so that in some sense they become poorer. We use this measure of economic development in this module because it helps to determine the nature of any country's urban areas as well as the environmental effects of urbanization.

In order to fully understand urbanization, we also need to understand the concept of **population density**, or the number of people per unit of area living in a defined area such as a city or country. For example, a city that occupies 20 square miles and contains 1,000 people has a much lower population density than a city of the same size that contains 100,000 people.

Another important concept is **urban growth**—the *rate* of increase of an urban population, usually given as a percentage such as 10% per year. Urban areas grow when the birth rate of the population exceeds the death rate and when people move there from rural areas or from other countries. In some parts of the world, people have had to move to cities because they were forced off their rural lands by drought, famine, war, or political or racial persecution. Others move to cities in search of jobs. In more recent times, people are also drawn to some cities by the attractions of better health care, better schooling, and greater varieties of entertainment and other pastimes.

Let's **REVIEW**

- Define *urbanization* and define and distinguish between an *urban area*, or *city*, and a *rural area*.
- Define and distinguish between *more-developed* and *less-developed countries*.
- What is *population density*?
- Define *urban growth* and list some of the reasons why people move to urban areas.

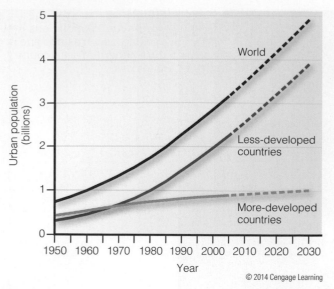

FIGURE 3.3 Urban populations have grown dramatically around the world, but more so in less-developed nations than in more-developed countries. This graph also shows projections to 2030 for the growth of urban populations of the world as well as those of less-developed and more-developed countries. (Data from United Nations Population Division)

Current Trends

One of the major trends in urbanization is that it continues to increase dramatically (Figure 3.3). Between 1850 and 2012, the percentage of the world's people living in urban areas rose from 20% to 51%, and experts such as those at the UN Population Division project that it could reach 70% by 2050. These percentages vary among countries. In 2012, the percentage of people living in urban areas in more-developed countries, on average, was 75% (79% in the United States) compared to a 46% average in less-developed countries.

A second important trend is that most urban growth is now taking place in the cities of less-developed countries (Figure 3.3). Experts project that about 88% of urban growth through 2050 will happen in these rapidly growing, overcrowded, and stressed urban areas (Figure 3.4).

A third trend in urbanization is that the number of large urban areas in the world is growing. In 2012, more than 400 cities had a million or more people each, and 22 *megacities*, 19 of them in less-developed countries (Figure 3.5) each had 10 million or more people. Soon there will be a number of *hypercities*, each holding more than 20 million people. Two examples are Tokyo, Japan, with 36 million people and Mexico City, Mexico (Figure 3.1a), with 21 million people.

CONSIDER this

- Every week, about 1 million people are added to the world's urban populations.

- Within the next 15 years, China will add more people to its cities than the number of people now living in the United States.

A fourth major trend is that poverty is increasing in the urban areas of many less-developed countries. According

FIGURE 3.4 This crowded street is in Delhi, India, which has a population of about 17 million people.

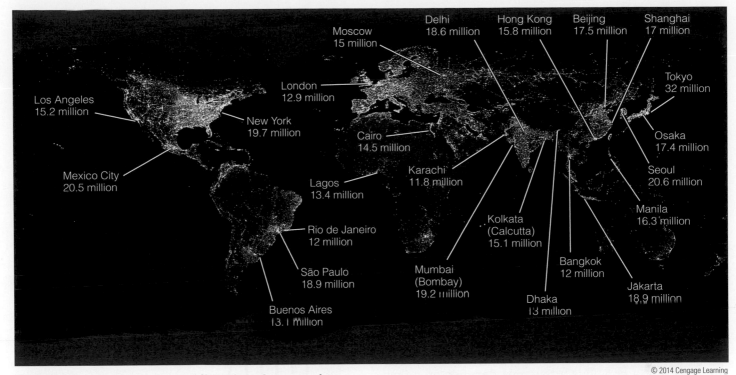

Moscow 15 million
Delhi 18.6 million
Hong Kong 15.8 million
Beijing 17.5 million
Shanghai 17 million
Los Angeles 15.2 million
London 12.9 million
Tokyo 32 million
New York 19.7 million
Cairo 14.5 million
Osaka 17.4 million
Mexico City 20.5 million
Karachi 11.8 million
Seoul 20.6 million
Lagos 13.4 million
Manila 16.3 million
Rio de Janeiro 12 million
Kolkata (Calcutta) 15.1 million
São Paulo 18.9 million
Mumbai (Bombay) 19.2 million
Bangkok 12 million
Jakarta 18.9 million
Buenos Aires 13.1 million
Dhaka 13 million

© 2014 Cengage Learning

FIGURE 3.5 We can see the world's major urban areas by using satellites to photograph them at night. Shanghai, China, shown in the module-opening photo, has about 17 million people. (Data from National Geophysics Data Center, National Oceanic and Atmospheric Administration, and United Nations)

to the United Nations, at least 1 billion people in these countries (more than 3 times the current U.S. population) live in densely populated slums within cities or in ramshackle shantytowns on their outskirts (Figure 3.6). Some urban population experts project that there could be 2 billion people—more than 6 times the current U.S. population—living in such conditions by 2030.

Let's REVIEW

- Summarize four current trends in urbanization.
- How many people in less-developed countries live in slums and shantytowns?

What Are the Problems?

Benefits and Drawbacks of Urbanization

City living can provide a variety of benefits. Cities are the sites of businesses and industries that provide jobs for millions of people. Most colleges and universities are located in urban areas. Cities are also centers of commerce, technological advances, arts and entertainment, and economic and political power.

On average, urban dwellers live longer than do rural residents and their children are more likely to be able to go to school. Urban area residents also tend to have better access to health care, family planning, and social services than do people in rural areas. Many people living in rural areas, especially young people, migrate to cities in hopes of gaining these benefits.

FIGURE 3.6 This shantytown is in Lima, Peru.

Photochris/Dreamstime.com

The growth of cities can also benefit the environment. For example, with higher concentrations of people and material goods, cities can more easily support recycling programs than can small towns and rural areas. They can also support mass-transit systems such as buses and light-rail systems, and this helps to reduce energy use and the pollution that results from high concentrations of motor vehicles used daily. In addition, where urban growth is managed in such a way that the human population is more concentrated in the city and not spread out across a vast area around the city, we can preserve larger areas of land for the benefit of wildlife, ecosystems, ecosystem services, and people.

Despite their many benefits, urban areas also have major harmful impacts on the environment and on human health, and these impacts are growing rapidly. In fact, many environmental analysts argue that the great majority of the world's urban areas are currently not environmentally sustainable for the long term.

Urban areas occupy only about 2% of the earth's land area. But a populous urban area depends on a flow of resources from other areas to meet most of its residents' needs and wants, while a portion of its pollution and wastes flows into the air and water and onto the land around it. Therefore, the area of land and water needed to provide any set of urban dwellers with their food, water, and other vital resources, and to absorb their wastes and pollutants is typically much larger than the area of the city in which they live (Figure 3.7). In other words, an urban area can have an *ecological footprint* (see Module 1, Figure 1.23, p. 24) that extends far beyond its physical boundaries.

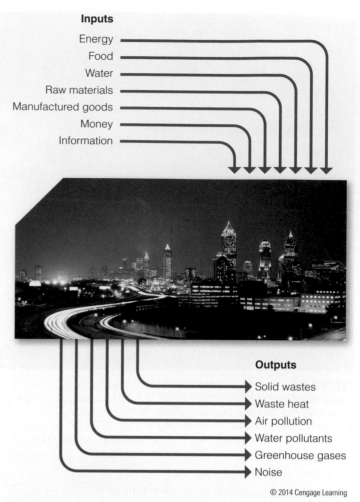

FIGURE 3.7 Most urban areas depend on rural areas to meet their resource needs and to receive their outputs of pollution and wastes.
Photo credit: Jeremy Woodhouse/Getty Images.

CONSIDER this

Ecological footprint experts Mathis Wackernagel and William Rees estimate that supplying the 7.6 million people living in the Greater London area of the United Kingdom with the resources they need requires gathering and importing resources from an area that is 58 times larger than the city.

KEY idea

Urban areas have large ecological footprints because they depend on a flow of resources from areas far beyond their borders to meet most of their needs and their surrounding environments receive some of the pollution and wastes they produce.

Let's REVIEW

- What are three ways in which people benefit from living in cities?
- What are three ways in which the environment can benefit from urban growth?
- Explain why most of the world's urban areas are not sustainable in the long run.

We now consider more closely some of the harmful environmental and human health impacts of urbanization.

Urban Sprawl

One of the major urban problems that can result when a city is located in a large land area where it can grow outward is **urban sprawl**—the spreading of low-density housing and other forms of development around city centers. For example, between 1970 and 2010, the population of

FIGURE 3.8 These photos show how urban sprawl occurred around Las Vegas, Nevada, between 1984 and 2010.

Clark County, a desert area that includes the city of Las Vegas, Nevada, increased sevenfold from 277,000 to more than 2 million, and the city's area grew dramatically (Figure 3.8).

Why is urban sprawl a problem? Again, we can use the concept of the ecological footprint. As a city sprawls, both its footprint and its overall environmental impact grow.

For example, in many of the world's cities surrounded by urban sprawl, it is difficult to get around without a car. This makes for a lot of engines burning gasoline, polluting the air, and adding climate-changing carbon dioxide to the atmosphere. In fact, there is a whole set of sprawl-related problems stemming from the dominance of the car in most suburbs as we discuss in the next section.

By contrast, these and other impacts are not as large in more compact, centralized urban areas where mass-transit options are available. This is because, compared to central cities, suburbs have lower-density housing typically located farther from the workplaces, stores, and other facilities that suburban residents use, so car use per person is far higher on average because people need to drive for purposes such as shopping, employment, and general transportation. Also, homes and land plots are typically larger in suburbs, so it takes more energy and generates more pollution to heat and cool homes and to care for the land. In other words, the per capita ecological footprint (see Module 1, Figure 1.23, p. 24) of a suburban resident is typically larger than that of a city dweller.

Another impact of urban sprawl is the conversion of large natural areas and croplands to housing and commercial developments, parking lots, streets, and highways. Undisturbed natural areas absorb rainwater, filter and purify it, and then release it slowly to help control flooding. These areas are also critical to replenishing underground water supplies, or *aquifers*. Undisturbed natural areas and crop fields also provide food resources and wildlife habitats. When these natural areas are paved over, the ecological services they provide are degraded or lost.

On the other hand, many people enjoy living in suburban areas. These areas provide less crowded living and more spacious single-family homes. Many of these areas also have newer public schools and health-care facilities, and lower crime rates. However, in terms of the environmental impacts of urbanization, urban sprawl is a key factor to consider. Figure 3.9 (p. 62) summarizes some of the environmentally harmful results of urban sprawl.

FIGURE 3.9 These are some of the undesirable impacts of urban sprawl.

Photo credits (left to right): Alexander Roberge/www.clker.com; Sami Sarkis/Getty Images; artwork by Patrick Lane; Federal Emergency Management Agency (FEMA).

Energy, Air, and Climate

Increased energy use and waste

Increased emissions of air pollutants and climate-changing carbon dioxide

Land and Biodiversity

Loss of cropland

Loss and fragmentation of forests, grasslands, wetlands, and wildlife habitat

Water

Increased use and pollution of surface water and groundwater

Increased runoff and flooding

Economic Effects

Decline of downtown business districts

More unemployment in central cities

© 2014 Cengage Learning

Let's REVIEW

- What is *urban sprawl* and why is it a problem?
- Give three examples of the harmful environmental impacts of urban sprawl.
- What are three benefits that many suburban areas provide for their residents?

Urban Transportation

In countries such as the United States, Australia, and Canada, many cities are located in areas where there is plenty of land available for outward expansion. As the populations in these cities grow, urban sprawl is the usual result (Figure 3.10a), and residents of most of these vast urban areas depend on the automobile (Figure 3.10b) to get from one place to another.

Most American urban areas are car-centered. According to the U.S. Department of Transportation, the United States has about 4.5% of the world's people, but Americans own almost 33% of the world's motor vehicles and use about 43% of the world's gasoline. In all U.S. urban areas taken together, people use their cars for 98% of all transportation.

The American car culture has provided benefits to a huge number of people while helping to sustain the country's economy for many decades. However, Americans are now learning about the downside of the car-dependent culture. For example, motor vehicles are the world's largest source of outdoor air pollution and the fastest-growing source of carbon dioxide (CO_2) emissions. Evidence indicates that CO_2 and other greenhouse gases (see Module 1, p. 15) are helping to warm the atmosphere, which could lead to disruption of the earth's climate.

FIGURE 3.10 Los Angeles, California (a), is a sprawling, car-centered urban area that experiences heavy traffic congestion (b) on most days of the year.

Another problem is that about half of all U.S. urban land is dedicated to motor-vehicle-related uses such as streets, highways, and parking lots. All of these uses involve paving over or otherwise covering land that, in its natural state, absorbs the precipitation that helps to replenish aquifers. Thus, a great deal of this precipitation runs off paved and covered surfaces, adding to flooding and pollution of surface waters in these areas and reducing the recharge of some aquifers.

Traffic congestion is another growing problem in urban areas (Figure 3.10b). It results in high levels of air pollution in urban areas, enormous emissions of carbon dioxide, millions of hours of wasted time for drivers and their passengers stuck in traffic jams, and higher levels of stress for all involved.

In addition, motor vehicle accidents in the United States kill about 33,000 people a year, on average, and injure another 5 million. Worldwide, about 1.2 million people are killed every year in such accidents and about 15 million people are injured.

Let's **REVIEW**

- What percentage of the world's motor vehicles do Americans own and how much of the world's gasoline do they burn? What percentage of all U.S. urban transportation involves cars?
- What are four problems resulting from the widespread use of cars in urban areas?

Resource Use and Depletion

Because cities have large ecological footprints, they use resources from far beyond their borders. As cities grow, so does this resource use, and in some cases, cities can degrade or deplete soil, water, wildlife, minerals, and other natural resources in the surrounding areas.

For example, as a city grows, its total demand for water increases. This can lead to extensive pumping of aquifers and withdrawal of surface waters for use as drinking water supplies and for the watering of lawns, gardens, and golf courses. This can reduce the supplies of locally available water resources.

A case in point is the city of Las Vegas, Nevada (Figure 3.8). This city, along with other urban areas of the southwestern United States, gets much of its water from the Colorado River. This river has been dammed in several places to create reservoirs that supply water to about 30 million people in seven states. It also provides irrigation water to grow crops in many of the dry southwestern states, much of which goes to feed the people of the growing southwestern cities. So much water is now being taken from the Colorado River for these uses that very little of it reaches the Gulf of California where the river flows into the sea (Figure 3.11).

This situation has been made worse by a long-term *drought*, a condition under which an area does not get enough water from rain or snow for a long period of time. Drought is a global problem, affecting at least 30% of the earth's land (excluding Antarctica) and hundreds of cities, according to atmospheric scientist Aiguo Dai and his colleagues. Currently, more than 30 countries, most of them in Africa and the Middle East, suffer from water shortages that prevent them from meeting the needs of their populations. Water supply experts working with the UN project that by 2050, 60 countries, many of them containing densely populated and rapidly growing cities, are likely to be short of water.

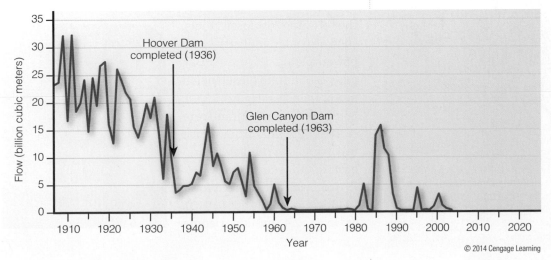

FIGURE 3.11 Since 1905, the flow of the Colorado River has dropped sharply because of increasing use of its water for agriculture and for the growing urban areas of the Southwest, combined with a prolonged drought. (Data from U.S. Geological Survey)

© 2014 Cengage Learning

In many urban areas, much of the water supply comes from aquifers. As cities in some parts of the world have grown, groundwater has been pumped faster than it can be naturally replenished by rainfall and snow melt. For example, the U.S. Geological Survey estimates that, on average, groundwater is being withdrawn in the United States about 25% faster than it is being recharged and much faster in parts of the dry western half of the country, and this is affecting some major urban areas (Figure 3.12). Much of this water is used to irrigate crops in these dry areas, but the growing demand for water in some of the area's cities is adding to the problem.

Many urban areas have met their water needs by moving massive amounts of water overland from relatively water-rich areas. One example is the system of *aqueducts*, or the combination of troughs, pipes, and canals used to transfer water from Northern California to comparatively dry Southern California. This system helped cities such as San Diego and Los Angeles to grow rapidly, and it encouraged the production of irrigated crops to help supply those urban areas with food.

However, this system and others like it have drained whole mountain lakes, destroying their ecosystems. In California, even the vast San Francisco Bay and Sacramento River Delta ecosystems have been seriously degraded with losses of wildlife habitat and the natural services provided by these systems. When these ecosystems are degraded or destroyed, the human communities that depend on them also suffer.

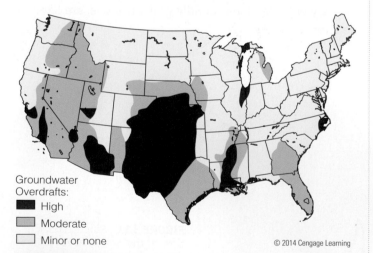

Groundwater Overdrafts:
- ■ High
- ▨ Moderate
- □ Minor or none

© 2014 Cengage Learning

FIGURE 3.12 In some areas of the United States, aquifers are being depleted because water is being removed much faster than it is replenished, partly because of urban water demand. (Data from U.S. Water Resources Council and U.S. Geological Survey)

Forests are another resource that have been exploited and sometimes depleted in order to meet the needs of urban areas. Over the centuries, billions of trees have been cut for use as firewood and building materials, and forests have been cleared to make way for the growth of cities as well as for crop fields and grazing land. These forests had provided wildlife habitat and ecosystem services such as natural water storage and purification, and absorption of climate-changing carbon dioxide. These important services are eliminated or severely degraded when forests are cleared.

Another resource that can be impacted by rapid urban growth is farmland containing fertile topsoil that gets paved over or covered by housing developments. Non-renewable mineral resources—the metals used to make building materials and consumer goods for urban populations—and nonrenewable energy resources such as coal and oil can also be impacted by urbanization.

Let's REVIEW

- What are the two major sources of water that can be depleted by a growing demand for water from urban populations?
- What are two problems that arise from transferring water from water-rich areas to drier urban areas?
- What are three other examples of resources that can be depleted by growing urban populations?

Pollution and Wastes

Just as urban areas require huge inputs of resources, they also create huge outputs of pollution and wastes because of their high population densities and high rates of resource use per person. Most of the world's air pollution, water pollution, and solid wastes are generated in urban areas.

The biggest share of the urban outputs of outdoor air pollutants are produced by the burning of fossil fuels, such as coal in industrial and power plants (Figure 3.13a), and gasoline and diesel fuel in motor vehicles (Figure 3.13b). Some of these pollutants are carried by winds from dense urban areas to rural areas and even to other countries.

According to the World Health Organization (WHO), at least 2.4 million people

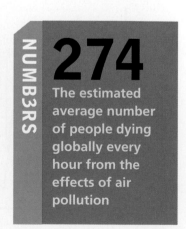

NUMB3RS

274

The estimated average number of people dying globally every hour from the effects of air pollution

FIGURE 3.13 Urban areas can experience high levels of air pollution from factories and coal-burning power plants (a), and from high concentrations of motor vehicles (b).

worldwide, most of them living in urban areas, die each year from the effects of prolonged breathing of polluted air—an average of more than 6,500 deaths per day. According to a 2007 World Bank study, in China, air pollutants contribute to the deaths of about 656,000 people a year. In addition, according to the U.S. Environmental Protection Agency, very small airborne particles of soot and other pollutants, produced mostly by coal-burning power plants, contribute to the deaths of at least 24,000 Americans every year.

Another pollution problem generated primarily in urban areas is *acid deposition*—sometimes called *acid rain*. It is formed from the emissions of coal-burning industrial and power plants as well as those of motor vehicles, and it consists of acidic particles of sulfur dioxide, nitrogen oxides, and other compounds formed from these chemicals. From the smokestacks of coal-burning plants, these pollutants travel long distances and return to the earth's surface in areas downwind from the plants—washed out of the air by rain and snow or simply falling to the surface. Acid deposition can severely degrade forests (Figure 3.14), soils, and aquatic systems, and damage human-made materials like paints and structures such as statues and buildings.

Urban areas are also major contributors to the growing levels of carbon dioxide and other greenhouse gases that are helping to accelerate the warming of the earth's atmosphere. Most of the world's leading climate scientists project that this warming will lead to rapid and widespread climate change during this century.

Large quantities of *water pollutants* are discharged by factories (Figure 3.15, p. 66) located in urban areas, and in some cities, untreated sewage often pollutes urban waters. Sediments such as eroded soil and chemicals including solvents, fertilizers, and pesticides run off of urban

FIGURE 3.14 These trees have been damaged by acid deposition.

FIGURE 3.15 This river flowing through a Chinese city suffers from uncontrolled pollution by thousands of factories.

construction sites, streets, parking lots, lawns, and golf courses, and pollute streams, lakes, and coastal waters near urban areas. This pollution has ruined the drinking water supplies for millions of people, especially in the urban areas of less-developed countries.

CONSIDER this

According to UN estimates, factory wastes and sewage pollute more than two-thirds of India's water resources as well as 54 of the 78 rivers and streams monitored in China (Figure 3.15).

Another form of pollution that is far worse in cities than in rural areas is *noise pollution*—any form of sound that can damage or interfere with one's hearing. Noise from industrial motors, motor vehicle traffic, and horns and sirens can add to city dwellers' levels of stress and affect their health, and it can make them less productive at work. Prolonged exposure to loud noises can cause permanent hearing damage.

Still another form of pollution that is growing worse in urban areas around the world is *light pollution*, the increasing levels of artificial light that illuminate urban areas (Figure 3.5). This form of pollution threatens some animal species. For example, the babies of some endangered species of sea turtles hatch from seashore nests at night and, in order to survive, must scamper into the sea as quickly as they can. Coastal urban lighting distracts many of these hatchlings, leading them off course and making them more likely to be eaten by predators. Light pollution also hinders astronomers and others living in and near cities who want to study and enjoy the night sky.

People and industries in urban areas also contribute by far the largest share of the world's growing mass of **municipal solid waste (MSW)**, commonly called *garbage* or *trash*, from urban homes and workplaces. It consists of paper and cardboard, cans, bottles, plastics, metals, glass, wood, and all other discarded solid materials. A special type of MSW is electronic waste, or e-waste—discarded cell phones, computers, TV sets, and other electronic devices that contain various toxic chemicals such as mercury and lead as well as valuable metal resources that could be recycled. It is the fastest growing category of MSW.

Because urban areas have a relatively high population density, MSW can build up quickly and must be disposed of somewhere. In several of the more-developed countries, most MSW is buried daily in *sanitary landfills* (Figure 3.16a), which are designed and built to help control the spread of diseases and other hazards associated with solid wastes. In the urban areas of less-developed countries, most of these wastes are put into *open dumps* (Figure 3.16b) where their hazards are uncontrolled. This can present a threat to people who live in shantytowns located in or near these dumps and who scavenge in them for goods that they can use or sell. E-waste and other toxic wastes in these dumps can lead to groundwater and surface water pollution. Fires in such dumps add to air pollution.

CONSIDER this

- Every year, urban areas in the United States generate enough municipal solid waste to fill a convoy of garbage trucks that would encircle the earth's equator almost eight times.

- Americans throw away enough office paper to build a ten-foot-high wall across the country from New York City to San Francisco, California.

FIGURE 3.16 In sanitary landfills **(a)**, wastes are covered regularly with soil or some other material. However, in open dumps such as this one in Manila, the Philippines **(b)**, wastes pile up and decay, creating health hazards for children and families who live in or near the dumps.

Let's REVIEW

- Why do urban areas generally have higher levels of pollution and wastes than rural areas?

- What are three problems related to urban air pollution? About how many people die each year largely as a result of air pollution?

- What are three types of urban water pollutants and what are their typical sources?

- What are two other types of pollution that are worse in urban areas than in rural areas?

- What is *municipal solid waste (MSW)* and why is it a problem, especially in some urban areas? What is e-waste? What happens to MSW in many urban areas of less-developed countries and what problems result?

Urban Poverty

Poverty, or the set of conditions that people endure when they are not able to provide for their basic needs, is not solely an urban problem. However, it can be made worse by the nature of urban life in some parts of the world, and it can contribute to the worsening of other urban problems such as pollution and depletion of natural resources (see *For Instance*, p. 68).

In the less-developed countries, where urban population density tends to be higher than it is in more-developed countries, at least 1 billion urban dwellers live in poverty, in crowded and unsanitary conditions (Figure 3.17). According to a 2006 UN study, that number could reach 1.4 billion—more than four times the U.S. population—by 2020.

Such conditions arise in *slums*—areas dominated by tenements and rooming houses where as many as 10 people might live in a single room—and in *shantytowns*, or collections of shacks on the outskirts of cities (Figure 3.6). Usually, people living in such places find it very hard to get enough food and clean water for healthy living. Some have to build their own shelters out of found scraps of metal, plastic, cardboard, and wood (Figure 3.18, p. 68). Many end up living in abandoned buildings or junked cars. In the face of such challenges, many poor people have worked hard to create livable communities, but they still face serious health threats.

FIGURE 3.17 Untreated sewage water runs through the Mathare Valley shantytown—one of the largest of such settlements in Africa—near Nairobi, Kenya.

FIGURE 3.18 This mother and her child live in a hut built in a poor area of Mumbai (formerly known as Bombay), India.

Life in shantytowns and slums can be hazardous living for many other reasons. They are often located near factories, airports, or open dumps (Figure 3.16b), which means people living there are exposed to air and water pollution as well as to toxic materials. Many of these settlements are in locations especially prone to landslides, flooding, or earthquakes.

These dangerous conditions, especially the lack of clean water and the presence of open sewers (Figure 3.17) can expose poor urban residents to *infectious diseases*—sicknesses caused by bacteria, viruses, or parasites that invade and multiply within a person's body. For people living in crowded, poverty-stricken urban areas, the most threatening of these diseases are flu, HIV/AIDS, tuberculosis, malaria, and diarrheal illnesses such as cholera.

Some of these diseases can spread rapidly in crowded urban areas, as they can be transmitted through the air, water, and food, or through body fluids such as feces, urine, blood, and droplets sprayed by sneezing and coughing. For example, one of the most difficult health problems in some areas of the world is the rapid spread of tuberculosis (TB), a disease that destroys lung tissue. The WHO estimates that this highly infectious disease kills about 1.4 million

NUMB3RS

3,560

The number of people killed, worldwide, every day by tuberculosis

people per year—80% of them living in urban areas of less-developed countries. Without treatment, each person with active TB can infect 10–15 other people.

FOR INSTANCE...

Mexico City—A Struggling Urban Area

Mexico City is a densely populated, rapidly growing city with a lot of problems. It has about 20.5 million people, and at least 400,000 more move into the city every year, mostly from poor rural areas.

The air over Mexico City is highly polluted because of three major factors. One is the presence of millions of cars and hundreds of polluting factories. Until recent years, motor vehicle exhaust in the city was not well regulated and is therefore a major source of pollutants, and older vehicles still have no pollution control equipment.

Second, the city enjoys a sunny climate, and on most days, as sunlight interacts with pollutants from cars and factories, a type of pollution called *photochemical smog* forms (Figure 3.19). A third factor is geography; the city's air pollution builds to dangerous levels because the city is located in a bowl-shaped valley surrounded on three sides by mountains—conditions that trap air pollutants at ground level. Traffic congestion adds to this problem. Many of the city's inhabitants, especially children, suffer from respiratory diseases. According to some health officials, breathing the city's air is roughly equal to smoking three packs of cigarettes per day.

At least 3 million people in Mexico City have no sewer services. Human waste ends up in gutters, ditches, and vacant lots. When it dries, winds pick it up and create a *fecal snow* that spreads across parts of the city. The *Salmonella* bacterium that causes food-borne illnesses and the bacterial form of hepatitis, an infection of the liver, are spread in this way, especially to children playing outdoors. According to some health officials, Mexico City's air and water pollution together play a key role in the deaths of about 100,000 of its people every year, or 11 every hour.

Mexico City also suffers from frequent water shortages. About two-thirds of the city's water comes from underground and as this groundwater is being depleted, the shortages are becoming more frequent. Most of the rest of the city's drinking water is transported from a source located 78 miles away, which requires a lot of energy for pumping, which in turn adds to the city's air pollution.

The city's high population density has contributed to consistently high unemployment and poverty, which has

FIGURE 3.19 Mexico City suffers from high levels of air pollution.

led to a high crime rate, made worse by a growing illegal drug trade. More than one of every three of Mexico City's residents live in slums called *barrios* or in shantytown settlements with little or no clean water or electricity.

The city has made progress toward dealing with some of its problems. By 2010, the percentage of days in which air pollution standards were violated each year had dropped from 50% to 20%. City officials have banned cars in the central zone, phased out the use of leaded gasoline, and now require that all cars built after 1991 be equipped with air pollution controls. The city has also replaced old taxis, delivery vehicles, and buses with cleaner burning models. In addition, it has expanded its subway system, planted more than 25 million trees to help absorb pollutants, and set aside land for green space.

Let's **REVIEW**

- What is *poverty* and how is it affected by urban life? About how many people live in poverty in the urban areas of less-developed countries?

- What are some of the hazards faced by poor people who live in urban areas of less-developed countries?

- Why are infectious diseases a problem for people living in crowded, poverty-stricken urban areas? What are five of the most threatening infectious diseases for these people?

- Describe Mexico City's major problems and the progress made in dealing with some of them.

What Can Be Done?

Controlling Sprawl with Smart Growth

According to many environmental scientists and urban planners, urban growth in itself is not necessarily something to be avoided. In fact, it can be beneficial to the environment when it concentrates more people in relatively small areas, allowing for more land to be preserved for sustaining biodiversity and vital ecosystem services. The problems arise when urban growth makes cities less sustainable and less livable over time, as is the case with most of the world's urban areas.

One approach to more environmentally sustainable urban development is called **smart growth**—a set of strategies that includes reducing dependence on cars, controlling and directing sprawl, and reducing wasteful resource use in urban areas. These strategies involve the use of energy-efficient mass-transit systems; neighborhoods containing a mix of houses, stores, and other services; and more compact, higher-density housing. Other goals of smart growth are to reduce traffic and congestion and to make neighborhoods into more pedestrian-friendly and bicycle-friendly places to live.

Scientists and planners who argue for the use of smart-growth strategies accept the fact that urban growth will take place. However, they seek to use various tools to channel growth into areas where it can improve abandoned or blighted areas, or at least avoid damaging environmentally sensitive and important land and waterways.

A commonly cited example of a city using successful smart-growth strategies is Portland, Oregon. Many of Portland's residents live and work in mixed-use neighborhoods that include high-density housing as well as single-family homes, stores, professional offices, and businesses. Portland's city planners and administrators developed extensive, easy-to-use bus and light-rail transit systems, walking paths, and bike trails. They also enacted strict boundaries for growth around the city and promoted the renewal of abandoned areas within those boundaries instead of their expansion into outlying areas.

These measures helped to make Portland highly livable and attractive; thus, while it has grown rapidly, it has not sprawled out as other American cities have. Since 1975, its population has grown by 50%, while the area of land it occupies has grown by only 2%. Figure 3.20 lists some of the smart-growth tools used by Portland and other cities.

Smart Growth Tools

Limits and Regulations
Limit building permits
Draw urban growth boundaries
Create greenbelts around cities

Zoning
Promote mixed use of housing and small businesses
Concentrate development along mass transportation routes

Planning
Ecological land-use planning
Environmental impact analysis
Integrated regional planning

Protection
Preserve open space
Buy new open space
Prohibit certain types of development

Taxes
Tax land, not buildings
Tax land on value of actual use instead of on highest value as developed land

Tax Breaks
For owners agreeing not to allow certain types of development
For cleaning up and developing abandoned urban sites

Revitalization and New Growth
Revitalize existing towns and cities
Build well-planned new towns and villages within cities

© 2014 Cengage Learning

FIGURE 3.20 Some urban areas are using a variety of smart-growth tools to help control urban sprawl.

Photo credits (top to bottom): Bobby Mikul/www.publicdomainpictures.net; Rich Gribble/USDA Natural Resources Conservation Service; U.S. Environmental Protection Agency.

Improving Urban Transportation

Some cities are taking measures to reduce congestion, air pollution, and the other problems that stem from heavy motor vehicle use within their boundaries. For example, some city governments, such as that of London, England, have raised parking fees and charged tolls for using roads, tunnels, and bridges leading into their cities. Such fees are often higher during peak traffic times.

Another way to cut the number of cars in a city is to encourage the use of car-sharing networks, as is commonly done in Germany, Austria, Italy, Switzerland, Ireland, and the Netherlands. Members of these networks typically can reserve a car in advance or call the network at any time to see if a car is available. In Berlin, Germany, car sharing has cut car ownership by 75%. According to the Worldwatch Institute, car sharing in Europe has reduced the average driver's annual carbon dioxide emissions by 40–50%. Car-sharing networks are gradually growing in several U.S. cities as well.

Many cities around the world are encouraging the use of alternatives to motor vehicles. They promote the use of *bicycles*, by building systems of bike lanes and paths and bike storage areas (Figure 3.21a); *light-rail systems* (Figure 3.21b); *bus systems* (Figure 3.21c); and *high-speed rail systems* that connect urban areas (Figure 3.21d). For example, Western Europe, Japan, and China have high-speed trains, also called *bullet trains*, that travel between cities at up to 190 miles per hour.

High-speed rail, light-rail, and bus systems all use much less energy per passenger mile than do cars (Figure 3.22). Thus, commuters in cities that employ these mass-transit systems burn less gasoline, save money, and help to reduce overall pollution and climate-changing carbon dioxide emissions.

Bicycling and walking account for about a third of all urban trips in the Netherlands and in Copenhagen, Denmark, compared to about one of every 100 trips in the United States. Paris, France, has made almost 21,000 bikes available for rental at 1,450 rental stations throughout the city. However, because of maintenance and repair costs, and the incidence of bike thefts, the system has been losing money, so as with most transportation systems, this one faces challenges. Montréal, Canada, also has a bike-sharing program as do Barcelona, Spain, and a growing number of other cities around the world.

In many of the world's poorest cities, most people cannot afford to own a car. For example, 95% of the residents of Mumbai, India, must walk or use buses or trains to get to work. Only 1.6% commute to work by car, compared

FIGURE 3.21 (a) Bikes parked in the center of Amsterdam, the Netherlands; (b) light-rail system in Nice, France; (c) electric-powered bus in Budapest, Hungary; (d) bullet train in Milan, Italy.

to 75% in the United States. Thus, in the area of transportation, some of the world's poorer cities have smaller ecological footprints and more sustainable transportation systems.

Let's REVIEW

- What is *smart growth*? Describe five smart-growth tools that some cities are using. How has Portland, Oregon, applied these tools?

- List some ways in which cities are trying to discourage car traffic and encourage alternatives to conventional car use in cities.

- What are four alternatives to car use that are available in many urban areas?

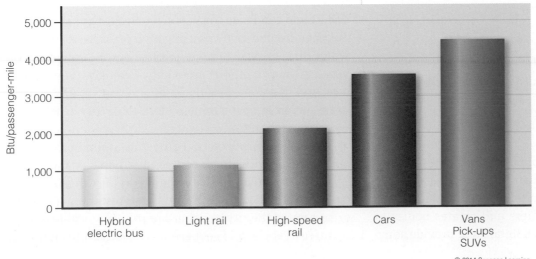

FIGURE 3.22 This chart uses units of energy called *British thermal units (BTUs)*. It compares the BTUs per *passenger-mile* (one mile traveled by one person) used by various transportation systems. For example, on average, a car uses 3 times as many BTUs as the best of today's light-rail systems would use to transport you one mile. (Data from Todd Litman, Victoria Transport Policy Institute, 2011)

Cutting Resource Use and Waste

In order to become more sustainable, cities will have to shrink their ecological footprints, or at the very least, stop them from growing further, according to most environmental analysts who study urban areas.

For example, according to these analysts, one of the most important ways to maintain and increase urban water supplies is to reduce the unnecessary waste of this precious resource. Geologist Mohamed El-Ashry of the UN Foundation has estimated that we lose about 66% of the water used in the world and half of the water used in the United States, largely through evaporation, water leaks, and inefficient use. Much of this evaporation loss results from irrigation, but a large amount of it occurs in the watering of urban lawns, gardens, and golf courses.

El-Ashry also estimates that we could meet most of the world's projected future water needs by cutting the global total water loss to 15%. Much of this could be accomplished with the use of water-saving technologies such as rain barrels and various systems for recycling used water, called *gray water*, from showers and dishwashers. Gray water can be used for watering lawns and for several other purposes. We can also use highly efficient watering systems, toilets, showers, and dishwashers to cut water losses.

An important way to encourage savings is to make water more expensive to users. Currently, most of the municipal, state, and national governments that control water supplies charge far too little for water, according to water resource experts such as Sandra Postel, founder of the Global Water Policy Project, and this undervaluation of water does not encourage its conservation or efficient use. Some governments in water-short regions are raising their water prices.

FIGURE 3.23 China's Three Gorges Dam on the Yangtze River created a reservoir that flooded an area that could fit roughly between the cities of Los Angeles and San Francisco, California.

ecological services. China used this approach, building the world's largest dam-and-reservoir system—the Three Gorges Dam—on the Yangtze River (Figure 3.23), and it plans to build more dam-and-reservoir systems on some of its other major rivers.

However, dam-and-reservoir systems have their drawbacks. They can restrict the flows of rivers and severely degrade their ecosystems, as in the case of the Colorado River in the United States (Figure 3.11). Dammed reservoirs also flood and destroy forests and croplands, and they force people to move. The Three Gorges Dam reservoir displaced about 1.4 million people from their homes and land.

Many people see the oceans as a potentially huge source of water for coastal urban areas. The key is to find an affordable and environmentally acceptable method of *desalination*, any process in which the salt from ocean water is removed. One way to desalinate water is to heat it, collect and condense the vapor, and leave the salts behind. Another method involves using high pressure to force saltwater through a membrane to filter out the salt. However because these methods are energy-intensive, very expensive, and in some cases damaging to coastal ecosystems, desalination is now supplying less than half of 1% of the world's water needs.

Other resource depletion problems around urban areas can be solved when the resources are managed more carefully. For example, the city managers of Curitiba, Brazil (see *For Instance*, p. 79), halted the deforestation of wooded areas around the city by using a permitting system for cutting trees. People who get permits can cut limited numbers

Some analysts argue that building dam-and-reservoir systems to provide water in urban areas is more sustainable than pumping it from underground. It relieves the need to pump over-taxed aquifers and creates bodies of water that can support wildlife habitats and provide some

of trees only if they plant and tend two trees for every one they cut. The city also enlisted volunteers who planted millions of trees where forests had been cleared.

Let's REVIEW

- Describe how water in urban areas is wasted and how such losses could be cut. How are water prices connected to water use, and how can they be used to cut water losses?
- Summarize the benefits and drawbacks of building more dam-and-reservoir systems, and of using desalination to supply water to urban areas.
- How has Curitiba, Brazil, controlled deforestation in wooded portions of its urban area?

Dealing with Pollution

The problems of poor urban air quality, acid deposition, and mounting greenhouse gas levels in the atmosphere leading to climate disruption can all be addressed with the broad strategies of *pollution reduction* and *pollution prevention*.

An important way for national and state governments to encourage pollution reduction is to enact and enforce strict air pollution control laws for industries and motor vehicles. Since 1970, the United States has had success with this approach by measuring and reducing the levels of several major air pollutants. These reductions came about because scientists and engineers found ways to capture pollutants before they were emitted into the air. This involves collecting pollutants in devices attached to industry smokestacks and motor vehicle exhaust systems.

While these systems have helped to reduce air pollution, many scientists point out that such "end-of-pipe" solutions have limits. They say that a more effective general approach is to find "front-of-pipe" solutions that prevent the production of pollutants in the first place—a pollution prevention approach.

One such solution is to rely less on the use of fossil fuels, especially coal and gasoline, and to shift to less environmentally harmful energy resources such as wind, solar energy, flowing water (hydropower), heat from under the earth's surface (geothermal energy), and biofuels produced from plant wastes and from sustainable crops like switchgrass.

Another important way to reduce energy waste and air pollution is to improve the energy efficiency of power plants, industrial processes, homes and other buildings, and motor vehicles. Whenever less fuel is burned for any of these uses, there will be that much less pollution

generated. For transportation purposes, urban dwellers can cut their pollution by walking and using bicycles and energy-efficient urban mass-transit systems where they exist. Homeowners can heavily insulate their homes and plug air leaks. Energy-efficiency efforts can also save money by reducing heating, cooling, and driving expenses.

As with air pollution, urban water pollution has been decreased by the enactment and enforcement of pollution control laws in many countries. The United States and other countries have had strict laws in place since the 1970s to control water pollution from urban construction sites, factories, and other urban sites. However, similar to air pollution control, efforts to check water pollution usually involve trying to collect pollutants before they get into waterways. Many scientists argue for putting much greater emphasis on preventing water pollution by finding ways to avoid creating the pollutants in the first place.

A promising way to prevent water pollution from urban sewage treatment systems is to work with nature by using wetlands to treat wastes. Some communities have used natural wetlands while others have created artificial wetlands to filter sewage from wastewater. Some of these systems make use of large tanks that hold bacteria, algae, and a variety of plants to treat the wastes (Figure 3.24). They also make use of solar energy, which cuts the use of fossil fuels and prevents more pollution.

FIGURE 3.24 This ecological wastewater treatment system is located in the city of Providence, Rhode Island. Biologist John Todd is demonstrating the ecological sewage treatment process that he invented.

The most effective ways to reduce the harmful impacts of urban air and water pollution involve preventing the creation of the pollutants.

Urban flooding is one of the major causes of water pollution, and there are many ways in which cities have tried to control flooding. Several cities have built levees and dams along the rivers that run through them, but this is just a way to funnel floodwaters downstream where they often flood other areas.

A better approach to urban flooding, according to many analysts, is to prevent it by protecting urban-area forests and wetlands that absorb heavy rains and snowmelts and then release those waters slowly. Many urban areas have instead paved over or developed these areas, eliminating this natural flood-control service. But some of these degraded natural areas can be restored (see *Making a Difference*, below). Cities can also limit future development in these natural areas.

We can also prevent noise and light pollution by the use of technological means, many of them quite simple. When they are required to do so by laws or by market forces, the manufacturers of cars, industrial motors, leaf blowers, and other gasoline-powered machines have found ways to muffle their noise. City streetlights can be shaded and aimed at the ground in order to keep them from lighting up the skies. Interior building lights can be similarly shaded and tinted office windows also help to soften the effects of interior lighting. Turning off interior lighting at night wherever people do not need it also prevents light pollution and saves energy.

Handling Municipal Solid Waste Problems

Cities use many different methods to handle their solid waste. In urban areas of the United States, most waste is buried in sanitary landfills that are designed to reduce some of the hazards of waste disposal (Figure 3.25) such as air pollution, water pollution, and groundwater contamination. Other more-developed countries burn most of their MSW in incinerators, some of which send hazardous pollutants into the air, although newer incinerators have better pollution controls.

In urban areas of many less-developed countries, people dump their solid wastes, often including hazardous wastes, into open dumps (Figure 3.16b). As countries become more developed and wealthier, they tend to move toward the use of sanitary landfills and incinerators.

However, even the best waste-management technologies have a hard time keeping up with rising outputs of MSW as both the human population and the global rate of resource use per person grow. Many environmental scientists and analysts say that the only really effective way to

MAKING A difference

Restoring an Urban-Area Stream

John Beal was a Boeing Company engineer whose doctors told him in 1980 that he had only a few months to live due to a severe heart ailment. As part of his effort to prove them wrong, he began walking for exercise, and one of his walking routes was along a small stream called Hamm Creek that flowed near his home in Seattle, Washington.

Beal knew that this stream had been severely degraded over the years. He remembered when evergreen trees stood along the creek's banks and salmon spawned in its waters. By 1980, the trees were gone, the creek was badly polluted, and the salmon no longer returned to spawn.

Beal decided that Hamm Creek needed his attention and that helping to restore the creek would be a good way to spend his remaining time. Eventually, he would haul away many truckloads of trash and work with various companies that were polluting the creek, convincing them to stop the pollution. Beal then set about planting trees along the stream's banks. His efforts, along with those of several people who joined him, helped to restore natural ponds that fed the creek, as well as waterfalls and salmon spawning beds in the creek.

Beal was to learn that by helping to lengthen the life of Hamm Creek, he would also lengthen his own life. Today, Hamm Creek's water runs clear, salmon have returned to spawn, and its banks are again lined with trees. John Beal lived for another 27 years after launching this project, and he greatly enjoyed the natural area that he had taken the lead in restoring.

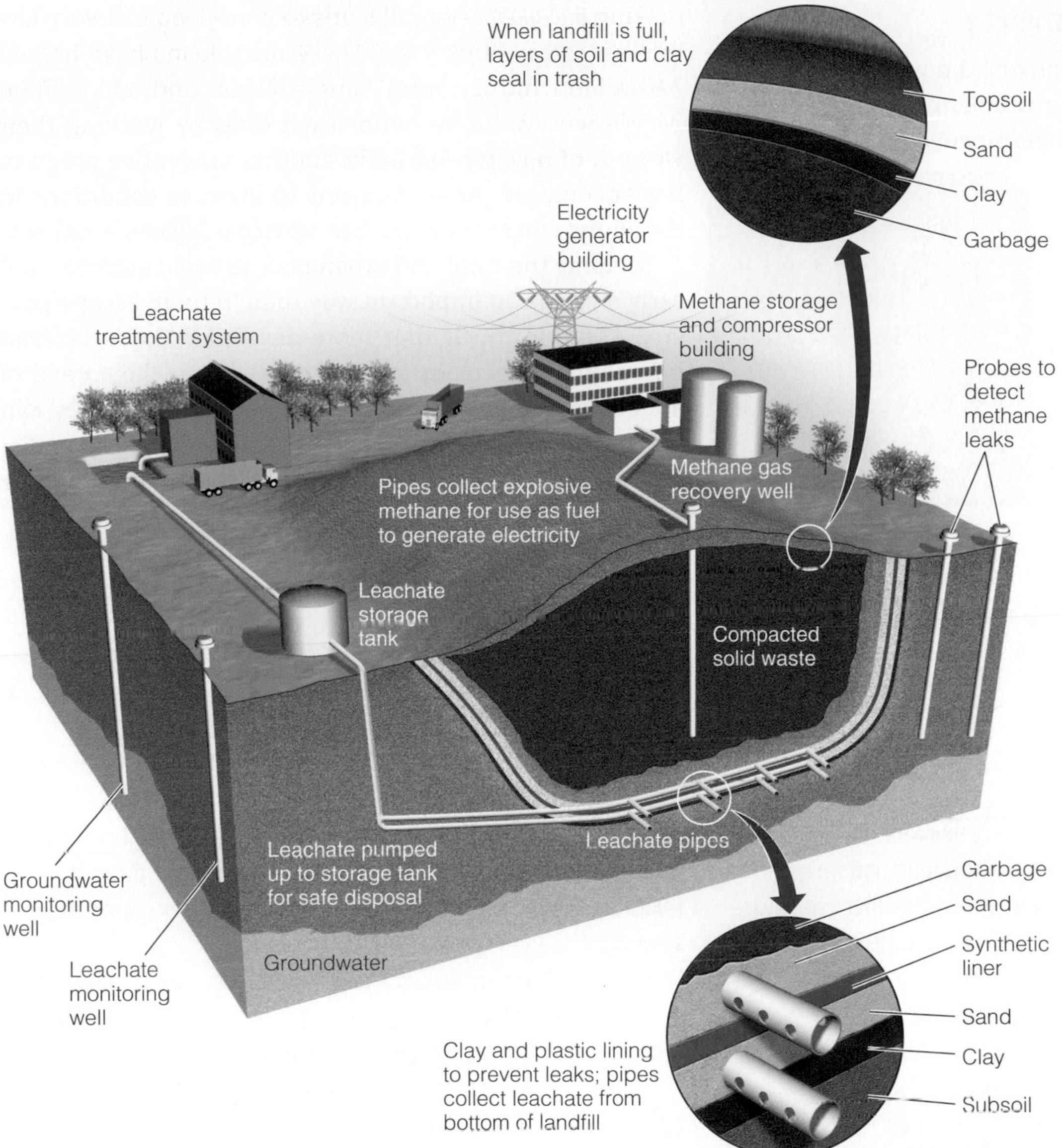

FIGURE 3.25 This diagram shows some of the features of sanitary landfills that can reduce the environmental impacts of dumping and burying solid waste.

When landfill is full, layers of soil and clay seal in trash

Topsoil
Sand
Clay
Garbage

Electricity generator building

Methane storage and compressor building

Probes to detect methane leaks

Leachate treatment system

Pipes collect explosive methane for use as fuel to generate electricity

Methane gas recovery well

Leachate storage tank

Compacted solid waste

Leachate pumped up to storage tank for safe disposal

Leachate pipes

Groundwater monitoring well

Leachate monitoring well

Groundwater

Clay and plastic lining to prevent leaks; pipes collect leachate from bottom of landfill

Garbage
Sand
Synthetic liner
Sand
Clay
Subsoil

© 2014 Cengage Learning

deal with MSW is to reduce it, mostly by preventing its production in the first place. Experts' estimates for how much MSW could be reduced in this way go as high as 80%.

One important strategy for reducing the production of solid waste includes reducing both the materials used to produce many products and the packaging used to market and deliver those goods. Another approach is to focus on making products that last longer and are easy to reuse, repair, recycle, or compost.

Some governments have passed laws to force these reductions. For example, many countries and states have laws that require payment of a small deposit on recyclable glass, plastic, and aluminum containers, which makes them more likely to be recycled. Some European countries require manufacturers to take back and recycle the electronic products and motor vehicles they produce.

While such regulations have been effective, many analysts call for greater investments in education to persuade manufacturers and consumers to use the three Rs of resource use—reduce, reuse, and recycle—in order to slow the growth of urban solid waste.

Let's REVIEW

- Explain the benefits of pollution reduction and pollution prevention, and distinguish between end-of-pipe solutions and front-of-pipe solutions.
- How can we reduce urban air and water pollution? What are some ways to reduce flooding in urban areas?
- How can cities deal with noise and light pollution?
- What are three ways to reduce the outputs of solid wastes in urban areas?

Dealing with Urban Poverty

Urban poverty has plagued many of the world's cities for centuries, and it is a growing problem. However, some cities have made progress in fighting poverty.

A number of governments and financial institutions have addressed urban poverty by focusing on rural poverty, which is one of the drivers of the migration of poor people to overcrowded, impoverished urban areas. In Bangladesh, for example, the Grameen Bank has been making *microloans*—small loans to poor people at very low interest rates—since 1983. These microloans have helped 7.6 million mostly rural Bangladeshis, and 100 million people worldwide, to better their lives by working their way out of poverty. Similarly, another innovative program has encouraged young Kenyans to improve conditions in the slums where they live (see *Making a Difference*, below).

Helping the rural and urban poor to avoid sickness and early death is an important way to help them escape poverty. There is much that more-developed countries and international agencies can do to help slow the spread of infectious disease in urban areas. For example, they can help to fund research in poorer countries on various diseases and the immunizations that prevent them. Some

NUMB3RS

100

The number of millions of people using microloans to work their way out of poverty

MAKING A difference

Bob Munro's Soccer Team

The shantytown in Mathare Valley near Nairobi, Kenya, represents poverty at its worst. Estimates of its population range from 200,000 to 600,000 people, all living on 3 square miles of devastated land. Most of its shacks, housing an average of 10 people each, are about the size of two double-bed mattresses. It has no regular roads, no electricity, and no closed sewers.

Most of Mathare Valley's inhabitants get up at 4 a.m. daily to stand in line at a community standpipe for a bucketful of water—one person's supply of water for the day. Many of them then move to another line to visit one of the latrines that are each shared by dozens of people at a time. Others, however, skip this line, and as a result, open gutters in the shantytown become sewers (Figure 3.17) full of human wastes that can spread infection.

In 1978, Bob Munro, a Canadian development agency worker, had an idea for a new way to help the people of Mathare Valley. Using their love of soccer, he told the town's youngsters that he would supply a soccer ball and provide a safe place to play if they would do something in return. He promised that for every hour that each of them spent digging drainage ditches, clearing away garbage, or improving their environment in other ways, they could play an hour of soccer.

Munro's scheme has been a huge success. Mathare Valley youths, armed with rakes and shovels, now clear away more than a ton of garbage on some days. Since this program started, it has served 120,000 boys and girls, all of whom became members of the Mathare Youth Sports Association (MYSA). In addition to soccer, these youngsters also play field hockey, basketball, and several other team sports. By mid-2007, there were more than 1,300 teams with 18,640 youngsters playing a dozen sports (Figure 3.A).

Munro long ago handed over the day-by-day running of MYSA to local leaders. His program is now being used in a number of other slums in Kenya, Uganda, Tanzania, and Botswana. Since 1978, Mathare's population has doubled, but its environmental squalor has been reduced, largely because of Bob Munro's idea.

Courtesy of MYSA SHOOTBACK

FIGURE 3.A These young people from the Mathare Valley shantytown are members of the Mathare Youth Sports Association.

Vestergaard Frandsen

FIGURE 3.26 Four young men in Uganda demonstrate the *LifeStraw*™, designed by Torben Vestergaard Frandsen. It is a personal water purification device that helps people avoid waterborne diseases.

private companies are also providing affordable tools to help people avoid these diseases (Figure 3.26).

Some cities have programs that designate land for shantytown settlements and supply them with clean water and bus service for workers to travel to and from their jobs. In some urban areas of Peru and in Nairobi, Kenya, experience with such programs has shown that once the residents knew they could stay in one place, they made improvements in their communities, including the establishment of schools and day care centers.

Let's REVIEW

- What is a microloan and how have microloans helped to reduce urban poverty? How can a focus on rural poverty benefit overcrowded cities?

- What are three ways to prevent or reduce the incidence and spread of infectious diseases in the urban areas of less-developed countries?

- Describe the conditions in Kenya's Mathare Valley shantytown and the program used to encourage young people to improve such conditions.

Making Urban Areas More Sustainable and Livable

In a number of cities around the world, officials, planners, and residents are working hard to make their urban areas more sustainable and livable, using well-tested policies, tools, and technologies. Many of these people envision what can be called an **ecocity** or a **green city**— a city that puts its highest priority on minimizing its ecological footprint and improving the quality of life for its inhabitants.

Some cities have made great strides toward these goals (see *For Instance*, p. 79). Examples of sustainability features commonly found in these cities are networks of bicycle paths (Figure 3.27a, p. 78); user-friendly systems for recycling and composting (Figure 3.27b); energy systems based on renewable energy sources such as wind and solar power (Figure 3.27c); and houses heated largely by the sun (Figure 3.27d).

Cities that are pursuing the ecocity model distinguish themselves from conventional cities in several ways. New neighborhoods in these cities are designed to be more pedestrian-friendly. Instead of having wide, multilane streets and highways, they have mostly narrow streets with wide sidewalks and paths for walking and biking. These cities also rely more on affordable, energy-efficient mass transit, neighborhoods with a mixture of housing and businesses, and compact development.

Ecocity developers try to preserve surrounding croplands and protect and restore nearby natural wildlife habitats and wetlands. Trees and plants adapted to the local climate and soils are planted throughout ecocities to provide shade, beauty, and wildlife habitat, and to reduce air pollution, noise, and soil erosion.

Green cities work toward sharply reducing pollution and the waste of energy and matter resources. In addition to encouraging their residents to reduce the use of these resources, ecocities also have strong programs to encourage reuse, recycling (Figure 3.27b), and composting. Garden trimmings, waste food, and even waste sewage can be converted to plant nutrients, as in natural systems, or

Elena Elisseeva/Shutterstock.com

ZOFotography/Shutterstock.com

Ulrich Mueller/Shutterstock.com

John Avenson/National Renewable Energy Laboratory

FIGURE 3.27 These are some of the features of more sustainable ecocities: **(a)** networks of bicycle paths are common; **(b)** most waste materials are reused, recycled, or composted; **(c)** wind turbines and solar cells such as these in Germany are widely used as a source of electricity; and **(d)** solar energy is used to heat water and interior spaces in businesses and homes such as this energy-efficient, passive solar home in Westminster, Colorado.

converted to biogas fuel. Some green cities recycle or compost 60% or more of all municipal solid waste.

Energy resources are carefully chosen and promoted in ecocities. To produce electricity, many make use of wind turbines, solar cells, (Figure 3.27c) and other locally available, renewable energy resources. As a result, homes and other buildings using these systems can often serve as miniature power plants by selling their excess electricity back to power companies. Many new buildings get much of their heat from direct solar energy (Figure 3.27d) or they use geothermal energy for heating and cooling.

In addition to striving for self-sufficiency in terms of energy resources, ecocity residents also try to be more self-sufficient in terms of their food supplies. Some cities have community urban gardens and farmers markets. Instead of maintaining grass lawns, some people fill their

lots with organic gardens and a variety of plants adapted to local climate conditions. People also make use of nearby organic farms as well as solar greenhouses and small gardens on rooftops, in yards, and even in window boxes.

Some examples of cities that are working toward becoming more environmentally sustainable and livable are Curitiba, Brazil (see the following *For Instance*); Bogotá, Colombia; Leicester, England; Stockholm, Sweden; Helsinki, Finland; Waitakere City, New Zealand; and the U.S. cities of Portland, Oregon; Olympia, Washington; Davis, California; and Chattanooga, Tennessee.

New York City also has plans to become one of the world's more sustainable large cities. Because it is a compact city with a large population and widespread use of public transit, the per capita carbon footprint for New Yorkers is already 71% smaller than that of the United States as a whole. The city's plan includes planting more than a million trees, establishing more bike lanes and pedestrian plazas, expanding its parks, improving the energy efficiency of its buildings, converting all of its taxis to hybrids or other energy-efficient vehicles, and modernizing its mass-transit system.

Curitiba, Brazil—One of the World's First Ecocities

Thanks to the vision and foresight of one of its former mayors, Curitiba ("koor-i-TEE-ba"), a city of 3.2 million people, has long been known as the "ecological capital" of Brazil.

In 1969, a group of planners led by former college professor and architect Jaime Lerner, who later served three times as mayor, envisioned Curitiba as a green city, one that would be driven by the needs of people, not cars. These planners decided to build an efficient mass-transit system, which is now known as one of the world's best and most energy-efficient *bus rapid transit (BRT)* systems. It makes use of state-of-the-art, low-polluting buses and customer-friendly bus boarding systems (Figure 3.28).

To encourage BRT use and to keep car traffic from clogging the inner city, the city managers created BRT express routes. Near these routes, only high-rise apartment buildings are allowed and the bottom two floors of each building must be devoted to stores where residents can shop. This gives a large number of residents easy access to the stores, buses, and other services, and it helps them to shrink their ecological footprints. In addition, bus sections can be hooked together to double or triple the lengths of the buses, giving them the capacity to carry as many as 270 passengers on the BRT routes.

The center of the city is a car-free zone, laced with pedestrian walkways, parks, and bicycle paths. As a result, Curitiba has lower emissions of greenhouse gases and air pollutants, less traffic congestion, less noise, and less energy use per person than do most cities of comparable size and population.

What helps to make Curitiba green, literally, are the 1.5 million or more trees that volunteers have planted throughout the city. City rules forbid the cutting of trees without a permit, and two trees must be planted for each tree that is cut down.

Recycling is a high priority in Curitiba. The city recycles roughly 70% of its paper and 60% of its metal, glass, and plastic. Most of these recovered materials are sold to the city's 500 or more major industries.

The city managers also believe in helping the poor to help themselves. People in need get help with caring for themselves and their children and finding a job. Some retired buses are used as soup kitchens, healthcare clinics, daycare centers that are free for low-income parents, and

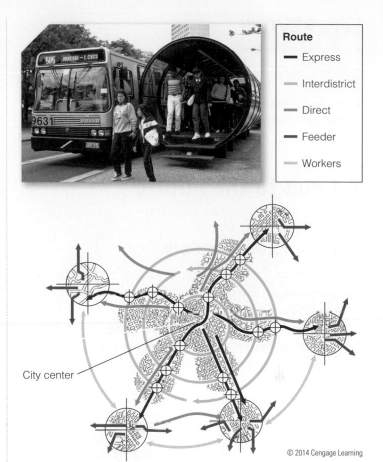

Route
- Express
- Interdistrict
- Direct
- Feeder
- Workers

City center

© 2014 Cengage Learning

FIGURE 3.28 Boarding a BRT bus in Curitiba, Brazil, is convenient and easy because of extra-wide bus doors and boarding platforms sheltered by large glass tubes where passengers can pay their fare before getting on the buses (see photo, top). The BRT system provides rapid transportation for large numbers of people with express routes laid out along five major spokes radiating out from the city center (see map, bottom).
Photo credit: John Maier, Jr./Peter Arnold

mobile classrooms where people can get free training for basic job skills.

Poor residents can also help to keep the city clean by collecting garbage. In exchange for filled garbage bags, they can get bus tokens, school supplies, and surplus food. Curitiba also runs a site-and-service program (see *A Look to the Future*, p. 80).

About 95% of Curitiba's citizens can read and write, 83% of its adults have at least a high school education, and all school children study ecology. Polls show that 99% of the city's inhabitants would not want to live anywhere else.

The population of Curitiba has grown fivefold since it began its sustainability programs, and this has put some stress on the city's systems. However, Curitiba continues to stand as an international model for ecocity development.

Let's REVIEW

- List eight characteristics of an *ecocity* or *green city*.
- How is New York City planning to become a more sustainable city?
- Summarize the ways in which Curitiba, Brazil, has become a more sustainable city.

A Look to the Future

Site and Service:
Hope for Slum Dwellers

Many cities in the developing world are exploding under the pressure of their growing populations, due both to births within these urban areas and to the inflow of migrants from the countryside. Every day, tens of thousands of poor people stream into these cities from rural areas, often overwhelming city services.

The majority of these immigrants head for shantytowns that now ring the outskirts of cities such as South Africa's Cape Town. There, an area called Cape Flats is crowded with tin shacks and suffers under the same conditions found in the Mathare Valley shantytown (see *Making a Difference*, p. 76). Mathare Valley itself, often referred to as Africa's oldest and worst slum, is one of 200 shantytowns located in and around Kenya's capital city of Nairobi. Similar slums exist in urban areas of Asia and South America.

The mushrooming growth of slums in the cities of less-developed countries is one of the biggest challenges of our time. These slums now have a total population of more than a billion people. Within the next three decades, that number could swell to 2 billion. Within the next decade, the slums of sub-Saharan Africa will likely have more people than the entire population of the United States.

So far, most national governments appear to have largely ignored this difficult problem. Many cities choose not to improve slum areas by adding roads, electricity, and other facilities, because they fear that this would attract still more poor people from rural areas. Some cities regularly bulldoze their slums. In any case, providing housing, water, and other basics in such areas would be far too expensive for most of these cities. With this lack of official support, hope is in short supply for most of the poor people trapped in city slums.

Enter a promising strategy known as **site and service**. Using this approach, a city government supplies a couple or a family with a piece of land—the *site*—big enough for a small house. The household members pay a low rent to the government in exchange for a document guaranteeing that they can stay on the site, and they are responsible for building the house. It usually amounts to a shack cobbled together from scrap lumber and other scavenged building materials, but given time, people are able to improve these basic shelters.

The *service* side of the strategy takes the form of roads, running water, streetlights, latrines, sewers, and other community essentials. An important part of this service side is security. The contract that renters sign typically includes guaranteed protection by the city against those who might resent the presence of the renters and try to do them harm—a threat that has arisen in some areas. By providing slum dwellers with secure, low-cost plots of land and access to basic services, site-and-service programs are giving these people a sense of hope and helping them to free up their creativity and energy in order to work themselves out of poverty.

For example, when some of the residents of the Mathare Valley shantytown wanted to use modern building materials such as concrete blocks and roof tiles, they found that they couldn't afford these materials, but this did not stop them. They learned how to build and operate simple brick and tile factories (large ovens, common in many less-developed countries). Thus, they manufactured their own building materials while creating jobs for several residents—something that would not have been possible without the site-and-service program.

In other cases, site-and-service programs have enabled people to work together to provide schools and day care centers for those residents who have jobs and for those who are looking for work. This, too, provides jobs for more people.

Site-and-service programs have now been established in urban areas in many countries, including Uganda, Tanzania, Indonesia, the Philippines, Peru, Ecuador, Venezuela, El Salvador, and Brazil (including the city of Curitiba; see *For Instance*, p. 79). Many analysts see these expanding efforts as a hopeful sign for cities of the future.

Let's REVIEW

What is a *site-and-service* program and how has it helped some slum dwellers?

What Would You Do?

We began this module by asking why we should care about urbanization. We learned that about one of every two people in the world and three of every four people in the more-developed countries live in urban areas, and within a few decades, about three-fourths of the world's people will be living in urban areas.

The big question is, how can we control the environmentally harmful impacts of urbanization in order to make urban areas more environmentally sustainable and more desirable places to live? As with most environmental issues, it boils down to what each of us does as an individual—how each of us can have a more sustainable and rewarding lifestyle in an increasingly urbanized world. This involves thinking about where to live, how to get from place to place, how to get enough food, water, and other resources, how to use those resources more sustainably, and how to reduce our production of wastes. All of these things together help to determine how big each of our ecological footprints is. Here are three ways in which people are dealing with this challenge:

Those who have a choice are thinking very carefully about where to live and that involves asking these kinds of questions:

- Do I prefer a rural area or a small, medium, or large urban area? Do I want to live in a central city, in a suburb, or in a developed rural area?
- Does the area where I want to live have a good mass-transit system as well as biking and walking paths, or will I be mostly dependent on a car? Does it have other ecocity features and how important are such features for me?
- How vulnerable to water shortages is the area that I'm considering, and will this change in the future?
- How vulnerable is the area to flooding, and will this change in the future?

Many people in more-developed urban areas are simplifying their lifestyles in order to lessen their own individual impacts and to live more sustainably by:

- Reducing car use and driving energy-efficient vehicles, using car-sharing networks, walking, biking, carpooling, and taking mass transit whenever possible.
- Getting as much of their energy as possible from the sun, wind, geothermal energy, and other renewable sources.
- Reducing their use and waste of water and energy.
- Cutting consumption of material goods and reusing and recycling more of what they do use.

People in urban areas are thinking about where their food comes from and what its environmental impacts are by:

- Growing more of their own food in urban gardens or buying more food from local sources.
- Making use of urban farmers' markets and community-supported agriculture programs that serve many urban residents.
- Replacing their lawns with gardens, native vegetation, and even edible plants that are adapted to local climate conditions.

KEYterms

city, p. 56
ecocity, p. 77
green city, p. 77
less-developed country, p. 57

more-developed country, p. 57
municipal solid waste (MSW), p. 66
population density, p. 57

poverty, p. 67
rural area, p. 56
site and service, p. 80
smart growth, p. 69
urban area, p. 56

urban growth, p. 57
urbanization, p. 56
urban sprawl, p. 60

THINKINGcritically

1. Do you believe that the advantages of suburban living outweigh the harmful effects of urban sprawl? Explain.

2. What might happen to suburban areas if the prices of oil and gasoline continue to rise? How might this affect your life?

3. Which two of the urban problems discussed in this module (pp. 60–69) do you think are the most serious? Explain. If it were up to you to lead the effort to solve those two problems, what are the first three steps you would take toward solving each of them?

4. Assume that you now get around mostly by using your own car or truck. What, if anything, would encourage you to rely less on your vehicle and to travel to work or school on foot, by bicycle, or on mass transit? Explain.

5. Should the United States develop a high-speed rail system connecting some of its major regions, as other countries are now doing? Explain. How would you pay the high construction costs for such a system? Would you instead spend the money on other priorities that you think are more important? If so, list two such priorities and explain why they are more important.

6. If you live in an urban area, how environmentally sustainable do you think it is, overall, on a scale of 1 to 10? Explain. List three steps that could be taken by the city's administrators and residents to make it more sustainable, and for each of these steps, explain how taking it would affect your life.

LEARNINGonline

Access an interactive eBook and module-specific interactive learning tools, including flashcards, quizzes, videos and more in your Environmental Science CourseMate, accessed through **CengageBrain.com**.

4

Food Resources

Why Should You Care about Food Resources?

There are at least three major reasons to care about food resources. First, we each need a nutritious diet to stay alive and to lead a healthy and productive life.

Second, about 1 billion people—or more than 3 times the number of people now living in the United States—do not get enough food to eat, and about 227,000 more people show up at the global dinner table every day. There are now 7 billion people on the planet. By 2050, there may be 9.3 billion. If we cannot provide enough food for the 1 billion hungry people today, how are we going to feed 2.3 billion more people in the next few decades?

The third reason for caring is that we have developed a food production system that meets the basic food needs of about 86% the world's population, but has an enormous environmental impact. Scientific evidence indicates that food production activities today have more and greater environmentally harmful effects than any other category of human activities.

For example, largely as a result of intensive food production systems, soils in many areas of the world are eroding faster than natural processes can renew them (Figure 4.1a). Also, we are reducing the earth's vital biodiversity by cutting down diverse forests to grow crops and graze cattle (Figure 4.1b). In addition, to irrigate their crops, farmers in some parts of the world are pumping water from underground supplies faster than this groundwater can be replenished by rain and melting snow. Also, the large-scale runoff of synthetic fertilizers and pesticides used to produce food on huge farms is severely polluting surface waters in many areas of the world.

CONSIDER this

According to a 2010 study by the United Nations Environment Programme, agriculture uses 38% of the world's ice-free land, accounts for 70% of the freshwater we remove from surface waters and aquifers, produces 60% of the world's water pollution, and emits 25% of all human-generated greenhouse gases.

In this module, we explore the various systems used to supply the world with food, as well as the harmful environmental and health effects of using these systems. We also consider several ways to produce food more sustainably.

What Do You Need to Know?

Energy Flow and Chemical Cycling

Two natural processes, *energy flow* and *chemical cycling* (see Module 1, Figure 1.19, p. 18), sustain the earth's amazing variety of species and ecosystems (biodiversity), including our own species and our economies.

In the first of these processes, the sun continually sends an enormous amount of high-quality energy to the earth, a tiny portion of which flows through the planet's

FIGURE 4.1 Two examples of the harmful effects of food production systems are (a) severe erosion of the earth's vital topsoil, as on this farm in Iowa, and (b) replacement of biologically diverse tropical forests with vast croplands such as this soybean plantation in Brazil.

organisms, which use this energy to sustain their life processes. From there, this solar energy flows into the environment as lower-quality energy in the form of heat, and some of it is reflected or radiated back into space (see Module 1, Figure 1.15, p. 14). This high-quality solar energy cannot be recycled because as it is used, it is degraded into lower-quality energy in accordance with the second law of thermodynamics (see Module 1, p. 12).

Some of this energy from the sun is used by certain organisms to make **nutrients**, or chemicals that the earth's organisms need in order to survive and live healthfully. Unlike solar energy, nutrients are continually recycled. This is because the earth gets no new supplies of these chemicals, and over billions of years, the earth's life forms, in order to survive, had to develop ways to recycle them.

One of the ways in which nutrients are recycled is through feeding relationships. Scientists have classified the planet's organisms into different *feeding levels* or *trophic levels* according to how they get their nutrients. The major class of organisms at one such level is called *producers*, the organisms (mostly plants) that produce their own nutrients from compounds and energy they get from their environment (Figure 4.2a). All other organisms, called *consumers*, must get their food by feeding on or by decomposing other organisms (Figure 4.2b). Consumers are further classified as *herbivorous species* (those that eat plants), *carnivorous species* (those that eat animals), and *omnivorous species* (those that eat both plants and animals).

Decomposers are consumer organisms, primarily bacteria and fungi, that break down the wastes of live organisms and the remains of dead ones, thereby releasing nutrients and returning them to the environment for use again by producers. In other words, decomposers act as nature's nutrient recyclers.

Energy flows through ecosystems from one trophic level to another in a **food chain**—a sequence of organisms, each of which serves as a source of food or energy for the next (see Module 1, Figure 1.17, p. 16). Most organisms actually feed on more than one other type of organism and are, in turn, eaten or decomposed by more than one type of consumer. Thus, each is part of a number of different food chains that intersect in complex ways to form a feeding network called a **food web** (see Module 1, Figure 1.18, p. 17).

When energy flows from one feeding level to the next in a food chain or food web, much of it ends up in the environment as low-quality heat due to the second law of thermodynamics. Therefore, the more trophic levels there are in a food chain or web, the greater is the total loss of chemical energy.

KEY idea In every transfer of energy from one feeding level to the next, a large share of the energy transferred flows into the environment as low-quality heat.

Two of the earth's life-sustaining nutrient cycles are the *water cycle* (see Module 1, Figure 1.20, p. 19) and the *carbon cycle* (see Module 1, Figure 1.21, p. 20).

FIGURE 4.2 The earth's *producers* (a) include trees and other plants that live on land, as well as plants and algae that live in water. The earth's *consumers* (b) include this giraffe, feeding on the leaves of a tree, as well as many other animal species such as lions, which feed on giraffes, and humans.

Another chemical that is important to life on the earth is nitrogen (N), a crucial part of proteins, many vitamins, and DNA, a molecule that is vital to successful reproduction. Nitrogen gas (N_2) makes up 78% of the air we breathe, but plants and animals cannot just absorb N_2 from the air. It has to be converted to other forms that organisms can use as nutrients. One such form is the nitrate ion (NO_3^-), an important plant nutrient. Natural processes create nitrates and other useful chemical forms of nitrogen through the *nitrogen cycle*—the chemical cycle in which nitrogen moves through air, water, soil, and organisms over and over again. However, scientists have learned how to make nitrates that are used, in turn, to make *synthetic inorganic fertilizers* that farmers apply to their crops to help them grow.

Phosphorus (P) is another important nutrient that cycles through parts of the environment in the *phosphorus cycle*. The phosphate ion (PO_4^{3-}) is another important plant nutrient that scientists have figured out how to make for use in the manufacture of synthetic inorganic fertilizers.

Over time, scientists and farmers have learned how to use the water, carbon, nitrogen, and phosphorus cycles to greatly increase crop *yields*, or outputs per unit area of land, of the key crops that feed the human population. At the same time, these higher-yield farming practices are altering the vital chemical cycles in ways that cause environmental harm. In the long run, these harmful effects might actually cause a decrease in food production. We discuss these human impacts on the nutrient cycles in the next major section of this module.

Let's REVIEW

- Give three reasons why we should care about food resources.
- Describe how life on the earth is sustained by energy flowing from the sun.
- What are *nutrients*? What are feeding levels, or trophic levels, and what are the three major classes of organisms, based on these feeding levels? Distinguish among herbivorous, carnivorous, and omnivorous species.
- Define *food chain* and *food web*. What happens in every transfer of energy from one feeding level to the next?
- Describe how natural processes recycle the nutrients that sustain life. Name four major chemical cycles.

The Importance of Soil

Soil is a complex mixture of eroded rock, mineral nutrients, decaying organic matter, water, air, and billions of tiny organisms. This vital resource supports plant growth, which in turn provides food for us and other animals.

Soil begins to form when natural processes slowly break down rock from the earth's crust into fragments and particles. Then bacteria and fungi and, later, plants and animals further break down and enrich these materials to form soil. As it builds up, soil forms several layers (Figure 4.3). The top layer, called **topsoil**, eventually contains enough nutrients (from plant and animal matter that has been broken down by decomposers) to support plants, or producers. These producers, in turn, support consumers,

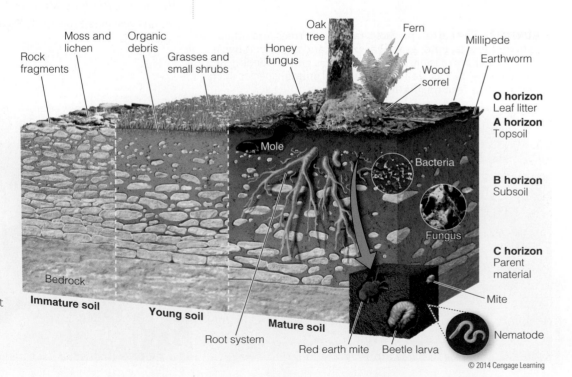

FIGURE 4.3 Over time, complex natural processes convert immature soil (left) to mature soil (right). *See an animation based on this figure at* www.cengagebrain.com.

Rock fragments — Moss and lichen — Organic debris — Grasses and small shrubs — Honey fungus — Oak tree — Fern — Wood sorrel — Millipede — Earthworm

O horizon Leaf litter
A horizon Topsoil
Mole
Bacteria
B horizon Subsoil
Fungus
C horizon Parent material
Mite
Nematode
Root system — Red earth mite — Beetle larva

Immature soil — **Young soil** — **Mature soil** — Bedrock and all of these organisms together form an ecosystem that is literally based on the soil.

It can take many hundreds of years for immature soil (Figure 4.3, left) to become mature soil (Figure 4.3, right) and for fertile topsoil to form. Topsoil is a renewable resource as long as human activities and natural events do not remove it faster than it can be renewed. But the renewal process is very slow, which means that topsoil can be depleted.

CONSIDER this It can take hundreds of years to build up just a half-inch of topsoil. But this amount of topsoil can be washed or blown away in a matter of weeks or months when we clear a forest or plow grassland and leave its topsoil unprotected.

FIGURE 4.4 This farmer is harvesting rice from a paddy in Thailand.

Most land-based life depends on the nutrients found in topsoil, including water and compounds of elements such as carbon, nitrogen, and phosphorus that support plant growth. Soil also plays a key role in purifying the water we use, by slowly filtering groundwater, and it helps to control the earth's climate by absorbing carbon dioxide from the atmosphere and storing it as carbon compounds. Thus, soil is one of the most important forms of natural capital (see Module 1, Figure 1.13, p. 13).

Types of Agriculture

Early agriculture was **traditional subsistence agriculture**, a form that relies on the labor of humans and animals, such as oxen, to produce enough crops for a farm family's survival, with little left over to sell or to store away (Figure 4.4). This ancient form of agriculture is the root of the first of three major food production systems that now feed the world (see *The Big Picture*, pp. 88–89). It is still practiced today by about 40% of the world's farmers, mostly in less-developed countries. It is used on about 75% of the world's farmland, but provides only about 20% of all food crops.

Over time, some farmers developed this form of agriculture into **traditional intensive agriculture**, another type of farming in which farmers use both human and animal labor, but also use greater amounts of fertilizer and water to increase their yields, with the goal of selling some of the food as a source of income.

Some traditional farmers grow a single crop in one field, a form of farming called *monoculture* (Figure 4.4), while others grow several crops in the same field during the same season—a practice known as **polyculture** (Figure 4.5). Because the variety of crops planted mature at different times, this practice can provide food throughout the year.

As agriculture developed, farmers increasingly used machines to replace human and animal labor and increase their yields. As a result, much of today's food is now produced by **industrialized agriculture**, or **high-input agriculture**, which uses heavy equipment, irrigation, synthetic fertilizers, and pesticides to produce monoculture crops

FIGURE 4.5 This plot of land in Australia has a number of different crops planted in the same area.

Where Our Food Comes From

Most of the world's population relies on three major types of food. The first type is *grains*, mainly wheat (see module-opening photo), corn (a), and rice (b), which are grown on croplands. The second type is *meats* from livestock that are raised on grass in rangelands and pastures (c), as well as in feedlots (d) and other confined areas where they are fed grain, fishmeal, and fish oil, usually mixed with growth hormones and antibiotics. The third type of food includes *fish* and *shellfish*, both those that are caught wild (e) and those raised by *aquaculture* or *fish farming*—the process of raising fish and shellfish in captivity. Some aquaculture operations use freshwater ponds, while others use underwater cages (f), usually in coastal or open ocean waters.

Tish1/Shutterstock.com

Marko5/Shutterstock.com

(a) Cornfield

(b) Terraced Rice Fields

(c) Cows and Sheep Grazing in a Meadow

(d) Cattle Feedlot

(e) Commercial Fish Catch

(f) Coastal Fish Farm

J van der Wolf/ Shutterstock.com

iStockphoto/Dave Hughes

Stockbyte/Thinkstock

Federico Rostagno/Shutterstock.com

FIGURE 4.6 In these two examples of industrialized agriculture, (a) a farmer is harvesting wheat using an expensive machine called a combine, and (b) specialized equipment is being used to irrigate a crop.

in very large quantities (see module-opening photo and Figure 4.6). This form of farming is typically very costly, with high usage of fossil fuels and other resources. Today, industrialized agriculture produces about 80% of the world's food on roughly 25% of the world's farmland, mostly in more-developed countries.

Industrialized agriculture has been the key component of a major global effort to multiply crop yields in a process called the *green revolution*, which has taken place in three phases, each of which helped to boost yields. First, scientists developed high-yield varieties of common crops such as wheat, corn, and rice. Second, farmers planted these varieties in large monoculture crops and used large inputs of water and synthetic fertilizers and pesticides. Third, farmers used *multiple cropping*, in which they planted more than one crop per year (one after another) on some plots of land. Since 1950, this high-input approach has raised crop yields dramatically in several of the more-developed countries, and more recently, in some less-developed countries such as Brazil, China, and India.

Let's **REVIEW**

- Define *soil* and summarize the process of its formation. What is *topsoil* and why is this renewable resource so easy to deplete?
- Define and distinguish between *traditional subsistence* and *traditional intensive agriculture*. What is *polyculture*?
- What is *industrialized* or *high-input agriculture*?
- What are the three major types of mass-produced foods?
- Describe the green revolution. What are its three phases?

Protecting Crops from Pests

An important part of industrialized crop production is the large-scale use of **pesticides**—synthetic chemicals designed to kill or control *pests*, those organisms such as weeds, insects, rats, and mice that we consider undesirable. Pesticides include *herbicides* (for killing weeds), *insecticides* (for killing insects), and *rodenticides* (for killing rats and mice).

In natural ecosystems, usually no single species can overwhelm others in a community, because each species' population is held in check by other species called *natural enemies*. They include predators, parasites, and disease organisms that feed on a particular organism, helping to keep its population from growing too large. For example, the world's 30,000 known species of spiders, including the wolf spider (Figure 4.7), kill many more crop-eating insects every year than the pesticides we use. In turn, birds eat the spiders to keep their populations from getting out of control.

Monoculture crop fields, on the other hand, do not contain the variety of organisms found in natural ecosystems. Thus, an insect that likes to eat a particular crop is less likely to run into a natural enemy in a monoculture field. It can then eat and reproduce more freely and its population is more likely to overrun and damage the crop, cutting its yield. To deal with this threat, farmers have to find ways to protect monoculture crops from pests. The most common method used in industrialized crop production is to spray fields with pesticides (Figure 4.8).

There are many different types of pesticides. Some, called *broad-spectrum agents*, kill many pests but also kill some species that are not pests. Other pesticides, called

FIGURE 4.7 This wolf spider helps to control pest populations by preying on insects such as this grasshopper. Wolf spiders do not harm humans.

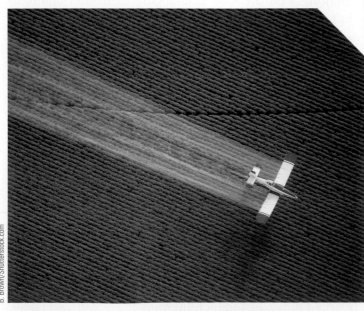

FIGURE 4.8 Modern agriculture involves spraying chemical pesticides on crops using airplanes and ground-spraying equipment.

narrow-spectrum agents, are effective against a narrowly defined group of organisms. Some pesticides are more *persistent* than others, meaning they last longer in the environment. For example, a once widely used chemical called DDT has been banned in most countries largely because of its harmful effects on wildlife and humans. But it is so persistent that it remains in the environment for years as a deadly poison threatening many insect species as well as their predators.

Industrialized Meat Production

In the early days of livestock production, beef cattle were raised in pastures and rangelands, pigs were reared in fairly large penned-in areas on farms, and chickens found their food in farmyards and fields.

While about half of the world's beef still comes from cattle grazing on grass in pastures and rangelands, the other half is produced through industrialized systems. One such system is the *feedlot*—a densely packed feeding area (see Photo d, p. 89) where cattle are fed corn and other grains, and sometimes fishmeal and fish oil. Other animals such as chickens, turkeys (Figure 4.9a, p. 92), and pigs (Figure 4.9b) are raised in large buildings called *confined animal feeding operations* where they are also fed grain or fishmeal. Some of these animals spend their entire lives in densely packed pens and cages within these operations.

Fish and Shellfish Production

The third major source of food produced through industrialized food production is fish and shellfish. The industrialized approach is now being used both for catching wild fish (see Photo e, p. 89) and for raising fish and shellfish in ponds and underwater cages (see Photo f, p. 89).

A population of wild fish or shellfish that is regularly harvested by commercial fishing operations is known as a **fishery**. These fishing operations are dominated by large factory ships in industrial fishing fleets that operate mostly on the oceans but also on many of the world's large lakes. Some factory ships, called *trawlers*, drag huge, heavily weighted nets designed to scoop up tons of fish at a time. Every year, thousands of trawlers scrape vast areas of ocean floor.

Fishing fleets also use high-tech equipment such as satellite global positioning systems (GPS), sonar fish-finding devices, spotter planes, and very long lines with thousands of large hooks attached. They also use refrigerated factory ships designed to process and freeze the fish they catch while still at sea.

FIGURE 4.9 Confined animal feeding operations are used widely for raising (a) turkeys, (b) pigs, chickens, and other animals.

In recent years, the fastest-growing form of animal food production has been **aquaculture** or **fish farming**—the process of raising fish and shellfish in confined areas. It is used to raise both marine and freshwater species. Some fish farmers use freshwater ponds (Figure 4.10), lakes, and reservoirs. Others raise fish in underwater cages suspended in rice paddies, saltwater lagoons, and estuaries, as well as in both coastal (see Photo f, p. 89) and deep-ocean waters. Led by China, which produces 70% of the world's farmed fish, aquaculture operations in 2010 produced nearly half of all fish and shellfish eaten by people.

FIGURE 4.10 This is an aerial view of some inland catfish aquaculture farms near Yazoo, Mississippi.

Let's REVIEW

- Define *pesticide*. How are pests controlled in natural ecosystems?
- Distinguish between broad-spectrum and narrow-spectrum pesticides.
- What is a persistent pesticide? Describe the two ways in which livestock are raised for meat production.
- Define *fishery* and *aquaculture*.
- What is the world's fastest-growing form of animal food production?

Crossbreeding and Genetic Engineering

During the 1800s, farmers and scientists began experimenting with the idea of changing their crop plants by influencing the plants' reproductive processes. These experimenters learned that they could select individuals of a species that had one or more of the characteristics they desired and pair them for reproduction in hopes of producing offspring with the desired characteristics. This process eventually became known as **crossbreeding**, or **artificial selection**.

For example, the numerous varieties of peppers that farmers produce today did not exist centuries ago. Over time, pepper farmers developed new types of peppers that were sweeter, hotter, or larger than the varieties they had traditionally grown. For instance, they had selected the sweetest peppers they could find and paired them to make their offspring sweeter. They repeated this process until the sweet pepper became a common crop. Cross-breeding has been used to produce many foods that we take for granted today, including numerous fruits and vegetables, and milk from cows that were selected and crossbred to produce more milk per day.

Developing new varieties of plants and animals through traditional crossbreeding takes a long time, typically 15 years or more. Partly for this reason, scientists have developed the field of **genetic engineering**, which involves changing an organism's genetic material by adding, removing, or changing segments of its DNA molecules to produce desirable characteristics or to eliminate undesirable ones. The resulting organisms are called *genetically modified organisms (GMOs)*.

While artificial selection is limited to species that are similar in their genetic makeup, genetic engineering enables scientists to transfer genes between different species that otherwise could not be crossbred and would not interbreed in nature. It takes about half as long to develop a new crop variety through genetic engineering as it does to use traditional crossbreeding. It also usually costs less in money and resources. Genetic engineering has been used to develop both crop plants that are better able to resist some insect pests (Figure 4.11) and varieties of cattle, pigs, and poultry that grow very rapidly.

Genetic engineers are working to develop new varieties of crops that are resistant to heat, cold, plant diseases, and drought. They also hope to develop crop varieties that can grow faster with little or no irrigation and with less fertilizer and pesticides. Some scientists see genetic engineering as a promising technology while others call for more testing to learn more about any risks it might present. At least three-fourths of the food products on U.S. supermarket shelves already contain some form of genetically engineered ingredient.

FIGURE 4.12 The production of organic fruits and vegetables is increasing rapidly.

Organic Agriculture

Industrialized food production creates large environmental impacts, as we discuss in the next section of this module. Partly for this reason, a growing number of farmers are using an alternative to industrialized food production known as **organic agriculture**, a form of farming that uses only organic (natural) fertilizers, natural pest-control methods, and no pesticides, synthetic fertilizers, genetically engineered seeds, synthetic growth hormones, or antibiotics.

In the United States, in order for a food to be certified as *100% organic*, it must meet all of the requirements stated in the definition given above. Organic agriculture is rapidly becoming more popular and, since 1990, has been the fastest-growing form of agriculture in the world (Figure 4.12). However, officially certified 100% organic foods are produced on less than 1% of the world's cropland and on only 0.6% of U.S cropland. We explore organic farming methods in more detail in a later section of this module, along with a comparison of its benefits and drawbacks.

Let's **REVIEW**

- Define *crossbreeding*, or *artificial selection*, and give two examples of how it has been used.
- Define *genetic engineering* and explain how it is being used. Briefly compare the advantages of artificial selection and genetic engineering.
- Define *organic agriculture*.

FIGURE 4.11 Both of these tomato plants were exposed to the same destructive caterpillars. The plant on the right was genetically altered to resist pest damage. The unaltered plant (left) suffered much more damage.

What Are the Problems?

Hunger, Malnutrition, and Overnutrition

The world's farmers produce more than enough food to supply everyone on the earth with a healthy, nutritious diet. As a result, most people in the United States and in the other more-developed countries have few worries about **food security**—the condition under which everyone in a population has daily access to enough nutritious food to live healthfully.

Meanwhile, about one of every seven people or about 1 billion people, most of them in less-developed countries, live with **food insecurity**—the condition under which the necessary amount of food to live healthfully is not easily or consistently available to everyone. It can lead to chronic *undernutrition*, or hunger, and *malnutrition*, or a shortage of key nutrients. Both undernutrition and malnutrition can prevent people from living healthy and productive lives.

According to many food security experts, the main cause of food insecurity is poverty, which leaves millions of people too poor to buy or grow enough food for themselves or their families. Other causes of food insecurity include wars (Figure 4.13), government corruption, degradation of land that could be used for food production, flooding, and prolonged drought and heat waves.

Many of the world's poor people who are not starving still suffer from chronic malnutrition. Their diets may be lacking in *macronutrients*, which include carbohydrates, proteins, and fats that help us build and maintain body tissues. For example, those who can afford only a diet of grains such as wheat, rice, and corn, suffer shortages of important proteins and fats. Other key nutrients are *micronutrients*, including vitamins such as A, C, and E, and minerals such as iron, iodine, and calcium that help to keep our organs functioning properly.

Chronic malnutrition can weaken people, make it harder for them to fight off diseases, and slow the physical and mental development of children. Undernourished and malnourished children are also more likely to die from infectious diseases like measles or from acute diarrhea—illnesses that rarely kill well-nourished children.

Malnutrition can cause other problems. Every year, as many as 500,000 children younger than age 6 go blind because they do not get enough vitamin A, and within a year of the onset of blindness, more than half of them die. Also, diets that lack *iodine*, a chemical element that helps to keep our thyroid glands functioning properly, have caused irreversible brain damage in 26 million

NUMB3RS

6 million

The number of malnourished children under age 5 who die each year from illnesses such as measles and diarrhea

FIGURE 4.13 These starving children were looking for ants to eat. They were victims of famine caused by a civil war in Sudan, North Africa, between 1983 and 2005.

Harmut Schwartzbach/Peter Arnold, Inc.

children, resulting in stunted growth and mental retardation for many of them.

While 1 billion people suffer from undernutrition and malnutrition, another 1.1 billion people suffer from **overnutrition**—the buildup of excess body fat that results when one takes in more food energy than he or she uses. People who are underfed, and those who are overfed and overweight, face similar health problems: increased risk of disease and illness, lower life expectancy, and lower energy and life quality.

In 2010, the U.S. Centers for Disease Control and Prevention (CDC), reported that two out of three American adults are overweight and one out of three are obese (20% or more above their ideal weight). In 2011, the CDC reported that about 17% of American children, ages 2–19 years, were obese—almost triple the percentage from 1980. In the United States, four of the top ten causes of death—heart disease, stroke, Type 2 diabetes, and certain forms of cancer—are diet-related diseases.

Let's **REVIEW**

- Define *food security* and *food insecurity*.
- Distinguish undernutrition from malnutrition. How many people in the world suffer from these conditions?
- Distinguish macronutrients from micronutrients and give two examples of each.
- What is *overnutrition*? How many people in the world are overweight?

Topsoil Erosion

Wind and flowing water can move soil from one place to another—a process called **soil erosion**. Some soil erosion is natural, but much of it is caused by human activities, and the erosion of vital topsoil has become a major problem in many parts of the world.

On land that is covered by natural vegetation, topsoil is usually not removed faster than it is renewed, except in the cases of flooding and severe windstorms. But when people remove vegetation by plowing up grasslands or clearing forests to make way for crops, buildings, roads, and cities, topsoil is often then left unprotected. Exposed topsoil can be washed (Figures 4.1a and 4.14a) or blown away (Figure 4.14b) much faster than natural processes can renew it.

Agriculture is among the leading causes of soil erosion. According to the UN Environment Programme (UNEP) and the World Resources Institute, topsoil is eroding faster than it can be renewed on about 38% of the world's cropland (Figure 4.15, p. 96).

Topsoil erosion removes the nutrients needed by plants and, in turn, by the animals (including humans) and the ecosystems that depend on these producer species. This *loss of soil fertility* is a big threat to food security for people in many countries.

NUMB3RS

90%
The percentage of American farmland that, on average, is losing topsoil 17 times faster than natural processes can renew it

FIGURE 4.14 Flowing water from rainfall is the leading cause of topsoil erosion, as seen on this farm in Iowa (a). On another farm in Iowa (b), wind is blowing topsoil off the fields.

Lynn Betts/Natural Resources Conservation Service

Lynn Betts/Natural Resources Conservation Service

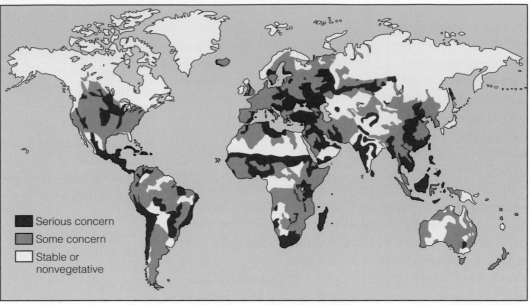

FIGURE 4.15 This map summarizes the extent of soil erosion around the world, according to a survey done by the World Resources Institute working with the UN Environment Programme.

Serious concern
Some concern
Stable or nonvegetative

© 2014 Cengage Learning

In the United States, a third of the country's original topsoil has been eroded away faster than it could be renewed. According to the U.S. Natural Resources Conservation Service (NRCS), in the state of Iowa, which is dominated by agriculture, half of the topsoil that took many centuries to form has disappeared after less than 200 years of farming (Figure 4.14).

Soil Degradation

Soil degradation occurs when topsoil loses some of its ability to support plant growth. One type of *soil degradation* is **desertification**, the drying out of soil due to a combination of prolonged drought and human activities that reduce or degrade topsoil.

Desertification can be moderate, in which case it can support some plant growth and can be renewed with some effort, or it can be very severe, supporting little or no plant growth and subject to severe erosion. In extreme cases, desertification actually leads to what we call a *desert*, which can expand and threaten nearby croplands and grazing areas (Figure 4.16), thus threatening food security in those areas.

According to the UNEP and the World Bank, desertification affects the food security of about 1 billion people and this number could grow to 1.8 billion people by 2025. Two countries whose food security is seriously threatened by advancing deserts are China, the most populous country in Asia, and Nigeria, the most populous country in Africa. Many scientists project that as the global climate changes, due primarily to projected atmospheric warming, drought

and desertification in various areas of the world will become more severe. This will likely limit the growth of food production in these areas and threaten the food security of billions of people.

About 70% of the water we withdraw from rivers and lakes and from underground sources of water is used to *irrigate*, or add water to soil on about 20% of the world's cropland. This has helped to produce about 45% of the world's crops. Irrigation water from surface and groundwater sources usually contains mineral salts that it has

FIGURE 4.16 In this part of the Sahel region of West Africa, desert sand dunes are moving in on cropland. This severe desertification resulted from a combination of prolonged drought and human activities, such as farming and the grazing of livestock, which destroyed much of the natural vegetation.

Voltchev-UNEP/Peter Arnold Inc.

FIGURE 4.17 This soil on a farm in Colorado once supported crops that were heavily irrigated. Because of high evaporation rates in this arid region, irrigation water quickly evaporates, leaving behind a crust of salts where the crops once grew.

picked up as it flowed through soil and over rocks. This slightly salty water is called *saline* water. Irrigation water that the soil does not absorb evaporates, leaving behind a thin crust of dissolved salts in the topsoil. A gradual build-up of these salts in the upper layers of soil that is repeatedly irrigated is called **soil salinization**. It is another form of soil degradation.

Salinization stunts crop growth, lowers crop yields, and can eventually kill plants and ruin the topsoil. This has become a severe problem in China and India, Asia's two most populous countries, as well as in Egypt, Pakistan, and Iraq. According to the UN, severe salinization has reduced yields on at least 10% of the world's irrigated cropland. Salinization also affects about 25% of irrigated cropland in the United States, especially in western states (Figure 4.17).

Overpumping of Groundwater

Aquifers, the bodies of water that are stored underground, are recharged by precipitation and surface water that seeps into the ground. But in many areas of the world, water is being pumped out of aquifers—primarily to irrigate crops—faster than they can be recharged. Overpumping of aquifers is a problem especially in the three countries that produce most of the world's grain—India, China, and the United States.

According to the U.S. Geological Survey, groundwater in the United States, on average, is being pumped four times faster than it can be replenished (Figure 4.18). Underlying the central part of the country is the world's largest known aquifer, called the *Ogallala aquifer* (the largest red area in Figure 4.18), which is used to irrigate vast areas of cropland. Some parts of the southern Ogallala are being pumped 10 to 40 times faster than they can be recharged. As a result, the total amount of farmland irrigated in this area has declined, along with crop yields. Such overpumping is also occurring in California's Central Valley, where half of the country's fruits and vegetables are grown on irrigated land (the long red area near the southern West Coast in Figure 4.18).

Overpumping of the Ogallala is a threat to food security and to the water supplies of the American Midwest. It also threatens *biodiversity*—the variety of plants and animals that help to sustain most ecosystems—in areas where the Ogallala flows onto the land and creates wetlands. Some of these vital wetland habitats are drying up as the Ogallala shrinks.

Let's REVIEW

- Define *soil erosion*. What are the two natural agents that drive it?

- Give three examples of human activities that can cause soil erosion and explain how they do so.

- What is the extent of soil erosion in the world and in the United States?

- What is *soil degradation*? Define *desertification* and *soil salinization*, and explain how each of these processes occurs and how each of them can affect crop production. Explain why the overpumping of aquifers for irrigation is a problem.

- What are three ways in which overpumping of the Ogallala aquifer is a threat?

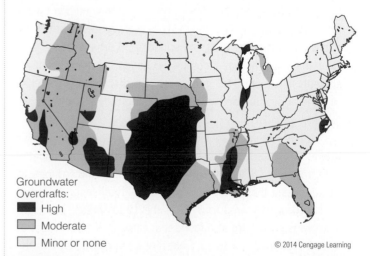

Groundwater Overdrafts:
- High
- Moderate
- Minor or none

© 2014 Cengage Learning

FIGURE 4.18 This map shows the areas of the continental United States that are most threatened by overpumping of aquifers, mostly for irrigation. *See an animation based on this figure at* **www.cengagebrain.com.**

Harmful Effects of Pesticides

Although the use of pesticides has provided real benefits for many people (which we discuss in the next major section of this module), it also has been environmentally harmful in several ways. For example, the USDA has estimated that about 98% of the insecticides and more than 95% of the herbicides that farmers spray on crops (Figure 4.8) end up as potential pollutants in the air, surface water, groundwater, and food crops. Some are toxic to nontarget organisms, including wildlife and humans.

Another problem is that persistent insecticides such as DDT can accumulate in the fatty tissues of organisms (including humans) that eat smaller organisms that have eaten the poisoned pests. In this way, these chemicals become *biologically magnified* to much higher levels in the bodies of consumers at high trophic levels in food chains and food webs.

The widespread use of synthetic pesticides can actually make some pest populations larger and stronger by promoting *genetic resistance* to the pesticides—a serious and growing problem. For example, insects breed rapidly and within 5 to 10 years, some species can develop immunity to widely used insecticides. As a result, scientists have to develop new insecticides to kill them, and after a time, the target insects can develop immunity to the new poisons. Since 1945, hundreds of species of insects, rodents, and weeds have developed genetic resistance to one or more pesticides.

Some pesticides also kill the natural enemies of the pests that the pesticides are supposed to kill (Figure 4.7). With these natural predators gone, the populations of other organisms that they helped to keep in check can grow and become new pests. One of every three of the most destructive insect species in the United States were once minor pests that became major pests after widespread use of insecticides.

Some pesticides harm birds, fish, and other forms of wildlife. In 1962, pioneering biologist Rachel Carson was

MAKING A difference

David M. Pimentel

Renowned ecologist David Pimentel (Figure 4.A) has dozens of scientific interests but he has applied his talents primarily to agriculture.

Pimentel is Professor Emeritus in the Department of Ecology and Evolutionary Biology at Cornell University. He has authored hundreds of articles and books, and has researched and taught for more than 50 years in the fields of ecology, biodiversity, crop science, pest management, soil science, and organic agriculture. Over the years, he has vigorously engaged his many graduate students in exploring the major problems that result from industrialized agriculture and looking for ecologically-based solutions to these problems.

When Pimentel, an expert on insect ecology, turned his attention to the use of pesticides in agriculture, he evaluated data from more than 300 agricultural scientists and economists. He concluded that about 37% of the U.S. food supply is lost to pests today, compared to 31% in the 1940s before pesticides were widely used. He also found that since 1942, crop losses to insects rose from 7% to 13%, despite a tenfold increase in the use of insecticides. Pimentel concluded that for every dollar spent on pesticides, the estimated harmful environmental and health

FIGURE 4.A David M. Pimentel

costs of pesticide use in the United States are valued at $5–10 in damages.

Pimentel also studied *integrated pest management (IPM)*, an alternative, ecological approach to pest management now practiced by many of the world's organic farmers. He concluded that by using IPM, we could cut by half the use of chemical pesticides on 40 major U.S. crops without reducing crop yields.

While the pesticide industry disputes these findings, many scientists and farmers see them as evidence that organic agriculture can compete with industrialized agriculture in feeding the growing human population. This is just one example of how Pimentel's work has spurred researchers and food producers to take creative and innovative approaches to solving the environmental problems associated with food production.

among the first scientists to sound the alarm on this issue. In some years, according to the USDA and the U.S. Fish and Wildlife Service, pesticides applied to cropland have killed as much as a fifth of the honeybee population, which plays a key role in pollinating important crops including apples, cranberries, cherries, blueberries, broccoli, and almonds. Pesticides are also blamed for the deaths of more than 67 million birds and 6–14 million fish every year. In addition, pesticides are a major threat to one of every three endangered and threatened species in the United States.

People also suffer harm from some pesticides. The World Health Organization has estimated that pesticides annually poison at least 3 million agricultural workers in less-developed countries and at least 300,000 people in the United States. Between 20,000 and 40,000 of these cases end in death every year. Household pesticides are also a major source of accidental poisonings and deaths of young children.

According to the National Academy of Sciences, residues from commonly used pesticides that end up in foods could be causing 4,000–20,000 cases of cancer per year in the United States. Scientists are also concerned about other possible harmful health effects of pesticides—especially in children—including birth defects, nervous system and behavioral disorders, and effects on the immune system. Pesticide manufacturers argue that people are generally not exposed to high enough doses of pesticides to suffer such adverse effects.

Some scientists contend that the benefits of pesticides do not outweigh these harms. Ecologist David Pimentel (see *Making a Difference*, at left) has studied pesticide use and found that between 1942 (when pesticides were not yet used widely) and 1997, U.S. crop losses to insects nearly doubled, despite a tenfold increase in the use of insecticides.

Let's REVIEW

- Describe five ways in which pesticides can be environmentally harmful.
- Describe David Pimentel's contribution to our understanding of the harmful effects of pesticides.

Water and Air Pollution

One of the reasons scientists are concerned about soil erosion from cropland is that eroded soil often ends up in nearby surface waters as sediment, which is one of the most widespread forms of water pollution. Such sediments

FIGURE 4.19 This lake in the state of New York has received excessive amounts of plant nutrients such as nitrogen and phosphorus from the surrounding land, much of it applied as fertilizer for crops. As a result, the lake's surface is covered with mats of algae and other aquatic organisms.

can smother aquatic plants, fish, and shellfish, and disrupt aquatic life in rivers, lakes, and coastal systems such as estuaries and coral reefs.

Another problem is the flow of nitrate (NO_3^-) and phosphate (PO_4^{3-}) plant nutrients in synthetic inorganic fertilizers from farm fields, lawns, and golf courses to nearby bodies of water. Because of such runoff, these aquatic systems can become overfertilized and experience explosive growths of algae (Figure 4.19).

For example, much of the volume of synthetic fertilizers used on the vast farmlands of the Midwestern United States runs off into hundreds of streams that eventually flow into the Mississippi River, which empties into the Gulf of Mexico. In the gulf, the bacteria that decompose these algae use most of the dissolved oxygen in the water, creating an *oxygen-depleted zone* (Figure 4.20, p. 100). Then, fish and bottom-dwelling organisms die off or abandon this zone, which is why it is sometimes called a *dead zone*, even though it still hosts hordes of bacteria.

Every year, at least 400 oxygen-depleted zones form in coastal waters around the world. This is an example of how humans can interfere with the earth's nutrient cycles, in this case, the nitrogen and phosphorus cycles. By adding nitrogen and phosphorus to the soil in the form of synthetic fertilizers in an effort to boost crop productivity, we affect those cycles and also disrupt aquatic systems, thereby interfering with the water cycle as well (see Module 1, Figure 1.20, p. 19).

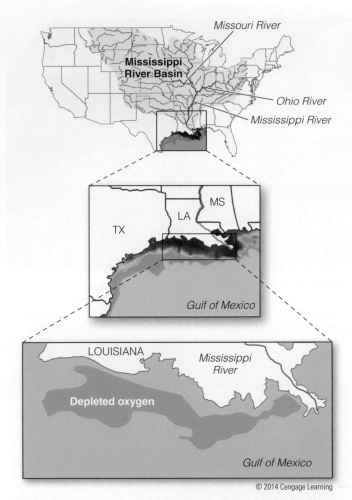

FIGURE 4.20 The oxygen-depleted zone in the Gulf of Mexico often covers an area larger than the state of Massachusetts (bottom image). The center map, based on a satellite image, shows the inputs of nutrients flowing as runoff from the land into the gulf during the summer of 2006. In the image, reds and greens represent high concentrations of phytoplankton and river sediment. (Data from NASA)

KEY idea Human activities can interfere with the major nutrient cycles—the chemical cycles that are vital to life on the earth—and disrupt agricultural systems.

Industrialized crop production also adds to air pollution, mainly from the emissions of machinery used to plant and harvest crops (Figure 4.6a), to pump irrigation water from aquifers and streams (Figure 4.6b), and to apply fertilizers and pesticides (Figure 4.8). These machines generally burn fossil fuels, emitting pollutants and climate-changing carbon dioxide gas into the atmosphere. The production and use of pesticides and fertilizers, and the processing of crops into foods also produce air pollutants.

Let's **REVIEW**

- What are two forms of water pollution that result from agricultural activities?
- What is an oxygen-depleted zone and how does it form?
- Give an example of how human activities can interfere with one or more of the earth's nutrient cycles. What are two sources of air pollution from agricultural activities?

Potential Hazards of Genetic Engineering

Genetic engineering holds promise for helping us to meet the challenges of increasing food security (which we discuss in the next major section of this module). However, it has some potential hazards that could limit its usefulness.

Critics of genetic engineering often point out that it involves mixing genes from widely differing species. They are concerned that this mixing could result in unintended and unpredictable results such as the creation of genetically modified (GM) organisms that could be harmful to certain species, to ecosystems, or to human health. As the use of genetically modified seeds increases, such harmful effects will become more likely.

For example, genes in the pollen of genetically engineered corn plants could spread onto a field of nonengineered corn. The GM corn could then mix with the natural crop, affecting its genetic identity and reducing the overall genetic diversity of the various strains of corn. In fact, this is already happening. A 2006 study by the Union of Concerned Scientists found that half of the non-engineered corn and soybean varieties tested contained DNA from GM varieties of these crops. Such contaminations are becoming more frequent, and they can ruin organic crops, which in order to be certified 100% organic, cannot contain genetically modified DNA.

Loss of Biodiversity

Many scientists argue that the growing loss and degradation of biodiversity—the earth's great variety of genes, plants, animals, and ecosystems—is a major threat to life on the earth. This is because biodiversity supports all life and human economies and provides the genetic variations that life as a whole needs to be able to adapt to major changes in environmental conditions. Think of biodiversity—the basis of one of the three scientific principles of sustainability (see Module 1, Figure 1.24, p. 26)—as a long-term genetic and ecological insurance policy.

Today, one of the most serious threats to the world's biodiversity is the cutting and burning of large areas of biologically diverse forest and grassland, mostly to grow crops or to graze cattle. Replacing these complex and diverse natural systems with simplified plantations devoted to growing a single crop such as soybeans (Figure 4.1b) represents a serious loss of biodiversity.

Another type of biodiversity loss is the elimination of thousands of varieties of vegetables, fruits, and grains. This is referred to as the loss of **agrobiodiversity**—the genetic variety of animal and plant species used to provide food. With the rise of industrialized crop production, farmers now rely on just a few varieties of each of the main types of food crops.

For example, in India, farmers at one time were using as many as 30,000 varieties of rice. Now, just 10 varieties are used to produce more than 75% of India's rice crop. Scientists believe that before long, almost all the rice produced in India will be of one or two varieties. Similarly, about 97% of all the food plant species used by U.S. farmers in the 1940s are no longer available to most farmers. By shrinking the gene pool for food crops, we may be reducing our ability to create new strains and to continue increasing crop yields in the future.

FIGURE 4.21 This whole area was once typical of rangeland used for grazing. Now, the land to the left of the fence has been overgrazed, while the land to the right of the fence has been lightly grazed by livestock.

> ## KEY idea
>
> The growing loss and degradation of the world's variety of genes, plants, animals, and ecosystems (its biodiversity) is a major threat to the well-being of life on the earth, to human economies, and to food production.

Let's REVIEW

- What are the potential problems that could result from more use of genetic engineering?
- What is one of the greatest threats to the world's biodiversity, resulting from agriculture?
- What is *agrobiodiversity* and how is it being threatened? Explain how losses of biodiversity and agrobiodiversity threaten life on the earth.

Environmental Impacts of Industrialized Meat Production

Meat production and consumption have increased with the growth of the human population and its rising affluence. This has led to higher environmental impacts.

One such result has been the degradation of much land due to **overgrazing**—the grazing of too many cattle in one place for too long. The resulting loss of protective vegetation (Figure 4.21) exposes topsoil to erosion by rains and wind. In 2008, according to the United Nations Food and Agriculture Organization (FAO), about 20% of the world's grasslands and pastures had been degraded in this way. The FAO also estimated that livestock production had caused well over half of the world's soil erosion and sediment pollution, and a third of its water pollution from the runoff of synthetic fertilizers.

The use of feedlots (see Photo d, p. 89) and other confined-area operations (Figure 4.9) to raise large numbers of livestock also has severe environmental impacts. One problem is the production of great amounts of animal wastes (manure). Piles of these wastes near feedlots often pollute the air with foul odors and release methane, a greenhouse gas that scientists believe is contributing to rapid climate change.

From a typical feedlot, only about half of these wastes are returned to the land as organic fertilizer. Much of the other half of this valuable resource is washed into streams, ponds, and lakes where it pollutes the water, and can contribute to the pollution of groundwater—another example of how agricultural activities can disrupt the earth's nitrogen, phosphorus, and water cycles.

> ## CONSIDER this
>
> According to the USDA, the amount of animal wastes produced by livestock in the United States is about 130 times the amount of wastes produced by the country's human population.

Feedlot systems also use large amounts of energy, mostly oil and natural gas, for feeding and watering livestock, and for hauling them to market. This results in air pollution, including emissions of the greenhouse gases carbon dioxide and methane. The FAO estimates that livestock production generates about 20% of all human-generated greenhouse gases.

Feedlots and other facilities for confined-area livestock production also use a great deal of water, which stresses already-strained water supplies in arid, drought-prone regions such as parts of the American West where many of these facilities are located. Also, according to the FAO, the cattle industry uses 37% of the world's synthetic pesticides, mostly on vast fields of crops such as soybeans and corn that are used to feed cattle. This adds to problems of air and water pollution as well as to human health concerns and biodiversity loss.

In addition, livestock production makes intensive use of antibiotics, which are added to animal feed not only to prevent the spread of diseases in crowded feedlots, but also to make livestock grow faster. A 2009 study by the Union of Concerned Scientists (UCS), along with several other studies, concluded that the widespread use of antibiotics in livestock production has played a role in the rise of genetic resistance among many disease-causing microbes. Thus, this overuse of antibiotics is reducing the effectiveness of those antibiotics in treating infectious diseases in livestock and in humans.

CONSIDER this About 70% of all antibiotics used in the United States and 50% of those used in the world are added to animal feed.

Environmental Impacts of Fishing and Aquaculture

Each year, commercial fishing fleets use a large array of factory ships and high-tech tools to try to find and catch more wild fish (see Photo e, p. 89) in the world's oceans and large lakes. As a result, according to studies by conservation biologists, 80% to 90% of the populations of the large, popular ocean seafood fishes, including tuna, swordfish, and marlin, have been depleted since 1950.

A 2006 scientific study of the global effects of overfishing found that the rate of overfishing by all nations together is unsustainable. As a result, overall, about 48% of the world's major fisheries are overexploited or depleted and this percentage could grow. Many scientists warn that this overfishing, along with other environmental impacts, is reducing not just the fish catch, but also the biodiversity of fish species.

Industrialized fishing operations have other harmful effects on marine ecosystems. For example, trawlers destroy or disturb large areas of the ocean floor every year. The total area disturbed annually is so large that marine scientist Elliot Norse has called bottom trawling "probably the largest human-caused disturbance to the biosphere."

Many analysts view aquaculture as a more sustainable alternative to industrialized fishing, but aquaculture has some potentially serious environmental impacts. One problem is that raising fish in cages in inland ponds and lakes, or in shallow coastal waters (Figure 4.22) can lead to the pollution of these waters with fish wastes and antibiotics that some fish farmers use.

Another problem is that coastal habitats and biodiversity are lost wherever fish farmers clear or degrade coastal wetlands. For example, mangrove forests that provide habitats for a variety of species are commonly cleared to create aquaculture ponds, especially for shrimp-farming operations.

Fish farming can also add to the overfishing problem when it involves raising carnivorous fish. These fish are often fed fishmeal or fish oil obtained from fish that are caught in the open ocean. As a result, populations of some of the species used to make fishmeal and fish oil are being depleted.

FIGURE 4.22 Raising large numbers of fish in offshore underwater cages can pollute coastal waters with fish wastes and with the antibiotics used in their feed to promote rapid growth.

Konstantin Karchevskiy/Shutterstock.com

Let's REVIEW

- Define *overgrazing* and explain why it is a problem. What is another environmental harm caused by the grazing of cattle?
- What are four problems resulting from feedlot meat production?
- How does trawler fishing threaten ocean life?
- What are three problems stemming from aquaculture?

What Can Be Done?

Producing More Food

One of the first answers to the question of how to produce more food has always been to convert more land to cropland. Some contend that we could more than double the global area of cropland by clearing tropical forests, cultivating marginal land, and irrigating more dry land.

However, by clearing all or most of the tropical forests, we would be eliminating the habitats of about two-thirds of the world's terrestrial plant and animal species. This would be a severe blow to the earth's vital biodiversity. Cultivating *marginal land*—land with poor soil fertility, steep slopes, or both—would be very expensive, requiring large amounts of water and fertilizers.

Also, because of projected climate change, much of the land now considered farmable will likely not be suitable for raising crops or livestock later in this century. Climate change projections indicate that much of the world's available and potential farmland near coastal areas will likely be flooded during this century by rising sea levels, while other potential cropland will be subject to longer and more intense drought.

The irrigation of more land is also a solution that is limited. The amount of irrigated land per person in the world has been declining since 1978 and is projected to fall much more by 2050. This is because the rates at which people are pumping water from many major rivers and aquifers has become unsustainable. Excessive water waste and population growth resulting in higher demand are making matters worse. Also, atmospheric warming is gradually melting many mountain glaciers that provide countries such as China and India with irrigation water, and glacial melting is projected to increase sharply during this century.

For these reasons, many food experts argue that instead of simply ramping up the systems we use now, we could shift to new or modified food production systems that have lower impacts on the environment. Some also argue for cutting the growth in demand for more food by greatly increasing efforts to slow population growth.

However, some believe that even while the population is still growing, we can increase food production while shifting to more sustainable food production systems. Next, we look at several aspects of such systems.

Let's REVIEW

- Why is it unlikely that we could cultivate all the world's land that could possibly be used for growing crops and livestock?
- Why is irrigating more land a limited solution?

Reducing Soil Erosion

One of the most important steps to take in shifting to more sustainable food production is to preserve the planet's topsoil. The key to this is greatly increased **soil conservation**—the various methods and techniques used to reduce soil erosion.

One of the oldest methods of soil conservation is *terracing*, in which a farmer creates a series of broad, nearly level terraces that run across the land's contours (see Photo b, p. 88). Each terrace retains water for its crops, slowly releasing it to the next level down, and this controls runoff and reduces soil erosion on steep slopes.

Another way to control runoff and soil erosion on sloping land is to use a combination of *contour planting* and *strip cropping* (Figure 4.23). Farmers work across the slope of the land rather than up and down, as they plow and plant their rows of crops. Like a small terrace, each row holds some water for its plants and slowly releases the

Ron Nichols/Natural Resources Conservation Service

FIGURE 4.23 Farmers can reduce topsoil erosion by planting along the contours of the land (*contour planting*) and on alternating strips of land (*strip cropping*) as was done on this farm in Wisconsin.

Pragati Biotechnologies

FIGURE 4.24 One way to reduce topsoil erosion is to grow crops in rows, or alleys, between rows of trees. Here, wheat is planted between lines of fast-growing eucalyptus trees in the state of Punjab, India.

excess water downslope to other plants. Crops are planted in alternating strips, and farmers harvest one crop at a time, with the remaining alternating crop strips left to catch runoff and reduce soil erosion.

On flat land, farmers often face the problem of wind erosion (Figure 4.14b), which can be reduced through *alley cropping*, or *agroforestry* (Figure 4.24). Farmers plant one or more crops together in strips, or alleys, between rows of trees or shrubs. These tree or shrub rows block some

wind and also provide shade, which reduces water loss by evaporation, and that further helps to reduce erosion by keeping the soil as moist as possible.

Similarly, some farmers plant trees around the edges of crop fields to block wind and reduce erosion. These plantings are called *windbreaks*, or *shelterbelts* (Figure 4.25). Like alley cropping, the use of windbreaks helps farmers to retain soil moisture in their fields, and it provides habitats for birds, pest-eating and pollinating insects, and other animals. It has also been shown to increase crop yields by 5% to 10%.

For centuries, farmers have regularly plowed and *tilled*, or turned over and loosened the topsoil in their fields before planting seeds. This opens the topsoil to wind and rain and increases the likelihood of soil erosion. Today, many farmers practice minimum-tillage farming, also called *conservation-tillage farming*, by using special tillers and planting machines that drill seeds directly into unplowed soil. However, because plowing and tilling also help farmers to control weeds, herbicides are needed with this minimum-tillage method. But these methods do help to reduce topsoil erosion and increase crop yields, while also reducing water pollution from sediment and fertilizer runoff.

In the United States, farmers used conservation tillage on about 41% of all cropland in 2008. The USDA estimates that by increasing this percentage to 80%, farmers could reduce topsoil erosion on U.S. farmland by at least half. It does have its drawbacks, as it requires costly machinery,

Federov Oleksiy/Shutterstock.com

FIGURE 4.25 Some farmers protect topsoil from wind erosion by planting strips of trees that serve as windbreaks along the edges of their crop fields.

does not work for all types of crops and soils, and requires more use of herbicides to control weeds.

Some scientists call for focusing soil conservation efforts on *erosion hotspots*—areas where the soil is highly erodible and where most of the world's soil erosion takes place. These hotspots occupy about one-tenth of the world's cropland.

Scientists say that, due to the difficulty of keeping these soils from being degraded and eroded, they should no longer be used for crops but instead should be planted with grasses or trees. In the United States, this is being done through the government-sponsored Conservation Reserve Program. More than 400,000 farmers have received payments through this program for retiring highly erodible cropland totaling an area larger than the state of New York. Since 1985, this program has helped to cut soil erosion on U.S. cropland by 40%.

FIGURE 4.26 This pile of decayed organic matter, or compost, can be used as a soil conditioner to help restore nutrients to topsoil.

Let's REVIEW

- Define *soil conservation*. What are five methods that have been used to reduce soil erosion? Briefly describe each.
- List the benefits and drawbacks of conservation-tillage farming.
- What are erosion hotspots and what are some scientists recommending we do with them?

Restoring Soil Fertility

The best way to maintain soil fertility is through soil conservation. However, it is also possible to restore some of the plant nutrients that are removed from soil by flooding, over-irrigation, wind, and continuous cycles of raising crops.

In order to restore soil, farmers have to add nutrients to it. One way to do this is to apply **organic fertilizer**—fertilizer made mostly out of plant and animal wastes and remains. The other major type of fertilizer is **synthetic inorganic fertilizer**, made of various combinations of nitrogen, phosphorus, potassium, and trace amounts of various minerals.

Organic fertilizers have been used for centuries. The most common type is *animal manure* produced by cattle, horses, poultry, and other farm animals. Another type, called *green manure*, is freshly cut or growing green vegetation that farmers plow into the topsoil shortly before planting a new crop. Another ancient form of fertilizer is *compost*, made from collected organic matter—including crop residues, tree leaves and twigs, and vegetable food wastes—that microorganisms break down into soil-like material that can be applied to crop fields (Figure 4.26).

Over time, the use of organic fertilizers has been replaced by the increasing use of synthetic inorganic fertilizers that now help to produce about one-fourth of the world's total crop yield. One problem is that unless these fertilizers are applied very carefully and are not used on sloping land near bodies of water, they can easily be washed off of cropland into bordering streams, ponds, or lakes that then become overfertilized (Figure 4.19).

Another traditional method of protecting soil fertility is *crop rotation*, in which farmers alternate the types of crops grown in a given field from one year to the next. Some crops such as corn and cotton remove nutrients from the soil, while others such as certain beans, peas, clover, and alfalfa add nutrients, particularly nitrogen. This method is good for protecting, and in some cases, restoring soil fertility.

Reducing Soil Degradation

Restoring land that is suffering from desertification (Figure 4.16) is difficult. One approach is to plant trees and grasses that will anchor topsoil and hold water, but this process takes time, and in areas suffering from drought, it is difficult to get such restoration started with little or no water. In some places, such restoration can be aided by using alley cropping (Figure 4.24) or windbreaks (Figure 4.25).

As with many environmental problems, the best solution to the problem of desertification is to use a preventive approach. Some desertification can be headed off by stopping or sharply reducing the overgrazing of grasslands and pastures, or by halting the clearing of trees or grasses for farming, especially in times of drought. Using low-till cultivation in place of conventional plowing and tilling that exposes topsoil to erosion by wind or rain can also slow the process of desertification.

Scientists and farmers know how to deal with soil salinization (Figure 4.17), but most methods are costly. For example, well-drained fields can be flushed with water to remove soil salts, and in fields that do not drain well naturally, underground drainage systems can be installed so that the salts can be flushed out without causing the soils to become waterlogged. Both methods are expensive and out of reach of most small-scale farmers in less-developed countries.

Again, prevention is the best approach. This might mean reducing irrigation and growing less water-intensive crops in areas subject to high evaporation rates and salinization. Farmers can also plant crops that can tolerate the salts fairly well, such as barley, cotton, and sugar beets.

> ## KEY idea
>
> The best way to deal with the problems of desertification and soil salinization is to prevent them from occurring.

Let's REVIEW

- Define *organic fertilizer* and *synthetic inorganic fertilizer*. What are three types of organic fertilizer?
- What is crop rotation and how does it help the soil?
- How can we prevent desertification?
- How can we prevent soil salinization?

Fighting Crop Pests

Using synthetic pesticides to help control crop pests has advantages and disadvantages (Figure 4.27).

Many scientists see crop pests as a problem that can be controlled with the use of ecological and biological approaches based on natural population control processes. They argue that in the long run, this is better than relying primarily on chemical pesticides that can lead to new pest populations (by killing natural pest enemies), larger populations of pests (through genetic resistance), pollution of air and water, and losses of biodiversity.

One approach is to import the natural enemies of the pests that have invaded a crop or garden. These enemies can include natural insect predators such as spiders (Figure 4.7), ladybugs (Figure 4.28), parasites, and bacteria and viruses that cause diseases in the pest populations. This

- Can increase food supplies
- Usually work quickly to rid crops of pests
- Can be safe, if used properly and in moderation

Pros

- Can promote genetic resistance in pest species
- Can kill natural enemies of pests and harm other species
- Can pollute the air and water and cause human health problems

Cons

Using Pesticides

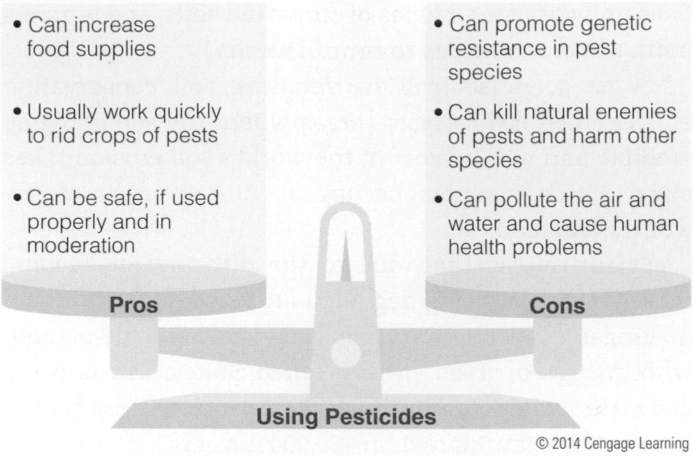

© 2014 Cengage Learning

FIGURE 4.27 *Weighing the pros and cons* of using pesticides to help protect crops from insect pests and weeds.

method, if used carefully, can be applied in such a way that it is nontoxic to other species, including humans. Such biological control, however, cannot always be done quickly enough or on a large-enough scale. Also, unless they are carefully evaluated, some of the natural enemies of pests can multiply and become pests themselves.

Another pest-control method is to use naturally occurring chemicals to trap or interfere with the biological functioning of pest insects. Scientists have found ways to lure the pests into traps by using their sex attractants, the chemicals that they use to attract a mate. The sex attractants of the pests' natural predators can also be used to draw the predators into crop fields that have been invaded by pest insects.

FIGURE 4.28 This ladybug beetle is preying on an aphid, a highly destructive insect pest. Ladybugs also eat plant mites and scale insects.

Steve Smith Photography/Shutterstock.com

Similarly, scientists have found ways to use insects' hormones to disrupt the normal life cycles of insect pests, preventing them from maturing, mating, and reproducing. Both of these natural chemical approaches can be applied without harming other species. But they are both costly and can take a long time to work.

Many pest control experts and farmers advocate using an ecological pest control program that treats each crop and its pests as parts of an ecological system. This approach, called **integrated pest management (IPM)**, uses a combination of cultivation techniques and biological and chemical pest control methods applied in a carefully coordinated way.

In such a program, crops are moved from field to field each year to stop pest invasions. Farmers watch for pests but do not act to kill them until the insects are numerous enough to pose the threat of economic damage to the crops. Then, farmers first use biological methods (natural predators, parasites, and disease organisms) and cultivation controls (such as the use of large machines to vacuum up harmful bugs).

If these methods do not work, farmers using IPM then turn to chemical controls, but they start with *biopesticides*—the naturally-based chemicals that plants and animals have used for centuries to repel their predators. They apply them in the smallest amounts possible and frequently switch from one chemical to another to slow down genetic resistance to these biopesticides. Farmers raising crops organically never use synthetic pesticides.

IPM has been used widely and successfully in several countries. After most pesticides were banned in Indonesia, pesticide use dropped there by 65% and rice crop yields actually rose by 15%. Sweden and Denmark have used IPM and cut their pesticide use by more than half. In Brazil, IPM has been used to reduce pesticide use on soybeans by as much as 90%, and Cuba, which uses organic farming to grow its crops, makes extensive use of IPM. According to a 2003 study by the U.S. National Academy of Sciences, these and other experiences show that IPM can reduce pest control costs by 50–65% without reducing crop yields or food quality.

However, IPM does not work in every situation. It requires more expert knowledge of pests, more training, and more time to work than does the use of conventional pesticides. Methods have to be carefully designed for each situation, depending on soil types, climate, growing season, and other factors. It can also be costly, at least up front, although long-term costs usually are lower than those of using conventional pesticides.

Using More Polyculture

Polyculture (Figure 4.29), which many regard as a more sustainable form of crop production, is a component of some IPM programs. It has been used for centuries in parts of South America and Africa, where some farmers still grow as many as 20 different crops together on small cleared plots in tropical forests.

Many farmers have learned that by using polyculture, they can reduce the chances of losing most or all of a season's crops to insect pests, storm damage, weather extremes, flooding, or other misfortunes. Even if they lose one or more polyculture crops to such damage, the diversity of crops in their fields helps to insure that some crops will survive. In other words, they are following nature's biodiversity principle of sustainability (see Module 1, Figure 1.24, p. 26).

Polyculture crops also use water and nutrients more efficiently because they have varying root depths that help to hold moisture in the soil and they tend to provide more shade, thus slowing evaporation of soil moisture. This cuts down on the need for fertilizing and irrigation.

Polyculture farmers usually don't use pesticides because their multiple crops provide multiple habitats for natural predators of crop-eating insects. Also, weeds have trouble competing with a richer mix of crop plants. The crops mature at different times, provide food throughout the year, and keep the soil covered for longer periods to reduce erosion from wind and water and to shade out some weeds.

Research shows that, on average, polyculture produces higher yields than most monoculture crops do. For

FIGURE 4.29 Polyculture, such as that being practiced on this plot of land, has many benefits.

example, a 2001 study by ecologists Peter Reich and David Tilman found that carefully controlled polyculture plots with 16 different species of plants consistently provided higher yields than did the plots that had one or only a few types of crops.

Let's REVIEW

- List four ways to work with nature in fighting crop pests.
- Define *integrated pest management*, explain how it works, and give an example of how it has been used successfully. What are its drawbacks?
- What are three advantages of using polyculture to raise crops?

More Sustainable Meat Production and Consumption

Meat production and consumption have huge environmental impacts. For example, according to a 2006 study by the FAO, the processes of meat production and delivery together generate 40% more greenhouse gas emissions than all of the world's cars, trucks, ships, planes, and trains combined. However, a Swedish study found that raising beef cattle on grass results in 40% lower greenhouse gas emissions and uses 85% less energy than fattening the animals for slaughter on grain crops in feedlots.

Raising livestock and poultry in pastures and other open-range areas is considered by most consumers to be more humane than raising these animals in feedlots and other confined areas. Many consumers also prefer organically raised livestock and poultry (Figure 4.30) because

organic production generally has a lower environmental impact. In addition, because organic farmers do not use antibiotics or feed supplements, consumers tend to believe that by eating organically raised meats, they are less likely to suffer harmful health effects.

The production of grains for feedlots also has a high environmental impact, but there are alternatives to growing grains for feed. For example, dairy farmers in India feed their cows mostly rice straw, wheat straw, corn stalks, and grasses gathered from roadsides. India is among the world's largest producers of milk and other dairy products.

One of the reasons why meat production has such high environmental impacts is that it is highly inefficient in terms of the food energy provided in meat, compared to the energy that goes into producing it (Figure 4.31). For example, well over a third of the world's grain harvest and about the same portion of the world's fish catch are fed to livestock. If all the grain harvested in the world today were fed to livestock, the amount of meat produced would be enough to feed just one of every three people on the planet, at the level of consumption of the average American.

This inefficiency problem is partly due to the loss of high-quality energy at each trophic level in a food chain as a result of the second law of thermodynamics (see Module 1, p. 12). Thus, one way to cut some of this energy waste and lower our environmental impact is to eat at lower trophic levels. That is, we can eat less meat or no meat and eat more fruits, vegetables, and protein-rich alternatives such as certain kinds of beans.

Also, some animals that are raised for meat eat less grain and other feeds than other animals do (Figure 4.31). Poultry and herbivorous farmed fish consume far less feed than do beef cattle, hogs, and carnivorous fish produced

FIGURE 4.30 These free-range chickens are being raised on an organic farm.

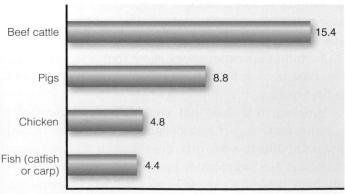

Beef cattle — 15.4

Pigs — 8.8

Chicken — 4.8

Fish (catfish or carp) — 4.4

© 2014 Cengage Learning

FIGURE 4.31 These bars represent the comparative amounts of grain consumed by various animals (in pounds) for each pound of body weight that they gain. (Data from U.S. Department of Agriculture)

through aquaculture. Many people are choosing to eat more of those grain-efficient meats and less beef, pork, and carnivorous fish, and there is some evidence that this provides health benefits as well as benefits for the environment.

More Sustainable Fish Production and Consumption

Some experts call for significant expansion of aquaculture, which already provides about half of the world's fish and shellfish. However, aquaculture will be a good way to produce more fish and shellfish only if it does not create other serious environmental problems.

Many scientists argue that in order to be more sustainable, aquaculture operations would have to be located far enough offshore or inland to avoid polluting coastal aquatic systems with their wastes. This would also avoid the ecological damage resulting from the destruction of coastal ecosystems to make way for fish farms. In addition, fish farmers would have to take all possible precautions to ensure that farm-raised fish do not escape to mingle with wild populations of fishes, which would reduce the genetic diversity of wild fish species.

Another way to make aquaculture more sustainable has to do with consumer choices. When consumers choose herbivorous species such as tilapia and carp, they are eating lower on the food chain. Carnivorous fishes, such as salmon, tuna, grouper, and cod eat other fishes. Raising such fishes involves feeding them smaller fishes, usually in the form of fishmeal, and it contributes to overfishing of the species used to feed these carnivores.

One approach that is more ecologically based than monoculture fish farming is *polyaquaculture*, in which farmers raise several species of fish and shellfish along with algae and seaweeds in coastal lagoons, ponds, or tanks. The wastes of some species become food for others. Polyaquaculture has been used for centuries, especially in Southeast Asia, and it applies both the biodiversity and chemical cycling principles of sustainability (see Module 1, Figure 1.24, p. 26).

KEY idea   By eating lower on the food chain— eating less red meat and more herbivorous fish, poultry, vegetables, and fruits—we cut the demand for grain, water, and energy resources used in industrialized meat and fish production, and we reduce the resulting harmful environmental impacts.

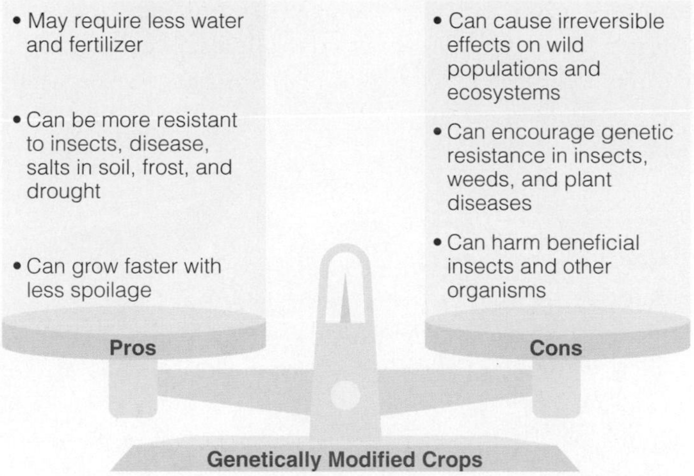

- May require less water and fertilizer
- Can be more resistant to insects, disease, salts in soil, frost, and drought
- Can grow faster with less spoilage

Pros

- Can cause irreversible effects on wild populations and ecosystems
- Can encourage genetic resistance in insects, weeds, and plant diseases
- Can harm beneficial insects and other organisms

Cons

Genetically Modified Crops

© 2014 Cengage Learning

FIGURE 4.32 *Weighing the pros and cons* of using genetically modified crops and foods.

Using Genetic Engineering

Some scientists see greater use of genetically modified crops as a way to reduce the harmful environmental effects of crop production while improving global food security by increasing crop yields. Others disagree and point to some of its potentially harmful side effects.

A 2008 assessment by 400 scientists concluded that, despite almost two decades of research and applications, genetically engineered crops have not significantly increased crop yields in the United States or elsewhere. Figure 4.32 lists the major pros and cons of using genetic engineering as a way to produce more of our food.

Let's REVIEW

- What is one way to lower the amount of greenhouse gases emitted during meat production?
- Why is it unlikely that everyone in the world could eat the same amount of meat as the typical American eats every year?
- What are three ways to make aquaculture more sustainable? Describe a polyaquaculture operation and explain its main benefits.
- List the benefits of eating lower on the food chain.
- Summarize the pros and cons of using genetic engineering to produce food.

Making the Transition to More Sustainable Food Production

Some scientists argue that we now have the knowledge and technologies to make a global transition to more sustainable food production systems that can meet the

Noam Armonn/Shutterstock.com

FIGURE 4.33 This lettuce crop in Israel was grown organically.

challenges of feeding more people every year with much lower environmental impacts than those of industrialized food production. One component of such a transition that is usually cited by these scientists is organic agriculture (Figure 4.33).

For example, agricultural scientists Paul Mader and David Dubois conducted a 22-year study in which they compared organic and conventional farming at the Rodale Institute in Kutztown, Pennsylvania. Later, ecologist David Pimentel (see *Making a Difference*, p. 98) and other researchers evaluated the Rodale study. Together, these researchers concluded that 100% organic farming tends to have the following advantages over industrialized farming:

- improves soil fertility, reduces soil erosion, and helps soils to retain more water during dry seasons and drought years;
- eliminates water pollution from synthetic fertilizers, pesticides, and antibiotics that are put into animal feed, because it uses none of these chemicals;
- sharply reduces air pollution and greenhouse gas emissions;
- benefits wildlife, especially birds, and supports biodiversity rather than degrading it; and
- uses almost a third less energy than industrialized agriculture uses for each unit of food produced.

According to these studies, however, yields of organic crops can be lower than those of conventionally raised crops, by as much as 20%. Another problem is that organically grown food costs more than conventionally grown food because organic farming requires more labor.

However, some economists point out that if the estimated costs resulting from the harmful environmental effects of industrialized agriculture were included in food prices, organically produced food would cost less than food produced by industrialized agriculture.

Some scientists and organic farmers are exploring the use of polycultures of *perennial crops*—crops that live for more than two years and do not have to be replanted each year. These experimenters are trying to find out whether or not such perennial polycultures can help us to make a transition to more sustainable agricultural systems (see the following *For Instance*).

Governments can play a role in both promoting food production and reducing its harmful environmental impacts. The major way in which they do this is by providing *subsidies*, or payments designed to help businesses survive and thrive, to farmers, fishers, and other food producers. However, these subsidies often promote unintended negative effects including the destruction of ocean-bottom habitats by fishing trawlers; depletion of ocean fish stocks; destruction of diverse tropical forests (Figure 4.1b); and overgrazing (Figure 4.21). Some experts call for governments to phase out harmful environmental subsidies for food production and to replace them with subsidies that promote more sustainable farming and fishing practices.

CONSIDER this Government farm subsidies, most of which support industrialized agriculture, account for about one-third of global farm income and amount to an average of more than $500,000 a minute in more-developed countries.

Governments can also help food producers to make the transition to more sustainable systems by funding research on sustainable organic and perennial agriculture as well as integrated pest management. In addition, government-sponsored demonstration projects can show farmers how more sustainable agricultural systems work. Governments could work together and set up an international fund to give farmers in less-developed countries access to new technologies for the more sustainable use of soil and irrigation water.

The distribution of food across and among countries also involves high rates of energy use and the emissions of CO_2 and various air pollutants that result. For those reasons, one way to make food production and consumption more sustainable is for consumers to rely more on locally grown food with an emphasis on certified 100% organic food. Consumers can also grow some of their own food in order to reduce their CO_2 emissions.

CONSIDER this

- In the United States, food travels an average 1,300 miles from farm to plate.
- A typical meal of food imported from other parts of the country or from other countries can easily account for 5 to 17 times more greenhouse gas emissions (from storing and transporting the food) than a similar meal of locally produced food would represent.

In the United States, the demand for locally grown food is increasing, as is the number of local *farmers markets* (Figure 4.34)—outdoor marketplaces in towns and cities, generally set up once a week, for bringing farmers and consumers together. Another option that is growing in popularity is *community-supported agriculture* (CSA)—programs in which individuals buy shares of a local farmer's crop and receive a box of fruits and vegetables each week during the summer and fall. Another way to eat locally grown food is to grow your own. In a growing number of cities, people are planting gardens in their yards, in vacant lots, and on rooftops.

FIGURE 4.34 These fresh organic vegetables are being sold at a farmers market.

FOR INSTANCE...

The Land Institute: Working Toward More Sustainable Agriculture

Plant geneticist Wes Jackson uses an ecological approach to agriculture at the Land Institute in Kansas. Jackson cofounded the Land Institute in the late 1970s with the ambitious goal of finding new ways to produce food by mimicking natural systems.

Researchers at the institute have been copying nature by growing diverse polycultures of edible *perennial plants*—those that can live and reproduce for two years or more. Jackson and his colleagues had observed that the prairie ecosystem sustained itself for many thousands of years before most of it was plowed under to create farmland. They also noted that, unlike croplands, the prairies did not need human inputs of water and fertilizers to keep producing. Thus, the institute's overall purpose is "to develop an agricultural system with the ecological stability of the prairie and a grain yield comparable to that from annual crops."

Land Institute scientists have experimented with perennials such as wheatgrass, a plant from which wheat was derived that produces a highly nutritious seed; Maximillian sunflower, which produces high-protein seeds not unlike soybeans; Illinois bundleflower (with seeds that could serve as livestock feed), a legume that can enrich the soil by adding nitrogen; and eastern gamma grass, a relative of corn with three times as much protein as corn.

One of the major challenges with perennial polyculture farming has been to improve crop yields. Annual crops generally have higher yields, however researchers in the state of Washington have found a perennial wheat variety that returned a yield 70% higher than that of most commercially grown annual wheat varieties.

Even with lower yields, some scientists argue that perennial crops could become an important supplement to the traditional annual crop yield while helping to reduce the harmful environmental effects of industrialized agriculture. With perennials, there is no need to till the soil every year, which reduces soil erosion and the resulting sediment pollution of rural streams, ponds, and lakes. Also, there is little or no need for synthetic fertilizers or chemical pesticides, so there is little or no resulting pollution from these sources.

Perennials need far less irrigation, because their deep roots retain more water than the shorter roots of annuals

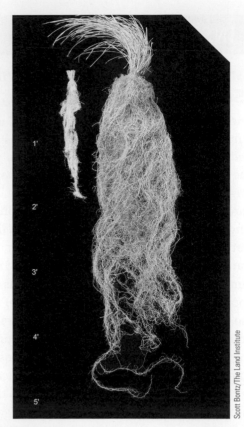

FIGURE 4.35 The roots of annual wheat crop plants (left) are less than half as long and massive as the roots of the big bluestem (right), a tallgrass prairie perennial plant.

can retain (Figure 4.35). The perennial plant is in the ground all year and because perennial crops require no annual plowing and planting, they are much better at preserving soil fertility. When these plants die, their abundant root systems stay in the ground, further enriching the soil.

In addition, perennial polycultures remove more carbon from the atmosphere, transferring it to the soil. This helps to lessen the problem of atmospheric warming due to increasing concentrations of carbon dioxide resulting from human activities such as the burning of fossil fuels. Finally, raising perennials requires far less energy than raising annual crops because of the lesser needs for running plowing, planting, and irrigation equipment. This means that perennial polyculture farming contributes less to air pollution than annual crop farming does.

Wes Jackson continues to work as the President of the Land Institute. He has stated that he believes the marriage of ecology and agriculture—to take place largely in farm fields bearing perennial polyculture crops—could become a robust and expanding reality in the foreseeable future.

Let's REVIEW

- What are six advantages that organic farming has over industrialized agriculture?
- What are three ways in which governments could help food producers to make a transition to more sustainable food production?
- What are some ways in which people can find more locally grown food? How does this help to lessen the environmental impacts of food production?
- Describe the work of Wes Jackson and other researchers at the Land Institute. What are three advantages of growing perennial crops?

A Look to the Future

The Challenge of Improving Food Security

Improving food security is important for the welfare of the human species and also for the well-being of the environment and many other species. About one out of every seven people on the planet is trying to survive on the equivalent of less than $1.25 a day, and about half of the world's people struggle to survive on less than $2.25 a day. Such poverty is a key reason why so many of the world's people face food insecurity, regardless of the amount of food we produce.

Because poverty is the root cause of most food insecurity, many scientists urge governments to create programs aimed at reducing poverty by helping the poor to help themselves. A central part of these programs would be a goal of reducing the population growth rate. Experience has shown that countries can reach that goal by providing family planning, education and jobs (especially for women), and small loans to poor people to help them buy land to grow their own food or to help them with starting businesses with which they can support themselves.

Some experts recommend focusing more on children in creating programs to end poverty and to reduce deaths and illness from hunger and malnutrition. Studies by the United Nations Children's Fund (UNICEF) indicate that one-half to two-thirds of nutrition-related childhood deaths could be prevented at an average annual cost of $5–10 per child with the following simple measures:

- Immunizing children against childhood diseases such as measles
- Encouraging breast-feeding (except among mothers with AIDS)

- Preventing dehydration from diarrhea by giving infants a mixture of sugar and salt in a glass of water
- Preventing blindness by giving children a vitamin A capsule twice a year, or by adding vitamin A and other micronutrients to their food

Food waste is a major problem. For example, according to environmental scientist Vaclav Smil, consumers in the United States waste 35–45% of all the food the country produces for domestic sale and consumption. It has been estimated that every year, Europe and North America together throw out enough food to feed the world's hungry people three times over.

The challenge of improving food security now and in the long term comes down to three urgent issues:

- How can we reduce hunger and malnutrition for the poor?
- How can we produce and distribute enough food for everyone while using more environmentally sustainable food production systems?
- How can we sharply reduce food waste?

Dealing with these issues will involve applying the three *scientific principles of sustainability* (see Module 1, Figure 1.24, p. 26) and the three *social science principles of sustainability* (see Module 1, Figure 1.28, p. 28) at the individual, local, national, and global levels.

Let's REVIEW

- What are three ways to try to slow population growth as part of an effort to reduce poverty and improve food security?
- Explain how we could lessen the number of childhood deaths stemming mostly from malnutrition.
- What are three key issues that we face now in our efforts to improve global food security?

What Would You Do?

Building more sustainable food production systems will depend to a large degree on our adopting more sustainable food consumption habits. There are a number of things that we as consumers can all do to make the food supply go farther without creating high environmental impacts. Here are several ways in which people are helping to make food production and consumption more sustainable:

Thinking about sustainability whenever they eat.
- Many people are sharply cutting their food waste by not buying or ordering more food than they can eat and by eating any leftovers before they spoil.
- People are also eating more meals from sources lower on the food chain. By eating less red meat (or none) and by choosing herbivorous fish over carnivorous fish, people are saving a great deal of energy and reducing the harmful environmental impacts of what they eat.
- Many have sharply reduced their exposure to pesticide residues by eating only 100% organic versions of fruits and vegetables.

Buying locally grown food.
- Consumers are finding farmers markets near where they live and making a habit of shopping there.
- Many have found food cooperatives and grocery stores that sell locally grown vegetables, fruits, and meats, and they add to the demand for these items by buying them.
- Others use community-supported agriculture (CSA) farms. Some people split a share with one or more friends.

Growing some of their own food.
- People are using books, periodicals, and the Internet to learn about organic gardening.
- Many have created organic gardens of their own. People who don't have land for a garden have used window boxes and other sorts of planters. (Note: it is important to have the soil tested in any urban area, as soils can be contaminated with lead and other toxic substances.)
- Others have used community gardens in the areas where they live. Friends often team up to help each other with the labor of gardening and to share the produce.

KEYterms

agrobiodiversity, p. 101
aquaculture, p. 92
artificial selection, p. 92
crossbreeding, p. 92
desertification, p. 96
fish farming, p. 92
fishery, p. 91
food chain, p. 85
food insecurity, p. 94
food security, p. 94

food web, p. 85
genetic engineering, p. 93
high-input agriculture,
   p. 87
industrialized agriculture,
   p. 87
integrated pest
   management (IPM),
   p. 107
nutrients, p. 85

organic agriculture, p. 93
organic fertilizer, p. 105
overgrazing, p. 101
overnutrition, p. 95
pesticides, p. 90
polyculture, p. 87
soil, p. 86
soil conservation, p. 103
soil degradation, p. 96
soil erosion, p. 95

soil salinization, p. 97
synthetic inorganic
   fertilizer, p. 105
topsoil, p. 86
traditional intensive
   agriculture, p. 87
traditional subsistence
   agriculture, p. 87

THINKINGcritically

1. Do you think the fact that about 1 billion people are going hungry is related to the fact that about 1.1 billion people are overweight? Explain. If you think these facts are related, how do you think the solutions to these two problems could be related?

2. Why do we need soil and why should we care about soil erosion? Some growers now produce food crops in greenhouses equipped with troughs that contain nutrient solutions that take the place of topsoil. Why can't we avoid using large areas of soil and instead grow most of the world's food in such greenhouses?

3. Explain how widespread use of a pesticide can (a) benefit some pest species in the long run, and (b) create new pest organisms.

4. Suppose that all of the grains and vegetables that you eat will someday come from one or two varieties of each plant species. In other words, suppose that someday there are only one or two varieties each of rice, wheat, broccoli, potatoes, and all of the other grains

and vegetables. Would this matter to you? Would it be important? Why or why not?

5. Keep track of the food you throw away for 3 days. Include food scraps, spoiled food, and all foods that you throw away for whatever reason. At the end of 3 days, tally up the total and estimate the weight of all the food you discarded. (You might even try to weigh the amounts you throw away and sum up the total weight after 3 days.) Compare your results with those of your classmates.

6. Given the choice between foods produced as sustainably as possible and foods that cost less but were not produced as sustainably, which would you choose? If you were to opt for the more expensive, sustainably grown food, how much more would you be willing to pay for it? Half again as much as the cheaper food? Twice as much? Three times as much? About where would you draw the line, and why?

LEARNINGonline

Access an interactive eBook and module-specific interactive learning tools, including flashcards, quizzes, videos and more in your Environmental Science CourseMate, accessed through **CengageBrain.com.**

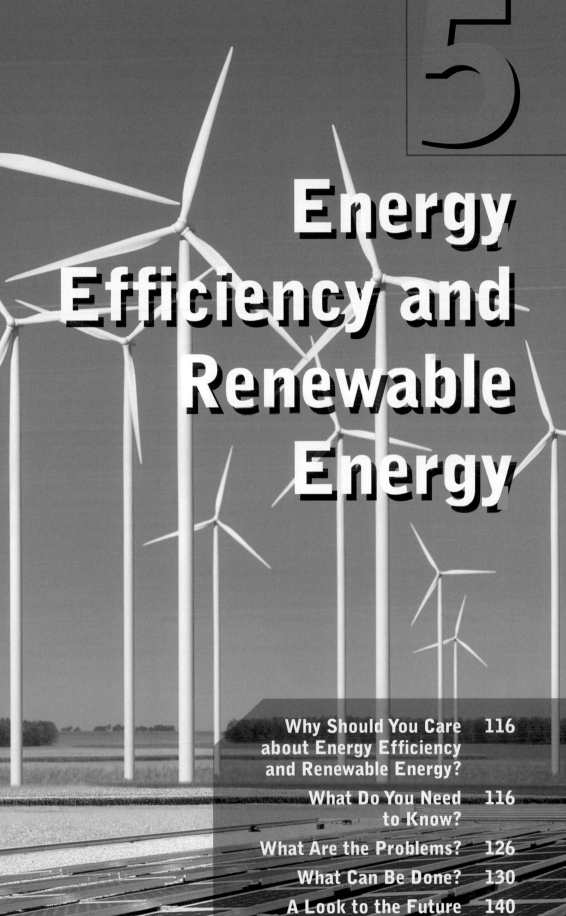

5

Energy Efficiency and Renewable Energy

Why Should You Care about Energy Efficiency and Renewable Energy?

Everything runs on energy. This includes all living things, as well as all the cars, houses, factories, lights, and appliances on which we depend. Every unit of energy we use costs us something, be it money or some other type of resource. So whenever we waste energy, we waste money or resources or both.

This is why we need to care about using energy more efficiently. Probably every one of us wastes energy in some way: by driving vehicles with a low fuel efficiency (getting less than 40 miles per gallon); living in poorly insulated houses; using energy-inefficient lighting and appliances; or leaving lights and computers on or motors running when we are not using them. Also, because our use of energy often has harmful effects on the environment, the more energy we waste, the greater is our ecological footprint (see Module 1, Figure 1.23, p. 24).

We should also care about the sources of the energy we use. The use of any energy resource has an environmental impact. But our most widely used energy resources—nonrenewable oil, natural gas, coal, and uranium (which fuels nuclear power plants)—have large harmful environmental impacts. By shifting during the next several decades to greater use of renewable energy from the sun (Figure 5.1), wind (module-opening photo), flowing water, and geothermal energy stored as heat in the earth's crust, we can greatly reduce these harmful environmental impacts.

When we rely more on renewable energy sources, most of which depend on energy from the sun, we are following one of nature's three scientific principles of sustainability (see Module 1, Figure 1.24, p. 26) that have sustained life on the earth for about 3.5 billion years. Some scientists argue that by following these principles, we will have a better chance of preserving our species, cultures, and economies than we will if we ignore them.

There is a great deal of evidence that certain human activities, in violation of these principles, are degrading the earth's life support system. In these activities, we are wasting huge amounts of energy and relying on environmentally harmful energy resources. The good news is that, by observing the earth's scientific sustainability principles, we can still enjoy many of the benefits of today's affluent societies without wasting so much energy and money, and we can reduce our harmful environmental impacts.

In this module, we look at how we can reduce our environmentally harmful waste of energy, and we consider our options for relying more on renewable energy resources.

What Do You Need to Know?

Renewable Energy from the Sun

A **renewable energy resource** is one that natural processes can replenish within a relatively short time. For example, solar energy is continually replenished, and firewood from

FIGURE 5.1 Devices called *solar cells* convert solar energy to electricity for (a) these apartments in Germany and for (b) this village in Niger, Africa.

FIGURE 5.2 Hydropower (a) is electricity produced by hydroelectric dams all over the world. Wood, a form of biomass (b), is another widely used renewable energy resource.

trees can be replenished within years. Other examples are energy from flowing water and heat stored in the earth's interior (geothermal energy).

By contrast, a **nonrenewable energy resource** is a resource that can be used up and not replenished on a human time scale. For example, supplies of *fossil fuels* (oil, coal, and natural gas) and of uranium are finite and will not be replaced for millions of years once we have used up our affordable supplies of these resources.

The most important renewable energy source, by far, is the sun. Without the sun's heat and light, which plants use to produce food, there would be no life as we know it on the earth. Like plants, we also use solar energy, for example, to heat water and to produce electricity (Figure 5.1).

Without the sun, we would have no wind with which to produce electricity (see module-opening photo), because wind is a moving mass of air created by differences in solar heating between the earth's equator and its poles, and by the earth's rotation. We would also not have flowing water, because the sun drives the water cycle (see Module 1, Figure 1.20, p. 19). It evaporates water, mostly from the oceans, into the atmosphere where some of that moisture then falls to land as rain and snow. Some of this precipitation flows into the rivers we use to produce electricity by means of hydroelectric dams (Figure 5.2a). Without the sun, we would also not have wood to burn (Figure 5.2b), because trees depend on sunlight for life. Thus, renewable wind energy, hydropower, wood, and other fuels made from plants (biofuels) are *indirect forms* of solar energy.

CONSIDER this A tiny fraction of the energy that the sun produces hits the earth, warms the planet, and keeps us alive. Without it, the earth would be a frozen planet.

Figure 5.3 shows the major types of energy that we use to supplement the direct input of solar energy that reaches earth. Nonrenewable energy resources provide

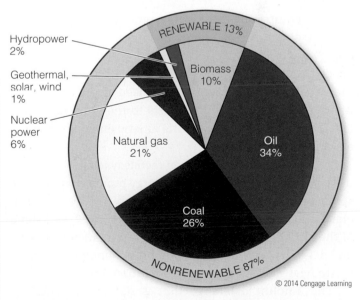

FIGURE 5.3 This chart show the mix of commercial energy resources used throughout the world in 2010 to supplement energy from the sun. (Data from U.S. Department of Energy, British Petroleum, Worldwatch Institute, and International Energy Agency)

about 87% of this supplemental energy—81% from burning fossil fuels and 6% from electricity produced at nuclear power plants, fueled mostly by uranium. Renewable energy resources, including hydropower, wind, solar energy, biofuels, and geothermal energy, supply the remaining 13% of our supplemental energy, often called *commercial energy*.

Let's REVIEW

- Why should we care about energy efficiency and renewable energy?
- Define and distinguish between *renewable* and *nonrenewable energy resources* and give an example of each.
- Why is the sun the most important energy resource? What are three forms of indirect solar energy?
- List four energy resources that we use to supplement the sun's energy.

Net Energy Yields

It takes energy to get energy. For example, producing a form of useful energy (*high-quality energy*, see Module 1, p. 11), such as wind-generated electricity, involves a number of steps, each of which requires energy. We burn coal to produce the steel that is used to make the towers and other parts for wind turbines (see module-opening photo). Then we burn diesel fuel to truck the large towers, turbines, and other parts to the wind farm site. After they are built, the turbines convert energy from the wind into electricity.

In each step of this energy production process, we convert one form of energy to another. The chemical energy in coal is converted to the high-temperature heat used to make the steel. The chemical energy in diesel fuel is converted to kinetic energy used by the trucks that haul the turbines and other parts to the wind farm site. Finally, the turbines convert the kinetic energy in wind to electricity.

Because of the *second law of thermodynamics* (see Module 1, p. 12), in each step of this energy production process, some energy is lost. Some of the high-quality chemical energy found in the coal and in the diesel fuel, and some of the kinetic energy in the wind, is automatically wasted and degraded to a lower-quality form of energy, primarily waste heat that ends up in the environment.

The amount of energy that is not lost in such an energy production process becomes useful to us and can be thought of as a **net energy yield**: the total amount of high-quality energy available from an energy resource minus the amount of high-quality energy required to make it available for our use. For example, suppose it takes 5 units of high-quality energy to produce 100 units of high-quality electricity from wind. This means that the net energy yield for the entire process is 95 units. Generally, the more steps we have to take in finding, processing, and using a source of energy, the lower its net energy yield will be. (This is similar to the decline in available high-quality energy with each transfer of energy among feeding levels in a food chain or food web; see Module 1, Figure 1.17, p. 16.)

Using data from several sources, including the U.S. Department of Energy, we can put the estimated net energy yields for various energy resources and energy systems into four general classes: high, medium, low, and negative (representing a net energy loss). Figure 5.4 shows such generalized net energy yields for energy resources and systems used to produce electricity, heat buildings, provide high-temperature heat for industrial processes, and move people and goods around. While these yields are general estimates, they do allow us to compare energy resources and systems. For example, Figure 5.4 shows that using wind to generate electricity produces a considerably higher net energy yield than do solar cells and nuclear power plants.

> **KEY idea** For any source of energy, the more steps we have to take to find, process, and use the resource, the lower its net energy yield will be.

Net energy yield is like the net profit from a business—the money that is left after the business has paid its expenses. For example, a business with $1 million in sales and $900,000 in expenses has a net profit of $100,000. Investors evaluate the success of a business, and decide whether or not to invest in it, by gauging its net profit, among other important factors.

According to many scientists and economists, we should apply the same reasoning when deciding what energy resources to use. A logical goal would be to use energy resources with high or medium net energy yields (Figure 5.4) as long as their harmful environmental effects are not too high. Generally, an energy resource with a low or negative net energy yield, does not provide enough energy to make it worth developing and using, unless the developers of that resource get outside financial help. Without such help, the energy resource would not survive in free-market competition with comparable energy resources that have medium or high net energy yields.

Electricity	Net Energy Yield
Reducing energy waste	High
Hydropower	High
Wind	High
Coal	Medium to high
Natural gas	Medium
Geothermal energy	Medium
Solar cells	Low to medium
Nuclear energy	Low
Hydrogen	Negative (Energy loss)

High-Temperature Industrial Heat	Net Energy Yield
Reducing energy waste (cogeneration)	High
Coal	High
Natural gas	Medium
Oil	Medium
Heavy shale oil	Low
Heavy oil from tar sands	Low
Direct solar (concentrated)	Low
Hydrogen	Negative (Energy loss)

Space Heating	Net Energy Yield
Reducing energy waste	High
Passive solar	Medium
Natural gas	Medium
Geothermal energy	Medium
Oil	Medium
Active solar	Low to medium
Heavy shale oil	Low
Heavy oil from tar sands	Low
Electricity	Low
Hydrogen	Negative (Energy loss)

Transportation	Net Energy Yield
Reducing energy waste	High
Gasoline	Medium
Natural gas	Medium
Ethanol (from sugarcane)	Medium
Diesel	Medium
Gasoline from heavy shale oil	Low
Gasoline from heavy tar sand oil	Low
Ethanol (from corn)	Low
Biodiesel	Low
Hydrogen	Negative (Energy loss)

© 2014 Cengage Learning

FIGURE 5.4 This is a generalized comparison of net energy yields for various energy systems. (Data from U.S. Department of Energy; U.S. Department of Agriculture; Colorado Energy Research Institute, *Net Energy Analysis*, 1976; *Encyclopedia of Earth*, 2007; Howard T. Odum and Elisabeth C. Odum, *Energy Basis for Man and Nature*, 3rd ed., New York: McGraw-Hill, 1981; and Charles A.S. Hall and Kent A. Klitgaard, *Energy and the Wealth of Nations*, New York: Springer, 2012)

Photo credits: Top left. Scott T. Baxter/Getty Images. Bottom left. National Renewable Energy Laboratory. Top right. Neil Beer/Getty Images. Bottom right. Alexander Roberge/www.clker.com.

In energy markets, such financial assistance usually comes in the form of government *subsidies*, or payments designed to help a business survive and thrive. In the United States and other countries, very large government subsidies have been provided to energy companies by taxpayers over the past several decades.

KEY idea

Net energy yield—the total amount of high-quality energy available from an energy resource minus the high-quality energy needed to make it available to consumers—is an important factor for comparing and evaluating energy resources.

Let's REVIEW

- Define *net energy yield* and give an example that illustrates it.
- Explain why an energy resource with a low or negative net energy yield must be subsidized to survive in the marketplace.

Energy Conservation and Energy Efficiency

Energy conservation is any reduction in the use or waste of energy. It can be as simple as choosing to walk or bike instead of using a car or turning down the thermostat at night.

However, we can accomplish energy conservation on a much larger scale using important strategies to improve **energy efficiency**—the measure of how much work we can get from each unit of energy we use. The lower the efficiency, the greater the amount of energy we waste. Note that by improving efficiency, we can use less energy and still enjoy the same benefits of using it. For example, most fuel-efficient cars can take us from place to place as quickly and comfortably as a gas-guzzling vehicle can.

Many of the major technologies that we rely on are highly inefficient. They waste a lot of energy and money, and have unnecessary harmful environmental impacts. For example, when we turn on an incandescent lightbulb, only 5–10% of the electricity it uses, and only 5–10% of the money we pay for this electricity, provides us with light. The other 90–95% of the electricity is converted to heat that ends up in the air around us. This is a major reason that these bulbs are being replaced by more energy-efficient and longer lasting compact fluorescent and LED lightbulbs. Improving the energy efficiency of lightbulbs and other energy technologies involves using our creativity and technology to do more work using less energy and, thus, spending less money.

Direct Solar Energy

As we explore various forms of renewable energy, it makes sense to start with the sun. Direct solar energy, which is essentially limitless, is by far our most abundant source of energy. We have learned how to use this energy resource in several ways—some simple and some more complex.

One of the simplest and cheapest ways is to use direct solar energy to cook food or sterilize water in solar cookers (Figure 5.5a) and solar ovens (Figure 5.5b). The use of these devices, especially in less-developed countries, reduces the need to collect and burn firewood, which can lead to depletion of forests and the habitats they provide for wildlife. These devices also reduce human exposure to dangerous air pollution from indoor open fires (Figure 5.2b) and leaky stoves. Such exposure contributes to the deaths of at least 1 million people a year, mostly poor people in less-developed countries.

Another approach is to provide much of the heat for a house by using a **passive solar heating system**, which absorbs and stores heat directly from the sun during daylight hours and slowly releases it throughout the day and night (Figure 5.6). Passive solar houses typically use brick, stone, adobe, or concrete walls and floors to store the sun's heat.

There is nothing new about passive solar heating. For more than 4,000 years, people have been orienting their dwellings toward the sun and using systems such as those described here. Today, we are rediscovering and improving these simple and proven technologies. Passive solar heating is a widely used application of the earth's solar energy sustainability principle (see Module 1, Figure 1.24, p. 26).

Another way to take advantage of direct solar energy is through an **active solar heating system** (Figure 5.7a), which uses a fluid, such as water or an antifreeze solution, that can absorb heat. As the fluid is pumped through tubes mounted inside of flat roof panels that face the sun (Figure 5.7b), it absorbs some of the sun's energy and carries it to a heat storage tank containing gravel, water, clay, or some other heat-absorbing substance. Or, it can be used to heat water that can then be stored in an insulated tank much like a conventional hot water heater. In China, at

FIGURE 5.5 This woman in India is using a solar cooker (a) with a curved mirror surface that concentrates direct solar energy on a pot of food or water. A solar oven (b) works in a similar fashion.

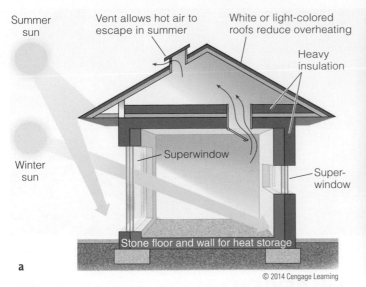

Summer sun

Vent allows hot air to escape in summer

White or light-colored roofs reduce overheating

Heavy insulation

Winter sun

Superwindow

Super-window

Stone floor and wall for heat storage

a

© 2014 Cengage Learning

Donald Aitken/National Renewable Energy Laboratory

b

FIGURE 5.6 This diagram (a) shows the basic principles of passive solar design, which were used to build this house (b) in Colorado.

least one of every ten households heats its water by using inexpensive rooftop solar water heaters, while in sunny Israel, nine of ten households do so.

A more high-tech and large-scale way to use direct solar energy is through a **solar thermal system**, which collects and concentrates direct solar energy to a temperature high enough to boil water and produce steam for generating electricity. In one system (Figure 5.8a, p. 122), huge arrays of curved mirrors collect and focus sunlight on fluid-filled pipes that run through the center of each collector. The concentrated heat in the fluid is used to produce steam that powers a turbine, which drives an electricity-producing generator.

In another type of system (Figure 5.8b), an array of computer-controlled mirrors track the sun and focus reflected sunlight on a central receiver, sometimes called a *power tower*, to boil water and produce steam that spins a turbine and generates electricity. We can store excess heat produced by these systems in molten salts and use it to produce electricity at night or on cloudy days (which are rare in most of the sunny desert areas where these systems are built).

FIGURE 5.7 This diagram (a) shows the basic principles of active solar heating system design. The panels shown in the photo (b) are part of an active solar water heating system.

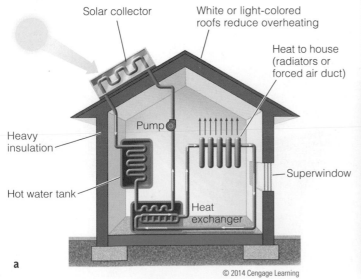

Solar collector

White or light-colored roofs reduce overheating

Heat to house (radiators or forced air duct)

Heavy insulation

Pump

Hot water tank

Superwindow

Heat exchanger

a

© 2014 Cengage Learning

AtominumeroUNO/Shutterstock.com

b

FIGURE 5.8 Curved solar collectors concentrate solar energy and use it to produce electricity in this solar thermal power plant (a) in the desert near Kramer Junction, California. In another type of system used at this plant built by Abengoa Solar in Seville, Spain (b), mirrors focus sunlight on a power tower to produce electricity.

Kramer Junction Company/National Renewable Energy Laboratory

Abengoa Solar

Another rapidly growing technology uses **photovoltaic (PV) cells**, commonly called **solar cells**, devices that capture direct solar energy and convert it to electricity (Figures 5.1 and 5.9). A typical solar cell is a small piece of material—mostly purified silicon (Si)—that is about the thickness of a human hair and has no moving parts. Each cell emits electrons when sunlight strikes it. When many solar cells are wired together, they can generate large amounts of electricity in solar cell power plants (Figure 5.9a). Newer, thin, and flexible solar cells (Figure 5.9b) can be incorporated into fabrics and other common materials.

Let's **REVIEW**

- Define *energy conservation* and *energy efficiency*.
- Define and distinguish between *passive* and *active solar heating* systems.
- Define *solar thermal system*. Describe two types of such systems.
- What are *photovoltaic (PV)* or *solar cells*? Explain how they can be used in solar cell power plants.

Hydropower

Hydropower is the energy produced by harnessing flowing water to generate electricity, usually by means of a dam-and-reservoir system (Figure 5.2a). The potential energy stored in the water behind the dam becomes kinetic energy as it flows through pipes built into the dam and spins turbines to generate electricity.

FIGURE 5.9 Individual photovoltaic (PV) cells, or solar cells, can be combined (a) to create solar-cell power plants. Newer solar-cells (b) are smaller, thinner, and more flexible.

Martin D. Vonka/Shutterstock.com

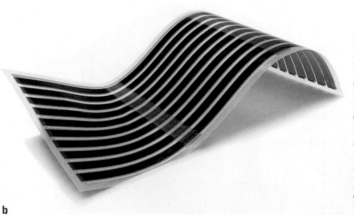

Konarka Technologies, Inc./National Renewable Energy Laboratory

Hydropower produces more electricity than any other renewable energy source. The world's leading producers of hydropower, in order, are China, Canada, Brazil, the United States, Russia, and Norway. According to the United Nations, only about 13% of the world's potential for hydropower has been developed. China plans to double its hydropower output between 2010 and 2020 and Canada and Brazil may also expand their use of hydropower. The United States has essentially reached the full capacity of its rivers to produce hydropower, while Norway gets 99% of its electricity from hydropower.

We can generate electricity at low cost on a much smaller scale by using micro-hydropower generators, which can be placed in a stream or river with little environmental impact. Other hydropower technologies attempt to capture and use the kinetic energy of waves and the daily ebb and flow of ocean tides. Devices with floating tubes are anchored near shorelines of countries such as Portugal to capture energy from the up and down action of waves, which is used to spin turbines and produce electricity. In a few places in the world, such as the Bay of Fundy in Canada, the tides move through very long, narrow formations in the coastal bluffs. This concentrates the force of the moving water, and as the tides flow in and out every day, their kinetic energy can be used to spin turbines and produce electricity. However, there are not many places on the earth where such systems are financially feasible.

Wind Power

One of the world's most rapidly growing sources of energy is **wind**, the movement of air masses driven largely by differences in solar heating across the earth's surface in combination with the earth's rotation. Like flowing water, wind is an indirect form of solar energy.

The kinetic energy of wind can turn the blades of a large wind turbine (see Module 1, Figure 1.9, p. 10, and Figure 5.10) and convert it into electricity that can be transmitted by power lines to users. Some wind turbines are as tall as a 40-story building. Their very long blades—some as long as a small jet passenger plane—capture higher altitude winds that tend to be steadier and stronger than winds near the earth's surface.

Wind prospectors have been mapping wind flows throughout the world, on land and at sea, to identify areas with favorable winds where clusters of wind turbines called *wind farms* could be built (see module-opening photo and Figure 5.10). Some prospectors believe that much of the future development of wind power will likely take place offshore because of the availability of stronger and more consistent winds. There has been some public opposition to wind farms, but for offshore wind farms, such opposition might be avoided if they are located far enough offshore to be out of sight for coastal residents.

FIGURE 5.10 Wind farms can be located in areas with favorable winds on land (see module-opening photo) or offshore as shown here.

FIGURE 5.11 In less-developed countries such as India, people in rural areas collect cow dung, dry it, and then burn it in their dwellings for heat and for cooking.

Biomass

Green plants that convert direct energy from the sun into chemical energy stored in their tissues are a form of *biomass* that we can burn to produce energy. **Solid biomass**, such as wood, crop residues, and animal manure (Figure 5.11), provides about 95% of the energy used for heating and cooking in the world's poorest countries. This use of wood is by far the largest form of renewable energy use in the world.

Also, wood chips are burned in some industrial processes and used for generating electricity because, as a fuel, wood chips can produce the high-temperature heat necessary for these purposes.

We can also convert solid biomass into liquid or gaseous fuels called **biofuels**. Two common liquid biofuels made from plants and plant wastes are *biodiesel* and *ethanol*, both of which are increasingly used to fuel motor vehicles.

Let's REVIEW

- Define *hydropower* and list three ways in which we can generate it.
- What is *wind* and how can we harness it to produce electricity?
- What is biomass? Distinguish between *solid biomass* and *biofuels* as energy resources, and give two examples of each.

Geothermal Energy

We can also tap into some of the heat stored within the rocks and fluid materials found below the earth's surface. This heat stored beneath our feet represents a gigantic energy resource known as **geothermal energy**.

We have learned how to extract and use geothermal energy to heat and cool buildings, and to generate electricity. For example, we can heat and cool our homes by using a *geothermal heat pump system* (Figure 5.12). It takes advantage of the temperature differences between the earth's surface and the ground below the frost line at depths of 3 to 10 feet, where the year-round temperature is typically 50–60°F. Such a system can be very energy-efficient, environmentally clean, and reliable.

In about 40 countries, engineers have tapped into deep reservoirs of hot water that is heated by geothermal energy. To extract heat from these *hydrothermal reservoirs*, a hole is drilled into the reservoir and a pipe is inserted to pump steam or hot water to the surface. The steam, or hot water converted to steam, spins turbines to produce electricity (Figure 5.13a). It can also be used to heat interior spaces and water within homes and to heat greenhouses in order to grow vegetables and other plants (Figure 5.13b). Iceland gets almost three-fourths of its energy and almost all of its electricity from hydroelectric power and geothermal energy, while California gets about 6% of its electricity from geothermal energy.

If we drill deep enough—3 miles or more—almost anywhere on the planet, we find hot, dry rock. We could also use heat from this source to boil water and make steam to generate electricity. So far, tapping into these deeper sources of geothermal energy is too costly, but scientists and engineers are trying to learn how to tap into this immense source of heat at an affordable price.

CONSIDER this According to the U.S. Geological Survey, tapping just 2% of the hot, dry-rock geothermal energy in the United States could produce more than 2,000 times the amount of electricity used annually in the country.

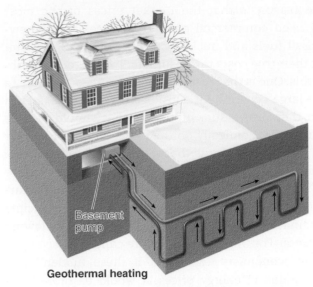

Geothermal heating

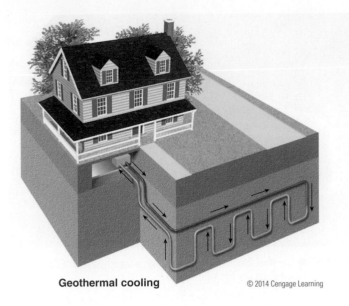

Geothermal cooling

© 2014 Cengage Learning

FIGURE 5.12 We can use a geothermal heat pump system almost anywhere. In winter, it heats a house by transferring heat from the ground into the house. In the summer, it cools the house by transferring heat from the building into the ground.

Hydrogen

Some scientists say that the fuel of the future is hydrogen gas (H_2). The problem is that H_2 is not found in the earth's atmosphere. However, we can use electricity to decompose water (H_2O) to produce H_2 gas (which we can then collect and burn) and oxygen (O_2) gas. Another way to produce H_2 gas is to strip it from the methane (CH_4) found in

natural gas and from gasoline molecules. However, these processes require more energy than the amount of energy that we can get from burning the H_2 fuel that is produced. That is, hydrogen fuel has a negative net energy yield (Figure 5.4).

Even so, some people see hydrogen as a promising energy resource. We have learned how to burn H_2 by combining it with O_2 in devices called *fuel cells*. While burning fossil fuels produces a variety of air pollutants and releases climate-changing CO_2 into the atmosphere, burning H_2 in a car powered by a fuel cell emits no air pollutants and no CO_2. The only thing coming out of the car's exhaust pipes is water vapor. For several years, major car companies have been developing and testing fuel-cell cars (Figure 5.14, p. 126) and buses that run on hydrogen.

FIGURE 5.13 This geothermal power plant in Iceland (a) generates electricity and heats a nearby body of water called the Blue Lagoon. This greenhouse in Alaska (b) is also heated by geothermal energy.

FIGURE 5.14 A fuel cell that runs on hydrogen gas powers this experimental car built by Honda.

Let's REVIEW

- Define *geothermal energy* and explain how we can use it to heat and cool a house, and to produce electricity.

- Distinguish between hydrothermal reservoirs and hot, dry rock as sources of geothermal energy.

- How can we make hydrogen into a fuel? What is the main problem with hydrogen fuel that keeps us from using it widely?

What Are the Problems?

Energy Waste

According to the U.S. Department of Energy, only about 16% of the commercial energy used in the United States is actually put to work. The other 84% is wasted. About 41% of this energy is automatically degraded to low-quality heat that ends up in the environment because of the second law of thermodynamics (Module 1, p. 12). The remaining 43% is unnecessarily wasted.

This wasted high-quality energy is a resource that can be used at a lower cost and with a lower environmental impact than any other energy resource. So why are we mostly ignoring this resource and wasting so much energy and money?

NUMB3RS

43%

The percentage of commercial energy used in the United States that is unnecessarily wasted

One reason is that the market prices of the most widely used commercial energy resources—nonrenewable fossil fuels and nuclear power (Figure 5.3)—do not reflect their true costs to consumers. Two factors contribute to this. One is that these energy resources continue to receive large government subsidies and tax breaks even though they are mature industries that should be able to compete economically on their own without taxpayer help. The second factor is that the harmful environmental and health costs of the production and use of these and all energy resources are not included in their market prices—a violation of the full-cost pricing principle of sustainability (see Figure 1.28, p. 28 and the following *For Instance*). For this reason, they are called *hidden costs*.

Many economists argue that if these hidden costs were included in energy prices, it would be more obvious to most consumers and businesses that conventional sources of energy are, in reality, very costly. They would learn that reducing energy waste by investing in energy efficiency could save them a lot of money while helping them to reduce their own impacts on the environment.

FOR INSTANCE...

The True Cost of Gasoline

Why do so many Americans buy fuel-inefficient vehicles that get less than 40 miles per gallon? Technology is not the problem. We have known for decades how to produce affordable, fast, comfortable, and safe cars that can get 50 to 100 miles per gallon. These fuel-efficient cars are widely used in Europe and other parts of the world.

One reason why many people keep buying gas-guzzling vehicles is that they have been conditioned by decades of effective advertising to crave vehicles with size, speed, and power. Automobile and oil companies understandably promote this because they make more money selling such vehicles and the fuels they burn. So for decades there has been a very strong and well-funded effort to convince consumers and politicians to favor fuel-inefficient vehicles, while there has been no comparable political effort to promote fuel-efficient vehicles.

As a result, U.S. fuel-efficiency standards are considerably lower than those of China and most European Union countries. This is beginning to change as the United States has increased its motor vehicle fuel-efficiency standards to be met by 2020. However, by 2020, the standards will again be lower than those of many other countries.

Another reason for the popularity of gas-guzzlers is that most U.S. consumers do not realize that the true cost of gasoline is much higher than what they pay at the pump. This is because there are hidden costs, including the costs associated with protecting oil company assets in foreign countries; pollution control expenses; higher medical bills and health insurance premiums because of illnesses from air pollution; and taxpayer-funded government subsidies and tax breaks for oil companies, car manufacturers, and road builders. All these costs are paid by the consumers at some point.

The International Center for Technology Assessment has estimated that the hidden costs of gasoline for U.S. consumers are at least $12 per gallon. So, assuming a pump price of around $4 per gallon, the real cost of gasoline in the United States is about $16 per gallon. Because most consumers are not aware of this, they have little incentive to drive readily available fuel-efficient cars that would save them money and reduce threats to their health and to the environment.

Another problem is that most fuel-efficient vehicles are currently more expensive than many of the less-efficient vehicles. At this point, most consumers cannot afford to buy the most fuel-efficient models. However, as demand for such vehicles increases, it's likely that their prices will come down.

Let's REVIEW

- How much of the commercial energy used in the United States is put to work? What happens to the rest?
- What is one reason why energy waste continues? What are two factors that contribute to this?
- What are three reasons why many Americans keep buying gas-guzzling vehicles?

Drawbacks of Solar, Hydropower, and Wind Energy Systems

Using direct solar energy to heat houses and water passively (Figure 5.6) has a medium net energy yield (Figure 5.4). This helps homeowners to lower their heating bills, which in turn helps them to pay for their investments in these passive systems fairly quickly. However, active solar heating systems (Figure 5.7) are more costly than passive solar systems and also require more maintenance and repair.

Another problem is that some areas have more exposure to sunlight than others (Figure 5.15). Also, many houses and other buildings are either not oriented to take advantage of direct solar energy or they are blocked by trees or other structures from receiving enough direct sunlight.

Solar thermal power plants that are built in sunny desert areas (Figure 5.8) have ample access to sunlight. However, they have a low net energy yield (Figure 5.4) and high

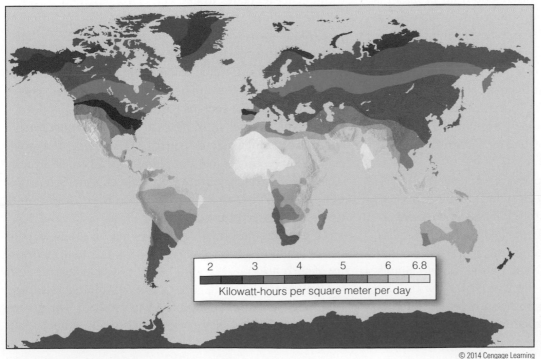

FIGURE 5.15 Global access to direct solar energy varies. An area with more than 3.5 kilowatt-hours per square meter per day (see scale) is a good place for passive and active solar heating systems as well as the use of solar cells to produce electricity. (Data from U.S. Department of Energy)

© 2014 Cengage Learning

costs, and usually need to be subsidized. Adding to these costs is the fact that updated electric power grids will have to be in place in order to distribute electricity from most of these plants to consumers.

The chief problem with solar cells (Figure 5.9) is their high cost. In fact, they are so expensive and valuable that thieves have been stealing solar-cell panels from roadside emergency call stations and from the roofs of some homes. Also, lack of access to sunlight in some areas (Figure 5.15) limits the use of solar cells, and, as with solar thermal systems, updated electrical grids would be necessary for solar-cell power plants to be widely used. Using solar cells to produce electricity has a low to medium net energy yield (Figure 5.4), although their net energy yields are increasing with new designs.

Flowing water is the most widely used form of renewable energy. We have used only a small portion of the world's potentially available hydropower and it has a high net energy yield (Figure 5.4). However, hydropower has some serious environmental impacts. When a large dam is built, land is flooded, wildlife habitat is degraded or destroyed, and people living in the area to be flooded are forced off their land. Then, flooded vegetation in the reservoir behind the dam begins to decay, especially in hot tropical areas, and emits methane, a potent *greenhouse gas*—one of the gases that are contributing to rapid atmospheric warming and projected climate change.

Also, hydropower from these dams is not renewable forever, because their reservoirs eventually fill with silt deposited by the river to the point where the hydroelectric plants no longer work. Another potential problem is that some of the rivers used to produce hydropower will likely see decreased flows of water as the mountaintop glaciers that feed them melt due to atmospheric warming projected for the rest of this century.

Many analysts contend that the use of wind to produce electricity has fewer drawbacks than any other source of energy except for energy savings from energy-efficiency improvements. Also, wind power has a high net energy yield (Figure 5.4).

However, wind power also has some disadvantages. For one, electricity generated by wind farms has to be routed overland through transmission lines, which are not always welcomed by people living along the power line routes. Also, some people object to the sights and sounds of wind turbines, and every year in the United States, 40,000–100,000 birds and bats die in collisions with large wind turbines. In some cases, another problem is land disturbance resulting from construction of wind farms.

Another problem is that winds die down from time to time. This means that wind power suppliers must rely on backup sources such as natural gas-fired, coal-fired, and nuclear power plants. Every wind farm must have a system that causes one of these backup sources to kick in whenever the wind dies. Scientists and engineers are working on ways to store electricity generated by the wind for use during such calm periods. Also, an updated electrical grid could enable wind farms across a region to serve as backup sources for one another, a topic that we discuss later in this module (see *For Instance*, p. 138).

Let's **REVIEW**

- What are the key problems associated with the use of **(a)** passive solar heating, **(b)** solar thermal power plants, and **(c)** solar cells?
- What are three problems related to the use of hydropower?
- What are four drawbacks of wind power?

Drawbacks of Biomass as an Energy Source

Solid biomass, especially wood, is renewable only when it is not used faster than nature can replenish it. In 2010, about 2.7 billion people in 77 less-developed countries were relying mostly on wood, charcoal made from wood, dried animal manure, or dung, and other forms of biomass for fuel. Populations in these countries are growing and so are firewood shortages, as people burn wood or convert it to charcoal fuel faster than it can be replenished.

Some countries have tried to deal with depletion of their biomass resources by planting grasses or fast-growing trees such as poplars (Figure 5.16a) in *biomass plantations*. In some tropical countries, vast plantations of oil palms (Figure 5.16b) are being planted and used to produce liquid biofuels such as biodiesel.

However, biofuel farming can become unsustainable and thus nonrenewable. For one thing, the repeated growing and harvesting of biofuel plantation crops eventually depletes soil of its important nutrients, especially in tropical forest areas where soils are not rich in plant nutrients to begin with. In addition, clearing grasslands, tropical forests, and other old-growth forested areas to make way for these plantations degrades or destroys the rich biodiversity of such areas.

One reason some people promote biofuels is that, when burned, they are supposed to produce less greenhouse gases than fossil fuels. However, a 2007 study by Paul

FIGURE 5.16 Fast-growing poplar trees (a) can be used as a source of firewood. Plantations of oil palms such as these in Malaysia (b) are a source of palm oil that can be converted to biodiesel fuel for motor vehicles.

Crutzen, a Dutch atmospheric chemist who won the Nobel Prize in Chemistry, made this less certain. Crutzen estimated that intensive biofuel crop farming, including the use of fertilizers that release the greenhouse gas nitrous oxide, along with the burning of biofuels, would emit more greenhouse gases than would burning the amount of fossil fuels that the biofuels would replace.

Another problem with biofuels is that when food crops such as corn are diverted from the food supply to make biofuels, food prices rise. In 2010, because of a massive taxpayer subsidy, about 40% of the corn grown in the United States was converted to ethanol that was mixed with gasoline to fuel cars. Until the subsidies expired in 2012, they made it more profitable for farmers to grow corn to fuel cars than to raise corn as a food crop. The result was a sharp rise in corn prices, which led to riots and other forms of protest by poor people in countries such as Mexico where corn was a staple.

Corn is still used to produce ethanol in the United States. Its promoters argue that it will help the country reduce its heavy dependence on imported oil and that using it instead of gasoline will cut greenhouse gas emissions. However, environmental economist Stephen Polansky and other analysts have estimated that if the entire U.S. corn crop were made into ethanol every year, it would satisfy the country's current demand for gasoline for less than three months and thus would not do much to reduce the country's dependence on oil imports. Scientific studies also indicate that, when we factor in the fossil fuels used

to plant, harvest, process, and convert corn to ethanol, the net energy yield for this biofuel is low.

Brazil runs almost half of its motor vehicles on ethanol or a mixture of ethanol and gasoline, but instead of using corn, it produces fuel from sugarcane grown on large plantations. The process makes use of *bagasse* (Figure 5.17), a waste product created when sugarcane is crushed. This ethanol has a moderate net energy yield, compared with the low net energy yields for gasoline and corn-based ethanol.

Another alternative to corn ethanol is *cellulosic ethanol*, which is produced from the cellulose found in plant leaves,

FIGURE 5.17 Bagasse is the sugarcane residue used to make ethanol in Brazil.

FIGURE 5.22 This photograph, taken by a camera that can sense infrared energy, or heat, is called a *thermal scan*. It compares energy losses in a home (shown in red and yellow) before (left) and after (right) it was insulated.

a homeowner can reduce energy losses by up to 30%. In the near future, *smart houses* will save even more energy by making use of light, thin sheets of aerogel insulation, made with the use of nanotechnology, and windowpanes that automatically get darker or lighter to control inputs of light as well as inputs and outputs of heat.

Homeowners can use more energy-efficient heating and cooling systems and appliances. For example, the most wasteful and expensive way to heat a building is with electric heat (Figure 5.4, Space Heating), while a modern natural gas furnace can be as high as 94% efficient.

Water heaters and other appliances also vary greatly in their efficiency. For example, a natural gas-fired *tankless instant water heater*, which is about the size of a typical suitcase, heats water on demand instead of storing large amounts of heated water. It uses from one-quarter to one-third less energy than a conventional tank water heater uses. (Electric instant water heaters, however, are nowhere near as energy-efficient as the natural gas-fired ones.) Front-loading clothes washers use less than half as much energy and only about two-thirds as much water as top-loading washers use.

Lighting is another important area in which homeowners and businesses can save money and energy. Replacing incandescent lightbulbs with CFLs and LEDs (Figure 5.18) would cut U.S. household energy bills and reduce air pollution and greenhouse gas emissions from coal-fired electrical power plants. Another way to save lighting energy is to focus efficient lighting on work areas instead of lighting

whole rooms. Still another is the use of motion detectors to turn lights on or off automatically in the presence or absence of people in a room.

We can also save some of the energy and money we use to keep televisions, stereo systems, and other devices on standby mode. Consumers can plug their electronic devices into a smart strip that senses when a device is not being used and turns off its standby feature.

CONSIDER this

- The EPA has estimated that if all U.S. households were to use the most efficient frost-free refrigerators now available, the amount of electricity saved every year would equal that generated by 18 large coal or nuclear power plants.

- Energy analyst Amory Lovins (see *Making a Difference*, at right) has estimated that all of the TV sets, computers, and other devices in American homes that are left on standby together use 5% of the country's energy and cost American consumers more than $4 billion per year—an average of more than $11 million a day.

Let's REVIEW

- What are the two quickest and least expensive ways to make a home or other building more energy efficient?

- What are the most energy-efficient ways to **(a)** heat a home, **(b)** provide hot water, and **(c)** provide lighting?

MAKING A difference

Amory Lovins: A World-Renowned Energy Expert

Since he began his energy research in 1976, Amory Lovins (Figure 5.A) has demonstrated that by improving energy efficiency and using a combination of renewable energy sources, we can reduce pollution, slow projected climate change, preserve biodiversity, avoid conflict over oil supplies, and ultimately save the world at least $1 trillion a year.

In 1976, Lovins published a groundbreaking paper in *Foreign Affairs* titled "Energy Strategy: The Road Not Taken." He distinguished between a "hard energy path," based on fossil fuels and nuclear power, and a "soft energy path," based on more efficient use of energy and a diversity of renewable energy sources such as solar, wind, biofuels, and geothermal energy.

At the time, critics called him an impractical dreamer proposing technological breakthroughs in energy efficiency and renewable energy that were simply not feasible. Now, more than 35 years later, his pioneering ideas are the centerpiece of global, national, and corporate energy strategies largely built around his soft-path strategy. Lovins is cofounder, chairman, and chief scientist at the Rocky Mountain Institute (RMI) in Snowmass, Colorado. This independent, nonprofit think tank has a full-time staff of more than 90 scientists, designers, and engineers who carry out cutting-edge research and consultation on energy efficiency, energy resources, energy policy, building and business design, and transportation.

In the mid-1990s, RMI designed an ultralight, extremely energy-efficient, hydrogen-powered *hypercar* that is the model for most of the prototype hydrogen fuel-cell cars now being tested (Figure 5.14). RMI also developed cutting-edge designs for hybrid and plug-in hybrid motor vehicles (Figure 5.19).

Lovins's analyses and ideas have proved to be so innovative, pioneering, and useful that he and the RMI are regularly consulted by 60 governments, 100 utility companies, and 100 major corporations around the world. *Newsweek* magazine has described Lovins as "one of the Western world's most influential energy thinkers." He has won most of the major global environmental awards, published more than 450 papers, written 31 books, and taught at 9 major universities, yet he finds time to write poetry and music, and to play the piano.

Lovins's energy-efficiency philosophy is embodied in his home and office, a large, superinsulated, solar-heated, solar-powered, and partially earth-sheltered building. Built in 1984 in Snowmass, Colorado, where winters are extremely cold, the building gets 99% of its heat and 90% of its electricity from solar energy. It is so energy efficient that the body heat generated by people inside makes a significant contribution to the heating of the building.

In 2011, Lovins published *Reinventing Fire*, which describes how we can make the transition to an energy future built around

strategies for improving energy efficiency and shifting to renewable energy resources. Today, if you ask Lovins what his chief goals are, he will tell you he wants to help businesses to make their profits in sustainable ways as well as to help all of us in making the world a secure, just, life-sustaining, and enjoyable place to live.

FIGURE 5.A Amory Lovins

Renewable Energy Resources

In an earlier section of this module (What Do You Need to Know?) we introduced several forms of renewable energy resources. We considered various applications of direct solar energy, as well as indirect forms of solar energy, including wind and hydropower. We also considered geothermal power and hydrogen as a fuel. We briefly discussed some of the benefits of each of these resources.

The use of each of these forms of energy presents some problems, as well as benefits for people and for the environment. We discussed some of those problems earlier in this module. Energy analysts have worked for years on weighing these benefits and problems in order to evaluate these renewable energy resources, both in comparison to each other, and in comparison to nonrenewable energy resources like oil and coal. Here, we challenge you to make your own comparisons and evaluations of the most widely used renewable energy resources. To do so, consider what you have read above along with the summaries of pros and cons for these resources that appear in *The Big Picture* on pages 136–137.

Evaluating Renewable Energy Resources

Every energy resource has its benefits and drawbacks. The key question to ask in considering each resource is, do its benefits outweigh any harms that will result from using it? If the answer is *no*, then use of the resource should be in serious doubt.

Energy analysts like Amory Lovins (*Making a Difference*, p. 135) ponder this question every day.

Now it's your turn. The five renewable resources represented here are the most widely used or fastest-growing sources of renewable energy. For each of them, weigh the pros and cons, decide whether or not the pros outweigh the cons, and explain your reasoning. Then, rank these energy resources in order from most to least useful.

Bryan Busovicki/Shutterstock.com

- Low-cost electricity with high net energy yield

- Low emissions of air pollutants and CO_2

- Large untapped potential

- Severe land disturbance and displacement of people

- High greenhouse gas emissions of methane from reservoirs in tropical regions

- Disruption of downstream aquatic ecosystems

Pros   **Cons**

Hydropower

- Medium net energy yield for some systems

- Little or no direct emissions of air pollutants and CO_2

- Easy to install, move around, and use in expandable arrays

- Expensive for the near term

- Need a backup or storage system when sun does not shine

- Solar cell power plants can disturb desert ecosystems

Pros   **Cons**

Solar Cells

Ascension Technology, Inc./National Renewable Energy Laboratory

- Little or no direct emissions of air pollutants and CO_2

- Moderate land disturbance

- Energy is free

- Low net energy yield and high cost

- Need a backup or storage system when sun does not shine

- Centralized system subject to sabotage and disruption

Pros | **Cons**

Solar Thermal Systems

Altangoa Solar

- Low-cost electricity with high net energy yield

- Widely available and relatively easy to harness

- Low environmental impact and low CO_2 emissions

- Backup or storage system needed when winds die down

- Can kill some birds and bats if not located carefully

- Aesthetically unpleasant for some people

Pros | **Cons**

Wind Power

Yegor Korzh/Shutterstock.com

Jim Barber/Shutterstock.com

- Moderate net energy yield (except for low net energy yield for corn and soybean biofuels)

- Potentially renewable and sustainable

- Can help reduce overall greenhouse gas emissions if produced sustainably

- Contributes to global climate change if produced unsustainably

- Clearing natural areas to make biofuel crop plantations degrades biodiversity

- Biofuel crops can compete with food production and raise food prices

Pros | **Cons**

Biofuels

137

The Promise of Wind Power

According to Amory Lovins (see *Making a Difference*, p. 135), Lester R. Brown, founder and head of the Earth Policy Institute, and many other energy and environmental experts, energy efficiency and wind power have more important advantages and fewer major disadvantages than any other energy resources. Between 1998 and 2011, the amount of electricity produced by tapping into wind energy grew by more than 20-fold (Figure 5.23), and in 2012, showed no signs of leveling off.

Since 1995, the cost of wind-generated electricity has fallen by more than half, making wind increasingly competitive with natural gas and coal for producing electricity. When the harmful environmental effects of using wind, natural gas, and coal are included in cost estimates, wind is easily the least costly. Even without these costs added, wind power may soon rank as the world's least expensive large-scale energy resource.

CONSIDER this

In 2004, Stanford University engineers estimated that tapping just one-fifth of earth's estimated wind energy potential could generate more than seven times the global demand for electricity.

The United States leads the world in total electricity generated by wind, followed closely by China and Germany. China is building wind farms at a fast pace, but the United States has the world's highest potential for wind energy development (Figure 5.24). According to the U.S. Department of Energy (DOE), just four states—North Dakota, South Dakota, Kansas, and Texas—have enough wind resources to satisfy the entire U.S. demand for electricity. In addition, a 2009 study by the DOE estimated that wind resources off the U.S. coasts, especially the East and Gulf Coasts, could also meet the entire country's demand for electricity.

Developing these land- and sea-based wind resources would allow the country to gradually phase out the use of environmentally damaging coal-burning power plants and costly, aging nuclear power plants. This would greatly reduce air pollution and greenhouse gas emissions.

Texas, known for its oil production, now produces more wind power than any other state. If it were a nation, Texas would be the world's fifth-largest producer of wind power. If the wind farms now planned for Texas are built, the state will be generating enough electricity from wind power by 2025 to supply up to 75% of the residential electrical needs of its population.

Now, farmers, ranchers, investors, wind developers, and the U.S. government are realizing the financial, environmental, and economic benefits of developing wind power. For example, without investing any money of their own, farmers and ranchers located in windy parts of the Midwestern states can get $3,000–$10,000 a year in royalties for each wind turbine placed on their land, and they can still use that land for growing crops or grazing cattle. Also, the DOE estimates that by 2030, the use of wind power in the United States could generate 500,000 jobs. As a result, wind power was the fastest growing source of electricity in the United States between 2007 and 2011 (Figure 5.25).

Critics point out that winds die down from time to time. Wind-power developers have dealt with this partly by building taller turbines with long blades that can capture winds at higher altitudes, which tend to blow harder and more steadily than do winds closer to the ground. Another solution to this problem is to locate more wind farms offshore (Figure 5.10) where winds also blow more steadily.

A key to the growing use of wind power in the United States, as well as the increased use of electricity from solar-thermal and solar-cell power plants, is the urgent need to build a *smart electrical grid*—an interactive, super-efficient national electrical grid connecting all wind farms, thermal solar plants, and solar-cell plants. A smart-grid system could distribute the electricity produced by strong winds offshore and in Midwestern states (as well as electricity

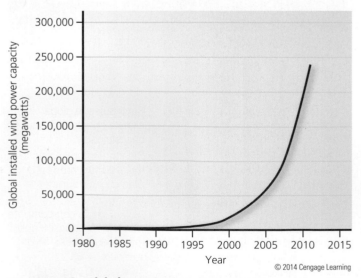

FIGURE 5.23 Global cumulative installed wind power capacity, 1980–2011. (Data from Global Wind Energy Council, Earth Policy Institute, Worldwatch Institute)

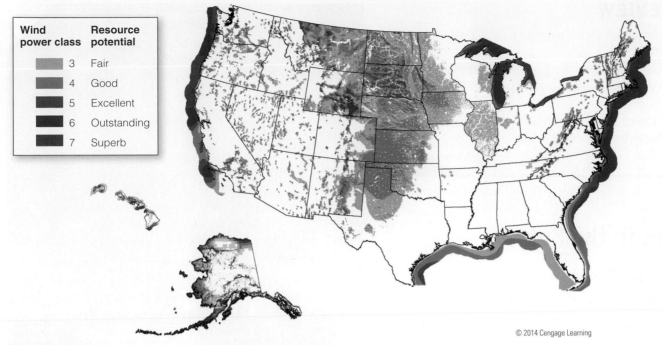

FIGURE 5.24 This map shows the land and offshore areas of the United States that have the strongest wind-power potential. (Data from the U.S. Department of Energy)

produced by solar power plants in southwestern states) to electricity consumers throughout the country. In addition, with a smart-grid system, when winds die in one area, the power plants in the rest of the country could make up the difference.

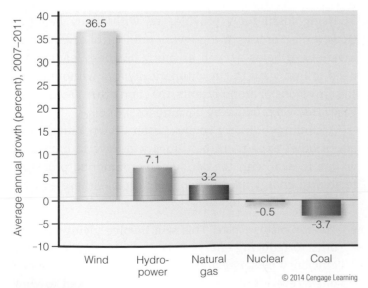

FIGURE 5.25 Of the top five sources of electricity, wind power had by far the largest average annual growth between 2007 and 2011. (Data from Energy Information Agency and Earth Policy Institute)

Critics of wind farms point out that many thousands of birds and bats die in collisions with wind turbines—a problem that engineers and scientists are trying to solve. Studies indicate that older wind turbines—built in the 1980s along some bird migration routes—are responsible for most of these bird and bat deaths. The large blades on newer turbines rotate more slowly, and the towers do not have surfaces that make good nesting and resting sites, as some older towers have, and these changes have reduced the number of bird deaths. Also, developers are being careful to locate new wind farms in areas away from known bird migration routes.

Wind power proponents also point out that wind farms are nowhere near the biggest cause of bird and bat deaths. According to the National Audubon Society, more than 1.4 billion bird and bat deaths result from collisions with cars, trucks, and buildings and other human structures (including coal-fired and nuclear power plants) and from encounters with domestic cats. The number of birds and bats killed in collisions with wind turbines is a fraction of this total.

In weighing the arguments of critics and proponents of wind power, several analysts, including Lovins and Brown, have concluded that wind power is an energy solution in which investors, consumers, job-seekers, and the environment all win. Use of this resource represents an application of several of the principles of sustainability (see Module 1, Figures 1.24, p. 26, and 1.28, p. 28).

- Describe the United States' potential for using wind to produce its electricity.
- How are wind developers dealing with the fact that winds sometimes die down? How would a smart grid help to deal with this problem?
- How is the wind industry dealing with the problem of bird and bat deaths caused by wind turbines?

A Look to the Future

The Energy Efficiency/Renewable Energy Revolution

For well over a century, the world has met most of its energy needs by burning carbon-based fossil fuels without putting a priority on reducing unnecessary energy waste. What will the energy resource mix of the future look like?

Among those who make educated projections about our energy future, there seems to be general agreement that we will continue to use fossil fuels in large amounts for the near term. However, many experts warn that there is an urgent need to deal with three serious environmental problems caused mostly by our heavy dependence on fossil fuels.

- *Outdoor air pollution*, primarily resulting from the burning of coal in power and industrial plants (Figure 5.26), kills at least 1 million people worldwide every year and as many as 68,000 people annually in the United States.
- *Climate disruption* projected for this century is largely due to the burning of fossil fuels, which adds carbon dioxide (CO_2) to the atmosphere faster than the natural carbon cycle can remove it (see Module 1, Figure 1.21, p. 20). This is resulting in atmospheric warming that is causing rapid climate change.
- *Increasing ocean acidification*, a process in which much of the CO_2 that we add to the atmosphere dissolves in the ocean and reacts with ocean water to form carbonic acid. If ocean waters become too acidic, they will slowly dissolve not only the world's coral reefs (Figure 5.27), which are vital aquatic centers of biodiversity, but also the shells of many species of ocean life that form vital links in the ocean's food webs. Some scientists contend that, in the long run, ocean acidification could be our most serious environmental problem.

FIGURE 5.26 Coal-burning industrial and power plants emit large amounts of pollutants and greenhouse gases.

Using fossil fuels provides us with many benefits. However, according to a growing number of scientists in a variety of fields as well as many business leaders, we need to find ways to cut, dramatically and without delay, the emissions of air pollutants, CO_2, and other greenhouse gases that we produce by burning fossil fuels.

According to Amory Lovins, Lester Brown, and a number of other experts, three things will have to happen if we are to deal with the serious environmental problems related to energy use. As soon as possible, these experts argue, we will have to:

- *Cut our energy waste.* This will involve giving top priority to reducing energy waste and improving energy efficiency by squeezing every bit of work that we can out of every unit of energy we use. We need to view the energy we are wasting as the quickest, easiest, cheapest, and least environmentally harmful source of energy that we can tap.

FIGURE 5.27 Coral reefs are threatened by ocean acidification. The organisms that build the reefs have a harder time doing so in acidic waters.

Pawe? Borówka/Shutterstock.com

- *Shift to renewable energy resources.* We have already begun phasing in a mix of renewable energy resources, including solar energy, wind, hydropower, biomass, and geothermal heat, and we need to fully implement this mix as soon as possible. These forms of energy produce little in the way of the air and water pollutants, greenhouse gases, and long-lasting, dangerous radioactive wastes that we are producing daily by relying on fossil fuels and nuclear power.
- *Reduce environmental harm.* We must find and quickly implement ways to lessen the harmful environmental impacts of our use of nonrenewable fossil fuels (especially coal) and costly and potentially dangerous nuclear power.

Many energy experts project that the large, centralized coal-fired and nuclear power plants that we now depend on will become less important over the next several decades. These experts foresee a gradual shift to a mix of smaller and more dispersed power systems such as rooftop solar water heaters, passive and active solar heating systems, energy-efficient houses and other buildings, regional and neighborhood wind turbines, small turbines that burn natural gas, and perhaps small stationary hydrogen fuel cells for houses and businesses (Figure 5.28, p. 142).

We can supplement such decentralized, localized systems with some centralized power sources, such as large land-based and offshore wind farms and large solar-cell and solar-thermal power plants that distribute electricity over a new smart grid. However, the decentralization of power sources will give localities and individual consumers more control over the energy resources they use. It will be similar to the shift, starting in the 1960s, from dependence on large centralized computers to widely dispersed home and laptop computers and handheld devices that now give people immense computing power.

This type of system could make greater use of locally and regionally available renewable energy resources. Windy areas would depend more on wind power. Those people living near streams and rivers could use microhydropower generators, and those living over high-quality geothermal energy resources could rely on them more than on other resources.

A decentralized energy system would also provide us with more physical security than we get from our current centralized system. A terrorist attack, a major accident, or a natural disaster such as an earthquake or tsunami could take out or severely damage any one of the big power plants, refineries, and pipelines on which we now depend. Relying on a diverse, smaller, and more dispersed mix of locally and regionally available renewable energy resources would greatly reduce the chances of disruption from such events.

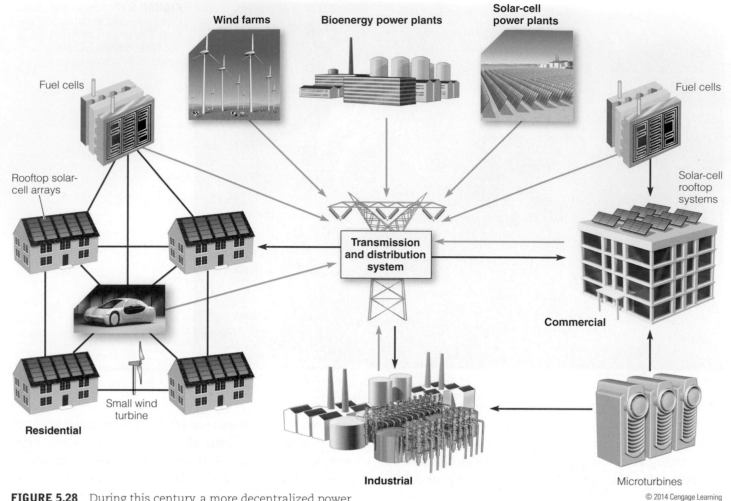

Wind farms **Bioenergy power plants** **Solar-cell power plants**

Fuel cells

Rooftop solar-cell arrays

Fuel cells

Solar-cell rooftop systems

Transmission and distribution system

Small wind turbine

Residential

Commercial

Industrial

Microturbines

© 2014 Cengage Learning

FIGURE 5.28 During this century, a more decentralized power system is likely to replace much of the current system of large, centralized coal-fired and nuclear power plants.

Finally, a decentralized system would provide much better economic security for most nations and for the world. Many economists argue that the renewable energy industries will be the future engines of growth for most of the world. The production, installation, and maintenance of certain types of renewable energy systems have already created millions of jobs worldwide. If consumers eventually save a lot of money and companies make a lot of money by using renewable energy resources, this will only help the national and global economies.

KEY idea A decentralized energy supply system, relying more on a mix of regionally and locally available renewable energy resources, would reduce dependence on imported oil and environmentally harmful coal-fired and nuclear power plants, while providing better physical and economic security through protection from energy supply interruptions.

Governments have powerful tools available for encouraging the use of one energy resource over another. They can provide money for research and development of a resource, give subsidies and tax breaks to developers and consumers, and enact laws and regulations that favor the use of particular energy resources. Governments have used these tools for decades to help the oil, gas, coal, and nuclear power industries to develop those resources. Many energy analysts and economists say it is now time to shift from providing such benefits to the fossil-fuel and nuclear industries to providing them to the renewable energy and energy-efficiency industries.

Governments and businesses alike have one more important tool—namely, education. When consumers understand the important benefits and risks of using various energy resources, they can make wiser and less environmentally harmful choices. Ultimately, what happens in the realm of energy will be up to individuals. The challenge of providing energy in the future is an immense economic and environmental opportunity to develop a more sustainable energy economy.

Let's REVIEW

- What are three serious environmental problems caused by our dependence on fossil fuels? List three steps that we will have to take, according to the experts, in order to deal with these problems.
- What are the major advantages of shifting from the use of a centralized energy distribution system based on large power plants to a decentralized system based on locally and regionally available renewable energy resources? How could this provide most nations with better overall physical and economic security?
- How can governments encourage or discourage the use of a particular energy resource? How can governments and businesses use education to help us find the best mix of energy resources in the future?

What Would You Do?

Just about all of the major solutions to our energy problems depend on what individuals do and on the choices we all make in our personal lives. What each of us does in almost every aspect of our lives each day helps either to solve these problems or to make them worse. More and more people are making this sense of personal responsibility a guiding principle in their lives. Here are some examples of how they are applying this principle in doing their part to help solve our energy and environmental problems:

Wasting less energy in getting from one place to another.

- Whenever possible, many people are walking and using bikes, buses, and other forms of mass transit. Others are sharply reducing their transportation needs by finding jobs that allow them to work at home, in our electronically interconnected world.
- When buying cars, many shoppers are choosing from only those models that will get 40 miles per gallon or better in fuel efficiency.

- Every year, more people join one of a growing number of car-sharing networks that allow them to use a car whenever they need one without having to deal with the expense of owning a car.

Wasting less energy where they live and where they work.

- Energy-conscious homeowners are making their houses as leak-free and well insulated as possible, and encouraging the owners of the buildings where they work to do the same.
- Programmable thermostats that can be set to adjust a house's indoor temperature up or down as needed are becoming more popular.
- Many people have cut their use of common appliances. For example, they hang their clothes up to dry instead of using a clothes dryer, one of the most energy-intensive appliances used in most homes.

Thinking about how their daily choices can add to energy problems or contribute to major energy solutions.

- When their electric utility companies offer electricity generated by renewable resources such as wind and sun (usually for just a few cents or dollars more per month), many people are choosing this option; some find ways to save electricity so that they can offset the slight increase in their electric bills.
- When making purchases, especially large ones, many shoppers are considering where the items were made and how the energy used to make them was supplied. Their goal is to choose items that were manufactured with the use of renewable energy.
- Buying locally has become steadily more popular. Transporting food items and other products across and between countries takes huge amounts of energy. By buying locally produced items, even if they have to pay a few cents more per item, people are saving energy and supporting their local economies.

KEYterms

THINKINGcritically

1. If we can have a positive net energy yield, using 50 units of energy to produce 100 units of energy, for example, why is this not a violation of the second law of thermodynamics? Explain.

2. Do you think we should publicize the net energy yield for every energy resource and use it to help people compare resources and evaluate their usefulness? Explain.

3. Should we phase out or at least stop developing energy resources that have low or negative net energy yields? Explain. List three candidates for such a phaseout.

4. Pick two renewable energy resources. For each of them, what do you think are the two biggest benefits of using them on a large scale, and what do you think are the two biggest drawbacks? Based on this analysis, do you think we should expand the use of the two energy resources you selected or phase them out? Explain.

5. Identify the three things that you regularly do that waste the most energy. Are you willing to stop or sharply reduce these activities, or to change them in a way that would reduce energy waste? You might try answering this question now, and then, for a month or more, keeping a diary of your activities and evaluating your energy waste and also your progress in conserving energy. Then review your diary to see if your answer to the question above has changed.

6. Imagine that you could climb into a time machine and send yourself forward in time about 50 years. Describe the mix of energy resources that you think people will likely be using. Is this the mix of energy resources that you think they should be using? Explain.

LEARNINGonline

Access an interactive eBook and module-specific interactive learning tools, including flashcards, quizzes, videos and more in your Environmental Science CourseMate, accessed through **CengageBrain.com**.

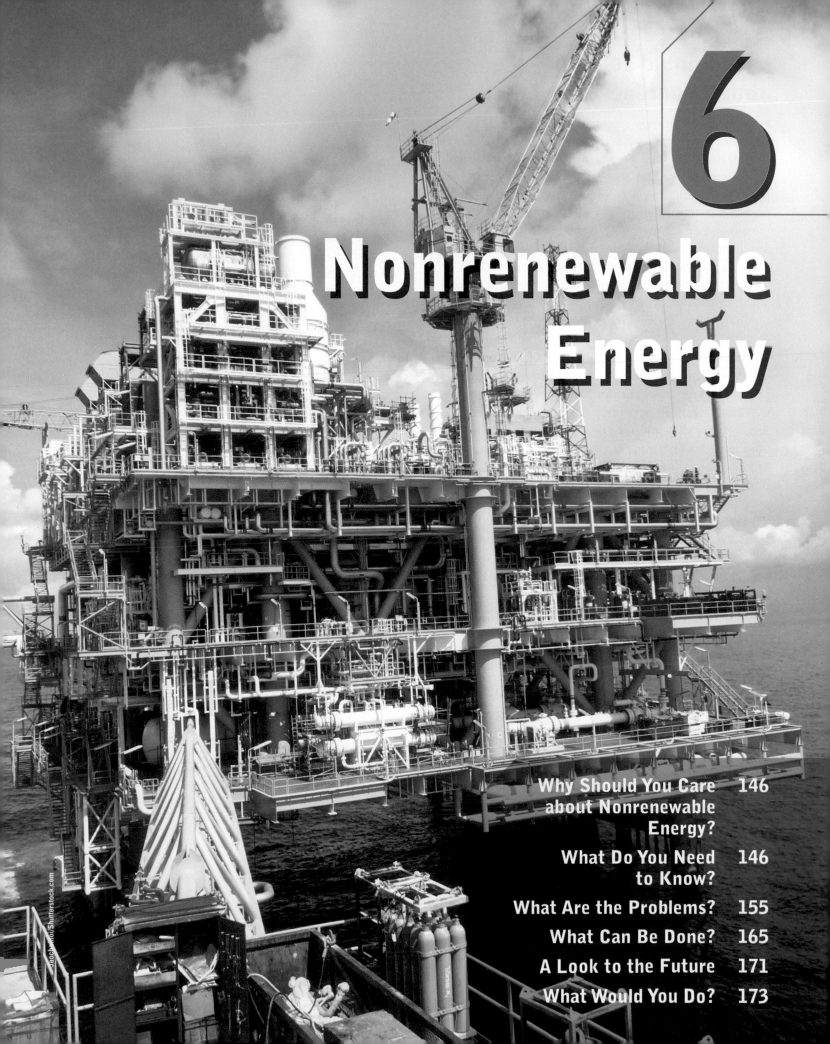

Nonrenewable Energy

Why Should You Care about Nonrenewable Energy?

Nonrenewable energy resources are those that exist in a fixed quantity, or *stock*, in the earth's crust. They include *fossil fuels*, or oil, natural gas, and coal, and the uranium that serves as fuel for nuclear power plants. One reason for caring about these resources is that they cannot be renewed by natural processes on a human time scale, and we depend on them greatly. Once the affordable deposits are depleted, it will take hundreds of millions of years for them to accumulate again in the earth's crust.

Nonrenewable energy resources provide 87% of the world's *commercial energy*—energy that is bought and sold on the market for heating, cooking, transportation, and other uses—and 92% of such energy in the United States (Figure 6.1). We currently depend on them for an astounding variety of goods and services that have enabled billions of people to live comfortably.

However, these benefits have come at a high cost. In producing and using our energy resources, we have created large and harmful environmental impacts. This is a second reason for caring about our heavy dependence on nonrenewable energy resources. For example, most of the outdoor air pollution that kills at least 1 million people a year (about 68,000 in the United States) comes from the burning of fossil fuels, especially coal, which in 2010 provided about 45% of the electricity in the United States and 80% of that in China.

A third reason for caring about our use of these nonrenewable resources is that whenever we burn carbon-containing fossil fuels, we create carbon dioxide (CO_2), which is one of the *greenhouse gases*—those gases that help to keep the earth's atmosphere warm enough for us to survive. For several decades, we have been adding CO_2 to the atmosphere faster than the natural carbon cycle can remove it (see Module 1, Figure 1.21, p. 20). Computer models developed by climate scientists project that if atmospheric levels of CO_2 continue to increase, the average temperature of the atmosphere is likely to rise to the point where it could disrupt the earth's climate during this century. The likely results will include rising sea levels, heavy flooding in some areas, intense and prolonged drought in other areas, more intense weather events such as heat waves and hurricanes, and disruptions of food production and water supplies in some areas.

In this module, we look at the pros and cons of using fossil fuels and nuclear energy, and we examine ways to reduce their significant harmful environmental impacts.

What Do You Need to Know?

Net Energy Yields

It takes energy to get energy. For example, producing a form of useful energy (*high-quality energy*, see Module 1, p. 11), such as the intense heat we get by burning coal to make steel, involves a number of steps, each of which uses

FIGURE 6.1 These charts show the mix of commercial energy resources used in 2010, in the world (left) and in the United States (right). (Data from U.S. Department of Energy, British Petroleum, Worldwatch Institute, and International Energy Agency)

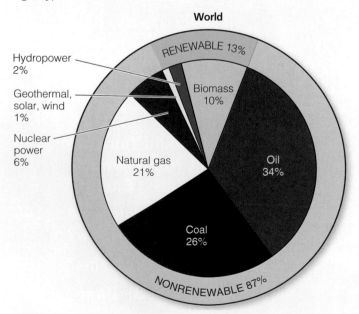

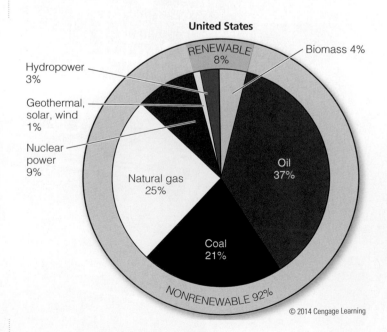

© 2014 Cengage Learning

energy. For example, coal miners typically use huge power shovels that burn diesel fuel to dig up the coal. Then more diesel fuel is used to transport the coal by train to the steel plant. Finally, the coal itself is burned to produce the steel.

In each step of this energy production process, we convert one form of energy to another. The chemical energy in diesel fuel is converted to kinetic energy used by machines to dig up the coal and to transport it to the steel plant. Then the chemical energy in the coal is converted to high-temperature heat used to make the steel.

Because of the *second law of thermodynamics* (see Module 1, p. 12), in each step of this energy production process, some energy is lost. Some of the high-quality chemical energy found in the diesel fuel and in the coal is automatically wasted and degraded to a lower-quality form of energy, primarily waste heat that ends up in the environment.

The amount of energy that is not lost in such an energy production process becomes useful to us and can be thought of as a **net energy yield**: the total amount of high-quality energy available from an energy resource minus the amount of high-quality energy required to make it available for our use. For example, suppose it takes about 4 units of high-quality energy from various sources to produce 100 units of high-quality energy in the form of the intense heat needed to make steel. This means that the net energy yield for the entire process is 96 units. Generally, the more steps we have to take in finding, processing, and using a source of energy, the lower its net energy yield will be. (This is similar to the decline in available high-quality energy with each transfer of energy among feeding levels in a food chain or food web; see Module 1, Figure 1.17, p. 16.)

Using data from several sources, including the U.S. Department of Energy, we can put the estimated net energy yields for various energy resources and energy systems into four general classes: high, medium, low, and negative (representing a net energy loss). Figure 6.2 shows such generalized net energy yields for energy resources and systems used to produce electricity, heat buildings, provide high-temperature heat for industrial processes, and move people and goods around. While these yields are general estimates, they do allow us to compare energy resources and systems. For example, Figure 6.2 shows that burning coal to provide high-temperature industrial heat produces a much higher net energy yield than does burning heavy oil for this purpose.

FIGURE 6.2 This is a generalized comparison of net energy yields for various energy systems. (Data from U.S. Department of Energy; U.S. Department of Agriculture; Colorado Energy Research Institute, *Net Energy Analysis*, 1976; *Encyclopedia of Earth*, 2007; Howard T. Odum and Elisabeth C. Odum, *Energy Basis for Man and Nature*, 3rd ed., New York: McGraw-Hill, 1981; and Charles A. S. Hall and Kent A. Klitgaard, *Energy and the Wealth of Nations*, New York: Springer, 2012)

Electricity	Net Energy Yield
Reducing energy waste	High
Hydropower	High
Wind	High
Coal	Medium to high
Natural gas	Medium
Geothermal energy	Medium
Solar cells	Low to medium
Nuclear energy	Low
Hydrogen	Negative (Energy loss)

High-Temperature Industrial Heat	Net Energy Yield
Reducing energy waste (cogeneration)	High
Coal	High
Natural gas	Medium
Oil	Medium
Heavy shale oil	Low
Heavy oil from tar sands	Low
Direct solar (concentrated)	Low
Hydrogen	Negative (Energy loss)

Space Heating	Net Energy Yield
Reducing energy waste	High
Passive solar	Medium
Natural gas	Medium
Geothermal energy	Medium
Oil	Medium
Active solar	Low to medium
Heavy shale oil	Low
Heavy oil from tar sands	Low
Electricity	Low
Hydrogen	Negative (Energy loss)

Transportation	Net Energy Yield
Reducing energy waste	High
Gasoline	Medium
Natural gas	Medium
Ethanol (from sugarcane)	Medium
Diesel	Medium
Gasoline from heavy shale oil	Low
Gasoline from heavy tar sand oil	Low
Ethanol (from corn)	Low
Biodiesel	Low
Hydrogen	Negative (Energy loss)

A net energy yield is like the net profit from a business—the money that is left after the business has paid its expenses. For example, a business with $1 million in sales and $900,000 in expenses has a net profit of $100,000. Investors evaluate the success of a business, and decide whether or not to invest in it, by gauging its net profit, among other important factors.

According to many scientists and economists, we should apply the same reasoning when deciding what energy resources to use. A logical goal would be to use energy resources with high or medium net energy yields (Figure 6.2) as long as their harmful environmental effects are not too high. Generally, an energy resource with a low or negative net energy yield does not provide enough energy to make it worth developing and using, unless the developers of that resource get outside financial help. Without such help, the energy resource would not survive in free-market competition with comparable energy resources that have medium or high net energy yields.

In energy markets, such financial assistance usually comes in the form of government *subsidies*, or payments designed to help a business survive and thrive. In the United States and other countries, very large subsidies have been provided to fossil fuel and nuclear energy companies by taxpayers over the past several decades.

> **KEY idea** Net energy yield—the total amount of high-quality energy available from an energy resource minus the high-quality energy needed to make it available to consumers—is an important factor for comparing and evaluating energy resources.

Nonrenewable Fossil Fuels

Coal, oil, and natural gas are referred to as **fossil fuels** because, like fossils, they are by-products of ancient decomposed plants and animals and were formed by heat and pressure within the earth's crust over hundreds of millions of years. Once these fuels are burned, we cannot reuse or recycle their high-quality chemical energy because it is automatically degraded to lower-quality energy in accordance with the second law of thermodynamics (see Module 1, p. 12).

Before we completely use up such a resource, we can deplete it to the point where it costs too much to find, extract, process, and use what is left. Once we use up the affordable supplies of these resources, we have to wait

around for several hundred million years for nature to replenish them.

Currently, we depend on nonrenewable fossil fuels for 81% of the commercial energy used worldwide and for 83% of that energy used in the United States.

Now, let's look more closely at each of these nonrenewable energy resources.

> **KEY idea** About 81% of the world's commercial energy and 83% of that energy used in the United States comes from burning nonrenewable fossil fuels, primarily oil, natural gas, and coal.

Let's REVIEW

- What are *nonrenewable energy resources*? What is commercial energy? Why should we care about nonrenewable energy resources?
- Define *net energy yield* and give an example that illustrates it.
- Explain why an energy resource with a low or negative net energy yield must be subsidized to survive in the marketplace.
- What are *fossil fuels* and why are they classified as nonrenewable energy resources?
- Of the commercial energy we use, what percentage is provided by fossil fuels in (a) the world and (b) the United States?

Oil

We use hundreds of products that are made from oil—from gasoline and heating oil to asphalt pavement to clothing to plastic drink bottles. **Petroleum**, or **crude oil**—also known as *conventional* or *light crude oil*—is a black, gooey liquid made mostly of compounds called *hydrocarbons* that contain atoms of carbon (C) and hydrogen (H). It is formed from the decaying remains of organisms that lived 100–500 million years ago. Some oil also contains small amounts of sulfur that form the air pollutant sulfur dioxide (SO_2) when oil is converted to gasoline and other products, and when oil and these other products are burned.

As crude oil formed over millions of years, it filled in pores and cracks in underground rock formations similar to the way that water saturates a sponge. We have learned how to find these ancient underground deposits of oil and to drill wells into the deposits and pump the oil to the surface. Today, using high-tech equipment, we can drill on land and at sea (Figure 6.3) as far down as 7 miles below the earth's surface, but at costs of millions of dollars per well.

FIGURE 6.3 Conventional crude oil can be pumped from deposits that lie under land (a) as well as from deposits under the sea floor (b).

Until recently, most of the world's oil has been pumped from large concentrated deposits that are located not too deep underground or not too deep beneath the ocean floor in fairly shallow water. As many of these fairly accessible deposits have become depleted, oil producers have looked for and shifted to deposits that are deeper underground or deeper beneath the ocean floor in much deeper water. To find and extract this oil costs more and takes more energy, which reduces the net energy yield for such oil.

Since 2008, oil production in the United States has increased after declining for more than two decades. Most of this new production comes from oil that is dispersed and held tightly in layers of shale rock. Oil producers are removing this oil by drilling horizontally through the shale rock layers and then pumping large volumes of a mixture of water, sand, and chemicals between the layers to fracture the rock and release the oil—an extraction technology called **hydraulic fracturing** or **fracking** (Figure 6.4).

Once crude oil is pumped out of a well, it is moved by pipeline, truck, or tanker ship to a *refinery* (Figure 6.5, p. 150, and module-opening photo) where it is pumped into a giant column and heated. Its components are removed at various levels from the column, depending on their boiling points, and used to make different products. Some of these products, called **petrochemicals**, are used as transportation fuels, including gasoline and diesel fuel, and as lubricants such as grease. Others are used to make plastics, paints, pesticides, fabrics, certain medicines, and many other products.

Conventional oil is the most widely used commercial energy source. In 2010, it provided 34% of the commercial energy used in the world and 37% of that energy used in the United States.

To estimate future supplies of conventional oil, geologists and economists look at **proven oil reserves**—known deposits from which conventional light crude oil can be pumped using current technology at a cost that will allow for making a profit. In 2010, about 77% of the

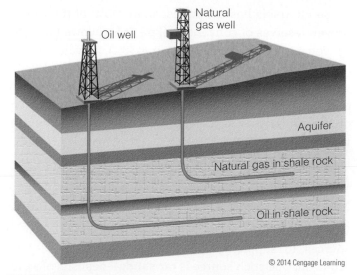

© 2014 Cengage Learning

FIGURE 6.4 Energy producers use horizontal drilling and hydraulic fracturing (fracking) to remove conventional oil and natural gas that is dispersed and held tightly in shale rock found in the United States and in several other countries.

FIGURE 6.5 An oil refinery separates oil products according to their boiling points. The photo shows an oil refinery in Anacortes, Washington.

Photo credit: Natalia Bratslavsky/Shutterstock.com.

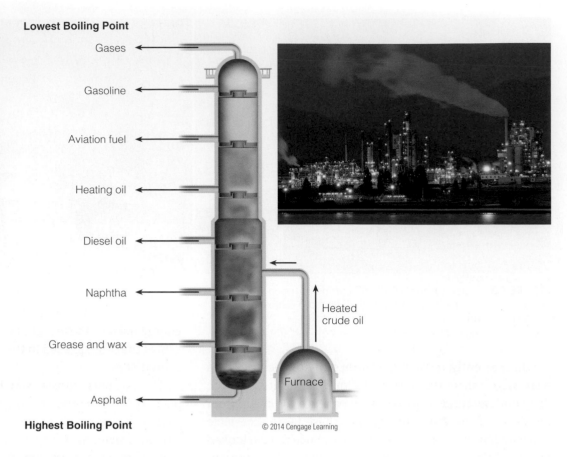

Lowest Boiling Point

Gases

Gasoline

Aviation fuel

Heating oil

Diesel oil

Naphtha

Heated crude oil

Grease and wax

Furnace

Asphalt

Highest Boiling Point

© 2014 Cengage Learning

world's proven oil reserves were owned and controlled by the governments of the 12 countries that belong to the Organization of Petroleum Exporting Countries (OPEC): Algeria, Angola, Ecuador, Iran, Iraq, Kuwait, Libya, Nigeria, Qatar, Saudi Arabia, the United Arab Emirates, and Venezuela.

Saudi Arabia has the largest share (19%) of the world's proven reserves of conventional oil. It is followed in order by: Venezuela (15%), Iran (10%), Iraq (8%), Kuwait (7%), the United Arab Emirates (7%), and Russia (6%). In order, the world's three largest oil users—the United States, China, and Japan—have only 2%, 1%, and 0.003%, respectively, of the world's proven oil reserves, which helps to explain why they import oil.

Heavy Oils from Tar Sands and Oil Shale

We can also use heavy oils from other underground sources to supplement supplies of conventional light crude oil. One such source is **tar sands**—deposits of sandstone rock containing a heavy, gooey substance called *bitumen* (Figure 6.6). Developers of this energy source dig up the rock, extract its bitumen, and convert it to heavy oil.

About three-fourths of the world's known tar sands are found under forests in northeastern Alberta, a province of Canada, and these deposits are now being developed.

Another possible source of heavy oil is **oil shale**, a type of rock (Figure 6.7, left) that contains a mixture of hydrocarbons called *kerogen*. When this rock is crushed and

FIGURE 6.6 This is a sample of tar sand.

iStockphoto/AdShooter

FIGURE 6.7 This is oil shale rock (left) and the heavy shale oil (right) extracted from it.

heated in water to release the kerogen, the end result is **shale oil**, a heavy oil that must be extensively processed to be useful (Figure 6.7, right). This unconventional heavy oil is different from the conventional oil that we can remove from between layers of shale rock through fracking, as described above. Shale oil is contained in the rock.

The United States has almost three-fourths of the world's oil shale reserves, lying deep in rock formations found in the states of Colorado, Wyoming, and Utah. According to the U.S. Bureau of Land Management, there could be eleven times as much heavy oil in these reserves as is stored in Canada's tar sand reserves. This would also equal about four times the amount of conventional light oil in Saudi Arabia's reserves.

Let's REVIEW

- Define *petroleum,* or *crude oil.* What is *hydraulic fracturing,* or *fracking?*
- What is a *refinery?* Define and give three examples of a *petrochemical.*
- What are *proven oil reserves?* What country has most of the world's conventional crude oil reserves? What percentage of these reserves does the United States have?
- What are *tar sands* and where are most of them located? Define and distinguish between *oil shale* and *shale oil.* Where are most of the world's oil shales located?

Natural Gas

Natural gas is a mixture of gaseous compounds containing mostly methane (CH_4) that is widely used as a fuel for heating and cooking, among other uses. As oil deposits

formed underground hundreds of millions of years ago, natural gas often collected above the deposits. When oil producers drill for oil, they can also remove this natural gas and transport it to consumers via pipelines. In those areas where gas pipelines have not been built, however, the companies often just burn off the gas.

Other deposits of natural gas that are not associated with oil deposits have also been tapped as a source of energy. Some natural gas is held tightly between layers of underground shale rock. Producers are using fracking to free up and extract this natural gas, just as they have been for removing conventional oil from between layers of shale rock (Figure 6.4). The use of this form of natural gas production is increasing rapidly in the United States.

We burn natural gas in furnaces, cooking stoves, and water heaters. We can also burn it in gas turbines (Figure 6.8) in power plants to produce electricity. And natural gas can be used to fuel cars and trucks.

For most people who use natural gas in their homes and businesses, a pipeline supplies the gas. Natural gas producers also remove propane and butane from natural gas and store these gases in pressurized tanks as **liquefied petroleum gas (LPG)**. These tanks can then be taken to homes and buildings where natural gas is not piped in and refilled as needed.

It is too expensive to pipe natural gas overseas between continents. However, it can be liquefied and stored, at a very low temperature and under high pressure, as **liquefied natural gas (LNG)**. In this very flammable state, it is often transported overseas on refrigerated tanker ships. Upon arrival, it is heated and converted back to a gaseous state and then distributed through pipelines. This process greatly decreases the net energy yield.

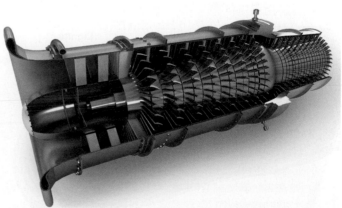

FIGURE 6.8 This is a natural gas turbine, similar to those used in some power plants to generate electricity.

The world's largest proven reserves of natural gas are in Russia (with 24% of the world's reserves), followed by Iran (with 16%) and Qatar (14%). The United States, with only about 4% of the world's proven reserves, is by far the largest consumer of natural gas.

CONSIDER this The United States has about 4% of the world's proven natural gas reserves but annually consumes about 22% of all the natural gas produced in the world.

Let's **REVIEW**

- What is *natural gas* and where is it found? What country has the largest supply of conventional natural gas?
- How is nonconventional natural gas lying between layers of shale rock removed?
- Define and distinguish between *liquefied petroleum gas (LPG)* and *liquefied natural gas (LNG)*.
- What percentage of the world's annual production of conventional natural gas does the United States use each year? What percentage of the world's conventional natural gas reserves does the United States have?

Coal

Some of the ancient plants that were buried 300 to 400 million years ago and subjected to very high heat and pressure over many of those years became **coal**—a rock-like

FIGURE 6.9 Coal forms in stages over millions of years. Each of these types of coal gives off a certain amount of heat when burned, along with carbon dioxide, sulfur dioxide, and small particles of carbon (soot) and toxic mercury.

Photo credits (left to right): Jon Sullivan/www.public-domain-image.com; U.S. Geological Survey; U.S. Geological Survey; U.S. Geological Survey.

material made mostly of carbon and containing varying concentrations of impurities such as sulfur and toxic mercury (Figure 6.9).

Coal is burned in power plants to generate electricity (Figure 6.10). It is first crushed and then fed into a boiler that produces steam, which spins a turbine to generate electricity. Coal-burning plants produce about 80% of China's electricity, 45% of the electricity used in the United States, and 41% of the world's electricity. Coal also fuels various processes used to make iron, steel, and other products.

Coal is the most abundant fossil fuel. In 2010, the three countries with the largest percentages of all proven coal reserves were the United States (28%), Russia (18%), and China (13%). The three largest users of coal in 2010 were, in order, China, the United States, and India. At current rates of use, the U.S. Geological Survey (USGS) estimates that the world's coal supply could last for a thousand years or more. Estimates for how long the U.S. supply will last vary between 100 and 250 years, depending on rates of use. Because of its dependence on coal for most of its energy, China is rapidly depleting its limited coal reserves and has started importing coal from the United States, Australia, and Indonesia. It is now the second-largest coal importer, after Japan.

Let's **REVIEW**

- What is *coal* and how is it formed?
- What percentage of the electricity used is produced by burning coal in **(a)** China, **(b)** the United States, and **(c)** the world?
- In order, what three countries have the world's largest coal reserves?

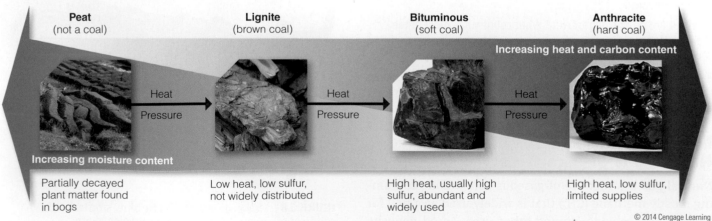

Peat
(not a coal)

Lignite
(brown coal)

Bituminous
(soft coal)

Anthracite
(hard coal)

Increasing heat and carbon content

Heat / Pressure

Heat / Pressure

Heat / Pressure

Increasing moisture content

Partially decayed plant matter found in bogs

Low heat, low sulfur, not widely distributed

High heat, usually high sulfur, abundant and widely used

High heat, low sulfur, limited supplies

© 2014 Cengage Learning

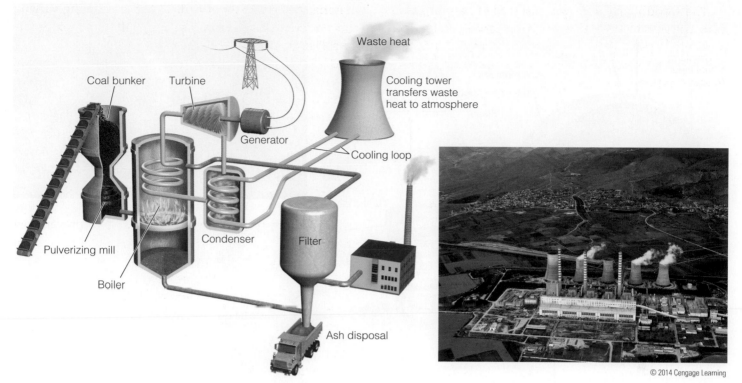

FIGURE 6.10 A coal-burning power plant is a fairly simple device. Power plants such as this one (photo) provide 41% of the world's electricity.

Photo credit: airphoto.gr/Shutterstock.com.

Nuclear Energy

Commercial nuclear power plants produce electricity by using the energy released from **nuclear fission**, a reaction that takes place when neutrons bombarding the nuclei of certain heavy atoms (such as uranium) split the nuclei and, in the process, release energy (Figure 6.11). For each nucleus that is split, two or three more neutrons are released, and each of them can trigger more fission reactions in what is called a *chain reaction* that releases an enormous amount of energy.

Each nuclear plant contains at least one *nuclear reactor*—a highly complex and costly system designed to perform a relatively simple task: to boil water to produce steam that spins a turbine and generates electricity (Figure 6.12, p. 154). The *reactor core* is designed to contain and control the nuclear fission reaction, which is the source of the heat used to boil the water.

The energy produced by most nuclear reactors comes from the fission of nuclei of the element uranium (U). The core of a typical nuclear reactor contains large numbers of uranium fuel pellets enclosed in cylinders called *fuel rods*. Each pellet is about the size of a pencil eraser and contains more energy than a ton of coal. *Control rods* are

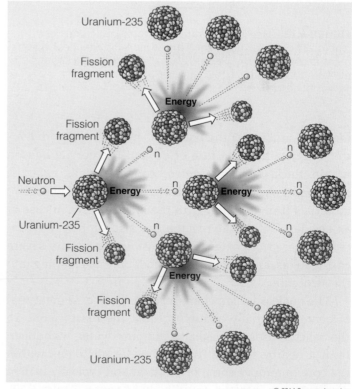

FIGURE 6.11 Nuclear fission is used to generate electricity in power plants.

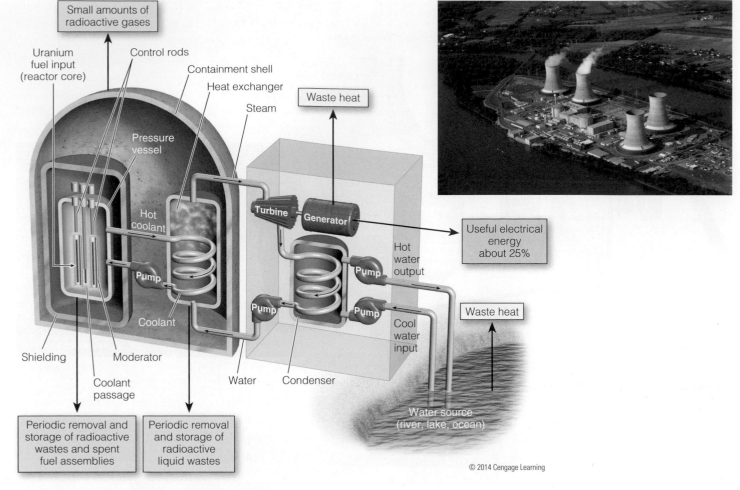

Labels in figure:
- Small amounts of radioactive gases
- Uranium fuel input (reactor core)
- Control rods
- Containment shell
- Heat exchanger
- Steam
- Waste heat
- Pressure vessel
- Hot coolant
- Turbine
- Generator
- Useful electrical energy about 25%
- Pump
- Hot water output
- Pump
- Coolant
- Pump
- Pump
- Cool water input
- Waste heat
- Shielding
- Moderator
- Water
- Condenser
- Water source (river, lake, ocean)
- Coolant passage
- Periodic removal and storage of radioactive wastes and spent fuel assemblies
- Periodic removal and storage of radioactive liquid wastes

© 2014 Cengage Learning

FIGURE 6.12 This is a diagram of a water-cooled nuclear reactor. The photo shows the Three Mile Island nuclear power plant near Harrisburg, Pennsylvania, the scene of a serious nuclear accident in 1979.

Photo credit: Robert Llewellyn/SuperStock.

moved in and out of the reactor as needed to absorb neutrons and control the rate of fission. The core is housed within thick, steel-reinforced, concrete walls called a *containment shell*, which is designed to keep radioactive materials from escaping into the environment in case of an accident within the reactor. It is also supposed to protect the core from external damage due to storms, severe earthquakes, and airplanes or other objects crashing into the structure.

The nuclear fission chain reaction gives off a great deal of heat—potentially enough to melt the reactor core. So the reactor uses a *coolant*, commonly water withdrawn from a nearby source. This cool water circulates within the core to remove some of the heat and typically is then pumped back into the same body of water (Figure 6.12). In some plants, the heat is discharged to the atmosphere from giant cooling towers (see photo in Figure 6.12).

The cooling system and other built-in safety features of a reactor greatly reduce the chances of a serious accident. However, if one or more of these safety features should fail, a massive heat buildup could cause an explosion or a *meltdown* of the reactor that could release large amounts of dangerous radiation and radioactive materials into the environment.

After several months of use, the fuel within a reactor is used up and must be replaced. Used or spent fuel rods are extremely hot and highly radioactive, so they must be safely stored away for at least 10,000 years in specially constructed containers. (We discuss this process in more detail later in this module.)

In evaluating nuclear power as an energy resource, we need to consider the entire process of using nuclear power. Operating the power plant is only one part of that larger process, which we call the **nuclear fuel cycle**—the cradle-to-grave process that includes fueling, operating, and dismantling a nuclear power plant and storing its highly radioactive parts (Figure 6.13).

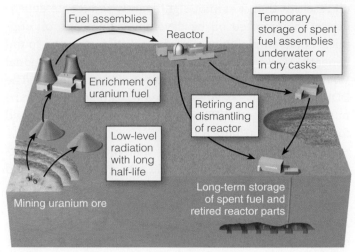

FIGURE 6.13 The nuclear fuel cycle begins with the mining of uranium ore and ends with the storage of radioactive wastes.

The nuclear fuel cycle starts with the mining and processing of uranium ore to obtain the uranium, which is then put through a process called *enrichment* to increase its concentration of fissionable material to the point where it can be used as a nuclear fuel. Beyond the operation of the plant, the nuclear fuel cycle also includes the safe storage of intensively radioactive spent fuel rods for thousands of years. The last stage of the cycle involves retiring the worn out plant (usually after 40–60 years) by removing all of its radioactive fuel and parts, and storing them safely for thousands of years.

Let's **REVIEW**

- What is *nuclear fission*? What is a nuclear chain reaction?
- Explain how a water-cooled nuclear reactor works.
- What is the *nuclear fuel cycle* and why is it important?

What Are the Problems?

Concerns about Conventional Oil Supplies

One question about our heavy dependence on oil is, how long will the affordable supplies of this nonrenewable resource last? There is still a lot of oil in the ground but we use a huge amount of it every year and the rate of use is rising rapidly.

No one knows when the world might run out of affordable conventional oil because the supply is influenced by a complex mix of at least five factors that can change with time: the demand for oil, the technology used to make it available, the rate at which we can remove the oil, the cost of making it available, and its market price. However, some analysts project that we might run out of affordable oil during this century. We could take it as a warning that 54 of the world's 64 leading oil fields have begun to decline in production, and not enough new oil fields have been found to make up for this decline.

Some experts believe that, with more exploration, we can find enough new oil fields to keep satisfying the world's demand for conventional oil. Others point out that these analysts are ignoring the rapid growth in world oil consumption, especially as China and India are becoming more affluent and ramping up their oil use. By 2010, China was importing 63% of its oil.

The U.S. Department of Energy (DOE) has estimated that if global oil use should grow at a rate of about 2% per year, the world's largest known proven crude oil reserves, located in Saudi Arabia, would last for only about 7 years. In addition, the remaining estimated reserves under Alaska's North Slope—the largest ever found in North America—would last for only about 6 months. At this annual growth rate of 2%, we would have to discover new proven oil reserves equivalent to those of Saudi Arabia every 5 years. Many oil geologists consider this to be highly unlikely.

FOR **INSTANCE...**

U.S. Consumption and Production of Oil

In 2010, the United States used about 22% of the world's oil but produced only about 9% of all the oil produced in the world. For the long term, many analysts are concerned that the United States has only about 2% of the world's proven reserves of conventional oil.

Between 1995 and 2008, oil production in the United States declined and the country imported oil to make up the difference between oil production and oil consumption. Since 2008, however, U.S. oil production has increased mostly because of high oil prices and the fact that horizontal drilling along with hydraulic fracturing have made it profitable to extract oil from between layers of shale rock (Figure 6.4). This plus a weak economy and more efficient motor vehicles helped to reduce U.S. oil imports from 60% of the fuel used in 2005 to 49% of that in 2010.

Geologists estimate that increasing domestic supplies of crude oil by opening up all of U.S. public lands and coastal waters to oil exploration would probably meet no more than 1% of the country's current annual demand for oil, and it would involve high production costs, decreasing net energy yields, and a large environmental impact.

Many American business and political leaders are urging the development of domestic sources of oil in order to further lessen U.S. dependence on imported oil. One major reason for this, they argue, is that huge amounts of wealth have gone out of the country to oil-producing nations, especially Canada, Saudi Arabia, Nigeria, Venezuela, and Mexico.

Some energy analysts say that production of conventional oil from shale rock will go up sharply over the next few decades and could lead to a new oil production era for the United States. However, no one knows how long this conventional oil from shale rock will last if production increases dramatically as projected.

Other experts warn that the United States uses so much oil and has so little of the world's remaining proven oil reserves that it cannot significantly reduce its dependence on imported oil by developing domestic oil supplies. For example, the amount of oil that might be found beneath Alaska's environmentally fragile Arctic National Wildlife Refuge (ANWR) is a subject of much debate. The DOE and the USGS have estimated that the oil likely to be pumped from the ANWR would meet current U.S. demand for about 1 year (Figure 6.14).

In 2008, the DOE estimated that making this short-term supply of oil available could take 10 to 20 years and would lower the price of gasoline at the pump by no more than 6 cents a gallon. Critics of proposals to drill in the ANWR argue that just getting drilling equipment to the site, as well as the drilling and transporting of ANWR oil, would likely cause serious and irreversible harm to some of the fragile ecosystems in this important wildlife refuge.

Let's REVIEW

- What are five factors that affect global oil supplies? Summarize the debate over whether there will be enough oil to meet the global demand.
- Explain why the United States has to import much of the oil it uses.
- Summarize the arguments for and against expansion of U.S. supplies by means of drilling for new and undeveloped domestic oil reserves.

Environmental Costs of Using Oil

The use of oil as an energy resource has cost the world a great deal in terms of harmful environmental effects. Air pollution is one of the major resulting problems. When oil is refined and when oil, gasoline, and diesel fuel are burned, they produce harmful air pollutants that have presented a major health threat for people and other living things in various areas of the world during much of the past 100 or more years.

For example, when gasoline is burned in motor vehicle engines, chemicals released into the air react under the influence of sunlight to form a mixture of harmful pollutants called *photochemical smog* (Figure 6.15). Photochemical smog creates an irritating red-brown haze in the air over big cities, especially those located in sunny areas. It can suppress plant growth and worsen breathing problems for people who have asthma and bronchitis.

A related problem is that, in the atmosphere, air pollutants resulting from the burning of oil products help to

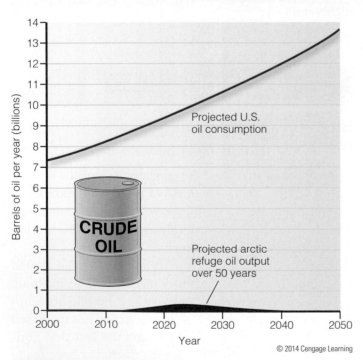

FIGURE 6.14 The amount of oil that might be pumped from the Arctic National Wildlife Refuge is a small fraction of the projected U.S. demand for oil. (Data from U.S. Department of Energy, U.S. Geological Survey, and Natural Resources Defense Council)

Nataliya Hora/Shutterstock.com

FIGURE 6.15 Thick photo-chemical smog over the city of Santiago, Chile, is caused mostly by motor vehicles burning gasoline and diesel fuel.

form acidic compounds that return to the earth as *acid deposition*, commonly called *acid rain*. These acidic chemicals can harm crops and trees (Figure 6.16), leach vital plant nutrients from the soil into nearby bodies of water, and make lakes too acidic to support most species of fish. Acid rain also damages stone building materials and statues as well as paint finishes.

Petr Vopenka/Shutterstock.com

FIGURE 6.16 Due to the effects of acid deposition, these spruce trees in the Appalachian Mountains in North Carolina have died off.

Another major problem is that using oil frequently causes water pollution. A lot of media attention is given to large spills of oil, such as the 1989 *Exxon Valdez* spill in Alaskan waters and the 2010 explosion and oil well blowout at the British Petroleum drilling rig in the Gulf of Mexico (Figure 6.17a, p. 158). However, the biggest overall source of oil pollution is the thousands of smaller sources, such as oil refinery leaks and small spills and leaks on the land bordering oceans and other bodies of water. Every time a coastal area resident pours any amount of a used oil product, including paints and paint thinners, down a drain, it can contribute to this problem.

Oil in ocean waters can coat the feathers of seabirds and the fur of marine mammals. Unless these animals are rescued and cleaned up, the oil can kill them by destroying their natural heat insulation and buoyancy (Figure 6.17b, p. 158).

Possibly the greatest threat from the burning of oil, gasoline, and diesel fuel is its projected effects on the global climate. A great deal of evidence indicates that the enormous amount of carbon dioxide emitted into the atmosphere due to the burning of these carbon-rich fuels over many decades is contributing to rapid atmospheric warming that threatens to disrupt the earth's climate during this century. Currently, the use of oil, gasoline, and diesel fuel for transportation accounts for about 43% of global CO_2 emissions, according to a variety of scientific studies, especially those conducted by several thousand scientists for the International Panel on Climate Change.

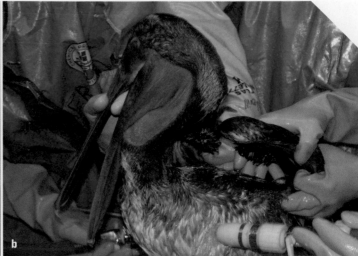

U.S. Coast Guard

U.S. Fish and Wildlife Service

FIGURE 6.17 In April 2010, an explosion on the British Petroleum (BP) *Deepwater Horizon* drilling platform (a) in the Gulf of Mexico killed 11 crew members and, over several months, released a huge volume of oil into the gulf's waters, injuring and killing an unknown number of fish, mammals, and birds. Many of these animals, including this brown pelican (b), were rescued.

Problems Caused by Use of Heavy Oils

Canada is strip-mining its vast supply of tar sands (Figure 6.6) to produce a heavy oil, and the development and use of this resource is causing major harmful environmental effects (Figure 6.18).

Producing heavy oil from tar sands involves clearing forests, diverting rivers, strip-mining sandstone rock and processing it to extract the gooey bitumen. This substance is then heated in giant cookers fueled by natural gas to make a form of heavy oil that can be refined into petroleum products. Because of the great amount of energy required for this process, this form of oil production has a low net energy yield.

Tar sand oil production destroys wildlife habitat, gouges the land, and pollutes the air and water. It also creates a toxic sludge and adds large quantities of climate-changing carbon dioxide (CO_2) to the atmosphere. The U.S. Environmental Protection Agency (EPA) estimates that CO_2 emissions from the production and use of tar sands oil will amount to 82% more per barrel than those resulting from the production and use of conventional oil and gasoline.

FIGURE 6.18 Producing heavy oil from the massive tar sands project in Alberta, Canada, has had an extensive environmental impact on the land, water, and air.

Jim Wark/Airphoto

Producing heavy oil from oil shale (Figure 6.7) has an even higher environmental impact and a lower net energy yield than producing oil from tar sands. At great expense, the rocks containing the shale oil must be dug up, crushed, and heated. This process requires large volumes of water and the oil shale exists primarily in areas (such as parts of the western United States) that are already short of water. Also, compared to conventional oil, shale oil production emits 25% to 50% more CO_2 per barrel of oil produced and causes much greater degradation and pollution of land, air, and water.

Let's REVIEW

- What is photochemical smog and how is it created? Describe its harmful effects.
- What is acid deposition and how does it occur? Describe its harmful effects.
- How does use of oil result in water pollution? What is the biggest overall source of oil pollution of water?
- How is the use of oil related to projected climate change?
- Describe the process involved in exploiting Canada's tar sands and the resulting environmental impacts.
- Describe the environmental impacts of producing heavy oil from oil shale.

Problems Caused by Use of Natural Gas

Of all the fossil fuels, natural gas has the lowest environmental impact. It releases much less CO_2 per unit of energy when burned than any other fossil fuel emits, and exploiting it results in much lower levels of land degradation and air and water pollution. However, the use and production of this resource has its problems.

First of all, when oil producers burn off, or *flare*, the natural gas located underground over oil deposits (Figure 6.19) they waste an important energy resource, pollute the air, and add climate-changing CO_2 to the atmosphere.

The growing use of liquefied natural gas (LNG) could lead to serious problems, especially when it is transported across the oceans. More than a third of the energy gained from burning LNG is used to prepare, transport, and deliver it, and then convert it back to gaseous methane. These processes all result in air and water pollution, CO_2 emissions, and degradation of ecosystems. This also means that LNG has a low net energy yield. If an LNG facility or tanker were to explode accidentally while it was in port (or to be deliberately blown up by terrorists), it would create a massive fireball.

FIGURE 6.19 Here, the natural gas found in a pocket above an oil deposit is being burned off and wasted because it costs too much to collect it or to build a pipeline to transport it.

Production of natural gas by horizontal drilling and fracking of shale rock deposits (Figure 6.4) is increasing very rapidly in parts of the United States. A growing number of residents in areas where such deposits are found question whether the benefits of exploiting this resource outweigh the costs. Fracking results in the scarring of land and involves the use of large amounts water, often in areas where water shortages are already a problem.

In some cases, groundwater supplies have reportedly been contaminated by the process of fracking. Natural gas producers claim that there is no solid evidence of such contamination, but many residents of areas where fracking occurs have claimed that it does.

Opponents of fracking say that, regardless of current evidence, eventually, the rapid expansion of the use of fracking will very likely lead to contamination of groundwater and surface waters. They argue that this will result from leaks in well casings and valves used in fracking, or from the leaching of toxic wastewater brought to the surface with the natural gas, or both. Opponents point out that the use of fracking for a single natural gas well can produce 1 million gallons of wastewater that often contains cancer-causing chemicals such as benzene and radioactive elements such as radium.

Some of this wastewater can be pumped deep underground and stored in disposal wells. However, there is some evidence that this practice could trigger earthquake activity that could allow the wastewater to migrate into underground drinking water supplies. Critics also point out that the major federal law that protects U.S. drinking water does not regulate fracking, because the natural gas industry pressured lawmakers who wrote the 2005 Energy Policy Act to exempt natural gas companies from such regulation.

In addition, natural gas producers are not required to list the chemicals they use in fracking or to test the wastewater flowing from gas wells for toxic and radioactive chemicals. Nor do they have to monitor and control the air pollution from methane (CH_4, a greenhouse gas more potent than CO_2) released by leaks from any type of gas production. As the use of natural gas from shale rock grows rapidly, opponents call for new laws to protect drinking water, and to encourage the recycling of mining wastewater.

In 2011, geoscientist David Hughes and ecologist Robert Howarth conducted separate studies, and both estimated that the overall emissions of the greenhouse gases CO_2 and methane from the entire process of producing and distributing natural gas could be much higher than scientists had thought. Earlier, lower estimates were based on only the CO_2 emissions from the burning of natural gas.

Hughes and Howarth pointed to large methane emissions that occur when natural gas is released into the atmosphere at oil well sites. They also looked at the flaring of natural gas (Figure 6.19), which emits CO_2. They raised concerns as well about the large quantities of CH_4 emitted into the atmosphere through leaks in natural gas wells (especially the rapidly rising number of fracking wells) and natural gas pipelines. They warn that unless these emissions of CH_4 and CO_2 are greatly reduced, a sharp rise in the use of natural gas over the next several decades could accelerate projected climate change. More research and data will be needed to test these claims.

Let's REVIEW

- What are three ways in which the impacts of burning natural gas are lighter than those of using other fossil fuels?
- Why is the natural gas found underground above oil deposits often burned off and wasted?
- What are some problems that could result from greater use of liquefied natural gas?
- Describe the controversy over the use of fracking to remove natural gas from shale rock deposits.
- Summarize the concerns about overall emissions of greenhouse gases during the production and delivery of natural gas.

Problems Resulting from the Use of Coal

Coal is the most abundant of the three conventional fossil fuels. It is also by far the most environmentally harmful fossil fuel we use. Every part of the process of using coal, from mining it to burning it to storing the resulting toxic wastes, has a severe impact on some part of the environment.

Coal is mined in several different ways. One method of *surface mining*, called *strip-mining* (Figure 6.20a), uses huge machines to tear away the vegetation, soil, and rock that overlie deposits of coal that are not very deep underground. This results in massive rows of waste rock called *spoils piles*. Another surface mining method is *mountaintop removal* (Figure 6.20b), practiced especially in the Appalachian Mountains of the eastern United States. Explosives are used to remove the tops of mountains so that huge machines can get at the coal that lies beneath the surface. The resulting waste rock and dirt end up in the valleys just below the mountaintops. This process buries forests and streams and destroys whole ecosystems.

Coal is also mined underground with use of shafts and tunnels. This *subsurface mining* has nowhere near the harmful environmental impacts on land, water, and air that surface mining has. However, acidic chemicals can leak from abandoned coal mines and contaminate nearby streams and groundwater. Also, the environment can be affected when the land above a mine collapses into the mine's shaft or tunnels—a process called *subsidence*. It can damage buildings, power lines, water supply pipes, sewer lines, and gas pipelines. Additionally, subsurface miners are subject to explosions, fires, cave-ins, and lung diseases from spending years inhaling coal dust and other air pollutants.

Burning coal in power plants (Figure 6.10) and industrial plants that have air pollution controls produces a toxic slurry, or *sludge*—the material produced in the process of collecting pollutants from a plant's exhaust before it leaves the smokestacks. This sludge is usually stored in holding ponds, and toxic liquids from these ponds can contaminate the groundwater lying beneath them.

In addition, these ponds can leak or even rupture, which is what happened on December 22, 2008, near the community of Kingston, Tennessee, when an earthen dam holding back a coal slurry pond broke open. The resulting flood of sludge covered part of a rural community up to 6 feet deep in places. It also spilled into two nearby streams, killing fish and contaminating the water with high levels of arsenic and other toxic metals.

FIGURE 6.20 These photos show two methods for producing coal: (a) strip mining, as is being done here at a site in Arizona, and (b) mountaintop removal, as in this operation in West Virginia.

Another serious and far-reaching environmental problem with burning coal in power and industrial plants is the harmful chemicals these plants emit into the atmosphere. The use of coal causes more air pollution per unit of energy provided than the use of any other energy resource. It also emits chemicals that are the major contributors to harmful acid deposition (Figure 6.16), and it releases toxic mercury and some radioactive materials.

According to the World Bank, the World Health Organization, the EPA, and a 2011 study by the Harvard Medical School's Center for Health and the Global Environment, the air pollution caused by coal-burning power and industrial plants annually kills several hundred thousand people worldwide and at least 13,000 people in the United States. Millions of other people suffer from severe lung problems, heart attacks, and other health problems related to this source of air pollution.

Coal-burning plants can be fitted with expensive and complex pollution control devices to reduce their contributions to air pollution, but most plants throughout the world do not have these devices (Figure 6.21). Even in the United States, a large number of older coal-burning power and industrial plants have not been required to meet U.S. air pollution standards for new facilities.

Another major problem with the use of coal is the release of huge amounts of carbon dioxide (CO_2) into the atmosphere. According to the DOE, burning coal—which is made mostly of carbon—to produce electricity releases over 3 times more CO_2 per unit of energy than burning oil, 5 times more than burning natural gas, and 17 times more

than the nuclear fuel cycle. According to a 2007 study by the Center for Global Development, about 25% of all human-generated CO_2 emissions and 40% of those in the United States come from coal-burning power plants.

Since 2008, the U.S. coal industry has mounted an intensive and effective publicity campaign to promote the idea of *clean coal*, which most environmental analysts say will never be a reality. We can burn some types of coal more cleanly by requiring expensive air pollution control systems, but CO_2 emissions will always result from the burning of coal, and strip-mining and mountaintop removal will always degrade large areas of land, water, and other resources. According to many energy experts, the idea of clean coal is a myth created by public relations firms to obscure the fact that coal is the world's dirtiest fuel.

FIGURE 6.21 This black smoke contains pollutants emitted by a coal-burning industrial plant with inadequate air pollution controls.

Most of these harmful environmental impacts of using coal are not included in the market prices of the electricity produced by burning coal. If the environmental costs were included and current government subsidies and tax breaks for coal were removed, as called for by the full-cost principle of sustainability (Module 1, Figure 1.28, p. 28), some analysts say that burning coal would be the most expensive way to produce electricity. As a result, these analysts contend, it would be phased out and replaced by cleaner energy resources such as solar energy and wind power.

CONSIDER this

The burning of coal:

- Is responsible for about 25% of the world's CO$_2$ emissions from fossil fuels and 40% of such emissions in the United States.

- Is responsible for about one-third of the toxic mercury emitted into the atmosphere by human activities.

- Emits pollutants that cause more than $100 billion a year in health costs in the United States, or an average of roughly $300 per year for every person in the country.

Let's REVIEW

- Describe the use of strip-mining and mountaintop removal for mining coal. What are two harmful results of subsurface coal mining?

- What are five environmental problems arising from the use of coal?

- Why do a number of energy experts argue that clean coal will never be a reality?

KEY idea   The burning of fossil fuels is the source of most of the world's outdoor air pollutants and climate-changing emissions of carbon dioxide.

Problems Caused by the Use of Nuclear Power

In the 1950s, proponents of nuclear power projected that by 2000, there would be at least 1,800 nuclear power plants supplying 20% or more of the world's commercial energy. Instead, in 2010, just 435 nuclear power plants produced only 6% of the world's commercial energy.

In 2008, the DOE projected that the percentage of the world's electricity produced by nuclear power would grow little or decline gradually as this century progresses because at least 300 older, worn-out nuclear plants would have to be retired about as fast as new ones could be built to replace them. Included among those older plants are all of the aging nuclear power plants that provide 20% of U.S. electricity and most of Japan's nuclear power plants that will have to be retired during the next two decades.

Five problems have stood in the way of the original, ambitious plans for this energy resource: the low net energy yield of the nuclear fuel cycle, safety issues, long-lived radioactive wastes, the very high costs of the nuclear fuel cycle, and the risk of nuclear power technology being used to help develop weapons-grade nuclear material and thus to expand the threat of nuclear war. Let's look more closely at these problems.

Low net energy yield. Producing electricity with the nuclear fuel cycle (Figure 6.13) is very inefficient. About 9% of the energy available in nuclear fuel is lost as waste heat to the environment when the fuel is mined, processed, and hauled to the power plant. Another 65% of the fuel's energy content is lost as waste heat to the environment because a typical nuclear power plant is only about 35% efficient. Experts estimate that at least another 8% is lost as waste heat in managing and storing the radioactive wastes produced at a typical plant during its 40- to 60-year lifetime. That brings the total energy lost as waste heat in the nuclear fuel cycle to about 82% of the total energy content of the fuel, which makes for a low net energy yield (Figure 6.2).

When we also include the large amounts of energy needed to run the machinery (mostly from fossil fuels) to dismantle a nuclear power plant at the end of its life and to transport the materials to a safe nuclear waste storage site, some analysts estimate that on a global scale, the conventional nuclear fuel cycle will require more energy than it produces.

Safety issues. Modern nuclear reactors have multiple-level safety features and backup systems that are designed to prevent accidents, releases of radioactivity, and meltdowns. This is why major nuclear power plant accidents are very rare. But having these extremely important safety features is a major reason why a nuclear reactor is so expensive to build, maintain, and retire at the end of its useful life, and why the net energy yield for the nuclear fuel cycle is so low.

Despite safety precautions, major nuclear accidents can happen. The first of these occurred in 1979 when one of the reactors at the Three Mile Island nuclear plant in Pennsylvania (see photo in Figure 6.12) had a partial meltdown of its core. No lives were lost but the owners of the plant suffered a $1 billion loss and public confidence in nuclear power dropped sharply.

A much more serious nuclear accident took place in 1986 when a reactor at the Chernobyl nuclear plant in Ukraine exploded, suffered a partial meltdown, and burned for 10 days. It released large amounts of radiation that spread through the atmosphere to many other countries. More than 350,000 people had to abandon their homes and were not able to return for two decades, and then only at their own risk.

In the course of the accident, the intense radiation that was released killed 56 people, half of them heroic workers and firefighters who rushed to the scene to put out fires and assess the damage. Estimates of the eventual number of deaths due to this accident, mostly from cancers caused by exposure to radiation, vary from 9,000 (from the World Health Organization) to as many as 212,000 (from the Russian Academy of Medical Sciences). We will never know the actual death toll because of poor medical records and spotty evaluations of people exposed to the radiation released by the accident.

UN scientists concluded that the Chernobyl accident was caused by a poor reactor design (not used in most of the world's other nuclear reactors) and human error on the part of the plant's operators. This accident also weakened public faith in nuclear power.

By 2011, a carefully crafted public relations and political lobbying program pushed by the nuclear power industry had rebuilt some of the lost public support for nuclear power. However, on March 11, 2011, there was another major accident, this time at Japan's Fukushima Daiichi nuclear power plant. Two reactors and some of the stored radioactive spent fuel at the plant were damaged by explosions (Figure 6.22). Evidence suggests that there was a complete meltdown of the core of one of the reactors along with partial core meltdowns of two other reactors. An unknown amount of radiation was released into the atmosphere and coastal waters near the plant.

This accident occurred after a major earthquake caused a powerful *tsunami*, a series of enormous ocean waves. This event devastated several Japanese cities, towns, and fishing villages on the country's east coast, and killed more than 25,000 people. Flooding at the plant cut off the power needed to run the cooling systems that were designed to keep the reactors from melting down. More than 350,000 people within 19 miles of the reactors were evacuated and might not be able to return to their homes and businesses for many years.

Preliminary studies revealed that the power plant operators and regulators had not designed the plant's operating systems to survive such major flooding. Apparently, they also had not been careful enough in testing and upgrading the safety systems of these aging plants, and they did not have adequate plans for dealing with a major nuclear accident. In the United States, there are 23 old reactors with the same design as that of the ill-fated Fukushima Daiichi reactors.

Kyodo via AP Images

FIGURE 6.22 This is one of the reactor buildings damaged by the 2011 accident at Japan's Fukushima Daiichi nuclear power plant.

Radioactive wastes. Each stage of the nuclear fuel cycle (Figure 6.13) produces dangerous and long-lived radioactive wastes, especially *high-level radioactive wastes* in the form of intensely hot, radioactive spent fuel rods (Figure 6.23). These wastes must be stored safely for at least 10,000 years—the time required for them to decay to safe levels of radioactivity. Spent fuel rods are usually stored outside of the nuclear reactor containment buildings in *water-filled pools*—deep, steel-lined concrete basins filled with water (Figure 6.23a).

After about 5 years, when they have cooled to the point where they will not melt their containers, the highly radioactive fuel rods can be moved out of the pools and into more secure, thick-walled steel and concrete containers called *dry casks* (Figure 6.23b). However, dry casks are not consistently used in Japan or in the United States because they add to the high costs of nuclear power.

Some of these pools and casks are located inside protective nuclear reactor containment buildings and some are stored outside of these structures in more vulnerable buildings. In 2005, the U.S. National Academy of Sciences reported that the pools and casks at 68 of the 104 U.S. nuclear power plants were highly vulnerable to sabotage or terrorist attack. In addition, more than half of all Americans live within 75 miles of such a spent-fuel storage site, according to the Institute for Resource and Security Studies and the Federation of American Scientists.

A spent-fuel pool typically holds 4 to 10 times as much radioactivity as the radioactive core of a reactor holds. If the water in the pool drains because of an accident or an act of terrorism, the exposed rods can catch fire and release radioactivity into the atmosphere, which is apparently what happened in two of Japan's Fukushima Daiichi reactors (Figure 6.22). Water-filled pools and dry casks are intended to be temporary storage devices. No acceptable form of permanent storage has yet been devised for spent nuclear fuel. Later in this module, we discuss this issue further.

High costs. World Bank economists and many investors note that no conventional or proposed newer nuclear power plant can compete in today's energy market unless it is backed by large government subsidies, tax breaks, government guaranteed loans, or all of the above. This is because of the very low net energy yield of the nuclear fuel cycle and the high construction and maintenance costs that are necessary for preventing nuclear power plant accidents. Adding to these costs are the processes for safely storing radioactive wastes and retiring the worn-out nuclear power plants.

Nuclear power proponents argue that the cost of nuclear power is not high. However, most industry cost estimates look only at plant operations, not the entire

FIGURE 6.23 Radioactive spent fuel rods are stored first in (a) water-filled pools. Later they can be moved to (b) dry casks.

U.S. Department of Energy/Nuclear Regulatory Commission

U.S. Department of Energy/Nuclear Regulatory Commission

nuclear fuel cycle, and they fail to include the government subsidies, tax breaks, and construction loan guarantees that help shield this energy resource from free-market competition. Also, no insurance company will fully insure a nuclear power plant, so taxpayers must finance insurance for the nuclear power industry, which means the industry does not have to include full insurance costs in its estimates.

Spread of nuclear weapons. For several decades, the United States, Great Britain, France, and Russia have been selling commercial and experimental nuclear reactors as well as the technology for processing uranium fuel to many other countries. In 2010, John Holdren, Director of the White House Office of Science and Technology Policy, reported that 57 countries have used civilian nuclear power technology to develop nuclear weapons or the knowledge needed to build them. Some critics see this as an important reason to halt the building of new commercial or experimental nuclear reactors and nuclear fuel enrichment facilities throughout the world.

KEY idea   Except for the United States, the former Soviet Union, and Great Britain, every nation that has gained nuclear weapons or the knowledge and ability to build them has done so by using civilian nuclear power technology.

Nuclear Power, Imported Oil, and Climate Change. Proponents of nuclear power say that it will reduce dependence on imported oil and does not produce climate-changing greenhouse gases. Opponents say that these are misleading arguments.

Using nuclear power instead of oil to produce electricity does very little to reduce the use of domestic or imported oil for two reasons. First, most domestic and imported oil is converted to gasoline and diesel fuel for motor vehicles, which has nothing to do with nuclear power. Second, oil is burned to produce only about 1% of the electricity generated in the United States and similarly small amounts of the electricity produced in other countries.

According to the nuclear power industry and the popular press, using nuclear power will reduce the threat of climate change by drastically reducing CO_2 emissions. If all we had to worry about were a nuclear reactor's operations (Figure 6.12), the use of nuclear power would reduce overall CO_2 emissions by quite a bit, because a nuclear reactor emits no CO_2. However, almost every step in the nuclear fuel cycle produces some climate-changing CO_2. While

this amounts to only about one-sixth of the CO_2 emissions of coal-burning power plants, per unit of electricity generated, nuclear power is certainly not carbon-free. Still, by keeping the focus on nuclear plant operations and ignoring the entire fuel cycle, the nuclear power industry has mounted a successful public relations campaign to convince people that nuclear power is carbon-free.

CONSIDER this

- According to a 2007 report by the Oxford Research Group, a new nuclear reactor would have to be built every week for about 70 years in order for nuclear power to have an effect in slowing climate change.

- Currently, it takes 4 to 5 years to build a nuclear power plant, and globally, the number of reactors being built will barely replace those being retired.

Let's REVIEW

- What are the five major obstacles blocking the development of nuclear power?

- Why is the nuclear fuel cycle considered to be inefficient?

- Explain the general safety features of conventional nuclear reactors and describe the major nuclear power plant accidents of 1979, 1986, and 2011.

- How long must the spent fuel rods and other radioactive wastes be safely stored? Why are the water-filled pools now storing spent fuel rods a potential threat to public safety?

- Why are the costs of using nuclear power so high? Explain why the nuclear power industry requires large and continuing government subsidies and tax breaks.

- Describe the relationship between the nuclear fuel cycle and the spread of nuclear weapons technology.

- Explain why nuclear power (a) does not lessen the need for oil-importing countries to import oil, and (b) is not a carbon-free source of energy.

What Can Be Done?

Choosing an Energy Future

In order to deal with the environmental and health problems caused by our use of various energy resources, we need to evaluate each potential resource in terms of its net energy yield, its overall benefits, and its costs, including environmental and health costs. Before we turn to such an evaluation process (see *The Big Picture*, pp. 168–169), let's look at some of the possible solutions to the problems we have explored in this module.

Reducing Air Pollution Emissions

There are two basic approaches to dealing with air pollution caused by the burning of fossil fuels in power and industrial plants and in motor vehicles. One approach is to control or clean up the pollutants, and the other is to prevent or sharply reduce the production of the pollutants as summarized in Figure 6.24.

Environmental analysts argue that prevention strategies are by far the better options compared to control and cleanup strategies. They argue that if pollutants are not created, we don't have to deal with them, but if we produce them, efforts to control them or clean them up can be ineffective and inconsistent, and can ultimately fail as energy use increases.

Laws can be used to encourage either method of dealing with air pollution. Governments have passed laws and established regulatory agencies to identify key pollutants and control their inputs into the environment from the burning of fossil fuels and other sources. In the United States, Congress directed the EPA to establish air quality standards for six major pollutants. Between 1980 and 2008, the combined emissions of these air pollutants dropped by 54%.

However, there are still two serious air pollution problems related to the burning of fossil fuels in the United States. First, for almost three decades, roughly 20,000 older U.S. coal-burning power and industrial plants, oil refineries, and cement plants have not been required to meet the air pollution standards for new facilities. This gives the owners of these plants an incentive to keep them in operation instead of building more modern facilities. Second, the U.S. Congress has failed to raise mileage standards for new motor vehicles to match the much higher standards of Europe, Japan, and China.

Another approach, other than to regulate, is to use the marketplace to help reduce air pollution emissions, such as is done in *emissions trading*, or *cap-and-trade* programs. For example, coal-burning power plants in the United States are allowed to buy and sell pollution rights for emissions of sulfur dioxide (SO_2), which is a threat to human health and a major contributor to acid deposition. Each year, a plant gets rights to emit a certain quantity of SO_2. If a plant emits less than its allotted emissions, it earns pollution credits. Its owner can use those credits to offset excessive emissions at other plants it owns, or it can sell them to other utility companies or private parties. Between 1990 and 2006, this market-based approach helped to cut SO_2 emissions from U.S. power plants by 53% at a much lower cost than that of the regulatory approach.

According to critics, however, this system allows utility companies with dirtier power plants to buy their way out of reducing their harmful emissions and it results in high SO_2 levels in certain areas. They also point out that the success of any emissions trading system depends on setting the cap low enough to get a significant overall reduction in emissions and then lowering the cap every few years to promote innovative technologies for preventing and controlling the emissions.

Control and Cleanup	Prevention

Disperse emissions with tall smokestacks	Reduce coal use
	Remove sulfur from coal
Set and enforce strict emission standards	Reduce car use by walking, biking, and using mass transit
Remove pollutants from smokestack emissions	Improve vehicle fuel efficiency
Remove pollutants from vehicle emissions	

© 2014 Cengage Learning

FIGURE 6.24 These are the major ways to control or clean up (left), or to prevent (right), air pollution from coal-burning power and industrial plants, and from motor vehicles.
Photo credits: Dudarev Mikhail/Shutterstock.com (left); Hemera/Thinkstock (right).

Reducing Carbon Dioxide Emissions

The most obvious way to deal with the serious problem of atmospheric warming and projected climate change is to shrink our carbon footprint by sharply reducing our unnecessary waste of energy. (About 43% of the energy used in the United States is wasted unnecessarily, according to the DOE.) We could also shift from using high-carbon energy resources such as coal and oil to using a mix of low-carbon renewable energy resources such as sunlight, wind, flowing water, and heat stored in the earth's interior. There is a growing movement around the world to oppose the continued use of dirty fossil fuel resources, for a variety of reasons (see *Making a Difference*, at right).

MAKING A difference

Maria Gunnoe: Fighting to Save the Mountains

In the 1800s, Maria Gunnoe's Cherokee ancestors arrived in what is now Boone County, West Virginia. Her grandfather bought the land where she now lives. In 2000, a ridge above that land was blasted away for a coal-mining operation, and Gunnoe's land now sits below a 10-story-high pile of mine waste. With the soil and vegetation gone, rains run quickly off the ridge, and since 2000, her land has been flooded seven times. Her yard has been covered by toxic sludge and her well is contaminated, making her groundwater unfit for drinking, cooking, and bathing.

This story is common in West Virginia, where more than 500 mountaintops have been blown apart. The air around such sites is polluted with coal dust during mining operations, the valley streams are buried, and the groundwater is often contaminated by arsenic and other toxic chemicals. Nineteen medical studies have found that higher rates of cancer and birth defects in the region are likely caused by this pollution.

Gunnoe (Figure 6.A), mother of two and former waitress, decided to fight the powerful coal companies that were blowing up the mountains. Since 2000, she has worked tirelessly as an advocate for protection of mountain ecosystems and com-

munities. She has been threatened and her property damaged, but she has kept up the fight. In 2009, she received the Goldman Environmental Prize, considered the "Nobel Prize for the Environment."

Gunnoe's inspiring work has led to tougher regulations in some areas. In 2012, she testified before a congressional committee, saying: "When mountaintop removal is permitted near your home, you will soon be forced to leave what is the birthplace of your family and your children's birthrights as heirs to your family's land.... Why is it acceptable to depopulate our communities and culture, poison our water and air, and leave us to die in a post-mining wasteland for temporary jobs and energy?"

FIGURE 6.A Maria Gunnoe

However, this prevention approach presents us with serious scientific, political, and economic problems because we get 87% of the commercial energy (92% in the United States) that we use by burning fossil fuels. One solution is to use fairly abundant natural gas as a fuel to help us make the shift from using high-carbon coal and oil and nuclear power to using a mix of improved energy efficiency and low-carbon renewable energy resources. However, greater dependence on natural gas may not help to reduce greenhouse gas emissions as much as was once thought, unless we control the numerous leaks of methane from the natural gas production and distribution system.

Let's REVIEW

- List four air pollution cleanup/control strategies and four prevention strategies.
- Explain how emissions trading programs work and give an example.
- How can we reduce CO_2 emissions by shrinking our carbon footprint? What is one problem presented by this approach?

Dealing with Nuclear Wastes

After 60 years of research and development, we still don't have a scientifically and politically acceptable way to safely store intensely radioactive and long-lived nuclear wastes for thousands of years.

Scientists have suggested several ways to deal with nuclear wastes. For example, some suggest converting harmful radioactive wastes into less harmful materials, but despite several decades of research, no such process has been developed. Other scientists have suggested launching such wastes deep into space or into the sun. But the risk of an accident would be unacceptable, considering the number of rocket launch accidents that have occurred throughout the history of the space program. For example, the 1986 explosion of the space shuttle *Challenger* after it was launched made it clear that such an accident would scatter radioactive debris far and wide.

One way to reduce the production of nuclear wastes would be to reprocess the radioactive wastes from spent fuel rods and from military sources into usable uranium

Evaluating Nonrenewable Energy Resources

In evaluating the possible mix of current and future energy resources, we need to recognize that any energy resource has a mix of benefits and drawbacks, as summarized here. The key question for evaluating each resource is whether or not its major advantages outweigh its major disadvantages.

Energy experts ponder this question every day. Now it's your turn. The resources represented here are the most widely used sources of nonrenewable energy. For each of them, decide whether or not the pros outweigh the cons. Then, rank them in order from most to least useful.

Stephen Coburn/Shutterstock.com

- Ample supply for several decades

- High net energy yield (but decreasing)

- Low land disruption

- Releases air pollutants and CO_2 when produced and burned

- Oil spills can severely pollute water

- Relies on a vulnerable international supply system

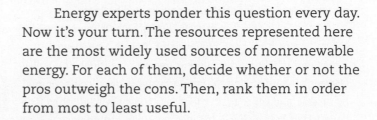

Pros   **Cons**

Conventional Crude Oil

- Large potential supplies in some countries (Canada and U.S.)
- Easily transported within and between countries
- Can provide domestic supplies for North American countries

- Low net energy yield

- Releases air pollutants and CO_2 when produced and burned

- Severe land disruption and high rates of water use

Pros   **Cons**

Heavy Oil from Tar Sands and Oil Shale

iStockphoto/AdShooter

Alexander Kalina/Shutterstock.com

- High net energy yield
- Emits less air pollutants and CO_2 than other fossil fuels when burned
- Ample supplies in the U.S. and in some other countries

- Low net energy yield for liquefied natural gas (LNG)
- Releases air pollutants and CO_2 when produced and burned and extraction can pollute groundwater
- Difficult and costly to transport from country to country

Pros | **Cons**

Conventional Natural Gas

- High net energy yield
- Very large potential supplies in many countries
- Inexpensive when environmental costs are not included

- Mining it severely disturbs land and pollutes water
- Emits large amounts of air pollutants and CO_2 when produced and burned
- Expensive when environmental costs are included

Pros | **Cons**

Coal

a.toni halim/Shutterstock.com

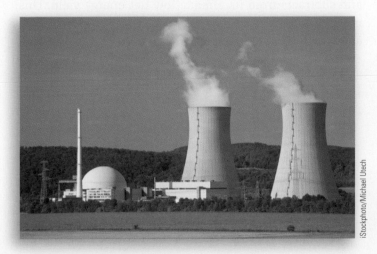

iStockphoto/Michael Utech

- Large fuel supply
- Low risk of accidents in modern plants
- Lower CO_2 emissions than coal

- Very low net energy yield and high overall cost
- Produces long-lived, intensely radioactive wastes
- Promotes spread of nuclear weapons

Pros | **Cons**

Nuclear Power Fuel Cycle

fuel. This expensive process involves extracting fissionable, bomb-grade plutonium from the wastes. However, if such plutonium were to fall into the wrong hands, it could be used to make nuclear weapons. This dangerous threat to national and global security is the main reason why the United States, after spending billions of dollars on research and development, abandoned the reprocessing option in 1977. However, France, Russia, and several European nations still reprocess their nuclear wastes.

According to most experts, the most acceptable option is to bury nuclear wastes deep underground in a geologically stable area using a specially designed vault that would contain the materials and not allow them to leak into the soil, groundwater, or atmosphere. This vault would also have to be able to withstand shocks from earthquakes and human-made explosions, and remain dry for at least 10,000 years. At this point, no such large-scale storage facility has been built and no geologists can even come close to guaranteeing such safety for 10,000 years, which is at least 400 human generations.

In 1985, the U.S. government approved plans to build such a facility in Nevada at a government-owned site known as Yucca Mountain, about 100 miles northwest of the city of Las Vegas. But after the U.S. government and nuclear industry spent more than $10 billion on evaluation and preliminary development of the site, the project was abandoned in 2011 for a combination of scientific and political reasons and a panel was established to evaluate other options.

Dealing with Older Nuclear Power Plants

In a nuclear reactor, after about 40 years of intense neutron bombardment along with high temperatures and pressures, many metal parts in the reactor core become brittle and corroded, and are subject to failure. Some of these parts can be replaced, but at great cost. More than half of the world's nuclear power plants are 40 years old or older, and this includes most of the U.S. plants, which have reached or will soon reach the expiration date of their original government-approved 40-year operating licenses.

However, under political pressure from utility companies, by 2011, the U.S. Nuclear Regulatory Commission (NRC) had extended the operating licenses for 66 of the 104 U.S. nuclear reactors from 40 years to 60 years and it was reviewing extension applications for 16 other reactors. This allowed plant owners to continue to make profits by running the old plants while avoiding the expenses of shutting down and dismantling the highly radioactive reactors. Opponents of these license extensions contend that they could increase the risk of nuclear accidents in aging reactors.

In 2011, the Associated Press (AP) reported the results of its one-year study of nuclear industry and NRC records related to the safety regulations for aging nuclear power plants in the United States. They found records of numerous incidences of cracked concrete and metal parts, clogged water lines, electrical problems, leaky valves and seals, and severe corrosion of some reactor components and underground pipes. They also found that the NRC had often allowed plant operators to delay repairs and inspections.

Several analysts have since called for much tighter regulations on aging plants along with a review of the extension of operating licenses for many plants. They have also called for a serious examination of the relationships between plant owners and nuclear industry regulators to ensure that regulations are actually enforced. Several experts have noted that ineffective regulation was apparently part of the problem that led to Japan's Fukushima power plant accident in 2011 (Figure 6.22).

As more than half of the world's 435 commercial nuclear reactors near the end of their 40- to 60-year lifespans, we face another radioactive waste issue—the problem of what to do with their radioactive parts. Each of the plants will have to be *decommissioned*, or retired from service, and taken apart (Figure 6.25). This is the final step in the cradle-to-grave nuclear fuel cycle (Figure 6.13). The remainder of the world's 435 plants will have to be decommissioned by the middle of this century. As with other forms of nuclear waste, we have yet to figure out how to safely store these tons of radioactive concrete, metal, and other materials for thousands of years.

Regardless of what method is used, decommissioning adds a great deal to the cost of the nuclear fuel cycle and subtracts from its already-low net energy yield. This makes nuclear power even less able to compete in the energy marketplace without large government subsidies.

Let's REVIEW

- What are four suggestions from scientists for dealing with highly radioactive wastes and what are the problems that could arise for each of them?
- Explain how governments have so far dealt with the problem of aging nuclear power plants. Define and explain the challenges of decommissioning nuclear power plants.

FIGURE 6.25 This nuclear power plant in Essex, UK, has been closed and is being decommissioned. Its radioactive parts will have to be stored somewhere for hundreds to thousands of years.

A Look to the Future

The Role of Nuclear Energy

There is considerable controversy over the future role of nuclear power in providing the world with a greater share of its electricity. It has some major advantages (see *The Big Picture*, p. 169) over coal but it also has some serious disadvantages.

In 2010, around the world, 63 new nuclear reactors were under construction. This is far from the number needed just to replace the aging reactors that will have to be decommissioned in coming years. Another 143 reactors are planned, but even if they are completed after a decade or two, they will not replace the 285 aging reactors that must be retired around the world over the next two decades.

For this reason, the International Energy Agency projects that the contribution of nuclear power to the world's electricity supply will decline over the next several decades as old plants are retired faster than new ones can be built and brought on line. This explains why the amount of electricity produced by the world's nuclear power plants has leveled off (Figure 6.26) and why the annual growth of electricity production from nuclear power decreased by almost 4% between 2007 and 2011.

To deal with this challenge, the nuclear power industry has proposed a new generation of smaller nuclear power plants that it argues would be safer, cleaner, and somewhat easier to build than conventional nuclear power plants.

The proposed designs are supposed to have built-in safety features that make explosions and radioactive emissions almost impossible. Most of these designs involve *high-temperature, gas-cooled (HTGC) reactors* that avoid some of the safety problems and the threat of meltdowns associated with conventional water-cooled reactors.

One of the proposed new reactor designs now being promoted by the nuclear industry is the *pebble bed modular reactor (PBMR)* (Figure 6.27, p. 172). In a PBMR, the fission reactions take place within a pile of fuel pellets referred to as the *pebble bed*. Heat from this fissioning fuel is removed by nonreactive helium gas that flows through the spaces

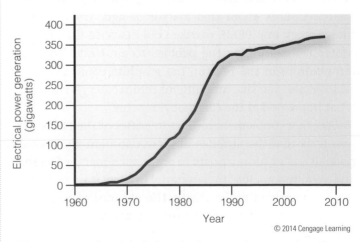

© 2014 Cengage Learning

FIGURE 6.26 The global electrical generating capacity of nuclear power plants rose steadily between 1960 and 1990 and then gradually leveled off. (Data from International Energy Agency and Worldwatch Institute)

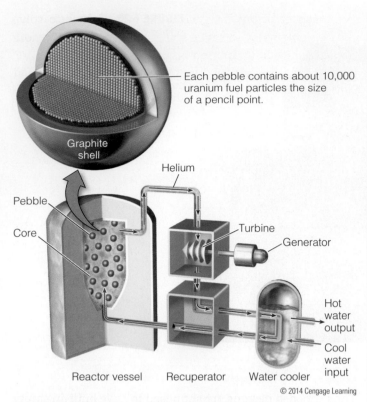

Each pebble contains about 10,000 uranium fuel particles the size of a pencil point.

Graphite shell

Pebble

Core

Helium

Turbine

Generator

Hot water output

Cool water input

Reactor vessel   Recuperator   Water cooler

© 2014 Cengage Learning

FIGURE 6.27 The *pebble bed modular reactor (PBMR)* is one of several new and smaller reactor designs that some nuclear engineers say should improve the safety of nuclear power and reduce its costs.

between the pebbles. The resulting hot helium gas, somewhat like steam, is used to spin a turbine that generates electricity.

Because the PBMR is configured so that it cannot melt down, proponents contend that it and other newer reactor designs eliminate the need for an expensive containment shell and emergency core cooling system. However, Edwin Lyman and several other nuclear physicists have serious doubts about this reactor design. They contend that a crack in a PBMR reactor could expose the graphite protective coating on the pebbles to air. Under the high temperature in the reactor, the graphite could catch fire and release massive amounts of radioactivity—similar to what happened at the Chernobyl reactor.

Another problem, according to critics of the PBMR proposals, is that the absence of a containment shell would make it easier for terrorists to enter such reactor facilities and steal the nuclear fuel material or blow it up to release large amounts of radioactivity. The absence of a containment shell would also make the reactor more vulnerable to earthquakes, airplane crashes, and bombs.

In addition, this technology would create about 10 times as much high-level radioactive waste per unit of electricity generated as a conventional nuclear reactor produces.

However, one advantage of the PBMR design is that its waste is less radioactive than that of conventional reactors, so the necessary storage time would be somewhat less.

Critics point out that Germany, South Africa, and the United States each built experimental pebble bed reactors but abandoned the technology because of technical problems, safety concerns, and high costs. China is building two commercial-size pebble bed reactors to evaluate this technology.

Some scientists have suggested replacing the current uranium-based reactors with new designs based on the element thorium. They contend that such reactors would be safer and cheaper to build, and would produce less radioactive waste. Other proposed new reactor designs use gravity rather than pumps to move cooling water, and they feature safety valves that automatically kick in without operator control. China plans to build some of these reactors to test this design approach.

However, none of these new designs eliminates the expense and hazards of long-term radioactive waste storage, power plant decommissioning, and the spread of knowledge and materials for the production of nuclear weapons. Also, each of them would still require massive government subsidies and loan guarantees.

Another possibility for nuclear power is **nuclear fusion**, in which two hydrogen nuclei are forced together at extremely high temperatures until they fuse to form a heavier nucleus (helium), releasing energy in the process. This reaction is the source of sun's energy that supports life on the earth. For over 60 years, scientists have been trying to produce this reaction with the goal of using controlled nuclear fusion to produce high-temperature heat and electricity.

According to some scientists, controlled nuclear fusion could provide an almost infinite amount of electricity and high-temperature heat. It could also be used to destroy toxic wastes and create hydrogen fuel by decomposing water. There would be no possibility of the type of accident or meltdown that can occur with nuclear fission. In addition, bomb-grade materials are not required for fusion energy, so there would be no risk of spreading nuclear weapons technology.

The problem is that after 60 years of research costing taxpayers many billions of dollars, scientists have not yet found a way to make controlled nuclear fusion work in such a way that it will generate more energy than it requires. In other words, so far, nuclear fusion has a negative net energy yield. Also, the estimated cost of building and operating a commercial fusion reactor (even with huge government subsidies) is several times the cost for a comparable conventional fission reactor. Unless there is some unexpected scientific breakthrough, it is unlikely that nuclear fusion will be a significant energy resource in the near future.

So where does this leave us? All experts agree that we will be using fossil fuels for many years to come. But many of them argue that during the next few decades, we would be wise to shift from burning coal to using less environmentally harmful natural gas, assuming the greenhouse gas emissions from the production and use of natural gas can be sharply reduced. We can also use nuclear fission, especially if newer-generation reactors turn out to be safe and affordable. But the number of increasingly expensive nuclear plants planned for the next few decades will barely replace the older plants that must be retired, also at great cost. In addition, we still don't have a scientifically and politically acceptable solution for long-term storage of our growing piles of nuclear waste.

As an alternative, one of our largest, least expensive, and most readily available sources of energy is the energy we waste—about 43% of all commercial energy used in the United States. We can also shift to a mix of low-carbon renewable energy resources such as solar energy, wind power, and the earth's internal geothermal energy.

Let's **REVIEW**

- Explain why experts expect little growth in the net amount of electricity produced by nuclear power plants during the next several decades.
- Define *nuclear fusion*. Explain why nuclear fusion would be an attractive energy resource and why it is not likely to be a significant resource in the near term.

What Would You Do?

Just about all of the major solutions to our energy problems depend on what individuals do and on the choices we all make in our personal lives. What each of us does in almost every aspect of our lives each day helps either to solve these problems or to make them worse. More and more people are making this sense of personal responsibility a guiding principle in their lives. Here are some examples of how they are applying this principle in doing their part to solve our energy and environmental problems:

Reducing their use of fossil fuels, especially coal.
- Many are reducing car use by walking, biking, car-sharing, and using mass transit.
- The use of plug-in hybrid and fully electric cars is growing. Ideally, the sources of the electricity used for recharging car batteries in the future will be mostly natural gas and low-carbon wind power, solar power, and geothermal energy.
- Some people are eating less meat, as it takes large inputs of fossil fuels to produce and deliver meat to consumers.

Cutting energy waste.
- Energy-conscious car buyers are choosing only models that get at least 40 miles per gallon.
- Many homeowners are heavily insulating their houses and plugging air leaks.
- Sales of energy-efficient windows, lighting, heating and cooling systems, and appliances have been growing steadily.

Thinking about their daily energy choices.
- It takes energy to make and transport anything, so energy-conscious people are buying locally made products and trying to avoid those that take a lot of energy to produce, deliver, and use.
- Many people are reusing more things, recycling as much as possible, and buying recycled and used items.
- Where electric utilities offer electricity generated by renewable sources such as wind and solar energy (usually for just a few cents or dollars more per month), many have chosen that option; some then look for ways to save electricity so they can offset any increases in their electric bills.

KEYterms

THINKINGcritically

1. Do you think that net energy yields should play a large role in the evaluation of energy resources for purposes of deciding which energy resources to subsidize or to invest in? If not, what are the other factors that you think should be more important in such evaluations? Explain.

2. Do you think the United States (or the country where you live) should phase out dependence on imported oil? Write up a three-point program that you would use to accomplish this. List two major challenges you would face in achieving this goal and explain how you could deal with each.

3. Do you think the United States (or the country where you live) should reduce its overall use of coal? If so, by what percentage would you reduce it, and what resources would you use in place of coal? What affect would this have on the price of electricity? If you do not think so, how would you deal with the harmful environmental and health effects of continuing to use coal? Explain your reasoning.

4. If you were in charge of the national energy policy, would you increase or decrease the use of nuclear power to produce electricity in the United States (or in the country where you live)? Explain. If you would cut its use, what resources would you use in place of nuclear power?

5. Suppose you were debating someone on the topic of clean coal. Give three arguments in favor of and three arguments to oppose the idea that coal can be made into a clean energy resource. Do the same exercise for the argument that nuclear power is a carbon-free source of energy.

6. Identify three practices from your daily life that result in your consuming energy from fossil fuels. If you were to decide to reduce your dependence on fossil fuels, what steps would you take? Write a short plan that includes these steps and explain your reasoning.

LEARNINGonline

Access an interactive eBook and module-specific interactive learning tools, including flashcards, quizzes, videos and more in your Environmental Science CourseMate, accessed through **CengageBrain.com**.

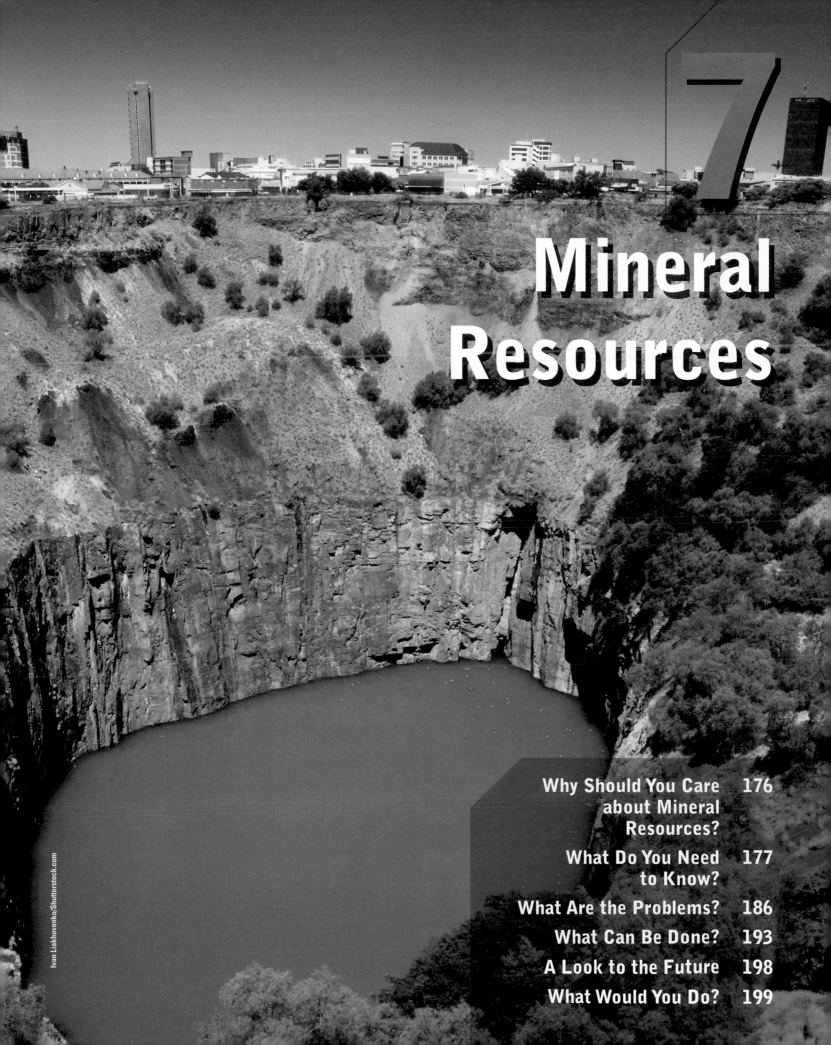

7

Mineral Resources

Why Should You Care about Mineral Resources?

About 8,000 years ago, some primitive human must have been drawn to pick up, from an eroded hillside or perhaps from the sandy bed of a shallow stream, a yellowish stone that was glinting in the sunlight. That person had discovered gold—the first metal to be used by humans, according to anthropologists.

Since that day, humans have discovered and learned how to extract more than 100 useful mineral resources—many of them metals such as gold and copper—from deposits in the earth's crust. These include sand that is used to make glass and bricks, and gravel that is used to make concrete and roadbeds. Another mineral resource is limestone, which is used as a building material and often is crushed to make a base material for roadways.

Today, almost everything that we use to support our lifestyles is made at least partly from mineral resources or produced by machinery and tools made from mineral resources. Look around you and see if you can find items that are not made from, or made with the use of, mineral resources.

Mineral resources have provided humanity with a great many benefits. However, the processes of using these resources—extracting them from the earth's crust, processing them for use in manufacturing the products we enjoy, and discarding these manufactured items after they are used—have a number of harmful environmental effects. Clearing forests or other vegetation that covers the ground where these minerals are buried eliminates large areas of wildlife habitat. Digging the minerals up disfigures the land in dramatic ways (see module-opening photo). And the use of mineral resources, from the mining of minerals to the disposal of manufactured products, can add greatly to air and water pollution and to emissions of climate-changing *greenhouse gases* such as carbon dioxide into the atmosphere.

In this module, we introduce you to mineral resources, examining how they are formed and how we extract them

FIGURE 7.1 Metals and other mineral resources are key components of (a) smart phones and tablet computers, (b) solar cells and wind turbines, (c) electric cars, (d) all aircraft, including this F-15 fighter jet, and thousands of other commonly used items.

from the earth. We then explore the environmental costs associated with the use of mineral resources. We also consider some ways to extend the lives of mineral supplies as well as ways to use mineral resources more sustainably. Finally, we look at how such solutions can be implemented on all levels, from the international arena down to each of our own personal, daily lives.

What Do You Need to Know?

The Earth's Major Geological Features

There are powerful natural forces at work under the earth's surface and aboveground. The scientific study of the structure of the earth and of the processes that take place in its interior and on its surface is called **geology**.

Billions of years ago, the earth was mostly a mass of hot liquid rock. As it slowly cooled, it separated into three zones (Figure 7.2). The extremely hot innermost zone, called the **core**, consists of a solid center, surrounded by a zone of hot, fluid rock. A thick zone around the core, called the **mantle**, makes up about 84% of the earth's volume. It consists almost completely of solid rock that is hot enough to be elastic and to be deformed by heat and pressure.

The earth's outermost zone is a thin layer of solid material called the **crust**. As the earth has aged, various physical and chemical processes affecting the crust have led to the formation on its surface of land masses (part of the *geosphere* in Figure 7.2), bodies of water (the hydrosphere), an envelope of gases (the *atmosphere*), and a variety of life forms (residing in the biosphere).

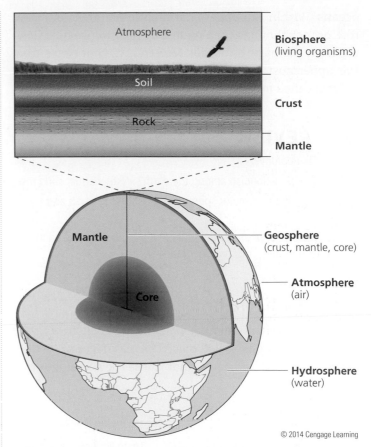

FIGURE 7.2 The earth is made up of four major spheres, containing rock, water, gases, and life forms.

Most of the processes that we will examine in this module take place within the earth's crust and upper mantle (Figure 7.3). The crust is made up of two major types of material: *continental crust*, which lies beneath the planet's land masses, and *oceanic crust* found beneath the planet's

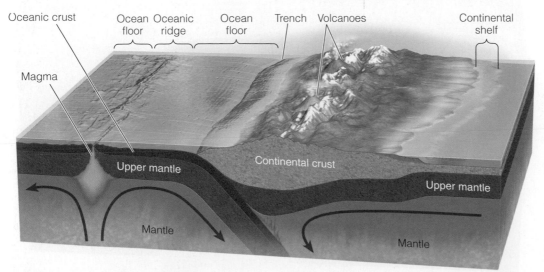

FIGURE 7.3 This diagram shows the key features of the earth's crust and upper mantle.

© 2014 Cengage Learning

oceans. Within the uppermost mantle are areas of *molten rock*, or rock that has been melted by the intense heat flowing from the earth's interior. Molten rock that seeps from the uppermost mantle into chambers within the earth's crust is called *magma*.

Let's **REVIEW**

- What is *geology*?
- Define and distinguish among the earth's *core, mantle,* and *crust*.
- What are the four major spheres that define the earth and its life?
- Distinguish between continental crust and oceanic crust and between molten rock and magma.

Major Geologic Processes

Within the earth's mantle, hot, elastic rock flows very slowly within loops, called *convection cells* (Figure 7.4). Geological research indicates that these powerful internal flows of heat and rock have broken the earth's crust into a number of gigantic rigid plates, called **tectonic plates** (Figure 7.5). In response to tremendous forces generated by the convection cells, these massive plates move around on top of the uppermost mantle. Tectonic plates move extremely slowly at an average rate of about 1 to 4 inches per year.

CONSIDER this You are riding on a gigantic surfboard—a tectonic plate that is moving at about the rate at which your fingernails grow.

As the tectonic plates move, some of them collide at *convergent plate boundaries*, while others move away from one another at *divergent plate boundaries*, and still others grind along against each other at their edges at *transform plate boundaries*. It is at these boundaries between plates, where immensely powerful forces are at work, that much geologic activity takes place.

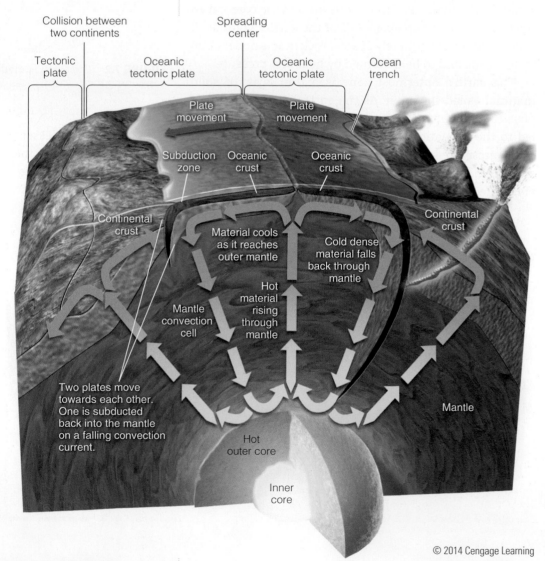

FIGURE 7.4 Convection cells of heat and elastic rock moving within the earth's mantle are powerful enough to have fractured the crust into several huge pieces, called *tectonic plates*. Geologists have identified the outlines of these major tectonic plates (Figure 7.5).

© 2014 Cengage Learning

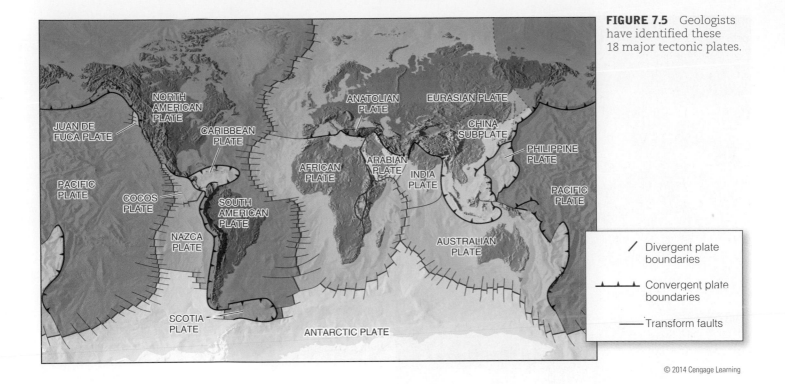

FIGURE 7.5 Geologists have identified these 18 major tectonic plates.

Legend:

/ Divergent plate boundaries

⊥⊥⊥ Convergent plate boundaries

— Transform faults

For example, when two plates grind against each other, they form what is called a *transform fault*, or simply a *fault*. Sooner or later, the plates get stuck and stop moving while the forces beneath them continue pushing, building up tremendous amounts of potential energy. When the adjoining plates break loose again, they shift suddenly and release a huge jolt of kinetic energy that shakes the crust. Such events, known as *earthquakes*, can be very destructive.

An earthquake on the ocean floor can generate a series of waves known as a *tsunami*. As these rapidly moving waves approach land, they slow down but build up in height and combine to form very high, massive waves that can cause great destruction. This happened in 2011 off the coast of Japan after a catastrophic earthquake (Figure 7.6a) caused a tsunami that greatly damaged the city of Fujitsuka (Figure 7.6b), leading to thousands of deaths and causing the second-worst nuclear power plant accident in history.

FIGURE 7.6 A transform fault **(a)** runs just offshore of Japan's northeastern coast. In 2011, a catastrophic earthquake centered just to the west of the fault line (see red circles) caused a massive and very destructive tsunami **(b)** that flooded Fujitsuka, Japan. In this satellite image, red represents vegetation.

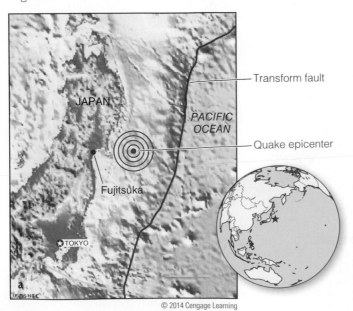

FIGURE 7.7 Volcanoes like this one erupt when magma that is pushed from the uppermost mantle toward the earth's surface under tremendous pressure breaks through the surface and flows out as lava.

The great forces produced by the movement of tectonic plates, combined with the flow of magma into chambers in the earth's crust can cause the formation of *volcanoes*: ruptures in the earth's surface that can emit gases, ash, and molten rock called *lava*. After lava flows out of a volcano, it cools to form a volcanic rock. Some volcanoes erupt slowly and emit mostly lava (Figure 7.7), while others erupt more explosively and eject large quantities of solids (mostly ash) and gases into the atmosphere.

CONSIDER this A large volcanic eruption occurred in 1991 when Mount Pinatubo in the Philippines exploded and threw enough ash and other particles into the atmosphere to darken the skies and cool the earth's atmosphere for 15 months.

The extremely slow movements of tectonic plates have caused masses of land to split apart and join together over many millions of years. Geological evidence indicates that about 200–250 million years ago, there was only one large mass of land (Figure 7.8a), a supercontinent called Pangaea. Since then, the movement of the planet's tectonic plates has split Pangaea apart and eventually resulted in the present-day arrangement of continents (Figure 7.8b).

The movement of the earth's continents on tectonic plates over long periods of time has played an important role in the distribution of the earth's species and in establishing the earth's biodiversity through long-term changes in the populations of various species. The joining of continents allows species to move into new areas and adapt to new environmental conditions. In the long run, this contributes to the formation of new species. When continents separate, a population can be split into two or more populations which then live under different environmental conditions. This can cause one species to gradually evolve into two or more new species. It can also cause the *extinction*, or disappearance, of various species.

a Supercontinent Pangaea

b Continents today

FIGURE 7.8 Over roughly 250 million years, the earth's continents have moved as they rode on their shifting tectonic plates. These movements long ago broke apart the supercontinent Pangaea (a) and the fragments eventually shifted to their present-day arrangement (b).

Let's REVIEW

- Define *tectonic plates* and explain how these plates were formed. Describe three types of boundaries between tectonic plates.

- What is a transform fault and how can it cause an earthquake? What is a tsunami?

- What is a volcano? Describe two major types of volcanic eruptions.

- Explain how the movement of tectonic plates has changed the sizes and locations of the earth's continents.

- Explain how the movement of tectonic plates has affected the distribution and diversity of the earth's species.

Minerals, Rocks, and the Rock Cycle

Over billions of years, the forces of heat and pressure under the earth's surface have forged a variety of materials in the planet's crust that have become important to humans. One major type of such materials is called **mineral**—any substance that occurs naturally in the earth's crust as a solid with a regularly repeating internal structure (a *crystalline* structure). Minerals occur in a variety of forms (Figure 7.9).

Almost all of what we call **rock** is a combination of one or more minerals found in the earth's crust. (There are a few exceptions, but for purposes of this module, we discuss only rocks that are made of minerals.) Some rocks contain only one mineral. For example, quartz (silicon dioxide, or SiO_2) is a mineral, but many people consider it to be a rock. Most rocks consist of two or more minerals.

There are three major types of rock. The type that we most often see is *sedimentary rock*—rock that is found on or near the earth's surface and is made of *sediments*, which consist of eroded rock particles and dead plant and animal remains. Such sediments accumulate in layers on land or underwater, and as these layers get thicker and heavier over thousands to millions of years, the resulting pressure, weight, and chemical reactions convert some of these sediments into rock. Examples of sedimentary rock are *sandstone* (typically formed from deposited layers of sand-sized particles of quartz) and *limestone* (formed mostly from the compacted shells, skeletons, and other remains of dead organisms).

Igneous rock forms on or below the earth's surface when magma flows up from the upper mantle (Figure 7.3) and eventually cools and hardens into mixtures of minerals in the crust or on the surface. Examples are *basalt*, which forms aboveground, and *granite*, which forms underground. Most of the earth's crust consists of igneous rock, but it is often covered by sedimentary rock.

Metamorphic rock forms when existing rock is converted to a new form of rock by high pressure, intense heat, chemically active fluids, or a combination of these agents. Two examples are *marble*, formed when limestone rock is exposed to pressure and heat, and *slate*, formed when mudstone and shale rocks are heated.

All of these types of rock are involved in the slowest of the earth's cyclical processes, called the **rock cycle**—the combination of physical and chemical processes, both on the surface and underground, that change rocks from one type to another. You can trace this cycle, which occurs over millions of years, in Figure 7.10 (p. 182). We should know and care about the rock cycle because over millions of years, it has formed most of the mineral resources that support the world's economies and our lifestyles.

In this simplified model, rock on the surface is eroded by wind and rain, and the resulting sediments build up and eventually are compressed into sedimentary rock. Over eons, some of this rock gets pushed down into the mantle where it can become magma that then rises toward the surface to become igneous rock. At some point, it can also be subjected to the chemical and physical processes that form metamorphic rock. After surface processes such as erosion strip off the materials covering

FIGURE 7.9 These are just a few of the hundreds of minerals found in the earth's crust, many of them used as semiprecious stones in jewelry.

Triff/Shutterstock.com

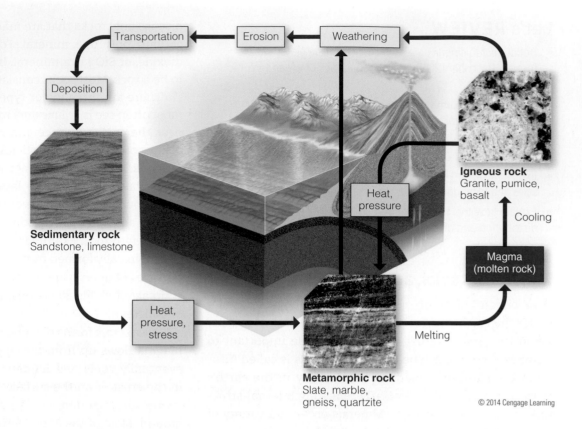

FIGURE 7.10 The *rock cycle* recycles the earth's major types of rock over millions of years and produces most of the nonrenewable mineral resources that we use.

Transportation

Deposition

Erosion

Weathering

Sedimentary rock
Sandstone, limestone

Heat, pressure, stress

Heat, pressure

Igneous rock
Granite, pumice, basalt

Cooling

Magma (molten rock)

Melting

Metamorphic rock
Slate, marble, gneiss, quartzite

© 2014 Cengage Learning

buried metamorphic and igneous rocks, these processes then start to erode and break apart the exposed rocks, and the process has come full circle.

> ## KEY idea
>
> Rocks found in the earth's crust are recycled very slowly, and the rock cycle produces deposits of many of the nonrenewable mineral resources that we use today.

Let's **REVIEW**

- Define and distinguish between *mineral* and *rock*.
- Define and distinguish among sedimentary, igneous, and metamorphic rock and give an example of each type.
- Describe the *rock cycle* and explain why it is important to us.

Mineral Resources

A **mineral resource** is a naturally occurring mineral deposit that we can extract from the earth's crust and use directly or convert into raw materials for use in the manufacturing of products. Because such minerals typically take millions of years to form in the earth's crust, they are classified as *nonrenewable resources* in contrast to renewable resources such as water and topsoil, which are renewed more rapidly.

> ## KEY idea
>
> The earth has fixed supplies of nonrenewable minerals that take millions of years to form.

We have learned how to locate, extract, and use more than 100 different nonrenewable mineral resources. Some of them are made of a single chemical element, such as copper (Cu), gold (Au), and diamonds, which are made of carbon (C). However, most mineral resources occur as *compounds*, or combinations of elements (see Module 1, p. 7–8), such as sodium chloride (NaCl), or table salt, and calcium sulfate ($CaSO_4$), or gypsum, used to make drywall.

Mineral resources can be classified as *metallic minerals*, such as copper and gold, or *nonmetallic minerals*, such as salt and gypsum. We cannot dig up and immediately use most mineral resources because they exist as components of rock. A body of rock that contains a useful amount of a mineral is called an **ore**. An ore with a mineral concentration high enough to be extracted at an affordable cost with current technologies is called **high-grade ore**.

a

b

FIGURE 7.11 Copper is extracted from a high-grade copper ore (a) and used to make a variety of products. Similarly, gold is removed from gold ore (b) and purified to make many valuable products.

A **low-grade ore** contains a much lower mineral concentration and is more difficult and costly to extract.

We depend on a number of metallic mineral resources. Copper, extracted and purified from copper ore (Figure 7.11a), is used for a variety of products including water pipes and electrical wire. Gold is extracted from gold ore (Figure 7.11b), purified, and used to make items such as jewelry. Aluminum is chemically extracted from an ore known as bauxite and is used to make beverage cans and structural material for use in building motor vehicles and aircraft. Without these and many other metallic mineral resources, our lives would be very different.

Widely used nonmetallic mineral resources include sodium chloride (NaCl), or table salt. Quartz (which is mostly silicon dioxide, or SiO_2) occurs in a number of different forms, including amethyst (Figure 7.12a) that is used as a gemstone in jewelry. Another common nonmetallic mineral resource is marble, which is used as a building material and to make statues (Figure 7.12b).

Life Cycle of a Mineral Resource

Mineral resources go through a *life cycle* that includes mining, processing, conversion to products, and disposal or recycling of the products (Figure 7.13, p. 184).

The first step in this cycle is to locate deposits of an ore that is of a high enough grade to be extracted profitably. Then, different mining techniques are used to remove the ore, depending on the type of mineral resource and its location. (We look at various mining methods in the next major section of this module.) The next step in the cycle is to extract the mineral from the ore by using heat or chemical solvents. If the mineral is metallic, the metal is processed in varying ways, depending on how it will be used, and converted into useful products such as steel girders and aluminum cans. Finally, the last step in a mineral's life cycle occurs when mineral products corrode or reach the end of their useful lives and they are discarded as solid waste. In some cases, a mineral's life cycle can be extended through recycling.

FIGURE 7.12 Amethyst (a) is cut into gemstones to make jewelry. Marble (b) is an important building material.

a

b

FIGURE 7.13 Each mineral resource has a *life cycle*.

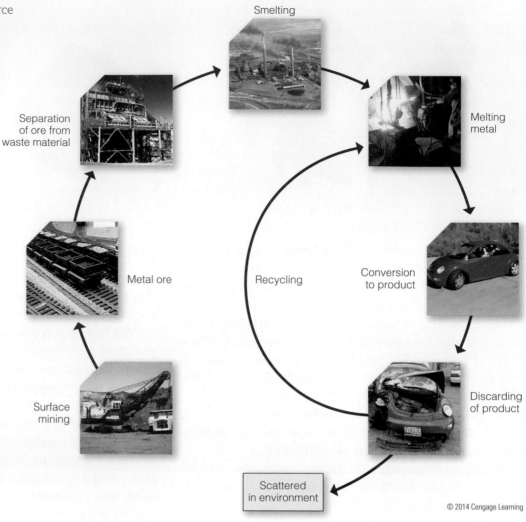

Smelting

Separation of ore from waste material

Melting metal

Metal ore

Recycling

Conversion to product

Surface mining

Discarding of product

Scattered in environment

© 2014 Cengage Learning

Each step in the life cycle of a mineral resource requires using large amounts of other resources such as energy and water. Each step also produces various types of wastes and pollutants that end up in the air, water, and soil, and in some cases, in our bodies, as discussed in the next major section of this module.

Let's **REVIEW**

- Define *mineral resource* and explain why mineral resources are nonrenewable.
- Distinguish between metallic and nonmetallic mineral resources and give two examples of each.
- Define *ore* and distinguish between *high-grade* and *low-grade* ores.
- Describe the life cycle of a mineral resource.

The Economics of Using Mineral Resources

Geologists estimate supplies of a mineral resource in terms of its **reserves**, or identified deposits from which we can extract the mineral profitably at current prices. The global physical supply of any mineral is fixed and nonrenewable on a human time scale. However, reserves are not fixed and can be increased by various means including the discovery of new profitable deposits, a rise in market prices that makes it profitable to extract deposits that were once too costly to extract, and the development of new technologies that make extraction less expensive.

For some mineral resources such as aluminum and iron, there are abundant reserves in the earth's crust. However, deposits of other important mineral resources such as platinum, cobalt, manganese, and chromium are more scarce. Also, deposits of mineral resources are distributed very unevenly among the world's countries.

As we use a nonrenewable mineral resource, we have to think about the rate at which we use it. Because there is a fixed supply, it will only last so long if we keep using it. We have never physically exhausted any reserve of a mineral resource because at some point, called **economic depletion**, it costs more than what the mineral is worth to find, extract, and process more of it. When we reach this point for any mineral resource, we have several options: develop more efficient and affordable mining technologies; use less and waste less of the resource; recycle or reuse products made from the resource; find a substitute for the resource; or do without it.

Technological advances can allow for the profitable mining of more of some mineral resources. For example, the high-grade iron ore from rich deposits in northern Minnesota was economically depleted shortly after World War II. However, by the 1960s, a new process had been developed for the mining of *taconite* (Figure 7.14), a low-grade but plentiful ore that had once been considered a waste rock in the iron-mining process. With that change, iron reserves in Minnesota were expanded greatly, and the taconite mining industry continues today.

We can also make reserves last longer by slowing the rate of their use. For example, by recycling millions of aluminum cans every year, in order to make new aluminum products, we have lessened the rate at which we extract aluminum ore from the earth to make those products. Another way to slow the rate of use of any mineral resource is to find a substitute material from which to make products. For example, by making more soft drink containers out of plastic, we have again reduced the need to mine aluminum for this purpose.

Another development that can extend the life of a mineral's reserve is a rise in the price of the mineral, due to shortages of the mineral or other economic factors. If a company can make more money by extracting and selling the mineral, it can afford to extract lower-grade deposits (Figure 7.14) that were too costly to extract at lower prices. Rising prices can also stimulate a search for new deposits, and they can spur the development of better mining and processing technologies. In addition, they can give buyers, who don't want to pay the higher prices, a reason to extend their supplies by reducing their use and waste of the resource, and by reusing and recycling more of it.

Another factor affecting mineral reserves is the use of government **subsidies**—payments or other forms of economic support designed to help an industry—and tax breaks for mining companies. These forms of support help an industry to keep its mining costs down. For example, the U.S. government gives subsidies and tax breaks to mining companies to encourage them to find and develop deposits of mineral resources. Several other countries, including China and Canada, also provide large subsidies to mining companies. We consider the benefits and drawbacks of such subsidies and other forms of financial assistance for mining later in this module.

iStockphoto/Phil Augustavo

FIGURE 7.14 The layers of reddish rock on this hillside in northern Minnesota are deposits of taconite.

For any mineral reserve, economists often assign an estimated **depletion time**—the number of years that it takes to use up a certain percentage—usually around 80%—of the reserve at a given rate of use. Experts sometimes disagree about depletion times, because they make different assumptions about the size of any reserve and about how fast it is being, or is likely to be, extracted.

Figure 7.15 shows three different general estimates of depletion times for a reserve of a nonrenewable resource such as copper or gold. Each is represented by a *depletion curve*. Curve A shows how long it would take to deplete the resource if it were simply mined at a fairly steady rate with no extensions of the reserve due to reuse and recycling of end products, improved mining technologies, unusual price increases, or new discoveries of the resource. Curves B and C show varying degrees of extension of the reserves due to these factors.

It can be difficult to expand reserves of mineral resources, partly because of the high financial risks involved. Experience shows that when geologists identify 10,000 possible mining sites, typically only about 1,000 of these sites are worth evaluating and only one ends up as a

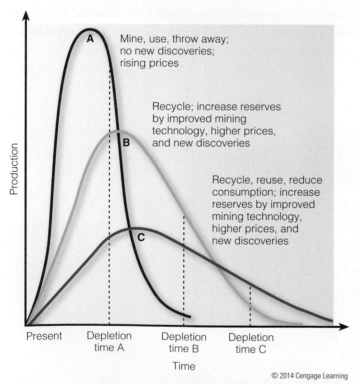

FIGURE 7.15 Different assumptions about the available supplies and rate of use of any nonrenewable mineral resource lead to different depletion curves. When 80% depletion occurs, as shown by the dashed lines, the resource is economically depleted because it costs too much to extract what remains.

productive mine. If there is a shortage of investment capital because of a lagging economy, expansion of reserves becomes even less likely.

Let's REVIEW

- Define *reserve* and *economic depletion*. Why is it important to think about the rate of use of any mineral resource?
- What are four ways to extend reserves of a mineral? What are three ways in which raising the prices of a mineral can extend its reserves?
- Define *subsidies*, and explain how government subsidies can affect mineral reserves.
- Define *depletion time* for a mineral resource and explain why experts sometimes disagree about depletion times.

What Are the Problems?

Environmental Impacts of Surface Mining

The exploitation of mineral resources has made life easier and more enjoyable for billions of people around the world by providing them with countless useful products as well as jobs and other economic benefits. However, there has been a steep price to pay for these benefits in terms of the environmental impacts of mining and processing mineral resources, and discarding their used end products.

First, the mining and processing of metals and other mineral resources require large amounts of energy and water. In arid, drought-prone areas such as the southwestern United States, this heavy use of water strains already over-taxed water supplies.

Second, mining operations are a major source of air and water pollution. The U.S. mining industry adds more toxic emissions to the atmosphere than any other industry. In some years, the industry has been responsible for almost half of all such emissions. And toxic runoff flowing out of unregulated mine sites often erodes topsoil and pollutes surface waters and groundwater.

A third major environmental impact of mining is disruption of land, especially from **surface mining**, any method used to access deposits found near the earth's surface, in which all vegetation, soil, and rock overlying the deposit are stripped away and the mineral deposit is then removed (see *For Instance*, p. 188). The soil and rock overlying a deposit that is to be mined are called **overburden**. Usually, these materials are deposited in piles called **spoils**. In the United States, surface mining is used

FIGURE 7.16 This copper mine near Salt Lake City, Utah, is the world's largest open-pit mine and has been in operation since 1906. The trucks used to haul out the copper ore appear to be tiny here, but each one is at least 23 feet tall.

to extract about 90% of the mineral and rock resources that are mined.

In one type of surface mining, called **open-pit surface mining**, machines are used to open an ever-widening pit from which the resource is removed (Figure 7.16). Such a pit can be miles in diameter and thousands of feet deep by the time the resource is depleted, and these pits are difficult if not impossible to restore. Because they contain little soil, it can take decades for vegetation to take hold. Thus, they often remain as huge eyesores subject to wind and water erosion (see module-opening photo). Precipitation running down through an open pit can wash toxic metals and other pollutants into underlying groundwater supplies.

Another form of surface mining is **area strip mining**, in which large, relatively flat areas of overburden are scraped away to expose the mineral resource, which is then extracted and hauled away. Gigantic machines, including 20-story-high power shovels and house-sized trucks, are used in this process. The end result is often a series of long, high spoils banks (Figure 7.17). Water and wind can erode the materials in spoils banks and pollute the air and nearby bodies of water. Regrowth of vegetation on spoils banks is quite slow because it takes a long time for soil and plants to become established on them.

In very hilly or mountainous terrain, other forms of surface mining are used. One is called **contour strip mining**, in which machines are used to cut broad terraces into hillsides, ripping up the overburden and extracting the mineral. The machines are then moved up the slope where they cut a new terrace and dump the overburden onto the terrace below. This process leaves a steep, exposed wall of dirt at the top of the hill and a series of highly erodible spoils banks laid out across its slope.

Perhaps the most extreme form of surface mining is **mountaintop removal mining**, in which entire mountaintops are cleared of trees and topsoil. Then, explosives and enormous excavating machines are used to expose the deposits of the resource. The massive amounts of spoils from such operations are typically dumped into the valleys below, destroying forests, burying streams, and polluting surface waters.

FIGURE 7.17 This spoils bank in Germany is left over from the mining of potash, a mineral salt that contains potassium used to make fertilizers.

Mineral deposits that are too far underground to be removed by surface mining are extracted by **subsurface mining**, a method that involves digging a deep vertical shaft and using explosives and excavating equipment to create tunnels and chambers to reach the deposit. From there, the metal ore is extracted with machinery and hauled to the surface (Figure 7.18).

Subsurface mining does not disturb nearly as much land as surface mining does, but under the surface, extensive shafts, tunnels, and caverns are created in order to remove a mineral resource. One of the major problems with this form of mining is the threats to the health and safety of the miners, who are subject to the risks of cave-ins, explosions, and fires every day. In addition, after years of inhaling mining dust, miners can contract black lung or other lung diseases that cause long-term breathing difficulties.

Regardless of the type of mining, the environmental impacts from mining an ore tend to get more severe as the ore's grade gets lower. The more accessible and higher-grade ores are usually exploited first, often with relatively low environmental impacts. As these deposits become depleted, miners sometimes go after lower-grade ores. The methods they use then require more energy, water, and other resources, and this increases land disruption, air and water pollution, and mining waste piles.

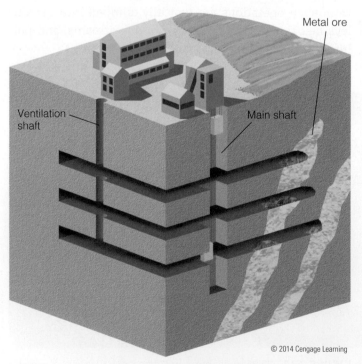

FIGURE 7.18 To reach mineral deposits, miners blast open shafts and tunnels, and remove huge amounts of rock.

FOR INSTANCE...

Some Environmental Impacts of Gold Mining

Gold mining can disrupt large areas of land because in most cases, a great deal of dirt and rock must be moved to find a few ounces of gold. For example, to obtain enough gold to make a pair of wedding rings, a typical mining operation would create a pile of waste rock and dirt that would weigh more than three mid-size cars.

This means that on many mining sites, especially in Australia and North America, mining companies have knocked down entire mountains of rock containing small concentrations of gold. This process, in addition to scarring the land (Figure 7.19), involves the use of highly toxic cyanide salts that are sprayed onto piles of crushed rock where they react with the gold, removing it from the ore. The solution drains from these piles into ponds, called *leach ponds*, where zinc is added to separate the gold from the cyanide solution. Then, acid is used to remove the zinc and what is left is a gold sludge.

These leach ponds attract birds and other wildlife searching for water, but they are extremely toxic. Where such ponds have leaked or overflowed, they have polluted groundwater and killed fish and other forms of life in nearby lakes, streams, and wetlands. Some ponds are equipped with liners and other systems intended to stop leaks, but these systems sometimes fail.

One such failure occurred at a gold mine in Romania in 2000, when heavy rains and snow washed out an earthen dam on one end of a cyanide leach pond. The pond released a slurry containing cyanide and toxic metals into the Tisza and Danube Rivers, wiping out large numbers of fish and other aquatic animals and plants. For months, government officials told people living along these major rivers not to eat fish from the rivers or drink water from their wells. Many businesses along the rivers had to close.

Gold mining has had harmful impacts in tropical forests and other tropical ecosystems as well. At such mining sites, gold miners often clear large areas of forest and pollute streams and groundwater with their toxic mining wastes. Some of these gold miners use water cannons to wash entire hillsides into collection boxes for gold removal. This technique, called *hydraulic mining*, has been banned in the United States, but it is still commonly used in tropical forests of Central and South America.

Small-scale gold mining has also resulted in mercury contamination in some areas of the world. Many small-scale

FIGURE 7.19 Strip mining at this gold mine in the Black Hills of South Dakota has disrupted a vast area of land. The green ponds in the foreground are toxic leach ponds.

miners in Latin America, Africa, and Asia illegally use mercury to remove gold from stream sediments. This is one of the two biggest human-related inputs of toxic mercury into the environment, second only to the burning of coal in power plants. In one area of Borneo, the mercury from illegal gold mining operations killed about 70% of the fish populations and poisoned miners and nearby villagers.

Let's REVIEW

- What are three major environmental impacts of mining?
- Define *surface mining, overburden,* and *spoils.* Distinguish between *open-pit surface mining, area strip mining, contour strip mining, mountaintop removal mining,* and *subsurface mining,* and describe the environmental impacts of each.
- How is the grade of ore being mined related to the severity of environmental impacts of mining?
- Summarize the major environmental hazards of gold mining.

Environmental Impacts of Processing Ores

Mining is just one part of the whole process of exploiting mineral resources. Once the ore is mined, it must be processed to remove the mineral. For example, some processing methods include breaking or crushing rock. This produces waste materials called *tailings*, which are usually heaped on the ground in large piles (Figure 7.20) or stored in ponds. Wind can blow particles of toxic metals off of tailings piles, and rainfall can wash them off, sometimes resulting in contamination of groundwater supplies and nearby wetlands, streams, and lakes. Tailings ponds can leak or overflow when flooded by heavy rainfall or rapidly melting snow.

In some processes, metals are separated from their ores through **smelting**—a process that uses heat and chemical reactions to decompose or drive off other chemicals in

FIGURE 7.20 The reddish dirt piles here are radioactive tailings at a mining site near Moab, Utah, where uranium was processed to make fuel for nuclear power plants.

an ore and leave the metal behind. Smelters can severely pollute the air and release climate-changing greenhouse gases including carbon dioxide. Unless they are equipped with effective pollution control equipment, smelters can also emit sulfur dioxide and tiny acidic particles that pollute streams and lakes, damage vegetation, and acidify soils in the surrounding and downwind areas of a smelter.

The processing of ore usually requires the use of large amounts of water. After processing, this water typically contains harmful pollutants such as sulfuric acid, arsenic, and mercury. Unless it is stored safely, this polluted water can get into drinking water supplies and into streams, lakes, and wetlands where it can kill fish and other forms of aquatic life. Even when it is stored safely in holding ponds, those ponds can leak or rupture.

Let's REVIEW

- How can tailings left from the processing of some ores affect the environment?
- Define *smelting*, and explain how it can harm the environment.
- How does the use of water in the processing of ores often harm the environment?

KEY idea Most of the processes involved in exploiting mineral resources result in the disturbance of land, the pollution of air and water, and the production of huge amounts of solid and hazardous wastes.

FIGURE 7.21 The metal tungsten is used in the manufacture of blades for wind turbines, which are increasingly used to generate electricity.

Suzlon Energy Ltd.

Problems of Unequal Mineral Distribution

The fact that mineral deposits are distributed unevenly around the globe means that many countries must import some of the minerals they need from areas where they are found. For example, the rare metal tungsten and mixtures of tungsten with other metals (alloys) are widely used in sawblades, milling tools, artillery shells, ballistic missiles, and wind turbine blades (Figure 7.21). According to the U.S. Geological Survey, China has 57% of the world's estimated tungsten reserves and Canada has 12%. Because the United States has only 4% of these reserves, it must import this important resource from other countries.

Another example of an unequally distributed mineral is lithium (Li), which is gaining importance because of the explosive growth in the use of lithium-ion batteries in electronic devices, in gasoline-electric hybrid cars,

and in new all-electric cars. Some analysts have suggested that the projected rise in use of these cars over the next few decades could make lithium one of the world's most important and valuable minerals.

According to estimates by experts at the U.S. Geological Survey, the four countries holding the largest identified lithium reserves are, in order, Chile, Argentina, Bolivia, and China. Several more-developed countries, including Canada, Australia, Japan, and the United States, each have just a tiny fraction of the world's lithium reserves. They will have to depend on the four countries with the large reserves for their supplies of lithium, much as they now depend on some Middle Eastern countries for much of the oil that fuels their cars.

The rates at which different countries use nonrenewable mineral resources also vary greatly. The more-developed countries such as the United States have seen high rates of mineral resource consumption for several decades. Now, rapidly developing countries like China, India, and

Brazil are following that pattern and increasing their rates of use for these resources.

This means that some countries that once had large deposits of key minerals have depleted a portion of their reserves and others are rapidly approaching that point. In the United States, high-grade ore deposits of iron, aluminum, and lead are now economically depleted, and the country depends on imports for all or most of its supplies of 24 key mineral resources. There is also growing concern about the availability of an important group of metallic elements and compounds, called *rare-earth metals*, which are essential to modern societies (see the following *For Instance*).

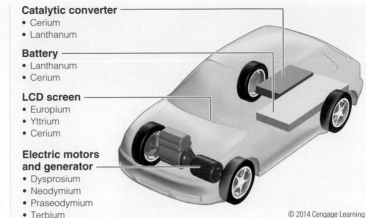

Catalytic converter
- Cerium
- Lanthanum

Battery
- Lanthanum
- Cerium

LCD screen
- Europium
- Yttrium
- Cerium

Electric motors and generator
- Dysprosium
- Neodymium
- Praseodymium
- Terbium

© 2014 Cengage Learning

FIGURE 7.22 The technology for hybrid and all-electric cars is highly dependent on rare-earth metals.

FOR INSTANCE...

Rare-Earth Metals

Rare-earth metals (or *rare earths*) are a group of 17 metallic elements and their oxide compounds that, because of their common set of unique properties, are vitally important in technologies on which we depend. For example, the rare-earth metal *lanthanum* is used to make nickel-metal-hydride batteries that power many electric vehicles and other devices. Another rare earth, *ytterbium*, is used to strengthen stainless steel. Rare earths are not all as rare as they were once thought to be. However, they are unevenly distributed around the globe and they are difficult to extract from most deposits because they are dispersed within the rock where they are found.

Most people are generally not well-informed about rare earths, but these elements are becoming more well-known for two reasons. First, they are used as key components in most of the world's vital technologies, making them among our most important mineral resources.

Here are just a few of many examples. Various rare-earth metals are essential components of most digital devices, including smart phones, digital cameras, and flat-screen TVs and computer monitors. We depend on rare earths for building medical scanning and imaging devices, fiber-optic cables, and jet engines. Rare earths are key components of military technologies, including missile guidance systems, aircraft electronics, radar, and satellites. Also, they are vital components of compact fluorescent and LED light bulbs, solar cells, wind turbines (Figure 7.21), and hybrid electric cars (Figure 7.22).

The second reason that rare earths are getting more attention is that the current supplies of rare earths are hard for many nations to come by, because they are not evenly distributed. According to the U.S. Geological Survey, China has about 33% of the world's rare-earth reserves, the United States has about 13%, and Russia and Australia each have 5–6%. About 97% of the available global supplies are produced and sold by China. Since 2010, China has been reducing its rare-earth exports to meet its own growing needs, and the price of Chinese rare earths has therefore risen sharply as the demand has exceeded the supply. For example, between July 2010 and February 2011, the price for a ton of rare earths rose by a staggering 778%, from $14,000 to $109,000. About 80% of China's supply of rare earths is extracted from the world's largest rare-earth mine near Baotou, a city in Inner Mongolia.

Large concentrated deposits of rare-earth metals are very scarce, so we rely mostly on dispersed deposits. It takes a lot of energy, money, and scientific expertise to find these deposits, extract the low concentrations of rare earths, and process the minerals into usable forms. Rare-earth deposits also contain radioactive elements, such as uranium and thorium, that can make workers sick and contaminate nearby water and soil. As a result, the mining and processing of rare earths have large and harmful health and environmental impacts, especially in China where the environmental controls on mining and processing have been lax.

The United States, Australia, Russia, Canada, and South Africa hold some undeveloped reserves of rare earths and have been scrambling to develop their own rare-earth mines and processing facilities to avoid dependence on China for these minerals. A large rare-earths mine operated by Molycorp Minerals in Mountain Pass, California (Figure 7.23), closed down in 2002, in part because of the expense of meeting pollution regulations for this very dirty mining process. China's rare-earth suppliers, with

FIGURE 7.23 Molycorp's rare-earths mine in the Mojave Desert near Mountain Pass, California, closed down in 2002 but is being modernized and may reopen in a few years.

China's lax environmental standards, did not have to pay such costs and could therefore beat the prices that Molycorp had to charge. In this way, China has driven most foreign mining companies out of the rare-earths market and now dominates the global supply of these immensely important and increasingly profitable resources.

Molycorp's California mine is being retooled to use more modern, less environmentally harmful mining technologies. Within a few years, this mine may be able to meet the U.S. demand for some of the important rare earths as it once did. Meanwhile, during this period of retooling, China will have an edge in selling finished products with rare-earth components, including electronic, solar, and wind-energy products, in the global marketplace.

One way to grow supplies of rare-earth metals is to salvage and grow these metals from the massive amounts of electronic wastes that we are producing in the form of discarded cell phones, computers, TV sets, and other electronic devices. However, this is difficult and costly because of the fairly small quantities of rare earths that can be recaptured from these products. Another option is to look for technologies that do not require rare earths. For example, companies that make batteries for electric cars could switch from nickel-metal-hydride batteries, which require the rare-earth metal lanthanum, to lithium-ion batteries. Still another option that scientists are working on is to find more accessible materials to substitute for rare-earth metals.

Environmentally Harmful Subsidies

Governments give subsidies and tax breaks to mining companies to encourage them to find and develop deposits of mineral resources. This financial assistance helps companies to lower their costs and therefore their prices—a result that can be called *artificially low pricing*. Thus, by lowering producers' costs and making the companies' end products more affordable for consumers, subsidies are intended to help a nation's economy grow, while also helping to free the economy from dependence on foreign supplies of minerals and other types of resources.

These are logical reasons for providing a company or industry with a financial advantage. In the United States, huge subsidies and tax breaks helped mining companies get on their feet and grow strongly during the 19th and 20th centuries. However, these subsidies continue and critics say they are now outdated and can promote environmentally harmful mining practices resulting from a shift to mining lower-grade ores. Critics argue that governments should, instead, subsidize the search for substitutes for increasingly rare minerals or for technologies that will allow us to exploit minerals while greatly reducing the environmental harm of mining and processing them.

CONSIDER this Each year, mining companies remove at least $4 billion worth of hard rock minerals from U.S. public lands, and they obtain these resources for a tiny fraction of their worth. The royalties that these companies are required to pay to taxpayers equal only about 2.3% of the value of these resources.

Let's REVIEW

- Why is lithium an important metallic mineral resource? Why is its unequal distribution a potential problem?
- Which four countries have the world's largest estimated supplies of lithium?
- What are *rare-earth metals,* why are they important, and why are they hard for most countries to obtain? Summarize the story of how rare earths have been supplied, globally, and how this might change.
- What was the intended effect of government subsidies and tax breaks for mining companies? Why do some critics argue that these forms of financial assistance are now outdated and environmentally harmful?

What Can Be Done?

Reducing the Environmental Impacts of Mining

Scientists are working on ways to extract mineral resources without severely degrading the environment around mining sites. One such area of research is *biomining*, the use of natural or genetically engineered microorganisms to remove desired metals from their ores. Because it takes place underground, this approach does not disturb the surrounding environment and reduces the air and water pollution associated with the smelting of metal ores. It also avoids the use of toxic chemicals and the resulting water pollution, such as that coming from gold mines.

The main drawback of this approach is that it is very slow, compared to conventional mining methods. Genetic engineers are seeking ways to speed up the process by altering the bacteria that are used, but there is concern that genetically altered bacteria might have harmful effects on ecosystems.

Some surface mining sites can be cleaned up and restored (see *Making a Difference*, p. 194). However, many sites, such as those shown in the module-opening photo and in Figure 7.16, will never be restored to their natural state because so much material has been removed. Also, restoring mining sites is very costly and, in reality, it is rarely done. It is estimated that the cost of cleaning up all the abandoned mining sites in the world would be in the trillions of dollars. Nevertheless, mine restoration can be done in some cases (Figure 7.24).

Governments can help to make our use of mineral resources more sustainable by halting subsidies that support unsustainable mining practices. Subsidies could instead be directed toward biomining and other efforts to lessen the impacts of mineral resource use, as well as toward efforts to find sustainable substitutes for mineral resources.

Expanding Mineral Reserves

All of the problems discussed above, from the environmental impacts of mining and processing mineral ore to the problems of unequal distribution of minerals, have something in common. They all stem from the increasing use of mineral resources—due to a growing population and rising rates of resource use per person—and the resulting over-exploitation of those resources. Finding solutions to these problems involves both expanding the supplies of mineral resources and reducing the demand for them. For this purpose, it would be helpful to find ways to use mineral resources more efficiently, by recycling and reusing them.

One way to expand reserves is to add lower-grade ores into the mix. That means finding ways to mine and process these ores profitably, as has been done, for example, in the case of taconite mining (Figure 7.14). Usually, this involves finding new or improved, less costly technologies for mineral extraction and processing. However, even with such improvements, the harmful environmental effects of mining and processing lower-grade ores can be worse than those of mining high-grade ores. They can include more land disruption, wastes, and air and water pollution, and more use of water and energy resources that are increasingly scarce in some areas.

FIGURE 7.24 This abandoned coal strip-mining site (a) near Lynnville, Indiana, was partially restored (b) as a wetland.

MAKING A difference

René Haller's Restoration Project

At a mining site outside of Mombasa in Kenya, Africa, the Bamburi Cement Company removed tons of limestone for its cement-making operation. By 1970, this open-pit mining operation had created several large pits. The company wanted to rehabilitate the degraded land to produce fruits and vegetables for its employees, and they hired Dr. René Haller (Figure 7.A), a Swiss expert in horticulture, landscaping, and tropical agriculture, to oversee the project.

By 1977, Haller had transformed many of the limestone quarries that scarred the site into a self-sustaining ecological oasis, which he called Baobab Farm—the site of a lush tropical forest area, wetlands, grassy pastures, lakes, and productive farms.

Baobab Farm now attracts about 100,000 tourists every year. Trees tower all around, and birds representing more than 220 species flit about while fish swarm in pools. Beyond the central area, cattle, sheep, and goats graze on grasslands that they share with antelope, buffaloes, ostriches, and giraffes. Visitors enjoy walking and biking on trails throughout the farm site.

The farm also includes a vineyard where grapes are grown on steep hillsides, and groves of trees where honeybees are raised. The rapidly growing tree plantations provide a sustainable supply of firewood, building materials, and vegetation for browsing livestock.

Over the course of Haller's decades-old experiment, he has developed an array of sustainable ecological farming practices that he and his colleagues have shared with local farmers. Haller's work has allowed these farmers to better their own lives while helping to protect their fragile tropical environment.

In 1997, René Haller was one of the first recipients of the Global 500 Roll of Honour award from the United Nations for his "outstanding environment achievements."

FIGURE 7.A Dr. René Haller sits with a giant tortoise on Baobab Farm, which he created on an abandoned limestone mining site in Kenya, Africa.

Another way to expand mineral reserves is to find them in ocean waters and under the ocean floor. The minerals bromine, magnesium, and sodium chloride (salt) are found in seawater and can be removed by various methods. There are other minerals dissolved in seawater, but the processes of removing them are too costly and energy-intensive to be profitable. Deposits of sand, gravel, sulfur, tin, copper, iron, tungsten, silver, titanium, platinum, diamonds, and phosphates (Figure 7.25) under the sea floor near the shorelines of several coastal areas are already being mined.

Ocean floor deposits will be more attractive as more land-based mineral deposits become depleted. For example, there is no substitute for the phosphates that are widely used in fertilizers to increase the yields for about 40% of the world's food crops. We may soon face phosphate shortages because yields from most of the world's phosphate mines are declining at a time when demand for this mineral is increasing as the world's population continues to grow.

FIGURE 7.25 The light-colored areas of this photograph are piles of tailings from a phosphate mine off the coast of Florida.

Mineral deposits are also found under the ocean floor in deeper waters, but not much has been done to extract them, mostly because of the high costs involved. Another problem is that because the open ocean is not owned by any country, proposals to mine the ocean floor lead to disputes among nations over rights to ocean-floor mineral deposits.

In particular, *hydrothermal ore deposits* are a possible source for some minerals. These deposits form when superheated mineral-rich water spurts from vents in volcanically active regions of the ocean floor. When this hot water mixes with cold seawater, black particles of various metal sulfides develop. They accumulate in naturally formed chimney-like structures, called *black smokers*, near the hot-water vents on the ocean bottom (Figure 7.26). These deposits are especially rich in minerals such as copper, gold, silver, lead, and zinc.

So far, it has been too costly to tap into these mineral deposits, but efforts are now underway. A Canadian company is building equipment that would be used to cut the black smokers apart on the ocean floor and lift the pieces to the surface where large processing ships would remove the minerals and dump the waste materials back into the sea. Japan is also venturing into deep-sea mining, sending robotic submarines to study areas off its coast that contain hydrothermal vents.

Some scientists are concerned about the effects of this mining on the ecosystems that surround these vents. The various species of exotic organisms in these ocean bottom ecosystems are supplied with food by bacteria that can produce nutrients from sulfur compounds in the water through *chemosynthesis*, a chemical process that does not require light as land plants do when they make nutrients through *photosynthesis* (see Module 1, p. 15). Because we have a limited understanding of how these systems work, scientists warn that by destroying them in order to obtain minerals, we would be destroying important sources of scientific information and possibly other valuable resources that do not exist anywhere else on earth.

Another potential ocean-floor mineral resource is the potato-sized lumps of manganese that cover areas of the ocean bottom in the Pacific, Atlantic, and Indian Oceans. These deposits form over millions of years, as metals slowly separate from seawater and sediments to accumulate in what are called *manganese nodules*. Some resource developers envision scooping them up with underwater mining machines or using giant vacuum pipes to suck them up from the ocean floor. Again, the effects of such operations on ocean-floor ecosystems are unknown.

The exploitation of ocean-floor mineral resources will likely become an industry in decades to come. Scientists are urging careful consideration of the environmental impacts of this exploitation along with the potential benefits of it. Figure 7.27 summarizes the major pros and cons of ocean-floor mining.

Black smoker

Sulfide deposits

Magma

Tube worms

White crab

White clams

© 2014 Cengage Learning

FIGURE 7.26 Hydrothermal ore deposits form on some areas of the ocean floor and are rich in certain minerals.

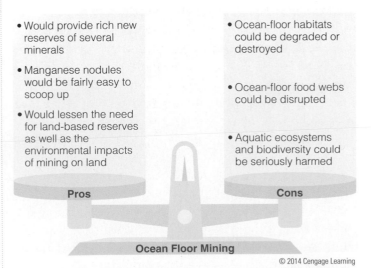

- Would provide rich new reserves of several minerals
- Manganese nodules would be fairly easy to scoop up
- Would lessen the need for land-based reserves as well as the environmental impacts of mining on land

- Ocean-floor habitats could be degraded or destroyed
- Ocean-floor food webs could be disrupted
- Aquatic ecosystems and biodiversity could be seriously harmed

Pros

Cons

Ocean Floor Mining

© 2014 Cengage Learning

FIGURE 7.27 *Weighing the pros and cons* of ocean-floor mining.

Let's REVIEW

- What is biomining and how can we use it to lessen the environmental impact of mining?
- What are two obstacles to restoration of mining sites?
- What role could government subsidies play in finding ways to use mineral resources more sustainably?
- What do the problems related to mineral resource use have in common?
- What are two general ways to expand mineral reserves, and what are the possible problems involved with using these strategies?
- Summarize the pros and cons of ocean-floor mining.

Reducing, Reusing, and Recycling

We can prolong the supply of nonrenewable mineral resources—especially valuable or scarce metals such as gold, iron, copper, aluminum, and platinum—by reusing or recycling the products that are made from these minerals.

NUMB3RS

58%
The percentage of aluminum drink cans recycled in the United States in 2010

For example, mining and processing aluminum ore produces 95% more water pollution and 95% more air pollution, and uses 97% more energy than does recycling aluminum beverage cans (Figure 7.28) and scrap aluminum. Cleaning up and reusing items such as glass bottles (made from the mineral silica, or sand) has an even lower environmental impact than does reprocessing them for recycling.

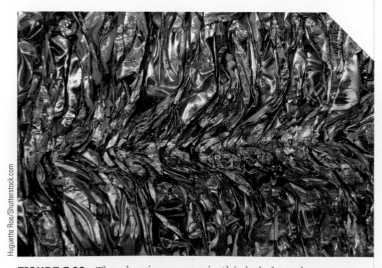

Huguette Roe/Shutterstock.com

FIGURE 7.28 The aluminum cans in this bale have been collected and crushed for recycling.

A growing number of analysts say that, of all the strategies for expanding mineral reserves and for using them more sustainably, the most important one is to cut our overall rates of use and waste of such resources. Some advise that we look at how nature uses resources and then mimic those processes (see the following *For Instance*).

We can encourage the reduction, reuse, and recycling of mineral resources by finding ways to include, in market prices, the harmful environmental costs of mining and processing mineral resources. This would result in higher prices for end products and it would spur producers and users of mineral resources to find ways to use them more sustainably. One way to do this is to levy taxes on mineral products. Also, governments can eliminate or cut mining subsidies and increase subsidies for recycling and reuse, and for finding less environmentally harmful substitutes for some minerals. Governments and businesses can also promote the education of consumers to help them cut their use and waste of mineral resources, and to reuse and recycle most of what they buy.

FOR INSTANCE...

Reducing Resource Waste by Copying Nature

Since the earth gets no new supplies of mineral resources (except for small amounts contained in meteorites), nature has developed ways to continually recycle all of the chemicals that the earth's life forms need for survival. In nature, the waste outputs of one organism typically become the nutrient inputs of another organism. This occurs primarily within the natural networks called *food webs* (Figure 1.18, p. 17). This explains why there is very little material waste in nature.

Thus, when we confront an environmental problem such as unnecessary resource waste, we can look at how nature has avoided this problem. This is an important part of a set of strategies called **biomimicry**, which generally involves two major steps. First, observe how natural systems have responded to changing environmental conditions over many millions of years. Second, try to copy or adapt these responses to human systems in order to help us deal with certain environmental challenges.

For example, some companies are making their manufacturing processes cleaner and more sustainable by redesigning them to mimic the ways in which nature deals with raw materials and wastes. They are trying to reuse or recycle most of the minerals and chemicals they use, instead of treating the used materials as waste and burying or burning them, or shipping them somewhere else.

In 1975, the 3M Company, based in Minnesota, which makes 60,000 different products, began a Pollution Prevention Pays (3P) program. It redesigned its equipment and processes to reduce pollution and wastes by using fewer raw materials for manufacturing. It also identified toxic wastes generated by its processes and recycled or sold those wastes as raw materials to other companies.

Between 1975 and 2008, the company's 3P program prevented a huge mass of pollutants—that 3M estimated to be roughly equal to the weight of more than 100 jumbo jet airliners—from reaching the environment. The company also saved more than $1.2 billion in waste disposal costs. Since 1990, a growing number of companies have adopted similar pollution and waste prevention programs.

Another way for industries to mimic nature is to develop *resource exchange webs*—similar to food webs in natural ecosystems. In these webs, the participants exchange waste outputs and convert them into raw materials. A resource exchange web, also called an *ecoindustrial park* or *industrial ecosystem*, has been developed in Kalundborg, Denmark (Figure 7.29), where an electric power plant, along with nearby industries, farms, and homeowners work together to reduce their outputs of waste

and pollution and to save money. There are more than 40 ecoindustrial parks in various places around the world (18 of them in the United States), and more are being built or planned.

Businesses gain many benefits from these industrial forms of biomimicry, just as 3M did. Biomimicry also stimulates companies to come up with new, environmentally beneficial chemicals, processes, and products that they can sell worldwide. Such companies tend to be viewed more favorably by consumers, based on their actual results rather than on advertising campaigns.

Let's REVIEW

- What are three major ways to prolong the supplies of nonrenewable mineral resources? Give an example of each.
- How can governments encourage the reduction, reuse, and recycling of mineral resources?
- What is *biomimicry*? What are the two key steps in applying this idea and how have some companies benefited from using this approach?
- Describe the industrial ecosystem used in Kalundborg, Denmark.

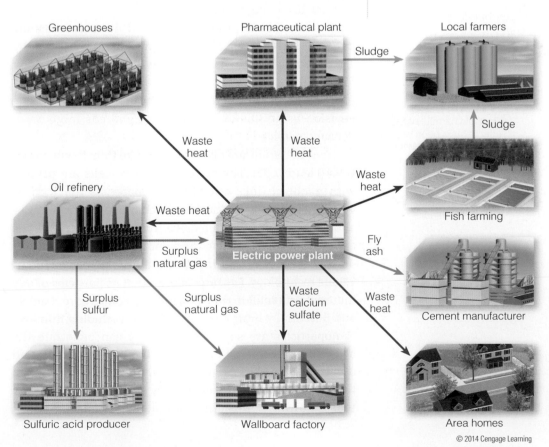

FIGURE 7.29 This *industrial ecosystem* in Kalundborg, Denmark, reduces waste production by mimicking a natural ecosystem food web.

© 2014 Cengage Learning

Try to Find Substitutes

According to some analysts, scientists and engineers will eventually find substitutes for key minerals when they become too scarce or when it costs too much to keep on mining them. For example, silicon and newer materials such as ceramics and high-strength plastics are now being used in place of some metals. Also, fiber-optic glass cables that transmit pulses of light are replacing copper and aluminum wires in telephone cables.

In 2004, for example, researchers developed a new form of the mineral graphite, which is composed of carbon atoms. Unlike the traditional form of graphite used as the lead in pencils, this new form can stretch like elastic, can conduct heat and electricity, and is 200 times stronger than steel. This revolutionary material, called *graphene*, could replace silicon transistors and be used to make lighter aircraft, more efficient batteries, and more advanced LCD and touch-screen displays. Prices are rising sharply as the demand for this new material grows.

Some analysts say that the next big advance in substitute materials will be in the form of **nanotechnology**—a rapidly emerging field of science and engineering that is focused on creating materials out of atoms and molecules. Scientists are working with the molecules of common materials such as silicon to build substitute materials that could replace scarce mineral resources. But nanotechnology, despite its promises, also has drawbacks (see *A Look to the Future*, which follows). Figure 7.30 compares the major pros and cons of nanotechnology.

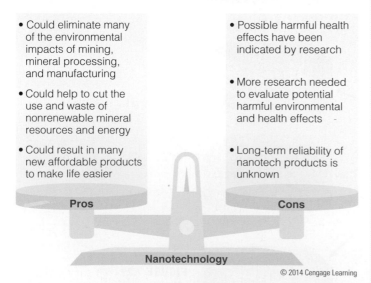

FIGURE 7.30 *Weighing the pros and cons* of nanotechnology.

A Look to the Future

The Nanotechnology Revolution

A *nanometer* is one billionth of a meter. It may be impossible to imagine, but consider this: if you could split a hair from your head lengthwise into 100,000 hairs, each of them would be about a nanometer across. Nanotechnology operates at the ultra-small scale of less than 100 nanometers.

NUMB3RS

1 million

The period at the end of this sentence is about 1 million nanometers wide.

Scientists envision arranging the atoms of elements such as carbon, oxygen, and silicon to create everything from medicines and solar cells to automobile bodies. Some also believe that we could use this technology to build substitutes for many minerals that are now becoming scarce.

The number of products in which nanomaterials are already used is more than 1,000 and is growing rapidly. These products include sunscreens, cosmetics, pesticides, food additives, odor-eating socks, and clothes with stain-resistant and wrinkle-free coatings. Nanotechnologists envision using this technology for hundreds more products and services.

Nanomaterials are also used to make thin, flexible solar cells (Figure 7.31). The circuits on these cells are printed with nanotech inks onto flexible materials that can be applied to almost any surface, including windows, walls, and even T-shirts. Some believe that these new materials are likely to revolutionize the solar power industry.

However, there are drawbacks to this potentially marvelous technology. Research shows that as particles of any material get smaller, they generally become more reactive and potentially more toxic to animals, including humans. Nanoparticles are so small that they can penetrate the body's natural defenses that normally keep out germs and other invaders, and in some cases, they can pass through the protective membranes of body cells. Thus, a chemical that is harmless at the macroscale may be hazardous at

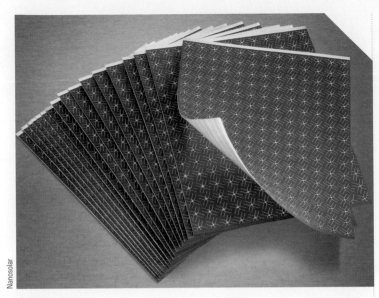

Nanosolar

FIGURE 7.31 The production and installation costs of these flexible, thin-film solar cells are lower than those of conventional solar cells.

the nanoscale when it is inhaled, ingested, or absorbed through the skin.

Laboratory animal studies show that nanoparticles can move from mother to fetus and from the nasal passages to the brain of a mammal. Other such studies have found that nanoparticles damaged the lungs, hearts, and blood vessels of test animals. Scientists note that such damage in humans could take years or decades to become apparent.

So far, products that include nanoparticles are mostly unregulated and unlabeled. Health and environmental analysts say that there is an urgent need to take three steps before nanotechnology becomes more widespread. First, carefully investigate the technology's ecological, economic, and health effects. Second, develop guidelines and regulations to control its use until we know more about its potentially harmful effects. Third, follow the strong recommendation of several scientists that we avoid releasing nanoparticles into the environment until more is known about their potential harmful impacts on human health.

Let's **REVIEW**

- Give two examples of substitutes that have been found for mineral resources.
- Define *nanotechnology* and give three examples of how it has been applied. What are the major pros and cons of nanotechnology?
- What are three steps recommended by scientists relating to the development of this technology?

What Would You Do?

At the beginning of this module, we asked why you should care about nonrenewable mineral resources. We learned that such resources are used to manufacture or supply most of the consumer goods that we use, and we saw that we waste large quantities of mineral resources. We also saw that in extracting, processing, and converting such resources into products and then discarding many of these items, we create a number of harmful environmental impacts.

Many concerned citizens and environmental leaders are finding ways to avoid adding to these environmental impacts on the personal level, and to help the world in extending its supplies of nonrenewable mineral resources by cutting their use and waste of these resources. Here are three key strategies that these people are now trying.

Reducing unnecessary mineral resource use.
- Buying fewer new mineral resource products and saving money by doing so.
- Donating unused tools and other metal items to second-hand stores and buying such items from these stores.
- Using e-mail lists, neighborhood groups, and lending libraries to borrow and share metal products such as tools.

Reducing unnecessary mineral resource waste.
- Using recycling centers to recycle rather than discard worn-out metal products.
- Buying products that are made of recycled materials as much as possible.
- When buying new metal products, spending more to buy higher-quality products that are built to last a long time.

Learning and teaching.
- Learning about where the materials used to make metal products came from and about the harmful environmental impacts of using these materials, using this information to help them know which products to avoid, and informing others of what they have learned.
- Finding out about local lending libraries and other ways to share metal products such as tools, informing others about them, or helping to form such groups where they don't exist.
- Helping to educate children, neighbors, and government representatives about the importance of using mineral resources as sustainably as possible.

KEYterms

area strip mining, p. 187
biomimicry, p. 196
contour strip mining,
    p. 187
core, p. 177
crust, p. 177
depletion time, p. 186
economic depletion, p. 185

geology, p. 177
high-grade ore, p. 182
low-grade ore, p. 183
mantle, p. 177
mineral, p. 181
mineral resource, p. 182
mountaintop removal
    mining, p. 187

nanotechnology, p. 198
open-pit surface mining,
    p. 187
ore, p. 182
overburden, p. 186
rare-earth metals, p. 191
reserves, p. 184
rock, p. 181

rock cycle, p. 181
smelting, p. 189
spoils, p. 186
subsidies, p. 185
subsurface mining, p. 188
surface mining, p. 186
tectonic plates, p. 178

THINKINGcritically

1. Explain what might happen if the earth's tectonic plates stopped moving.

2. Explain how you benefit from the rock cycle. In your explanation, include three specific products or services that you used this week that would not exist if there were no rock cycle.

3. For one week, keep a record of the products that you buy. Then analyze your list to determine what three mineral resources are most common in the products you bought. How could you reduce your use of these three mineral resources?

4. Do you think that the benefits of ocean-floor mining outweigh its drawbacks? (See Figure 7.27.) Explain.

5. How might the rapidly rising prices of rare-earth metals, with 97% of their production controlled by China, affect your life? Assume you are the president of your country and write a plan for dealing with shortages of these vital mineral resources.

6. How might nanotechnology affect the business of mining gold and other nonrenewable mineral resources? Do you think the potential benefits of nanotechnology outweigh its potential harms (Figure 7.30), or will the harms outweigh the benefits? Explain.

LEARNINGonline

Access an interactive eBook and module-specific interactive learning tools, including flashcards, quizzes, videos and more in your Environmental Science CourseMate, accessed through **CengageBrain.com**.

8

Species Extinction

iStockphoto/Geoff Kuchera

Why Should You Care about Species Extinction?

We share planet Earth with millions of other *species*, or life forms, each dependent on an *ecosystem*, or a natural system made up of species and nonliving elements such as soil or water. However, we would not survive for long were it not for **biodiversity**—the earth's great variety of species and ecosystems (see Module 1, p. 15). Scientific evidence indicates that eventually, all species die out, or become *extinct*. However, the *rate of extinction*—the number of species disappearing every year—has risen sharply as humans have dominated and degraded more and more of the planet to provide resources for the growing human population. Biodiversity expert Edward O. Wilson warns that "the natural world is everywhere disappearing before our eyes—cut to pieces, mowed down, plowed under, gobbled up, replaced by human artifacts."

If all species eventually become extinct, does it matter that the mere hundreds of Siberian tigers (see module-opening photo) remaining in the wild are likely to die out before the end of this century? The huge Sumatran corpse flower (Figure 8.1a) could become extinct because it is found only in the tropical rain forests of Sumatra, Indonesia, which are being rapidly cleared to grow crops such as soybeans. Its sickening smell, like that of the rotting flesh of a corpse, attracts certain flies and beetles that pollinate the flower. Will it matter if this flower becomes extinct?

Another animal that could soon go extinct is the orangutan (Figure 8.1b), which is in trouble because of the clearing of the forests in Indonesia and Malaysia where it lives, and because an illegally smuggled live orangutan can sell for as much as $10,000. Will it matter if this species becomes extinct in the near future? Does it matter that 25–50% of the earth's known land-based species could become extinct during this century, largely because of human activities?

Biologists say there are three reasons why these possible extinctions do matter and why we should care about extinctions caused mostly by our activities. First, the earth's biodiversity provides us with essential resources, including food and oxygen, ecological services such as natural pest control, and pleasure, because nature is beautiful and interesting. Second, through various activities, we are eliminating and degrading many biologically diverse ecosystems, such as tropical forests, that are potential sites

FIGURE 8.1 Human activities threaten the Sumatran corpse flower (a) and the orangutan (b) with extinction.

for the emergence of new species. These new species will be needed to replace species that will be going extinct and to maintain the planet's vital biodiversity. Third, many people share the ethical view that wild species have a right to exist as long as they can, regardless of their usefulness to us.

Many biologists regard the current rapid loss of the species and ecosystems that make up the earth's biodiversity, hastened by human activities, as the most serious and long-lasting environmental and ethical issue that humanity faces. In this module, we examine the causes of the rising rate of species extinction and we consider how we can slow this troubling loss of biodiversity.

What Do You Need to Know?

Species, Populations, and Communities

A **species** is a group of organisms that have similar genetic and other characteristics that distinguish them from other groups of organisms. Organisms of *sexually reproducing species* have the ability to mate and produce offspring that are, in turn, capable of mating and reproducing. Most organisms are part of a **population**, a group of individuals of the same species that live in the same space. Most populations live in a **community**, two or more populations of different species living and interacting in a certain space.

Scientists estimate that the earth has at least 14 million and possibly as many as 100 million species. So far,

they have identified almost 2 million different species, at least half of them insects (Figure 8.2), which play vital ecological roles. For example, the monarch butterfly (Figure 8.2a), along with bees and other butterfly species, pollinate flowering plants, while insects such as the praying mantis (Figure 8.2b) help control the populations of at least half the insect species that we view as pests. Some organic gardeners use praying mantises as a form of biological pest control instead of relying on chemical pesticides.

Scientists classify species as *producers* or *consumers* based on whether they produce their own nutrients or feed on other organisms or their remains (see Module 1, pp. 16–17). As organisms feed on or decompose other organisms, chemical energy and nutrients are transferred from one organism to another in *food chains* and *food webs* (see Module 1, Figures 1.17 and 1.18, pp. 16 and 17). Human activities that disrupt food chains and webs can threaten the existence of these species by altering or eliminating their sources of nutrients.

How Populations Change

One of the most important scientific theories (see Module 1, p. 6) is the theory of **evolution by natural selection**, which says that a population of any species can change when some individuals in the population have *traits*, or characteristics, that give them a better chance of surviving and reproducing under a particular set of environmental conditions than other individuals that do not have those traits. The theory also says that the trait in question must be a *genetic trait*, or one that can be passed on to an organism's offspring.

FIGURE 8.2 The monarch butterfly (a) pollinates various flowering plants. The praying mantis (b), shown here eating a monarch butterfly, helps control the populations of many other insect species.

For example, consider a population of hawks living on an island. The hawks that have better eyesight than other hawks on the island are more likely to find and capture their prey and to survive and reproduce. As a result, those hawks with the *beneficial trait* (better eyesight) are more likely to produce more offspring than are the hawks without that trait, and therefore, better eyesight will eventually become more common in this population of hawks. Scientists are still filling in some of the details of evolution by natural selection, but it is the only viable scientific theory that explains the history and diversity of life on earth.

Evolution by natural selection can lead to **speciation**, a long-term process in which one species splits into two or more new species. This can happen when two or more different populations of a sexually reproducing species remain separated for a very long time and evolve independently under different environmental conditions. Eventually, the genetic makeup of individuals from the separated populations may become so different that they will not be able to produce live, fertile offspring if they happen to come into contact with one another again. In such a case, one species has become two, and speciation has occurred (Figure 8.3).

KEY idea

Species evolve from other species through natural selection.

Let's REVIEW

- What is *biodiversity* and why is it important to life on the earth? Why should we care about species extinction hastened by human activities?
- Define *species, population,* and *community* and give an example of each.
- State the scientific theory of *evolution by natural selection.*
- Define *speciation* and explain how it can occur.

Ecological Niches

As species evolve, they develop patterns of living, most of which are different in important ways from the patterns of other species. Biologists refer to such a pattern of living as an **ecological niche**, or simply **niche**. It describes various aspects, including what the species does within its ecosystem, what it eats and what eats or decomposes it, how much water and sunlight it requires, and how much space it occupies. A species' niche includes its **habitat**, which is the physical place where a species lives.

Species vary greatly, according to types of niches. **Generalist species** have broad niches, which means that they can live under widely varying environmental conditions. For example, some generalists such as white-tailed deer (Figure 8.4a) can survive on a wide variety of foods. Deer can live on ground-cover plants and shrubs in a forest, as well as on garden produce or grains from farm fields. Generalists such as cockroaches can live for long periods of

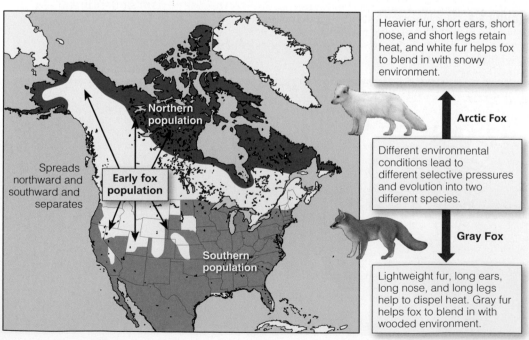

FIGURE 8.3 Scientists believe that two different species, the Arctic fox and the gray fox, arose from one common population of foxes in a process of speciation.

Heavier fur, short ears, short nose, and short legs retain heat, and white fur helps fox to blend in with snowy environment.

Arctic Fox

Different environmental conditions lead to different selective pressures and evolution into two different species.

Gray Fox

Lightweight fur, long ears, long nose, and long legs help to dispel heat. Gray fur helps fox to blend in with wooded environment.

© 2014 Cengage Learning

FIGURE 8.4 The white-tailed deer (a) is a generalist species. The giant panda (b) is a specialist species.

time without water. Other examples of generalist species are rats, raccoons, and humans.

On the other hand, **specialist species** are those species that have narrow niches, meaning that they require certain environmental conditions in order to survive. For example, the giant panda (Figure 8.4b) feeds almost exclusively on bamboo, which limits the areas where it can live. The corals that create spectacular coral reefs in tropical waters can thrive only within a narrow range of water temperatures. Many species fall in a range somewhere between highly specialized and broadly generalized species.

Other Ways to Classify Species

Biologists classify species in several ways, some based in part on their roles within ecosystems. Four general types of species are *native, nonnative, indicator,* and *keystone species*. Any given species may play one or more of these four roles.

Native species are those that have evolved to live within a particular ecosystem and often to be an important part of that ecosystem. For example, certain fishes such as lake trout (Figure 8.5) have lived in the Great Lakes of North America since the 1800s. They eat certain other species, helping to control those species' populations, and

perform other functions that help to keep the lake ecosystem functioning.

Any species that migrates into, or is deliberately or accidentally released into an ecosystem where it is not native, is called a **nonnative species**, also referred to as an *invasive species*. An invasive species can upset an ecosystem by interfering with, harming, or crowding out native species that help to keep the ecosystem functioning. For example, the sea lamprey (attached to the lake trout in Figure 8.5) is a parasitic species that preys on a variety of

FIGURE 8.5 This lake trout from one of the Great Lakes was the host for a parasitic sea lamprey that used its sucker-like mouth to attach itself to the fish and feed on its blood.

fish species. Since it invaded the Great Lakes in the early 1800s, it has multiplied and sharply reduced the numbers of lake trout and other native species of the Great Lakes.

Another role that biologists observe is that of an **indicator species**—a species that is especially sensitive to, and very responsive to, changes in long-term environmental conditions. Such a species can provide early warnings of damage to an ecosystem. For example, most trout species found in lakes and streams cannot survive if the water they live in falls above or below a certain temperature range. They also need clear, clean water with high levels of dissolved oxygen. When their populations decline, it is a strong sign of changes in temperatures, clarity, or pollution levels in their waters.

Many bird species, especially hummingbirds (Figure 8.6), often play the role of indicator species. They respond quickly to environmental disturbances such as air and water pollution, and damage to their habitats. Butterflies (Figure 8.2a) are also good indicator species because their association with various plant species makes them vulnerable to the loss or fragmentation of their habitat.

An important niche role is that of the **keystone species**—any species that, by filling its niche, has a large effect on the types and abundances of other species found in its ecosystem. In short, it plays a key role in the functioning of its ecosystem. For example, when the American alligator nearly disappeared from its natural habitat in the southeastern United States, the ecosystem it had

occupied for thousands of years began to unravel (see the following *For Instance*).

Other examples of keystone species are certain species of bees, butterflies, hummingbirds (Figure 8.6), and bats that play a critical role in pollinating flowering plant species. The loss of a keystone species can lead to population crashes and extinctions of other species that have evolved to depend on it for food or shelter or other benefits.

KEY idea

Each species has a role to play in its ecosystem as defined by its ecological niche.

FOR INSTANCE...

The American Alligator: A Keystone Species

The American alligator (Figure 8.7) is a keystone species because of the important role it plays in the wetland ecosystems of the southeastern United States where it lives.

However, the alligator has also been attractive to hunters for its hide, which has been used to make shoes, belts, and other items, and for its meat. In the 1930s, hunters freely killed alligators to harvest their meat and hides and for sport. By the 1960s, most of the alligators in the state of Louisiana and in the Florida Everglades had been eliminated.

Most of the people killing these alligators probably did not know about the important ecological roles that this keystone species plays in its wetland habitat. Alligators literally help to create their ecosystems by digging deep pits in the swamps, called "gator holes," which store freshwater during dry spells and serve as habitat for some species of fish, insects, snakes, and birds. They build large nesting mounds for themselves, but these mounds also become nesting and feeding sites for herons and egrets.

Alligators also eat a lot of gar, a predatory fish that feeds on game fish such as bass and bream. In this way, they help maintain populations of these game fish. As alligators move from their gator holes to nesting mounds, they help to keep freshwater ponds and coastal wetlands free of invasive shrubs and trees that would otherwise crowd out native species of plants and animals.

In 1967, the U.S. government classified the American alligator as endangered, which legally protected it from

FIGURE 8.6 Declines in hummingbird populations can indicate changes in environmental conditions such as the presence of pesticides in the air or on the flowers that these birds feed on and pollinate.

FIGURE 8.7 The American alligator is a keystone species in the natural communities found in the marshes and swamps of the southeastern United States, including the Florida Everglades.

being hunted. In 1987, the population of this keystone species had recovered enough for the U.S. Fish and Wildlife Service to remove it from its list of endangered species.

Let's **REVIEW**

- Define and distinguish between *ecological niche* and *habitat.* Distinguish between *generalist* and *specialist species,* and give an example of each.

- Distinguish between *native* and *nonnative species,* and give an example of each.

- What are *indicator species* and why are they important? Define and give an example of a *keystone species,* and explain why these species are important to their ecosystems.

- Summarize the story of the American alligator, explaining its importance to its ecosystem.

Species Interactions

Within their ecosystems, species share limited resources, and in filling their niches, they interact with each other in ways that help to keep these resources available and to keep their ecosystems functioning. These interactions also play a role in controlling the population sizes of species so that no one species can take over most of the food, habitat, and other resources. Ecologists have identified five basic types of species interactions.

The first of these is **interspecific competition**, in which members of two or more species try to make use of the same limited resources such as food, water, light, and space. Some interspecific competition results in a species becoming more efficient in making use of a resource, which can give that species an advantage over competing species. For example, the compass flower is a plant that has evolved to turn its leaves during the day to maximize the amount of sunlight that hits them. This helps the compass flower to keep from being shaded out by other plants.

Sometimes when populations of different species compete for the same resources over very long periods of time, they evolve genetic traits that allow them to reduce or avoid competition with each other. Typically, they do this by using shared resources at different times, in different ways, or in different places. The result is called *resource partitioning.* For example, some insect-eating bird species called warblers feed in different portions of the same trees and on different insect species (Figure 8.8).

A second major type of species interaction is **predation**, in which, a member of one species (the *predator*) feeds directly on all or part of a living organism of another species (the *prey*). Figures 8.2b and 8.9 (p. 208) show examples of such *predator-prey relationships.*

While predation harms individual organisms, it can benefit their species by helping to promote evolution by natural selection. Predators tend to kill the sick, weak, and

Blackburnian Warbler   Black-throated Green Warbler   Cape May Warbler   Bay-breasted Warbler   Yellow-rumped Warbler

FIGURE 8.8 In this example of resource partitioning among five species of insect-eating warblers in the spruce forests of Maine, each species spends most of its feeding time in a different part of each spruce tree (shaded here), and each consumes a different combination of insect species. (After R. H. MacArthur, "Population Ecology of Some Warblers in Northeastern Coniferous Forests," *Ecology* 36 (1958): 533–536)

© 2014 Cengage Learning

FIGURE 8.9 In this example of predation, an Alaskan brown bear (the predator) has captured and will eat this salmon (the prey).

Steve Hillebrand/U.S. Fish and Wildlife Service

aged members of a population because they are the easiest to catch. This leaves behind more offspring with traits that might help them avoid predation. Predation also helps to control the population sizes of prey species.

A third type of species interaction is **parasitism**, which occurs when one organism (the *parasite*) feeds on the body of, or otherwise makes use of the energy of, another organism (the *host*). The parasite is typically much smaller than its host and rarely kills its host. Some parasites, such as tapeworms, live inside their hosts. Others, like blood-sucking sea lampreys (Figure 8.5), attach themselves to the outsides of their hosts. Some parasites move from one host to another, as fleas and ticks do. Parasitism can benefit ecosystems by helping to control the populations of host species.

In a fourth type of species interaction, called **mutualism**, both species benefit by providing one another with food, shelter, or some other resource. For example, clownfish (Figure 8.10) usually live with sea anemones, whose tentacles sting and paralyze most fish that touch them except for the clownfish. The clownfish also feed on the leftovers of the anemones' meals. In turn, the clownfish fight off or eat some of the predators and parasites that threaten the anemones.

Another example of mutualism is taking place right now within each one of us. It involves armies of bacteria that live in our digestive tracts and help us digest our food.

They in turn get a safe habitat and a food supply from us, their hosts.

A fifth type of species interaction, called **commensalism**, benefits one species but has little or no effect on the other. For example, plants called *epiphytes* attach themselves to the trunks or branches of large trees in tropical

Khoroshunova Olga/Shutterstock.com

FIGURE 8.10 In this example of mutualism, a clownfish gains protection and food by living among deadly stinging sea anemones. At the same time, it helps to protect the anemones from some of their predators and parasites.

FIGURE 8.11 The rooting of this bromeliad, an *epiphyte* or air plant, to the trunk of a tree in Brazil's Atlantic tropical rain forest, is an example of commensalism.

and subtropical forests (Figure 8.11). They benefit by having a solid base from which to receive sunlight, rainwater, moisture from the humid air, and nutrients falling from the tree's upper leaves and limbs. Their growth apparently does not harm the tree.

Extinction

The central focus of this module is **extinction**, the process by which a species ceases to exist on the earth. Scientists study extinction by looking for information about early life on the planet as they analyze *fossils*—mineralized replicas of ancient bones, shells, leaves, and other items and the impressions of these things found in some rocks— along with the contents of ice cores drilled out of glaciers. They also study records of mammals and birds that have become extinct since humans evolved and compare this with fossil records of extinctions that happened before humans were here.

Scientists also measure the **rate of extinction**: the estimated number of species, or the percentage of known

species, that go extinct during a certain time period, typically a year. Throughout most years before humans evolved, species disappeared at a very low **background extinction rate**, which biologists estimate was about 0.0001% a year, or 1 species for every 1 million species on the planet, on average.

However, during the long history of life on the earth, drastic changes in environmental conditions caused by events such as ice ages and collisions between the earth and asteroids, have occasionally eliminated large numbers of species. Such an event, called a **mass extinction**, represents a sharp rise in the extinction rate above the background level. Evidence indicates that this has probably happened five times during the past 500 million years. There is growing evidence and concern that we might be entering into a new mass extinction caused by human activities—a threat that we will now explore in greater detail.

Let's **REVIEW**

- Define *interspecific competition* and give an example of it. What is resource partitioning?
- Define *predation* and give an example. How can predation benefit a prey species?
- Define *parasitism* and give an example. Define and give examples of *mutualism* and *commensalism*.
- Define *extinction* and describe two methods that scientists use to study historical evidence of extinction.
- Define *rate of extinction* and *background extinction rate,* and give the best scientific estimate for the latter. What is a *mass extinction*?

What Are the Problems?

The Rising Rate of Extinction

Extinction is a natural process. However, as the human population has grown and spread over the earth and used more and more of the planet's resources, an increasing number of species have lost their habitats and access to the resources they need. As a result, the estimated rate of extinction has risen steadily (Figure 8.12, p. 210).

Biologists Edward O. Wilson (see *Making a Difference*, p. 210) and Stuart Pimm, who together have studied extinction, have estimated that the annual rate of extinction is now about 1,000 times the background rate that existed before modern humans appeared on earth about 200,000 years ago. These scientists projected that during this

FIGURE 8.12 The number of species extinctions has risen as the human population has grown. (Data from U.S. Geological Survey and the U.S. Bureau of the Census)

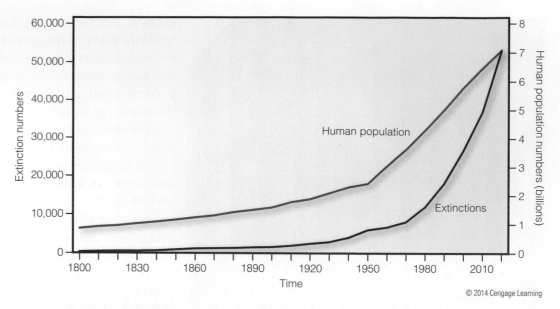

© 2014 Cengage Learning

century, the extinction rate is likely to rise to at least 10,000 times the background rate—largely because of the projected growth of the human population and the projected increase in rates of resource use per person during this century. This could lead to the extinction of 25–50% of all land-based species by 2100. In addition, they noted that

areas that contain most of the world's biodiversity, such as tropical forests and coral reefs, are among the planet's most threatened habitats.

KEY idea   Species are becoming extinct about 1,000 times faster than they were before modern humans appeared on earth. By the end of this century, the extinction rate is projected to be at least 10,000 times higher than the estimated background rate.

MAKING A difference

Edward O. Wilson: Biodiversity Guru

At age nine, Edward O. Wilson (Figure 8.A) became interested in insects. He has said: "Every kid has a bug period. I never grew out of mine." He is now one of the world's leading experts on ants and has applied the results of his ant research to the study and understanding of other social organisms, including humans.

Throughout his career, Wilson has widened his focus to include the nature and interactions of all of the world's species. As a result, he is viewed as one of the world's leading experts on biodiversity and its importance to life on earth. He is also working on Harvard University's *Encyclopedia of Life*, an online database of the planet's known and named species. In 1992, he published *The Diversity of Life*, a landmark book that describes the practical

issues and key principles of biodiversity for a wide audience of scientists and ordinary citizens.

Wilson has won the U.S. National Medal of Science and more than 100 other awards in the United States and in many other countries. He has written hundreds of scientific articles and 25 books, two of which won the Pulitzer Prize for General Nonfiction.

FIGURE 8.A Edward O. Wilson

Endangered and Threatened Species

An **endangered species** is a species whose total population is so small that it could soon become extinct. An example is the Siberian tiger shown in this module's opening photo. Figure 8.13 shows four other examples of endangered species. A **threatened species** (also known as a *vulnerable species*) still exists in numbers high enough to survive in the short term, but because its populations are declining, it is likely to become endangered in the near future. An example is the polar bear, which we discuss later in this module.

Some species have characteristics that raise their chances of becoming extinct. For example, the giant panda bear (Figure 8.4b) eats mostly bamboo, which severely limits its habitat area. The endangered blue whale and endangered sea turtles (Figure 8.13c) travel the oceans within fixed corridors, or *migration routes*, making them vulnerable to potentially harmful human activities such as industrialized fishing.

Species such as the Florida panther (Figure 8.13a) and the grizzly bear are vulnerable because they need large territories to find enough food. Other species, including popular ocean fish such as the orange roughy, are prone to extinction because of their low reproductive rates, which makes it difficult for the species to recover once their populations decline. The Siberian tiger (see module-opening photo), as well as other tiger species, are endangered because of shrinking habitat areas and the high market prices for their skins (Figure 8.14, p. 212) and bones. There are only about 3,200 tigers left in the wild, and extinction experts project that before the end of this century, they will probably be extinct.

FIGURE 8.13 These are just four of many endangered species: (a) a Florida panther, (b) a Venus flytrap, (c) a hawksbill sea turtle, and (d) a whooping crane.

Jo Crebbin/Shutterstock.com

David Huntley/Shutterstock.com

Cigdem Sean Cooper/Shutterstock.com

Al Mueller/Shutterstock.com

FIGURE 8.14 This tiger skin is worth thousands of dollars.

Biologists have identified six major causes of the endangerment of wild species. They can be summarized by using the acronym **HIPPCO**, which stands for **H**abitat destruction, degradation, and fragmentation; **I**nvasive species; **P**opulation growth and rising rates of resource use per person; **P**ollution; **C**limate change; and **O**verexploitation. Now, let's take a closer look at each of the HIPPCO factors.

Let's **REVIEW**

- Why do experts think the rate of extinction has risen steadily and will rise even more sharply during this century?

- How do current and projected rates of extinction compare with the background rate?

- Define and distinguish between *endangered* and *threatened species* and give two examples of each.

- What are two factors that can increase a species' chances of becoming extinct? List the six major causes of the endangerment of wild species.

Habitat Destruction and Damage

The biggest cause of species endangerment is *habitat loss*. For example, by clear-cutting a single forest (Figure 8.15), we eliminate the habitats for hundreds to thousands of

species. In this category, we also include *habitat degradation*, which occurs when natural habitat is damaged such that wildlife can no longer use it. For example, when the pollution in a stream reaches a certain level, the stream can no longer support fish populations.

This category also includes *habitat fragmentation*, which occurs when a large, intact area of habitat is partially destroyed and divided into smaller, isolated patches, typically by roads, crop fields, logging, and urban development. Species such as wolves, which need large areas of forest for mating, rearing young, and feeding, cannot survive in fragmented forests. Figure 8.16 shows the loss and fragmentation of a large area of tropical rain forest in Bolivia.

Many species, including the Indian tiger, the black rhino, and the African and Asian elephants (Figure 8.17), have been threatened by a combination of habitat loss, degradation, and fragmentation. Adding to this threat is the illegal hunting (poaching) of these species for their valuable skins and body parts.

Satellite images, ground level surveys, and other methods of analysis indicate that humans have taken over, disturbed, or polluted about 60–80% of the earth's land surface (not including the polar regions) and about 50% of the earth's waters. Some species are more affected by habitat loss, degradation, and fragmentation than are others. Among those most threatened by these factors are bird species (see *For Instance*, p. 214).

FIGURE 8.15 Extreme clear-cutting near Chiang Mai, Thailand, transformed a tropical rain forest with a diversity of species into this barren land.

FIGURE 8.16 These satellite images show the loss and fragmentation of tropical rain forest in the Brazilian state of Mato Grosso between 1992 (a) and 2006 (b). The forest (appearing in red) was cleared in order to grow crops, graze cattle, and build human settlements

NASA images created by Jesse Allen, using Landsat data provided by the United States Geological Survey and ASTER data provided courtesy of NASA/GSFC/METI/ERSDAC/JAROS, and U.S./Japan ASTER Science Team

African Elephant

Black Rhino

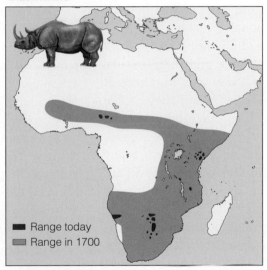

■ Range today
■ Probable range in 1600

■ Range today
■ Range in 1700

Indian Tiger

Asian or Indian Elephant

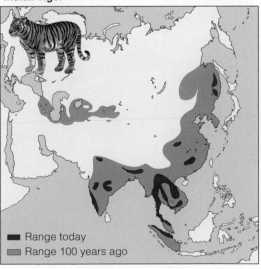

■ Range today
■ Range 100 years ago

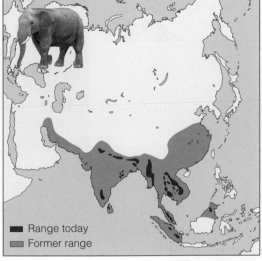

■ Range today
■ Former range

FIGURE 8.17 These maps show the severe loss of habitat for four endangered species: the African elephant, the black rhino, the Indian tiger, and the Asian elephant. *See an animation based on this figure at* **www .cengagebrain.com**. (Data from International Union for the Conservation of Nature and the World Wildlife Fund)

© 2014 Cengage Learning

Birds Are Losing Their Habitats

About 70% of the earth's known bird species are declining in numbers, and roughly 12% are threatened with extinction (Figure 8.13d), mostly because of the loss, degradation, and fragmentation of their habitats. About 75% of the word's threatened bird species live in forests. Many of these forests, especially in the tropical areas in Asia and Latin America, are rapidly being cleared or fragmented by roads and development (Figure 8.16). In North America, populations of many forest songbirds, including tanagers, orioles, and thrushes, which spend their winters in these tropical areas, have also declined sharply.

Among the world's *aquatic*, or water-based bird species, about 40% are in danger, many because of the draining and degradation of their wetland habitats. Other threats to these birds' habitats are oil spills, runoff of pesticides and fertilizers from farm fields, and other pollutants such as sediments and factory emissions.

Another threat to a number of migrating bird species are electrical power lines, cell phone and radio towers, wind turbines, and tall buildings. Every year, hundreds of millions of migrating birds are killed when they collide with such structures wherever they have been erected within bird migration routes.

The extinction of bird species can have a devastating ripple effect on the ecosystems of which they are a part. Birds help to control populations of rodents and insects that would otherwise devour many plants. They also help to sustain plant populations through pollination and dispersal of seeds. Without these free ecosystem services, many plants would become extinct, especially in tropical areas, and this would be followed by the loss of the specialized animal species that feed on those plants.

Let's REVIEW

- Distinguish among habitat destruction, degradation, and fragmentation, and give an example of each.
- Describe the decline of many bird species and explain how the extinction of bird species can affect ecosystems.

Invasive Species

The second biggest factor in the rising rate of species extinction is the deliberate or accidental introduction of harmful invasive species into ecosystems by humans. There are hundreds of examples of such introductions, and many of them have been beneficial. For example, deliberately introduced nonnative crop and livestock species now provide most of the U.S. food supply.

However, many introductions of nonnative species have caused serious ecological harm. For example, in the 1930s, the *kudzu* ("CUD-zoo") *vine* was imported from Japan and planted in the southeastern United States to help control soil erosion. That goal was achieved, but the vine grew so rapidly and was so hard to kill that it took over almost everything in its path, including gardens, trees, and abandoned cars and houses (Figure 8.18). It has overwhelmed many areas of the southeastern United States and is expected to spread northward as the climate gets warmer. So far, no one has found a good way to control the spread of kudzu.

Similar to kudzu, several deliberately or accidentally introduced species have taken over ecosystems where there are few or no natural predators, competitors, or parasites that can help to control their population growth. These invaders can crowd out or decrease populations of some native species and disrupt the ecosystem services they provide.

For example, in recent years African and Burmese pythons (Figure 8.19) and several boa constrictor species, acquired by people as pets, have ended up in Everglades National Park in Florida. Some of these snakes' owners

FIGURE 8.18 The rapidly growing kudzu vine has taken over this abandoned house in the state of Georgia.

FIGURE 8.19 Burmese python (shown here) and other python species, as well as boa constrictors, seize their prey with their sharp teeth, wrap them in their coils, and squeeze them to death.

found it hard to feed and manage them, and dumped them into the park's wetlands where the snakes have reproduced rapidly.

These snakes can be as big around as a telephone pole, grow to 20 feet in length, and weigh more than 200 pounds. Because of their huge appetites, they eat whatever they can catch, including raccoons, birds, pet dogs and cats, full-grown deer, and occasionally an American alligator (Figure 8.7)—the only species besides humans that are capable of killing these predators. These snake populations are slowly spreading to other areas and, by 2100, could be found in many wetlands in the southern half of the United States.

Population Growth

Human population growth is considered a major factor in the rising rate of species extinctions (Figure 8.12), mostly because humans use a large and growing portion of the earth's land and other resources to meet their needs and wants. This leaves other species with fewer resources—especially food, water, and space for habitat.

Population growth also affects aquatic species living in lakes, rivers, and oceans. About 45% of the world's people live on or close to the oceans' coasts, and this percentage is increasing. With this growing coastal population come more boats, road and housing construction, oil leaks and spills, garbage dumping, and other activities that add wastes and harmful chemicals to coastal waters. These activities threaten populations of some fish and other aquatic species, and can upset aquatic food webs.

Pollution

Pollution of water and air is a growing problem that threatens some species with extinction. For example, every year, pesticides kill more than 67 million birds and millions of fish. The U.S. Fish and Wildlife Service estimates that about one-fifth of the endangered and threatened species in the United States are at risk from pesticide poisoning. Also, oil pollution in the Gulf of Mexico adds to the threats that have endangered all species of sea turtles living in the gulf.

Since 2006, certain pesticides have been shown to play a role in the widespread loss of many honeybee colonies in the United States and in parts of Europe because of a mysterious problem called *colony collapse disorder*. Other factors, such as an invasive parasitic mite from Asia, are possibly involved in this disorder. Researchers are trying to learn why this is happening and what can be done to reduce this loss of honeybees, which pollinate almost a third of all U.S. food crops.

Let's **REVIEW**

- How can invasive species threaten native species and the ecosystems in which they live?
- Describe the harmful impacts of one deliberately introduced nonnative species and the impacts of one accidentally introduced nonnative species.
- Why is human population growth a factor in the rising rate of extinction?
- Give two examples of how pollution can threaten wild species.

Climate Change

A growing new threat to many of the world's species is climate change, triggered by an increase in the average temperature of the earth's atmosphere, which is projected to continue rising considerably during this century. A key factor in projected climate change is the growing emissions of carbon dioxide (CO_2) from the burning of carbon-containing fossil fuels, especially coal, oil, and gasoline. Carbon dioxide helps to control the earth's climate as part of the carbon cycle (Figure 1.21, p. 20). When average CO_2 levels rise, the atmosphere gets warmer and when they drop, the atmosphere cools. The problem is that our large-scale burning of fossil fuels is adding CO_2 to the atmosphere faster than the carbon cycle can remove it.

A second factor that affects climate change is the large-scale clearing of forests (Figure 8.15). This *deforestation* eliminates many of the trees and plants that take up much of the excess CO_2 from the atmosphere as part of the carbon cycle. This vegetation is being cleared faster than it can grow back, especially in tropical areas (Figure 8.16).

Projected climate change will likely cause more flooding in some areas and prolonged droughts in others. It will make some habitats unlivable for certain species faster than those species can move to other areas or adapt to the warmer, wetter, or drier conditions. It is already melting ice at the earth's poles and on many mountaintops. These changes are altering some land-based habitats and will likely raise sea levels and destroy areas of biologically diverse coral reefs and coastal wetlands during this century. In these ways, projected climate change is likely to contribute heavily to the rise in the rate of extinction.

Scientists at Conservation International project that by the end of this century, climate change could cause the extinction of one-fourth to one-half of all species of land animals and plants. Scientific studies indicate that polar bears (see the following *For Instance*) and 10 of the world's 17 penguin species are already threatened because temperatures in the polar regions where they live are rising much faster than in other parts of the world.

FIGURE 8.20 This polar bear is feeding on a seal that it hunted and killed on the floating ice of the Arctic sea.

KEY idea According to numerous studies and mathematical models, projected climate change during this century could result in the extinction of one-fourth to one-half of all known land animals and plants.

FOR INSTANCE...

Polar Bears and Climate Change

Globally, there are 19 populations of polar bears, totaling between 20,000 and 25,000 bears. More than half of these large white bears live in the Arctic regions of Canada, and the rest live in Alaska, Greenland, and parts of Northern Europe and Northern Russia.

During winter, when floating sea ice in the Arctic expands, polar bears swim from one ice patch to another, hunting for seals (Figure 8.20). As the temperature rises during the brief summer, this ice melts at the outer edges and the whole mass of ice shrinks. In these warmer months, the bears fast and live off the body fat that they have accumulated during winter. Now, atmospheric warming is causing the floating sea ice on which the bears hunt to melt at increasing rates, which gives them less time for hunting. As a result, polar bears in the Arctic could suffer a sharp drop in their populations.

Scientists are concerned that the summer sea ice may disappear during coming decades, in which case, the now-threatened polar bear would likely become an endangered species. According to scientists at the International Union for the Conservation of Nature (IUCN), the world's total polar bear population is likely to decline by a third or more by 2050. By the end of this century, the bears might be found only in zoos.

The Killing, Capturing, and Selling of Wild Species

A sixth major cause for rising extinction rates is overexploitation—the O in HIPPCO. With the rapid growth of the human population, hunters and fishers trying to help feed growing populations have depleted populations of many wild species.

Overhunting, much of it illegal, threatens gorillas, orangutans (Figure 8.1b), and chimpanzees. Some biologists project that by 2030, most of Africa's large ape species will likely be extinct in the wild and will be seen only in zoos or as stuffed specimens in museums.

Overhunting and other threats to wildlife have led to some legal protection of endangered and threatened species. However, some protected species are still illegally killed or captured for sale—a practice called *poaching*. Poachers sell the live animals and the hides, horns (Figure 8.21), and

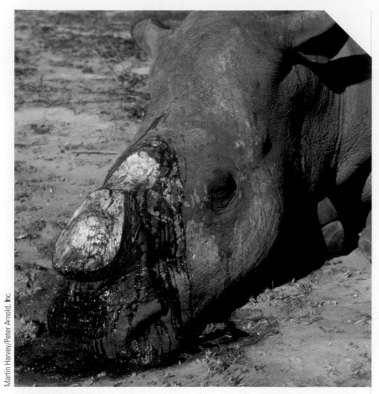

Martin Harvey/Peter Arnold, Inc.

FIGURE 8.21 This threatened white rhinoceros in South Africa was killed by a poacher for its horn. The horn can be worth many thousands of dollars because it is used to make dagger handles and is ground into powders that are used to make medicines.

other body parts that are highly prized, to collectors in the largely illegal wildlife trade. Few of the poachers are caught or punished, and many of the live wild animals transported between countries die in transit.

Poaching is playing a major role in pushing some species toward extinction. Tigers, for example, are highly valued for their furs (Figure 8.14), and for body parts that are used for medicinal purposes in China and other parts of Asia. According to experts, few if any tigers will exist in the wild by 2030 unless emergency actions are taken to control poaching and to preserve what is left of the world's tiger habitat (Figure 8.17, lower left).

CONSIDER this

- A coat made from the fur of the highly endangered Bengal or Indian tiger can sell for $100,000.
- The bones and other body parts of an illegally killed tiger are worth up to $70,000.
- There are about 3,200 wild tigers in the world, down from about 100,000 in 1900, mostly because of habitat loss and illegal hunting.

Wildlife biologists estimate that for every wild animal captured and sold in the pet trade, an estimated 50 others are killed or die in transit. Tropical birds are popular in this market, and at least 80% of these birds are imported legally or illegally from tropical forests, according to the IUCN. As a result of the pet trade, more than 60 bird species—most of them parrots—are classified as endangered or threatened.

The pet trade also affects coral reef ecosystems, especially in Indonesia and the Philippines. Divers go to the reefs and catch tropical fish by squirting poisonous cyanide into the water to stun the fish. This means that many fish die for each fish that is caught and sold. The cyanide solution also kills the coral animals that build and maintain the reefs. Other wild species that are being exploited and depleted because of the pet trade include amphibians and reptiles.

Several plant species, too, are in danger of being depleted, largely because of the buying and selling of these plants for use in decorating houses, offices, and lawns. Several species of orchids and cacti are endangered for this reason.

Let's REVIEW

- What are two ways in which climate change is likely to affect the rate of species extinction?
- Summarize the story of how polar bears survive and how they are threatened by climate change.
- Give three examples of overexploitation of plant and animal species. For each example, give a reason why it happens.

What Can Be Done?

Applying the Precautionary Principle

One strategy for dealing with the loss of species is to try to preserve and grow dwindling populations of wild species that are threatened or endangered. The other major strategy is to prevent the problem from occurring by protecting species' habitats before they become endangered. Increasingly, scientists are urging that we emphasize the latter prevention approach.

We could do this by applying the **precautionary principle**: When evidence indicates that an activity can harm human health or the environment, we should take precautionary measures to prevent or reduce such harm even if scientific research has not established some of its cause-and-effect relationships. Applying this principle to species extinction means taking precautions to protect the habitat of any species that could likely become threatened, even if we don't have conclusive evidence that this will happen.

Some conservationists argue that it is more important to protect the shrinking populations of already endangered species. Protecting habitat and protecting populations both cost money and other resources, all of which are limited, so these two efforts are competing for these resources. There can also be disagreement about which habitat areas and which species should get more or earlier protection. Thus, conservationists are debating about how best to spend their limited resources. Let's look more closely at ways in which we can slow the rate of extinctions caused mostly by human activities.

Protecting Habitat

One way to approach the problem of habitat destruction, degradation, and fragmentation is to establish a worldwide network of wildlife refuges, nature preserves, and other protected habitats. These areas can serve as reservoirs of biodiversity and centers for future evolution.

Setting aside land for wildlife refuges and nature preserves is quite difficult. Only 13% of the earth's land area (not including the polar areas) has been set aside as protected, and just 5% is strictly protected from human activities, even though conservation biologists call for strictly protecting at least 20% of the land. In other words, 95% of the earth's land is set aside for human use, which explains why so many of the planet's land species are increasingly threatened with extinction. Most politically and economically powerful mining and timber companies and land developers oppose even partially protecting 13% of the earth's land because of the valuable resources that might be on or under that land.

The U.S. government has established 548 partially protected National Wildlife Refuges. People can use these areas for limited purposes such as hiking and camping, and in some cases, hunting and fishing. Some of these refuges have been set aside specifically to protect certain endangered species. The trumpeter swan, Florida's key deer, and the brown pelican (Figure 8.22) have been protected in this way, and it has helped their populations to recover from near extinction.

Another approach, originally proposed by one of this book's authors, Norman Myers, is to identify and quickly protect **biodiversity hotspots**—areas of land that are especially rich in highly endangered plant and animal species found nowhere else on the earth. Biologists have identified 34 of these areas (Figure 8.23) where about 86% of the original wildlife habitat has been destroyed or severely degraded by human activities. Although these hotspots cover only about 2% of the earth's land surface, they are home to most of the world's endangered or critically endangered species. Myers and many other conservation biologists call for emergency action to protect these hotspots.

Another related approach to protecting habitat is to find out where endangered and threatened species live, and to pass laws designed to protect such areas from certain human activities. This is an important part of the U.S. Endangered Species Act (ESA). Other countries, including Canada, have passed similar laws.

FIGURE 8.22 The brown pelican (a) was an endangered species in 1903 when the Pelican Island National Wildlife Refuge (b) was established in Florida as the first U.S. national refuge dedicated to protecting an endangered species. The brown pelican was removed from the U.S. endangered species list in 2009.

Jeremy Woodhouse/Peter Arnold, Inc.

George Gentry/U.S. Fish and Wildlife Service

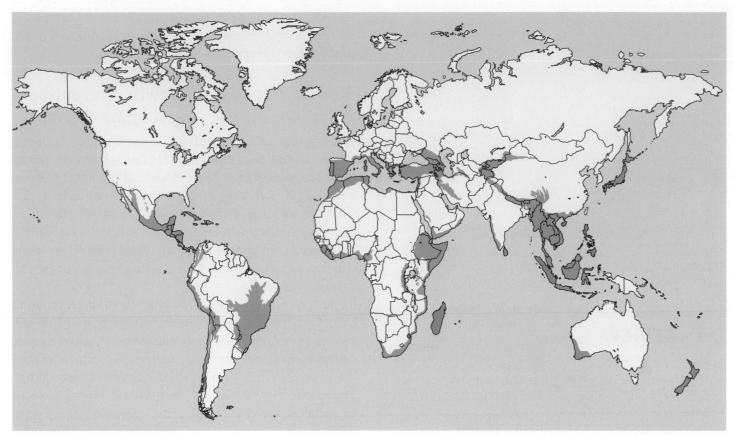

FIGURE 8.23 Conservation biologists have identified 34 biodiversity hotspots that are highly endangered centers of land-based biodiversity. They are in urgent need of protection. *See an animation based on this figure at* **www.cengagebrain.com**. (Data from Center for Applied Biodiversity Science at Conservation International)

The ESA directs the National Marine Fisheries Service and the U.S. Fish and Wildlife Service (USFWS) to identify and list endangered and threatened species and to designate and protect habitat to help those species in recovering their population numbers. The ESA also makes it illegal for Americans to sell or buy any product made from an endangered or threatened species or to hunt, kill, collect, or injure such species in the United States. This process has worked for a number of species, including the American alligator (see *For Instance*, p. 206), the gray wolf, the peregrine falcon, and the bald eagle.

Opponents of the ESA want to weaken or repeal it because it protects land that some people would like to use for various purposes that could harm protected species. They argue that the law puts the rights and welfare of endangered plants and animals above those of people. These critics have pressured the U.S. Congress to cut the funding for enforcement of the ESA and to have the law repealed. Some resent the fact that the law applies to private as well as public lands.

Most conservation biologists and wildlife scientists say that in spite of inadequate funding and other obstacles, the law has been effective. According to the USFWS, more than half of all species listed as endangered now have stable and recovering populations, and 99% are still surviving. This high success rate has been achieved on what many scientists say is a very small budget. Proponents would like to expand the ESA to establish a nationwide set of strictly protected reserves to try to prevent species from getting to the point where they need emergency protection. Figure 8.24 (p. 220) summarizes the major pros and cons of the Endangered Species Act.

CONSIDER this

The 2011 budget for the ESA was $25 million, which

- amounted to an average cost of less than 8 cents for every person living in the United States, and

- is about equal to the cost of eight 30-second Super Bowl TV commercials.

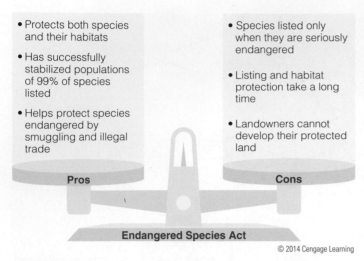

- Protects both species and their habitats
- Has successfully stabilized populations of 99% of species listed
- Helps protect species endangered by smuggling and illegal trade

- Species listed only when they are seriously endangered
- Listing and habitat protection take a long time
- Landowners cannot develop their protected land

Pros

Cons

Endangered Species Act

© 2014 Cengage Learning

FIGURE 8.24 *Weighing the pros and cons* of the Endangered Species Act.

Let's REVIEW

- What is the *precautionary principle*?
- What percentage of the earth's land should be strictly protected, according to conservation biologists? What percentage is strictly protected?
- How does the U.S. government protect habitat? Define *biodiversity hotspot* and explain the hotspots approach to protecting biodiversity.
- State the purpose of the U.S. Endangered Species Act and summarize its pros and cons.

Preventing Invasions of Nonnative Species

When a nonnative species is released into an ecosystem where conditions are favorable for its survival, it becomes next to impossible to reverse the impacts. So the best way to reduce this second-biggest cause of the rising rate of species extinction is to prevent the introduction of potentially harmful nonnative species into ecosystems in the first place.

This is a major challenge, but scientists argue that there are at least three ways to approach it. The first is to step up research efforts to learn more about how certain species become successful invaders and about what makes an ecosystem vulnerable to invaders. A second step would be to conduct new and regular ground surveys and satellite observations to detect and monitor species invasions while using computer modeling to predict how invasive species are likely to spread and what harmful effects they might have. A third major step would be to strengthen legal strategies for dealing with this problem by increasing the frequency of inspections of imported goods that might contain invasive species.

Slowing Population Growth, Preventing Pollution, and Slowing Climate Change

Slowing population growth is a complex and controversial topic, and government efforts to deal with this challenge are covered in another module of this book. It is also a topic that is very personal. Each of us needs to decide whether we believe that population growth should be slowed. We then must decide what our part in slowing it is—whether we choose to ignore it and produce as many children as we want or to take it seriously and to have few or no children.

So far, most pollution laws have focused on trying to clean up air and water pollution instead of trying to prevent it. For example, most air pollution and gases such as carbon dioxide that contribute to projected climate change are caused by the burning of fossil fuels. Thus, shifting away from these fuels over the next few decades would be a major prevention approach that would help to solve three major interconnected environmental problems: the outdoor air pollution that kills at least half a million people a year and damages forests and crops; climate change that is projected to disrupt ecosystems, food supplies, and the habitats of humans and other species; and the rapidly rising rate of extinction and loss of biodiversity discussed in this module.

Making such a shift would be difficult because of the immense political and economic power of the world's fossil fuel industry. However, a growing number of citizens and business leaders are calling for moving from dependence on nonrenewable fossil fuels to using a mixture of renewable energy sources such as the sun, wind, flowing water, and the geothermal heat stored naturally in the earth's crust. Equally important are efforts to cut energy waste, which contributes to pollution and projected climate change.

Many pollution threats to wild species come from our widespread and increasing use of chemical pesticides. This approach to pest control is likely to fail in the long run, because pest insects and weeds multiply so rapidly that they can develop resistance or immunity to a pesticide within a decade or so. Many scientists call for a different approach, known as *integrated pest management* (IPM) in which each food crop and its pests are viewed as parts of an ecological system.

With IPM, when pest populations reach an economically harmful level, farmers use various methods to control the pests. One is *cultivation controls*, such as altering planting times and changing the crops planted in a particular area in order to avoid the pests. Another is *biological controls*, including the use of natural predator insects, parasites, and disease organisms to kill the pests. Beyond those measures, farmers use conventional pesticides only as a last resort, and then very carefully and in small amounts.

In 1986, the Indonesian government banned most of the pesticides used on its rice crops and developed a nationwide program to help farmers switch to IPM. As a result, between 1987 and 1992, pesticide use dropped by 65% and rice production rose by 15%. IPM has also been used successfully in Denmark and Sweden. These countries have cut their pesticide use by more than 50%.

These and other programs show that by using IPM, farmers can cut pesticide use dramatically, without reducing crop yields. Thus, IPM is an important pollution prevention approach that helps farmers to keep producing while reducing their harmful effects on biodiversity. Scientists call for using it more widely, especially in areas where the habitats of threatened and endangered species can be degraded by the use of pesticides on nearby farms.

Let's REVIEW

- Why is prevention the best approach to dealing with invasive species? What are three ways to help prevent the spread of potentially harmful invasive species?

- How are population growth, pollution, and climate change interrelated as contributors to the rising rate of extinction?

- Explain how reducing our energy waste and use of fossil fuels, can help to slow the rate of biodiversity loss.

- What is integrated pest management, and how does using it reduce threats to biodiversity?

Reducing Overexploitation

To deal with over-hunting and poaching, one of the most promising strategies involves education. With the growth of ecotourism, people living near the habitats of endangered species such as sea turtles, tigers, and tropical birds are learning that these animals are worth more alive than dead (see *Making a Difference*, below). For example, conservation biologist Michael Soulé estimated that one male lion living to age 7 can bring in about $515,000 in tourist dollars to Kenya, while the money that a hunter can get for a lion's hide is about $1,000.

MAKING A difference

Pilai Poonswad Offered Poachers a Better Deal

Professor Pilai Poonswad (Figure 8.B) is a biologist teaching at Mahidol University in Thailand. In a rain forest near the university, the poaching of Rhinoceros hornbills (Figure 8.C) was a problem. The rare Rhinoceros hornbill is found in tropical and subtropical forest habitats in parts of Asia. It emits a loud honking squawk and uses its long bill to defend itself from predators such as snakes and monkeys. Its habitat is threatened by logging and agricultural development, and in some areas, local tribesmen kill it for food and to collect its feathers. It is also captured and sold as part of the illegal wildlife trade.

FIGURE 8.B Professor Pilai Poonswad

Dr. Poonswad visited the poachers in their villages and showed them why the Rhinoceros hornbill would be worth more to them as an ongoing attraction for tourists than as something they could sell just once. Some of the poachers chose to make a living by taking tourists into the forest to see these magnificent birds. Because they now have money and time invested in preserving the hornbills, they help in the effort to prevent poaching. The Rhinoceros hornbill's population was dropping steadily, but is now gradually recovering, thanks in part to Dr. Poonswad.

FIGURE 8.C The Rhinoceros hornbill

FIGURE 8.25 This endangered leatherback sea turtle became entangled in a fishing net and lines, and probably would have starved to death if it had not been rescued.

Technology can provide solutions to some overexploitation problems. For example, one type of overexploitation is the overfishing of the oceans by industrialized fishing fleets. It not only threatens many fish species, but also results in the accidental capture of endangered sea turtles. Some turtles get trapped in the large nets towed by fishing boats (Figure 8.25) or caught on one of the thousands of hooks that are dragged along behind some boats. However, several countries now require fishers to use turtle excluder devices, which allow turtles caught by the nets to escape before they drown. There are also special hooks that can catch fish, but are less likely to hook turtles. Use of these devices has reduced the number of sea turtle drownings.

Preserving Individuals of Certain Species

Another way to protect species from extinction is to preserve individual animals, plants, and seeds. Scientists are using about 100 *seed banks* (one underground in the frozen Arctic) to preserve genetic information from endangered plant species by storing their seeds in refrigerated, low-humidity environments. About one-third of the world's known plant species are being preserved in 1,600 botanical gardens and arboreta throughout the world. However,

these seed banks contain only about 3% of the world's rare and threatened plant species.

We can use zoos, aquariums, game parks, and animal research centers as shelters to help build up populations of endangered species for possibly returning them to their natural habitats in the future. While these facilities can help to educate the public about the importance of preventing human activities that hasten species extinctions, we cannot rely on them to preserve species from extinction, mostly because of limited funds and space.

Let's **REVIEW**

- Give an example of how we can use education to help reduce the overexploitation of species. Give an example of how technology can help in this regard.
- What are the limitations of seed banks, botanical gardens, zoos, and aquariums as places to preserve species? What roles can they play in preventing species extinction?

A Look to the Future

The Future of Evolution

The future of evolution should be regarded as one of humankind's most challenging issues. Currently, we probably are the world's greatest evolutionary force, mostly because we are destroying or degrading the habitats of many wild species and harming the species in other ways (HIPPCO). We have little knowledge about how this planet-wide experiment on the earth's life is likely to turn out.

We know that we could lose up to half of all species during this century, that this will be an irreversible loss, and that it could affect our own well-being. Perhaps we could get by without half of all mammals and other vertebrates (animals with backbones). However, if we lost half of all invertebrates (animals without backbones), especially insects that pollinate plants, we would be in trouble within just a few crop-growing seasons.

The mass extinction that we appear to be bringing about with unprecedented speed is the biggest of our environmental problems in terms of the likely duration of its impact and the numbers of people who will be affected. Most other environmental changes are potentially reversible, but once a species is gone, it is gone forever.

Of course, in the long run, evolution will replace the disappearing species with new ones that will occupy the abandoned ecological niches or create new niches. However, data on past mass extinctions indicate that this is likely to take millions of years. While hastening the extinction of many of the world's species, we are also severely degrading or destroying tropical forests, coral reefs, and wetlands, all of which have served as powerhouses of evolution in the past. This means that the natural recovery of biodiversity could take much longer.

Suppose that after this mass extinction is done, it takes the earth's life about 5 million years to recover, as scientific evidence suggests it has after some past mass extinctions. This would be 25 times longer than the estimated 200,000 years that our species has lived on the earth. This 5-million-year loss of biodiversity would affect the next 71,000 generations of humans. Yet, this issue is rarely discussed. How often do you hear leading scientists, ethicists, news reporters, or commentators talking about it?

Despite the severe lack of information about the effects of this essentially irreversible global experiment that we are conducting, we can venture a few guesses about what is likely to happen. For example, there will likely be rapid growth of opportunistic species such as weeds, cockroaches, rats, and flies. Another likely development is that, even if larger vertebrates such as elephants, rhinoceroses, bears, and big cats survive, our largest protected land areas will be far too small, and their populations will also likely be too small, for speciation among these animals in the foreseeable future.

What do such projections tell us about our efforts to protect species and their habitats? Our major strategy has been to look out especially for *endemic species*, those found in only one place, for species confined to small habitats, and for those with specialized niches.

However, the fossil record shows that endemic species often turn out to be evolutionary dead ends because, in general, they do not generate new species. So should we stop protecting such species? Should we spend less time and money on protecting species such as elephants and tigers that breed later in life and produce only a few offspring? Should we instead focus more effort on those insect species and other invertebrates that can produce many thousands of offspring within just a single season? Some biologists argue that this would make speciation and greater biodiversity in the future more likely.

Similarly, should we devote more attention to protecting the threatened evolutionary powerhouses such as tropical forests, wetlands, and coral reefs? Do they deserve to be protected more than other types of ecosystems such as northern forests, grasslands, and wetlands? There are no easy answers to these important questions that we need to confront. They represent areas of ongoing scientific research and debate.

What Would You Do?

This module began with a discussion of why we should care about species extinction. We have now explored this growing problem and some of the possible ways of dealing with it. As with all environmental issues, trying to lessen the human influence on the rate of extinction and the loss of biodiversity really depends on what individuals do—on the daily choices and decisions that each of us makes.

Many people are finding species extinction to be an issue that they want to do something about. Here are three key ways in which they are trying to help reduce the rising rate of extinction:

Considering how each of their choices will affect wildlife habitat.

- Trying to buy only wood products that have been certified as sustainably produced and avoiding products that were made from trees cut from old-growth forests or tropical forests.
- Trying to buy only food that was produced sustainably and avoiding food grown in areas where tropical and other forests have been cleared to make way for farms.
- Avoiding the use of all-terrain vehicles, speedboats, jet skis, and other machines that can degrade or fragment habitat or endanger wild species.

Making choices that will not further the spread of invasive species.

- Not buying wild plants or animals and not removing wild species from natural areas.
- Not dumping fishing bait or fish from aquariums into waterways, wetlands, or storm drains; not releasing wild species purchased as pets into wild areas.
- When camping, cleaning off recreational gear such as tents, canoes, and hiking boots when leaving the campsite.

Contributing as little as possible to the problems of pollution, climate change, and overexploitation of wildlife.

- When visiting or living near bodies of water, taking precautions not to add pollutants to the water.
- Minimizing emissions of climate-changing gases such as carbon dioxide by choosing low-carbon travel alternatives such as bicycles, mass transit, and fuel-efficient cars; using efficient lighting and appliances; and eating locally grown and organically grown foods.
- Choosing not to buy furs, ivory products, and other products made from endangered animal species.

KEYterms

background extinction
    rate, p. 209
biodiversity, p. 202
biodiversity hotspots,
    p. 218
commensalism, p. 208
community, p. 203
ecological niche, p. 204
endangered species,
    p. 211

evolution by natural
    selection, p. 203
extinction, p. 209
generalist species,
    p. 204
habitat, p. 204
HIPPCO, p. 212
indicator species, p. 206
interspecific competition,
    p. 207

keystone species, p. 206
mass extinction, p. 209
mutualism, p. 208
native species, p. 205
niche, p. 204
nonnative species, p. 205
parasitism, p. 208
population, p. 203
precautionary principle,
    p. 217

predation, p. 207
rate of extinction, p. 209
specialist species, p. 205
speciation, p. 204
species, p. 203
threatened species,
    p. 211

THINKINGcritically

1. Explain why we should care about the rising rate of
    species extinction. How do you think this loss of bio-
    diversity will affect your life? How might it affect the
    lives of any children and grandchildren you might
    have?

2. Explain why you agree or disagree that wild species
    should have a right to exist regardless of whether they
    are useful to us. Do we have an ethical obligation to
    help prevent our hastening of any species' extinction?
    Explain.

3. Suppose you had the power to save selected species
    from extinction. First, given the choice between sav-
    ing the American alligator and saving the polar bear,
    which would you choose and why? Second, from all
    the endangered species that you know of, which two
    would you choose to save from extinction (other than
    the alligator and polar bear), and why?

4. List three products that you used during the past week
    that, because of the way they were manufactured or
    delivered, might have contributed directly or indi-
    rectly to the rising rate of extinction. Are there ways in
    which you could have avoided using these products?
    Explain.

5. Do you know of any invasive species problems in or
    near the area where you live? If so, do some research
    to find out what is the nature of the problem—what
    other species have been affected and how, what the
    effects have been on the ecosystem, and what if any-
    thing is being done about the problem.

6. If you could talk to the poacher who killed the white
    rhinoceros shown in Figure 8.21, what would you say?
    Write out three questions that you would ask the
    poacher.

LEARNINGonline

Access an interactive eBook and module-specific interactive learning tools, including flashcards, quizzes, videos
and more in your Environmental Science CourseMate,
accessed through **CengageBrain.com**.

9

Land Degradation

Why Should You Care about Land Degradation?

Suppose you are living in an area where the land appears to be free of any major environmental problems. So why should you care about land that is being degraded in other parts of the world, such as mountaintops in West Virginia that are being blown apart for coal mining or forests in Oregon that have been cut down for timber? (See the module-opening photo.) There are several reasons.

For one, all areas of the world are connected through our life-support system (see Module 1, Figure 1.1, p. 2)—the earth's supply of air, water, and other resources that we all need. What happens on land in one area can affect other parts of the world through connections within this system.

For example, dust blowing off eroded lands in China forms part of a massive cloud of pollutants, called the Asian Brown Cloud (Figure 9.1) that stretches over much of Asia. It affects the air, water, and land of this vast area. In fact, satellites have tracked the spread of long-lived pollutants in this cloud from northern China across the Pacific Ocean to the West Coast of the United States.

A second reason for concern about land degradation is that most of us depend on land all over the world for the *natural resources* (see Module 1, Figure 1.13, p. 13, blue boxes) that help us to live and to run our economies. We use much of the earth's land to produce our food. We get building materials and paper from forests. Many of the medicines we depend on were derived from tropical forest plants, and our life-support system itself depends on the earth's great variety of plants and animals that live in land ecosystems.

A third reason is that land-based ecosystems supply us with vital *natural* or *ecosystem services*, such as climate control and the natural purification of air and water (see Module 1, Figure 1.13, p. 13, yellow boxes). Dense forests (see module-opening photo) absorb carbon dioxide from the atmosphere, which helps to keep atmospheric temperatures from soaring to levels that would drastically change the climate. This change would ruin some areas for farming, endanger certain water supplies, and result in rising sea levels and major flooding in some areas. Also, over many thousands of years, grasslands and forests have built up vital rich topsoil on which farmers depend for producing food crops. Healthy soil and land ecosystems also store and help purify much of the world's drinking water. As more land becomes degraded, these services will be severely strained.

FIGURE 9.1 The Asian Brown Cloud spreads over eastern China.

In short, because everything is connected, we need to be concerned about what happens on the land all around the planet, no less than we are concerned about what happens on the land where we live.

What Do You Need to Know?

The Earth's Land

Scientists have divided the **terrestrial areas**, or the land portions of the earth, into major types according to their characteristics. These characteristics include the types of soil in each area and the varieties of plants and animals that live in it. The major factor determining differences among the earth's land types is **climate**, the general

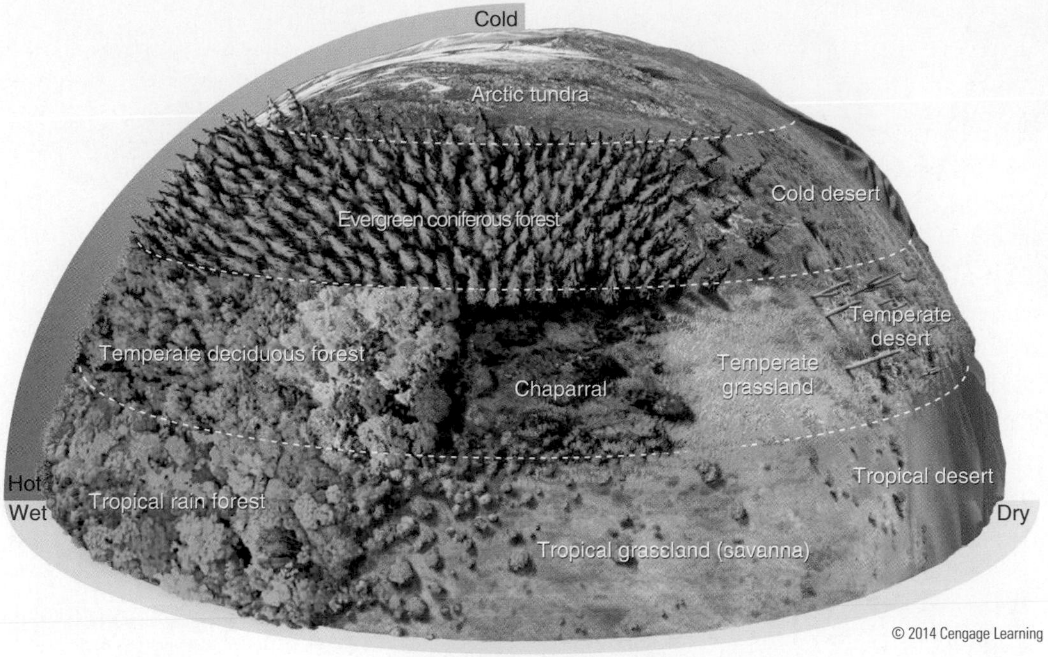

© 2014 Cengage Learning

FIGURE 9.2 Each different biome has certain dominant forms of vegetation, largely determined by its long-term average annual precipitation and average temperature.

pattern of atmospheric conditions in a given region over long periods of time, ranging from at least three decades to thousands of years.

The major types of land are called **biomes**, each being a region of some type of forest, grassland, or desert, with a distinct climate and a certain combination of dominant plant species adapted to the climate. (Animal species also vary across biomes, but biomes are usually described more in terms of their vegetation.) Figure 9.2 shows how the existence of different biomes is related to climate factors, especially average annual precipitation and average temperature. These climate factors, acting together over a long time, help to determine the type of desert, grassland, or forest biome in a particular area. The terms that you see in this figure describing biomes are explained later in this module.

In this module, we focus on four general types of biomes: forests, grasslands, deserts, and mountains. Figure 9.3 (p. 228) shows how scientists have divided the earth's land to reflect the characteristics of these major types of biomes. Each biome contains certain combinations of various life forms adapted to differences in climate, soil, and other environmental factors.

Figure 9.3 shows the biomes with clear boundaries, each covered with one general type of vegetation. In reality, a typical biome area is made up of patches of vegetation, each containing slightly different combinations of plant and animal life forms, but generally conforming to the biome type. Also, Figure 9.3 shows what these areas

were like before humans began covering much of the land with farms, towns, roads, and cities.

Let's **REVIEW**

- Give three reasons to care about land degradation.
- Define *terrestrial area, climate,* and *biome.* Explain how climate is related to biome type. What are the two most important climate factors in this relationship?

Soil

We start our examination of biomes by looking at soil, the foundation of terrestrial life on the earth. **Soil** is a mixture of rock particles, decaying plant and animal matter, water, air, and microscopic organisms. Different types of soil contain widely varying combinations of these components. Figure 9.4 (p. 229) shows how soils form and develop.

Soil is essential to life on land because it supplies many plants with most of their *nutrients*—the chemicals necessary for life processes. By supporting the growth of plants, soil also directly supports plant-eating species and indirectly supports meat eaters such as humans. Most soils that have developed over a long period of time, called *mature soils,* contain horizontal layers, or *horizons* (Figure 9.4). The numbers and types of horizons vary with different types of soils, but most mature soils have at least three of the possible horizons.

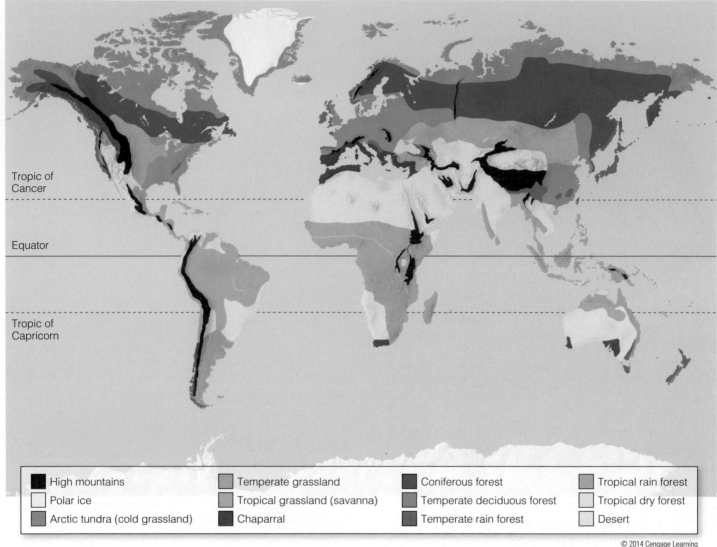

FIGURE 9.3 The earth has a variety of major biomes.

Legend:
- High mountains
- Polar ice
- Arctic tundra (cold grassland)
- Temperate grassland
- Tropical grassland (savanna)
- Chaparral
- Coniferous forest
- Temperate deciduous forest
- Temperate rain forest
- Tropical rain forest
- Tropical dry forest
- Desert

The roots of most plants and most of the soil's organic matter—the decaying plant and animal material that contains the soil nutrients—are concentrated in a soil's two upper layers. The uppermost layer, or the *O horizon*, contains leaves, sticks, and other matter that has not yet completely decayed. The second layer down is the *A horizon*, made of **topsoil**, a thin layer of soil rich in organic matter. The *B horizon* (*subsoil*) and the *C horizon* (*parent material*) are made up mostly of inorganic matter, including rock particles, sand, and gravel. The C horizon lies on a base of rocky material or bedrock.

In most mature soils, the top two layers host bacteria, fungi, earthworms, and small insects. These organisms help to break down decaying plant and animal remains, creating a nutrient-rich mixture of organic matter, called *humus*, an important component of topsoil. Moisture in the soil dissolves the nutrients in humus, which are then drawn up by the roots of plants and used to support plant growth. Plants, in turn, feed most of the earth's terrestrial species.

Soil can also hold water and gases, mostly nitrogen and oxygen gas, which are also used by plants to support their life processes. Certain types of soil act like a sponge, soaking up water from precipitation and releasing it slowly to plant roots and through evaporation into the air. In this process, water gets filtered by the soil and purified. Thus soil performs vitally important ecosystem services.

Biodiversity and Ecosystems

Some biomes have a greater variety of life forms than others do, but each of them supports a certain degree of **biodiversity**—the variety of species, the genetic materials

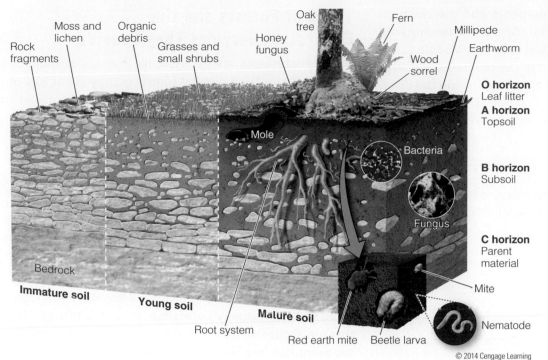

Rock fragments
Moss and lichen
Organic debris
Grasses and small shrubs
Honey fungus
Oak tree
Fern
Millipede
Earthworm
Wood sorrel
Mole
Bacteria
Bedrock
Root system
Red earth mite
Beetle larva
Mite
Fungus
Nematode
Immature soil
Young soil
Mature soil

O horizon Leaf litter
A horizon Topsoil
B horizon Subsoil
C horizon Parent material

© 2014 Cengage Learning

FIGURE 9.4 This diagram shows soil formation and development. Soil *horizons*, or layers, vary in their composition and other characteristics.

they contain, and the earth's forests, deserts, grasslands, oceans, lakes, rivers, and other ecosystems where species live. (It is the basis for one of the scientific principles of sustainability outlined in Module 1, Figure 1.24, p. 26.)

The earth's biodiversity plays a vital role in the formation and sustainability of biomes and it supplies us with food, fibers, energy, and medicines. Biodiversity also plays critical roles in preserving the quality of the air and water, maintaining the fertility of soils, decomposing and recycling waste, and controlling pest populations. Biodiversity is also greatly affected by *disturbances* to the earth's biomes—events or processes that seriously harm the structure or functioning of a natural system—which is the major focus of this module.

An **ecosystem** consists of one or more groups of organisms interacting among themselves and with nonliving matter and energy within a specified area. For example, a tropical grassland (savanna) ecosystem (Figure 9.5) includes plants, animals, and organisms that decompose the wastes and remains of other organisms, all interacting with one another, with solar energy, and with the chemicals in the air, water, and soil. The earth hosts a variety of *terrestrial ecosystems*, such as those we've introduced here, as well as *aquatic*, or *water-based ecosystems*, including oceans, rivers, lakes, and wetlands.

Ecologists classify the organisms in ecosystems, based on their feeding levels, as *producers*, *consumers*, and *decomposers* (see Figure 9.5 and Module 1, pp. 16–17). When one

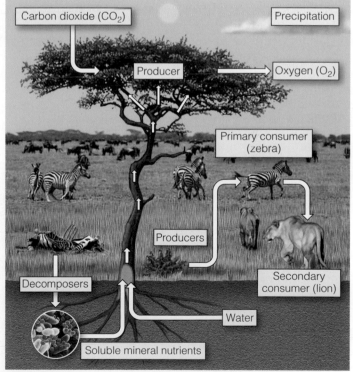

Carbon dioxide (CO₂)
Precipitation
Producer
Oxygen (O₂)
Primary consumer (zebra)
Producers
Decomposers
Secondary consumer (lion)
Water
Soluble mineral nutrients

© 2014 Cengage Learning

FIGURE 9.5 The principal living and nonliving components of an ecosystem in a tropical grassland interact and take part in the cycling of nutrients. (This is the basis for one of the scientific principles of sustainability, see Module 1, Figure 1.24, p. 26.) *See an animation based on this figure at* **www.cengagebrain. com**.

organism feeds on another, nutrients and the chemical energy they contain are transferred from one feeding level to another in *food chains* and *food webs* (see Module 1, Figures 1.17 and 1.18, pp. 16 and 17). The processes of energy flow from the sun and chemical cycling sustain the diversity of life on the earth (encompassing all three scientific principles of sustainability; see Module 1, Figure 1.19, p. 18 and Figure 1.24, p. 26).

KEY idea   The earth's variety of life forms and ecosystems are sustained by energy flow from the sun and by the cycling of nutrients.

Let's REVIEW

- Define *soil* and *topsoil*. Why is soil important to life on land?
- Describe how soil is formed and how different soil horizons compare.
- What is *biodiversity* and why is it important?
- What is an *ecosystem*?
- What two processes sustain life in ecosystems?

We now turn to an exploration of the major terrestrial ecosystems and biomes.

Types of Forests and the Ecosystem Services They Provide

Forest ecosystems are ecosystems that are dominated by trees. Natural and planted forests occupy about 30% of the earth's land surface (Figure 9.3), excluding the planet's polar ice-covered land. Ecologists classify natural forests into two major types, based partly on their age. The first type is an **old-growth forest**: a forest that has not been seriously disturbed by human activities or natural disasters for 120 to several hundred years, depending on the types of trees in the forest (Figure 9.6a). The second type is a **second-growth forest**: a stand of trees that has developed naturally over a number of years following the disturbance of an old-growth forest (Figure 9.6b).

A *planted forest*, often referred to as a **tree plantation** (also known as a *tree farm* or *commercial forest*), is a managed stand of trees, usually consisting of one or two species that are about the same age (Figure 9.7). Tree plantations are eventually harvested for timber to make lumber (see module-opening photo), for pulp to make paper products, or for their fruit (such as olives, Figure 9.7). Forest managers try to protect these plantations from diseases and insects until they can be harvested. Some experts suggest that tree plantations may eventually supply most of the world's demand for these commercial uses, which would help to protect the world's remaining old-growth and second-growth forests.

FIGURE 9.6 This old-growth forest (a) in California's Sequoia National Park is much older than this second-growth forest (b) in northern Wisconsin.

FIGURE 9.7 This is an olive tree plantation on the coast of the Mediterranean Sea.

Forests are extremely important to us for the ecological and economic services they provide (Figure 9.8). For example, they generate oxygen through photosynthesis while removing carbon dioxide from the atmosphere. By performing this ecosystem service, forests help to stabilize the earth's atmospheric temperature and regulate climate change as a part of the carbon cycle (see Figure 1.21, p. 20).

Let's REVIEW

- Define *forest ecosystems.*
- Define and distinguish among *old-growth* and *second-growth forests* and *tree plantations.*
- List three important ecosystem services and three important economic services provided by forests.

Tropical, Temperate, and Cold Forests

The earth's three main types of forest—*tropical*, *temperate*, and *cold*—are each characterized by a certain combination of precipitation level and annual average temperature (Figure 9.2). For example, *tropical rain forests* (Figure 9.9a, p. 232) generally have higher levels of precipitation (Figure 9.9b) and higher temperatures than cold forests have, so they have denser, more diverse vegetation that grows throughout the year. These forests are named for the fact that they grow near the equator (Figure 9.3) where the air is hot and humid and it rains frequently.

Tropical rain forests are among the most diverse ecosystems on the planet. Although they cover only about 2% of the earth's land surface, ecologists estimate that they contain at least half of all known terrestrial plant and

Ecosystem Services

Energy flow and chemical cycling

Wildlife habitat

Water storage and purification

Soil erosion control

Atmospheric carbon storage

Climate moderation

Economic Services

Jobs

Lumber

Pulp to make paper

Firewood

Livestock grazing

Recreation

FIGURE 9.8 Forests provide a number of important ecosystem and economic services.

Photo credits (top to bottom): PhotoLink/Getty Images; Photodisc/Getty Images; Photodisc/Getty Images.

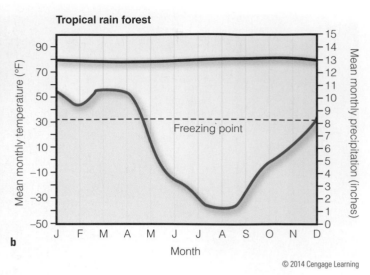

Tropical rain forest

Freezing point

© 2014 Cengage Learning

FIGURE 9.9 The dense uppermost layer of this tropical rain forest in Costa Rica (a) is called the *canopy*. The graph (b) shows the typical variations in annual temperature (red) and precipitation (blue) in a tropical rain forest.

animal species. Their vegetation is made up mostly of *broadleaf evergreen* trees and shrubs that are arranged in layers, according to their need for sunlight (Figure 9.10).

The highest tree tops in a rain forest form the uppermost layer, or *canopy*, which is so dense that it blocks most of the sunlight, resulting in only a dim greenish light reaching the forest floor on the sunniest days. Many of the plants growing in the lower layers have very large leaves to capture as much sunlight as they can.

Most of a rain forest's animal and decomposer species are also arranged in layers because they are highly *specialized*, meaning they each feed on or decompose only one or a few types of other species that live mostly in a particular layer. For example, bats, birds, and many insects live in the canopy layer where they feed on the fruits, flowers,

FIGURE 9.10 A tropical rain forest ecosystem is composed of layers of vegetation that support different species of plants, animals, and decomposers.

© 2014 Cengage Learning

and leaves of the trees. Because of this specialization and layering of life, a small area within a rain forest can host a great variety of species. Biologists have found that a single tree in a rain forest can support several thousand different insect species.

Unlike forests in other parts of the world, tropical rain forests do not have deep nutrient-rich soil. The wastes and remains of plants and animals that fall to the forest floor are decomposed quickly by the abundant decomposers, aided by the forest's hot, humid conditions. Then the nutrients from these broken-down organic materials, instead of remaining in the soil, are quickly taken up by the dense variety of plants in the forest. Thus, the soil does not build up a supply of nutrients.

Unlike tropical forests, *temperate deciduous forests* (Figure 9.11a) grow in areas that have moderate average temperatures and varying seasons. In such temperate areas, lying north and south of the equatorial tropical region of the earth (Figure 9.3), summers are warm and winters are cold enough to have several weeks of below-freezing weather. Precipitation is abundant and typically evenly divided between rain in warmer months and snow in winter (Figure 9.11b).

The dominant trees in this forest biome are *broadleaf deciduous trees*, including oak, maple, aspen, hickory, and beech. As winter approaches and temperatures drop, the trees' leaves turn to various shades of red, orange, and yellow (Figure 9.11a), and later drop to the ground. Deep layers of dead leaves and other decaying organic matter add nutrients to the soil, which unlike that of tropical

forests, becomes nutrient-rich. Each spring, new leaves appear and the trees begin producing again through photosynthesis (see Module 1, p. 15).

Cold forests (Figure 9.12a, p. 234), also called *northern coniferous forests*, *boreal forests*, or *taigas* ("TIE-guhs"), grow in regions lying between temperate areas and the Arctic tundra of northernmost North America, Asia, and Europe (Figure 9.3). They can also be found growing in the high altitudes of some mountain ranges. In these areas, winters are typically long and extremely cold, and daylight hours are few. Summers are short and relatively cool, and precipitation is low, compared to that in tropical and temperate forests (Figure 9.12b).

The dominant trees in boreal forests are *coniferous* (cone-bearing) *evergreen trees*, with leaves that stay on the trees year-round. They include pine, spruce, and fir, which have needle-shaped, wax-coated leaves, along with cedar, and hemlock. These trees have all adapted to be able to withstand the intense cold of winter when soil moisture is frozen.

Partly because of the short growing seasons and long, frigid winters with few hours of sunlight, a limited number of different species of plants can survive in these forests, and thus diversity is low, compared to tropical and temperate forests. Another reason for lower plant diversity is that the conifer needles that do drop off the trees accumulate in a deep layer on the forest floor. These needles release an acid into the soil and because of low temperatures and the waxy coating of the needles, decomposition is slow, so the soil is poor in nutrients. For these reasons, only certain shrubs and mosses that have adapted to such conditions can grow on the forest floor.

FIGURE 9.11 This temperate deciduous forest near Hamburg, Germany (a), is pictured during the fall. The graph (b) shows the typical variations in annual temperature (red) and precipitation (blue) in a temperate deciduous forest.

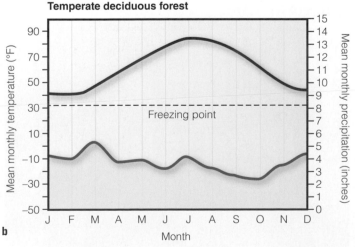

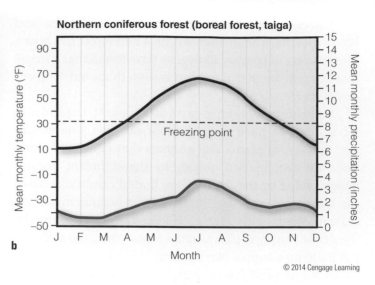

Northern coniferous forest (boreal forest, taiga)

FIGURE 9.12 This cold (boreal or northern coniferous) forest (a) is found in the Rocky Mountains of the western United States and Canada. The graph (b) shows the typical variations in annual temperature (red) and precipitation (blue) in boreal forests.

© 2014 Cengage Learning

Let's **REVIEW**

- Define and distinguish among tropical, temperate, and cold forests. What two factors characterize these different types of forests?

- How does a tropical rain forest support a high level of biodiversity? Why are its species arranged in layers?

- Why are temperate and cold forests generally less biologically diverse than tropical forests?

- How do soils in the three types of forests vary? Why is this so?

FIGURE 9.13 This tropical grassland, or savanna (a), is located in Kenya, Africa. The graph (b) shows the typical variations in annual temperature (red) and precipitation (blue) in a tropical grassland.

Tropical, Temperate, and Cold Grasslands

Grasslands are biomes where a combination of seasonal drought, grazing by large herbivores, and occasional fires prevent large numbers of shrubs and trees from growing, with the result that grasses dominate the landscape. Grasslands are found mostly in the interiors of continents in areas that are too dry for forests to thrive (Figure 9.3). The three main types of grassland—tropical, temperate, and cold—result from varying combinations of annual average temperatures and precipitation (Figure 9.2).

The most common type of *tropical grassland* is called *savanna* (Figure 9.13a), which is warm year-round and sees wet and dry seasons (Figure 9.13b). It is covered mostly with grasses but also has widely scattered clumps of trees. The animals that live on the savanna are grazing animals, including gazelles, zebras, and antelopes, and animals that

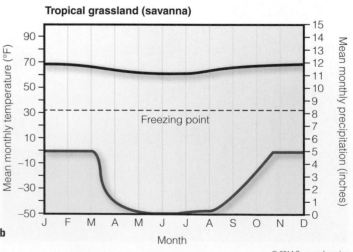

Tropical grassland (savanna)

© 2014 Cengage Learning

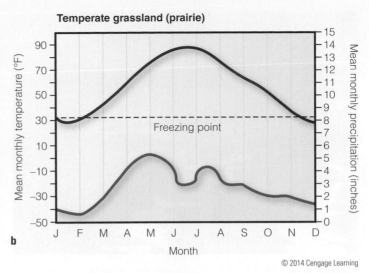

Temperate grassland (prairie)

Freezing point

Month

© 2014 Cengage Learning

FIGURE 9.14 This is a prairie (temperate grassland) (a) in the Midwestern United States, with wildflowers in bloom. The graph (b) shows the typical variations in annual temperature (red) and precipitation (blue) in a temperate grassland.

eat the leaves and twigs of trees such as giraffes, along with their predators, including lions and hyenas.

In a *temperate grassland*, or *prairie* (Figure 9.14a), winters can be long and cold and summers hot and dry. Annual precipitation is fairly light, falling unevenly through the year (Figure 9.14b). Winds blow on most days and evaporate moisture quickly, leading to dry conditions and regular wildfires during summer and fall. Because of these fires, trees cannot take hold on the prairies and grasses are the dominant vegetation.

The variety of grasses growing in this biome have evolved to withstand long dry periods and wildfires. These grasses have deep root systems that are not harmed by the fires and that help to hold moisture in the soil and to hold the soil in place, unless it is plowed up and exposed to erosion. The leaves and stems of prairie grasses die and decompose every year, adding organic matter to the soil, which becomes deep and rich in nutrients. Animal species on the prairies include burrowing animals like prairie dogs, plant-eaters such as grasshoppers, grazing animals including pronghorn antelopes, and predators such as coyotes and golden eagles, along with several other bird species.

Cold grasslands, or *Arctic tundra*, are flat, treeless plains (Figure 9.15a, p. 236) that lie to the north of the boreal forests and south of the Arctic ice cap (Figure 9.3). This biome endures long, dark, frigid winters with high winds and low precipitation (Figure 9.15b). During the summer months the sun shines almost around the clock and precipitation increases slightly but is still low.

The tundra's low-growing plants, including grasses, mosses, lichens, and dwarf shrubs, have just two months of summer to grow. Because of the brief growing season and cool-to-frigid conditions, trees and shrubs cannot grow in the tundra. Animals of this biome include a variety of insects, birds, grazing animals like the caribou, and predators such as the Arctic fox and the grizzly bear. Many of these species, including the birds and caribou, migrate to the tundra during the short summer to find food and raise their young before traveling south again as the frigid winter sets in.

The soil in the tundra develops very slowly because of the limited plant growth and low rate of decomposition. Because of the extreme cold, the deeper soil, called *permafrost*, stays frozen for long periods of time—2 years or more—and rarely thaws except during exceptionally warm summer months. During most summers, the permafrost layer holds water aboveground where it forms marshes, ponds, and other wetlands (Figure 9.15a). Mosquitoes, black flies, and other insects thrive in these wetlands, forming the base of the food web.

Because its soil forms so slowly and never becomes rich in nutrients, the tundra is a fragile biome. The soil and vegetation recover very slowly from damage caused by activities such as mining, oil-drilling, and construction of pipelines.

Let's **REVIEW**

- Define *grasslands* and name and describe the three major types of grasslands.
- Why are prairies dominated by grasses? Why is it that trees and shrubs cannot grow in the tundra?
- How does permafrost support the tundra ecosystem?

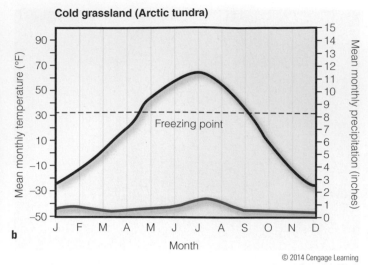

Cold grassland (Arctic tundra)

FIGURE 9.15 Arctic tundra (cold grassland) **(a)** is shown here during summer in Alaska's Arctic National Wildlife Refuge (ANWR). The graph **(b)** shows the typical variations in annual temperature (red) and precipitation (blue) in a cold grassland.

© 2014 Cengage Learning

Mountains

Mountains are high lands where dramatic changes in altitude, slope, soil, and vegetation are contained within relatively small areas (Figure 9.16). Mountainous land covers about one-fourth of the earth's surface.

Mountains, from their foothills to their peaks, host a variety of ecosystems, which can include tropical, temperate, and boreal forests, meadows, and tundra-like ecosystems called *alpine tundra*. Many mountaintops are covered with ice and snow, and for thousands of years, this has helped to keep the earth's climate at moderate temperatures by reflecting sunlight back into space. These ice-covered mountaintops also store large amounts of water. In the warmer weather, much of their snow and ice melts and flows into streams for use by humans and wildlife. Billions of people depend on this water for irrigation and drinking water supplies.

Mountains play an important role in supporting biodiversity because they hold most of the world's forests, which are habitats for a great number of species, many of them found nowhere else on earth. Mountains also provide habitats for animal species that migrate from lowland areas, driven by human expansion and development in those areas or searching for a cooler climate as the average atmospheric temperature warms.

NUMB3RS

4 billion
The number of people who depend on mountain ice and snow for some or all of their drinking water

However, some of these ecosystem services, including the reflective, cooling effect, are increasingly threatened. Because the earth's atmosphere is warming, much of the mountaintop ice is melting at increasing rates. As the ice melts, more mountaintop rock is exposed and, being darker than ice, it absorbs solar energy. As a result, warming increases, which melts more mountain ice in an escalating cycle of change.

Tropical, Temperate, and Cold Deserts

A **desert** is an area where average annual precipitation is very low and occurs unevenly throughout the year. Varying combinations of average annual temperatures and

FIGURE 9.16 Mountains such as these in Russia feature spectacular landscapes.

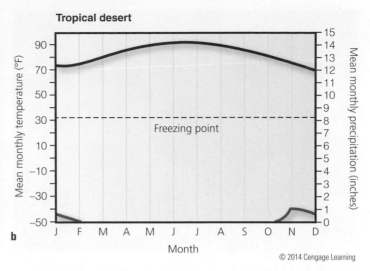

FIGURE 9.17 This *tropical desert* (a) is part of the Sahara Desert in southern Morocco. The graph (b) shows the typical variations in annual temperature (red) and precipitation (blue) in a tropical desert.

precipitation lead to the formation of tropical, temperate, and cold deserts.

The driest of all deserts are *tropical deserts* (Figure 9.17a), which are hot and dry throughout almost all of the year (Figure 9.17b). They are often quite rocky and winds blow sand into deep, shifting dunes that cover much of their surface. Biodiversity is very low in these deserts, with just a few species of plants and animals.

In *temperate deserts* (Figure 9.18a), precipitation is somewhat higher and more evenly distributed throughout the

year than it is in tropical deserts (Figure 9.18b). Because these deserts are located farther away from the equator, temperatures are also more variable, dropping lower during winter months. There is also more vegetation, although still dispersed, primarily various species of shrubs, cacti, and other plant species adapted to the dry conditions and temperature variations. Animal species living there include rattlesnakes, lizards, kangaroo rats, roadrunners, quail, and hawks.

Cold deserts (Figure 9.19a, p. 238) lie north and south of temperate areas, which means that winters are cold, summers are not quite as hot as in other types of deserts, and precipitation is low (Figure 9.19b). Vegetation is sparse, consisting mostly of low-growing plants adapted to the dry, colder conditions.

FIGURE 9.18 This is a *temperate desert* (a) in the state of Arizona, with saguaro cactus, a prominent species in this ecosystem. The graph (b) shows the typical variations in annual temperature (red) and precipitation (blue) in a temperate desert.

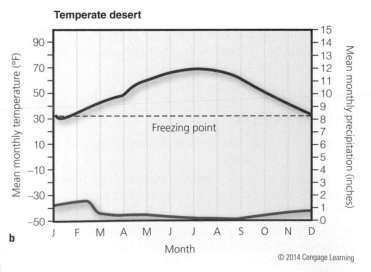

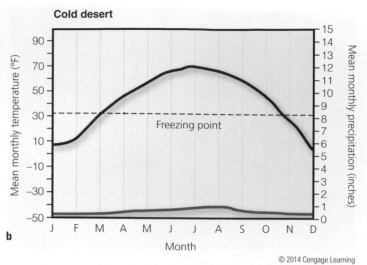

Cold desert

© 2014 Cengage Learning

FIGURE 9.19 This two-humped Bactrian camel (a) lives in Mongolia's Gobi Desert, a *cold desert*. The graph (b) shows the typical variations in annual temperature (red) and precipitation (blue) in a cold desert.

Desert plants and animals have evolved to be able to survive long, hot, and dry periods. For example, cacti can store water in their tissues and their tough outer casings prevent evaporation of this stored water. Some plants such as the mesquite tree drop their leaves during long dry periods to retain as much moisture as possible.

Many desert animals are ground-dwellers, small enough to hide during the day in shady spots such as crevices in rocks. They come out to feed only at night or early in the day. Camels and other larger animals (Figure 9.19a) can drink a lot of water whenever they can get it and store it in their fat for times when it is unavailable. Other animals such as the kangaroo rat do not need to drink water but get it from the foods they eat.

Of all the biomes, desert ecosystems are among the most fragile and can take as much as hundreds of years to recover from disturbances such as the tearing up of soils by off-road vehicles. This is because plant growth, decomposition, and nutrient cycling are slow in the desert, so soils are dry and fragile.

Let's **REVIEW**

- Define *mountain*. What are three important ecosystem services provided by mountains?
- Define *desert*, name and describe the three major types of desert, and explain where on the planet each major type is found.
- What are three examples of ways in which desert plants and animals have evolved to survive desert conditions?

Change within Ecosystems

Changing environmental conditions bring about changes in the makeup of ecosystems. Some of these changes in environmental conditions can have natural causes. For example, lightning starts some forest fires. However, human activities can also result in environmental changes when, for example, we clear forests and grasslands for agriculture. In response to such changes, there can be a gradual change in species composition in a given area—a process called **ecological succession**. During this process, some species arrive in an area and their populations grow, whereas populations of other species shrink and may even disappear.

Ecologists study two main types of ecological succession. *Primary ecological succession* takes place in areas where there is no soil and the very slow process of soil formation must take place before certain species of plants can arrive and begin the gradual process of building a plant community. These processes usually take many decades to thousands of years to occur. For example, when a volcano erupts, it can result in a layer of rock covering a large area where no plants will grow immediately. Primary succession can also take place in water-covered environments where there is no bottom sediment. For example, some commercial fishing methods can leave large areas of the ocean floor scraped clear of vegetation and bottom sediments.

The most common type of succession is *secondary ecological succession*, in which a series of ecosystems with different species develop in places already having a soil base (or bottom sediment in the case of an aquatic system). It takes place in an area where an ecosystem has been disturbed to some degree but where a soil or sediment

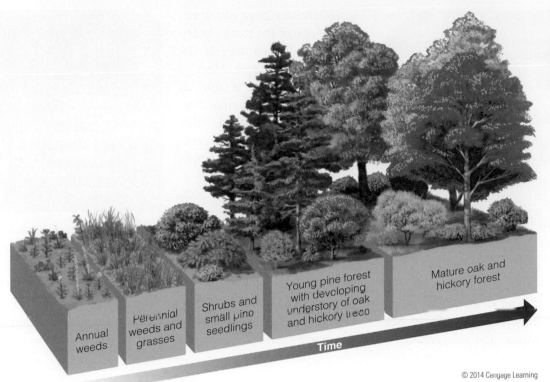

FIGURE 9.20 In this example of secondary ecological succession, plant communities became established in an abandoned farm field. It can take hundreds of years for such an area to become covered with a mature forest.

Labels in figure: Annual weeds · Perennial weeds and grasses · Shrubs and small pine seedlings · Young pine forest with developing understory of oak and hickory trees · Mature oak and hickory forest · Time

base remains. For example, a severe forest fire can leave a large area open and treeless in a matter of hours. Another common example is an abandoned farm field (Figure 9.20) where new vegetation can begin to grow within a few weeks from seeds already in the soil and from seeds blown in by wind or carried to the land by birds and other animals.

It was once thought that ecological succession followed an orderly, predictable process from early to later stages, resulting in mature ecosystems. However, after many decades of research, scientists now generally agree that they cannot predict the course of any case of ecological succession. It is a chaotic process involving an ongoing struggle and competition among all kinds of species for enough light, water, nutrients, and space. While general patterns of succession can be observed, most ecologists now recognize that ecosystems are generally in a state of continual disturbance and change.

Let's REVIEW

- Define *ecological succession*, and distinguish between primary and secondary ecological succession.
- Explain how secondary ecological succession typically works in an abandoned farm field.

What Are the Problems?

Changing the Face of the Earth

With great ingenuity and energy, we humans have made use of the earth's resources to provide an amazing array of material goods for ourselves. In the process however, we have changed the nature of many biomes as we have spread across the earth's surface. For example most of the vast, sweeping prairies that once covered the Midwestern United States (Figure 9.21, p. 240) no longer exist. They have now been replaced by farms, ranches, towns, and cities.

In supplying billions of people with goods and services over several generations, humans have had a very large environmental impact. We can think of this impact as our **ecological footprint**—the amount of biologically productive land and water needed to supply the people in a particular country or area with food and other renewable resources, and to absorb and recycle the wastes and pollutants produced by such resource use. In Module 1, Figure 1.23 (p. 24) shows that we would need an estimated 1.3 earths in the short term to sustain our use of renewable resources at 2008 consumption levels indefinitely, and two earths to sustain the levels projected for 2045.

One of the most widespread and most harmful of all the environmental problems that contribute to our

Fedorov Oleksiy/Shutterstock.com

FIGURE 9.21 Farm fields such as these have replaced most of the natural grasslands that once covered the Midwestern United States.

growing ecological footprints is **land degradation**. It is a catch-all label that includes deforestation, desertification, soil erosion, soil salinization, and overgrazing of grasslands by large and growing herds of livestock—problems that we explore in the rest of this module. Most of this degradation is due to human efforts to obtain firewood, grow crops, feed cattle, and mine the minerals necessary for producing the goods that we use.

According to the 2005 United Nations Millennium Ecosystem Assessment, 60% of the world's major terrestrial ecosystems are being degraded or used unsustainably, as the human ecological footprint gets bigger and spreads across the globe. This environmental destruction and degradation is increasing in many parts of the world. Figure 9.22 summarizes some of the human impacts on the world's forests, grasslands, deserts, and mountains.

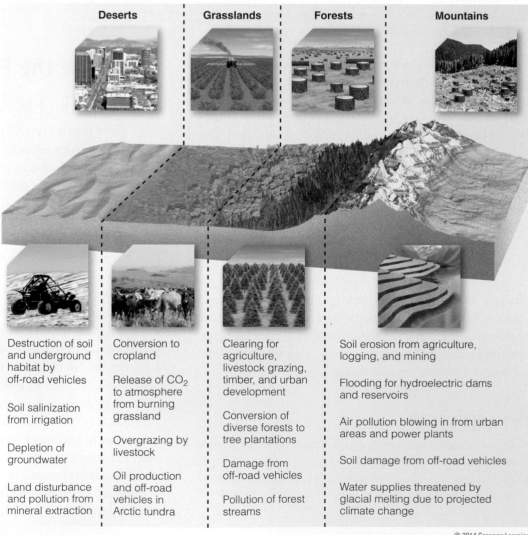

FIGURE 9.22 These are some of the major human impacts on the world's forests, grasslands, deserts, and mountains.

Deserts	Grasslands	Forests	Mountains
Destruction of soil and underground habitat by off-road vehicles	Conversion to cropland	Clearing for agriculture, livestock grazing, timber, and urban development	Soil erosion from agriculture, logging, and mining
Soil salinization from irrigation	Release of CO_2 to atmosphere from burning grassland	Conversion of diverse forests to tree plantations	Flooding for hydroelectric dams and reservoirs
Depletion of groundwater	Overgrazing by livestock	Damage from off-road vehicles	Air pollution blowing in from urban areas and power plants
Land disturbance and pollution from mineral extraction	Oil production and off-road vehicles in Arctic tundra	Pollution of forest streams	Soil damage from off-road vehicles
			Water supplies threatened by glacial melting due to projected climate change

© 2014 Cengage Learning

Let's REVIEW

- Define *ecological footprint*. What are some effects of expanding ecological footprints?
- Define *land degradation*. What are five major types of land degradation?

Deforestation

One of the most disruptive forms of land degradation is **deforestation**, the removal of large expanses of forest for agriculture, settlements, mining, firewood, or other uses. All forests naturally undergo change, with sections of any forest being damaged by natural fires and storms, but over periods of 100 to 200 years, they are generally restored by secondary ecological succession (Figure 9.20) if they are left undisturbed. However, many areas now undergoing deforestation for human purposes are being degraded much faster than natural processes can renew them.

Surveys by the World Resources Institute (WRI) indicate that over the past 8,000 years, human activities have resulted in the clearing of nearly half of the earth's original forests. Most of this loss has occurred since the 1950s. According to the WRI, if current deforestation rates continue, about 40% of the world's remaining intact forests will have been logged or converted to other uses by 2030, if not sooner.

Of special concern to ecologists is the loss of tropical forests. Climate data and biological research indicate that we have cut down or degraded at least half of these lush forests that existed for millions of years before the appearance of humans. The majority of tropical forest loss has taken place since 1950, as people have cleared the forests in order to grow crops, graze cattle, mine mineral resources, and build settlements (Figure 9.23). This form of land degradation is increasing as the world's population and resource use continue to grow.

CONSIDER this

Ecologists warn that without strong conservation measures, most of the world's tropical rain forests will probably be gone or severely degraded within the next 20 to 40 years.

The rapid loss of ancient tropical forests is reducing the earth's biodiversity by destroying or degrading the habitats of many of the unique plant and animal species found in these forests. Ecologists warn that perhaps a quarter of the world's species could become extinct in these forests during this century. Rapid tropical deforestation is

FIGURE 9.23 This satellite image reveals the loss of tropical rain forest, cleared for farming, cattle grazing, and settlements, near the Bolivian city of Santa Cruz between (a) June 1975 and (b) May 2003.

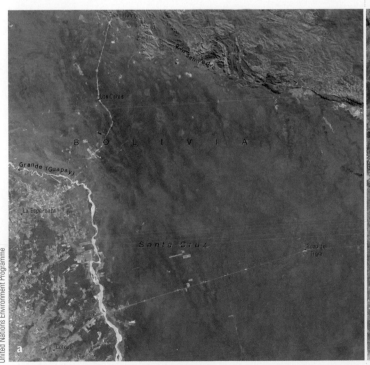

also drastically reducing the amount of vegetation that removes carbon dioxide from the atmosphere, which will help to accelerate projected atmospheric warming and the resulting climate change during this century.

For example, according to a 2005 study by forest scientists in the Amazon River basin of Brazil, the widespread use of fires to clear land is changing weather patterns by eliminating large tracts of forest that help to generate rain. The shade of the dense canopies is also eliminated so that the forest floor now heats up much more. These changes are converting large areas of forestland to savanna (Figure 9.13). Scientists project that, at the current rate of deforestation, most of the Amazon basin could become savanna by 2080.

In some cases, the clearing of trees (Figure 9.24) also dehydrates the forest's soil by exposing it to sunlight, and the dry topsoil can then be blown or washed away. In this situation, a deforested area can reach a point, beyond which hardly anything can grow back and it then becomes more like a tropical grassland or tropical desert area. In short, by eliminating large areas of trees faster than they can grow back, we change the nature of these forests and degrade the vital ecosystem services they provide, such as biodiversity, water storage, and climate moderation.

Similar to tropical forests, temperate forests (Figure 9.11a) have been extensively cleared for activities such as mining, ranching, farming, road building, and urban development. Worldwide, this biome has been disturbed by human activity more than any other, mostly to grow crops because of its fertile soils and moderate climate.

The news about forests is not all bad. In 2007, the UN Food and Agriculture Organization (FAO) reported that in several countries, including the United States, forest cover had increased or stayed about the same between 2000 and 2005. Some of this growth in forest cover was natural, due to secondary ecological succession (Figure 9.20) in cleared forest areas and abandoned croplands. Other cases of expanded forest cover were due to the planting of commercial tree plantations (Figure 9.7). However, because forest plantations have a low level of biodiversity, they cannot provide the ecosystem services that we get from old-growth and second-growth forests.

In the United States, by 1920, most of the old-growth forests had been cut down by farmers and loggers. However, most of the forest area that was cleared has now been regenerated as second- and third-growth forests in every region of the country, except in several western states. The country's total forested area is still increasing, and partially protected forests now make up about 40% of the total U.S. forest area. Most of this area is contained in the 155 forests that make up the National Forest System, managed by the U.S. Forest Service.

Let's REVIEW

- Define *deforestation*. Why is tropical deforestation a special concern to many scientists?
- Explain how large-scale cutting of tropical forests can change the nature of these forests.
- Describe the status of forests in the United States.

FIGURE 9.24 This area of tropical forest in Brazil's Amazon River basin has been deforested.

iStockphoto/luoman

Degradation of Forests

The clearing of forests and other types of forest degradation usually start when roads are built to gain access to the forest, usually resulting in a number of harmful effects (Figure 9.25). At the very least, a road usually fragments forest habitat, and this threatens biodiversity. Often, road construction can cause increased erosion and sediment runoff into streams and lakes. Logging roads also expose forests to invasions by insect pests, disease-causing organisms, and nonnative species. They also open undisturbed forests to logging, mining, farming, off-road vehicle use, and other disruptive activities.

Logging of forests is done in a number of ways, the most destructive of which is *clear-cutting*, in which loggers remove all or most of the trees from an area (see module-opening photo and Figure 9.26b). With the less intensive method called *selective cutting*, loggers leave many trees standing, selecting only certain intermediate-aged or mature trees to cut (Figure 9.26a). With this method, forests can be sustained indefinitely as long-term second-growth forests. Similarly, a variation of clear-cutting, called *strip cutting* (Figure 9.26c), is more sustainable than clear-cutting. Narrow strips of forest are clear-cut, but they can regenerate much more effectively than a large area of clear-cut forest can.

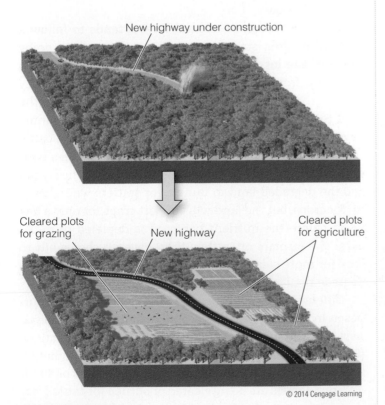

FIGURE 9.25 Building a road into a previously inaccessible forest can harm the forest ecosystem.

(a) Selective cutting

Clear stream

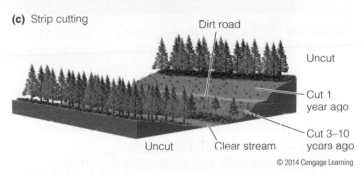

(b) Clear-cutting

Muddy stream

(c) Strip cutting

Dirt road

Uncut

Cut 1 year ago

Uncut   Clear stream   Cut 3–10 years ago

© 2014 Cengage Learning

FIGURE 9.26 This diagram shows three tree-harvesting methods: (a) selective cutting, (b) clear-cutting, and (c) strip cutting.

Scientists have found that removing all the trees from an area can lead to sharp increases in water runoff and soil erosion. This can, in turn, strip nutrients from the forest soils, which causes more vegetation to die and can leave barren ground that can suffer further erosion. More erosion also means more pollution of streams in the forest. All of these impacts damage or destroy terrestrial and aquatic habitats and further degrade biodiversity.

Fires can also degrade forests, although most temperate and boreal forests have evolved to withstand some level of seasonal fire damage and to benefit from fires in certain ways. There are two types of fires that affect forest ecosystems. *Surface fires* (Figure 9.27a, p. 244) usually burn only the leaf litter and low-growing plants on the forest floor. They spare most mature trees, and allow most wild animals to escape.

Surface fires can be beneficial to forests because they burn away flammable ground material that would serve as fuel for more destructive fires. They also aid ecological succession in several ways. When the forest floor litter and undergrowth burn, mineral nutrients important for new plant growth are released. These fires also release

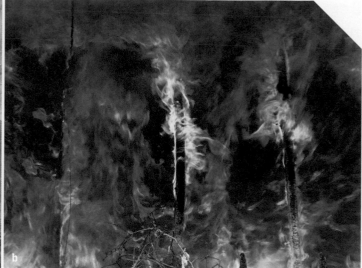

FIGURE 9.27 These photos show a surface fire (a) and a much more destructive crown fire (b).

seeds from the cones of lodgepole pines and help other tree species such as giant sequoia and jack pine to germinate and grow. Fires also help to control tree diseases and insects that can be harmful to forests.

The more dramatic and destructive fires that we often see in pictures and videos are called *crown fires* (Figure 9.27b). They become so hot that they generate their own winds and leap from treetop to treetop, engulfing whole trees in flame. They can kill most vegetation and many animals, destroy human structures, and leave the forest topsoil exposed to erosion. A crown fire can occur more easily in a forest that has gone for several decades without a surface fire. In such a case, dead wood, leaves, and other flammable ground litter has built up, making the forest a tinder box for an explosive fire.

Rising temperatures and expanding drought due to projected climate change will likely dry out many forest areas, making them more vulnerable to crown fires. These changes will also make some forest areas more suitable for insect pests, which could multiply and kill more trees. More dead trees will also provide fuel for any fires that get started in these forests.

Causes of Tropical Forest Degradation

Ecologists have identified a number of direct causes of tropical forest deforestation that stem from underlying causes. For example, population growth and poverty are underlying causes that force people to move into forests in search of food and firewood. Their activities—cutting and burning trees and building settlements—are direct causes of some deforestation and other forms of tropical forest degradation.

Other, more destructive direct causes are timber harvesting, cattle grazing, mining, and the establishment of vast plantations of crops such as soybeans and oil palms. One underlying cause of these activities is a demand for biofuels such as biodiesel produced from palm oil. Another is government *subsidies*, which are payments to timber, cattle, mining, and biofuel companies to help their businesses grow.

The degradation of tropical forests tends to follow a pattern. It often starts with foreign corporations doing much of the logging in a particular country, often with subsidies from that country's government. After they remove the best timber, the companies typically sell the land to ranchers who clear or burn away the remaining vegetation to make way for grazing cattle. After the cattle overgraze the land, within a few years, the ranchers typically move their cattle to another forest area. They often sell the degraded land to farmers or settlers who plow it up for crops, but the land will support crops for only a few years before the nutrient-poor soil is depleted of nutrients. The farmers or settlers then have no choice but to look for more cleared land. Together, these different uses degrade the land in a destructive cycle.

Different tropical forests are being used for varying purposes. In Indonesia, Malaysia, and other areas of Southeast Asia, forests are being replaced by oil palm plantations (Figure 9.28) that produce palm oil for cooking, cosmetics, and biodiesel fuel. In the Amazon and other parts of South America, forests are being cleared or burned for cattle grazing and for large fields of soybeans. In Africa, people are clearing plots for small-scale farming and for firewood.

FIGURE 9.28 The demand for palm oil has led to the planting of large oil palm plantations like this one in Indonesia. This has caused widespread tropical deforestation.

In Indonesia, every minute on average, an area of land equal to that of a football field is cleared and planted with oil palms.

Let's **REVIEW**

- How do new forest roads typically lead to forest degradation? Explain how clear-cutting can harm a forest.

- How do fires affect forests? How can a fire be beneficial to a forest?

- Name two underlying causes of tropical forest degradation. For each of these, name two direct causes of this degradation that are related to the underlying cause.

Desertification

Another threatening form of land degradation is **desertification** or loss of soil moisture to the point where the land loses some or all of its *productivity*, or ability to produce vegetation. It occurs primarily in areas where the climate is *arid*, or dry. Desertification does not literally mean "conversion of land to desert" except in some extreme cases such as that shown in Figure 9.29. In most cases, the land loses half or less of its productivity because of topsoil losses.

Desertification results from natural causes such as prolonged drought. However, many human activities have contributed to climate change that is now making drought conditions worse in some areas. The destruction of natural vegetation through human activities such as farming and overgrazing by livestock can also contribute to desertification.

The FAO estimated in a 2007 report that at least 70% of the world's arid lands used for agriculture are degraded and threatened by desertification (Figure 9.30, p. 246). As a result, this problem affects an area of the earth's land roughly equal to the area of China or of the United States.

According to a 2007 study by the Intergovernmental Panel on Climate Change, severe and prolonged drought

FIGURE 9.29 Expanding sand dunes threaten to engulf this agricultural area in the Sahel region of West Africa.

Moderate

Severe

Very severe

© 2014 Cengage Learning

will likely be a growing problem during the rest of this century, due to projected climate change. This will expand desertification in dry areas of the world and could lead to a sharp decline in overall food production, along with water shortages for as many as 3 billion people—more than 40% of the current world population. Many of these people could become environmental refugees searching their regions of the world for enough food and water to survive.

Overgrazed Rangelands

One of the direct causes of desertification is **overgrazing**, which occurs when too many animals graze for too long, exceeding the ability of a grassland area to support them.

Before humans began domesticating animals and managing herds of grazing livestock, grasslands supported wild herds of grazing animals. Eventually, people learned to manage herds on **rangelands**—temperate or tropical grasslands that supply vegetation for grazing (grass-eating) and browsing (shrub-eating) animals. Now, cattle, sheep, and goats graze on about 42% of the world's rangelands.

Rangeland grass is a renewable resource as long as it is not overused. Its blades grow from the base of the plant, not from its tip. So as long as only the upper half of the blade is eaten, a rangeland can be grazed repeatedly. Lightly grazed rangeland provides important ecosystem services such as soil formation and nutrient cycling. However, on overgrazed land, the grass's tissues are usually eaten down to the soil level. This can damage and kill grasses and expose grassland soil to erosion by water and wind (Figure 9.31). Overgrazing also compacts the soil, decreasing its ability to absorb and hold rainwater. When soil is exposed and degraded in this way, certain invasive plants like mesquite, sagebrush, and cheatgrass can take over a rangeland, making it unfit for grazing.

Let's **REVIEW**

- Define *desertification* and explain how it can lower the productivity of land. What effect will projected climate change have on this process?

- Define *overgrazing* and *rangelands*. What are two effects that overgrazing can have on land?

FIGURE 9.31 Overgrazed rangeland (left) is subject to a number of problems. Lightly grazed rangeland (right) can support grazing animals indefinitely.

USDA Natural Resources Conservation Service

FIGURE 9.32 These photos show the erosion of topsoil by (a) flowing water on a farm in Tennessee and (b) wind on a field in Iowa.

Soil Erosion

One of the biggest land degradation problems that we face is **soil erosion**—the movement of soil components, especially surface litter and topsoil, from one place to another by flowing water and wind (Figure 9.32).

In natural grasslands and other ecosystems that are covered by vegetation, the roots of plants help to anchor the soil, which then stores water and nutrients and releases them slowly. Although soil erosion occurs naturally, soil in undisturbed ecosystems usually is not lost faster than it forms. However, when soil is laid bare by activities such as the plowing of grassland and the clearing of forests, it can be washed or blown away much more quickly than it forms. When soil erosion caused by flowing water is severe enough, it forms gullies (Figure 9.33). It can take just a few years to destroy topsoil that took hundreds to thousands of years to form.

Soil erosion can limit food production if it becomes severe enough. This is because erosion carries away important plant nutrients, resulting in a condition commonly

FIGURE 9.33 Severe gully erosion has damaged this cropland in Bolivia.

FIGURE 9.34 Soil erosion is a serious problem in many areas of the world. (Data from UN Environment Programme and the World Resources Institute)

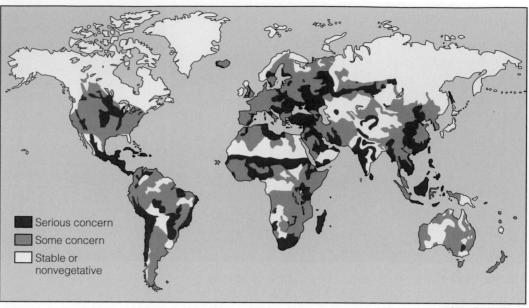

■ Serious concern
■ Some concern
□ Stable or nonvegetative

© 2014 Cengage Learning

described as *loss of soil fertility*. Making matters worse, soil erosion often results in the pollution of surface waters when vital plant nutrients in eroded soil end up as sediment in streams, rivers, and lakes.

Topsoil is a renewable resource, but it is renewed very slowly. This means it can be depleted by soil erosion. Topsoil is eroding faster than it forms on well over a third of the world's cropland, according to a joint survey by the UN Environment Programme (UNEP) and the World Resources Institute (WRI) (Figure 9.34). Globally, erosion rates are many times higher than natural replenishment rates—10 times higher, on average, in the United States and 30 to 40 times higher in China and India.

CONSIDER this Just half an inch of topsoil can take hundreds of years to form. But it can be washed or blown away by soil erosion in a matter of weeks or months when we plow grassland or clear a forest and leave its topsoil unprotected.

The UNEP/WRI study also reported that topsoil erosion has decreased global crop production by about 17%. This has only been made worse by the sharp increase in corn production for the purpose of making ethanol fuel for cars. Intensive corn planting for ethanol production in the United States is causing topsoil to be lost about 12 times faster than it can be formed on many of the most intensively used corn fields.

Soil Salinization and Waterlogging

Irrigation has allowed farmers in many parts of the world to increase their food production by expanding croplands into arid regions with limited rainfall. In fact, the world's area of irrigated cropland tripled between 1950 and 2008. About 20% of the world's cropland is irrigated, and this irrigated land produces about 45% of the world's food.

Similar to other aspects of farming, irrigation has an environmental cost. Most irrigation water contains various dissolved salts that are picked up as the water flows over or through soil and rocks. Some of this water is absorbed by the soil in an irrigated field, but some of it evaporates and leaves behind a thin layer of dissolved salts in the top layer of soil. The drier the climate, the more this evaporation occurs. Over several seasons of irrigating crops in dry climates, the soil accumulates growing amounts of salts in its upper soil layers—a soil degradation process called **soil salinization**. Excessive levels of salts can stunt crop growth, lower crop yields, and even kill plants and ruin the land (Figure 9.35).

According to United Nations estimates, by 2020, nearly one-third of the world's croplands could be degraded by soil salinization. This problem affects about one-fourth of the cropland in the United States. In 2012, the countries suffering the worst soil salinization were China, India, Egypt, Pakistan, Mexico, Australia, and Iraq.

To deal with soil salinization, farmers often try to wash the salts out of the topsoil by pumping more water onto the land. This can lead to another problem called **waterlogging**, in which salty, or *saline*, water accumulates

FIGURE 9.35 Because of high evaporation rates on this heavily irrigated land in the state of Colorado, its soil has become severely salinized and can no longer support the crops that once grew here.

underground and gradually saturates the soil. If this excessive saline water cannot be drained away fairly quickly, it will then surround the deep roots of plants, which can damage and eventually kill them.

CONSIDER this

Crop yields have been reduced on at least 10% of the world's irrigated cropland because of soil salinization and on at least another 10% of the world's irrigated cropland because of waterlogging.

FIGURE 9.36 Some of the effects of surface mining on the land are shown here. This open-pit copper mine (a) is located in Bisbee, Arizona, and this coal strip-mining site (b) is also found in Arizona.

Effects of Mining

Metals are mined to produce numerous products that are essential for many people, as well as a great number of products that people desire. However, mining, processing, and using metals can also disturb land, erode topsoil, and pollute the air, water, and soil.

Surface mining, in which land is stripped of its vegetation and soil, is much more destructive to land than is *subsurface mining*, which involves digging shafts and tunnels. The most obvious effect of surface mining is the scarring and disruption of the land's surface by open-pit mining and area-strip mining (Figure 9.36). These types of mining can also lead to severe soil erosion in areas that border mining sites.

Another form of surface mining, *mountaintop removal*, is used in several areas in the southeastern United States to mine coal. In these operations, huge volumes of waste rock and dirt are dumped into valleys below the mountaintops. This destroys whole forests and buries mountain streams. By 2010, this form of mining had destroyed 500 mountaintops, mostly in West Virginia (Figure 9.37, p. 250) and Kentucky.

Let's REVIEW

- What is *soil erosion* and why is it a problem? What is its effect on soil, in terms of nutrients in the soil?

- How serious is the global problem of soil erosion, according to UNEP? List three areas of the world that are severely affected by soil erosion.

- Define and describe the processes of *soil salinization* and *waterlogging* of soil.

- Explain how surface mining can degrade land.

FIGURE 9.37 This is a mountaintop coal mining site in West Virginia.

What Can Be Done?

Managing Forests More Sustainably

Researchers have come up with several suggestions for how to use forests in more sustainable ways. For example, they recommend that loggers rely more on selective cutting (Figure 9.26a) and strip cutting (Figure 9.26c) than on clear-cutting (Figure 9.26b). They also call for a halt to any clear-cutting on steep slopes where erosion can be highly destructive; no further logging in old-growth forests; and a sharp reduction of road construction in uncut forest areas. In addition, they argue that tree plantations should be located only on already deforested or degraded land.

Forest fire experts and ecologists have suggested several strategies for better forest fire management. One approach is to carefully set and control *prescribed fires*, or small, contained surface fires to remove flammable small trees and underbrush in forest areas with the highest risk,

thus helping to prevent destructive crown fires (Figure 9.27b). Another approach is to allow some fires on public lands to burn, as long as these fires don't threaten human structures, in order to remove flammable underbrush and smaller trees. Fire experts also recommend that homeowners thin out the forest within a zone of about 200 feet around their homes and try not to use highly flammable building materials such as wooden roof shingles.

Another important way to achieve more sustainable use of forests is to certify sustainably grown timber and sustainably produced forest products so that consumers, by purchasing these products, can choose to support more sustainable forestry. For example, the paper used in this book was made from wood grown in a certified sustainable manner, and it also contains recycled fibers.

Let's REVIEW

- What are three suggestions by researchers for how to use forests more sustainably?
- What are three ways to reduce the destruction of forests by fires?
- How can the certification of sustainably grown timber help us to achieve more sustainable use of forests?

Reducing Deforestation

One of the most important underlying causes of deforestation is the fact that forests are not valued for the important ecosystem services they provide (Figure 9.8, top). Factoring the estimated values of these services into the prices of forest products would help to reduce deforestation and forest degradation.

Another underlying cause of tropical deforestation is poverty. Many poor people are driven by poverty into the forests to grow food and find firewood. One way to lessen this form of deforestation is to restore degraded land by planting trees, partly for the benefit of the poor people in need of them, as Nobel Laureate Wangari Maathai did in Kenya, Africa (see *Making a Difference*, at right).

Another important way to reduce deforestation is to reduce the demand for wood. We can do this by cutting wood waste and by finding substitutes for wood in many applications. For example, shipping pallets, excess packaging, and junk mail are examples of wood-based materials that are often used once and then thrown away. Much of this material could be reused or at least recycled. A growing number of builders are working to salvage wood from demolished buildings—an important example of reuse.

People can choose building materials made of recycled plastic that are fashioned to resemble wood lumber. They

MAKING A difference

Wangari Maathai and the Green Belt Movement

When Dr. Wangari Maathai (Figure 9.A) was growing up in Kenya, Africa, she played along tree-lined streams in an area of the country that was lush with tropical forestland. By the mid-1970s, when Maathai was in her 30s, most of this forest had been degraded and largely replaced by plantations, and many streams had dried up.

In 1977, Maathai decided to do something about this environmental degradation and, at the same time, help poor women escape poverty. With the planting of a small tree nursery in her backyard, and with the goal of organizing poor women in rural Kenya to plant and protect millions of trees, she founded the Green Belt Movement. It has since grown to more than 50,000 members and has planted and protected more than 45 million trees.

Women of the Green Belt Movement help to form community tree nurseries and are paid a small amount for each seedling they plant that survives. While these women used to have to spend long hours finding and hauling daily firewood supplies, they now have firewood more easily available in the reforested areas of Kenya. With these benefits, many Kenyan women have escaped desperate poverty. The planted trees also have reduced soil erosion and improved the environment in other ways for people and wildlife. The Green Belt Movement has spread to more than 30 other African countries.

Before founding the movement, Maathai had been the first Kenyan woman to earn a PhD and she chaired the Department of Veterinary Anatomy at the University of Nairobi. In 2004, Maathai became the first African woman, and the first environmentalist, to be awarded the Nobel Peace Prize. The Nobel Prize Committee honored her lifelong struggle for environmental conservation, democracy, human rights, and women's rights. Shortly after learning that she had won it, Maathai planted a tree, declaring that it was "the best way to celebrate." Her message to the world was that for anyone, planting a tree can be a symbol of commitment and hope. With her death in 2011, the world lost one of its most important environmental leaders.

Martin Rowe

FIGURE 9.A Wangari Maathai (1940–2011) was the first Kenyan woman to earn a PhD and she became the head of an academic department at the University of Nairobi.

---

are equally strong and durable, if not more so. Similarly, paper need not be made only from trees. Fibers from fast-growing plants such as hemp and the woody annual plant called *kenaf* yield more paper pulp per unit of land area than tree farms do. According to the U.S. Department of Agriculture, kenaf-based paper products could replace wood-based paper products in the United States within 20 years. China uses rice straw and other agricultural residues to make much of its paper.

Let's **REVIEW**

- What are two underlying cause of deforestation and how can each be addressed?
- Describe the work that Wangari Maathai has done in her native country of Kenya.
- What are two ways in which we could reduce the demand for harvested trees?

Protecting Land

Reducing deforestation and other forms of land degradation ultimately means protecting more land. Only about 13% of the earth's land is officially protected within nature reserves, parks, wildlife reserves, and other areas. However, some conservation biologists argue that, because of budgetary limitations in many countries, only about 5% of the world's land is strictly protected from potentially destructive human activities. Conservation scientists call for full protection of at least four times that area.

The problem is that powerful economic and political interests oppose expanding protected areas to this extent. Some developers and resource extractors also oppose protecting lands that are currently off limits to them because these lands might contain valuable resources that would add to economic growth. However, ecologists and conservation biologists contend that protected areas support biodiversity and other ecosystem services on which all life and all economic activities depend.

Many biologists argue for using the *buffer zone concept* to design and manage protected areas. This means drawing a line around an inner core of such an area, and then drawing more concentric lines farther out from the center to create buffer zones around the inner core. In these buffer zones, local people can use resources, ideally in sustainable ways, without harming the inner core. The UN has used this principle to create 564 *biosphere reserves* in a global network within 109 countries. Some nations are establishing these reserves on their own (see the following *For Instance*). So far, however, most reserves fall short of their protection goals and receive too little funding.

The United States and other more-developed countries have set aside undeveloped lands by legally designating them as *wilderness*. Some critics are opposed to closing off these areas because they could be of economic value. However, wilderness advocates argue that two vitally important long-term needs are met by wilderness. One is to preserve biodiversity as a vital part of the earth's natural capital.

The other is to protect wilderness areas as centers for the recovery of biodiversity in response to usually unpredictable changes in environmental conditions.

KEY idea   An estimated 5% of the earth's land is truly protected from potentially destructive human activities, but conservation experts call for protecting at least 20% of the planet's land in this way.

In reality, few countries have the political or financial ability to set aside and preserve large areas of land for the protection of biodiversity and other ecosystem services. Some conservation biologists call for the international community to take emergency action to identify and quickly protect **biodiversity hotspots**—areas especially rich in plant and animal species that are found nowhere else and that are in danger of extinction. Of all conservation strategies, they argue, this one would protect the largest number of endangered species per unit of land area protected. Conservation biologists have identified a number of biodiversity hotspots (Figure 9.38) and are calling for their immediate protection.

FIGURE 9.38   These biodiversity hotspots (in green) have been identified by ecologists as important and endangered centers of terrestrial biodiversity. (Data from Center for Applied Biodiversity Science at Conservation International).

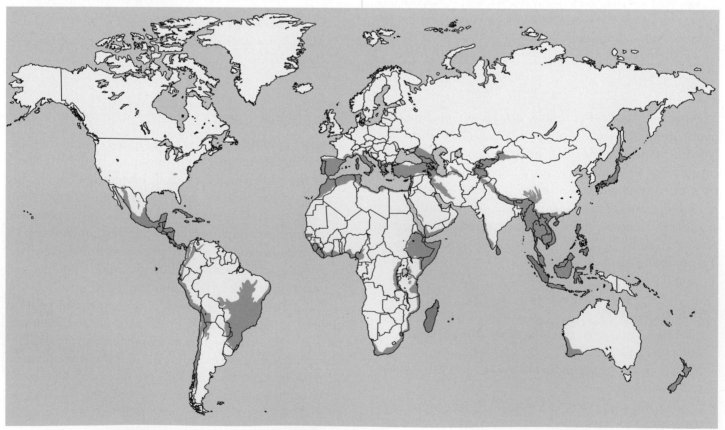

© 2014 Cengage Learning

One of this book's authors, Norman Myers, first proposed this idea in 1988. According to his research, no other strategy for biodiversity protection could achieve as much, at a comparatively small cost, as the hotspots strategy. So far, however, only about 5% of the world's identified biodiversity spots are truly protected.

FOR INSTANCE...

Protecting Land in Costa Rica

Costa Rica, like all Central American countries, was once covered by tropical forests. However, between 1963 and 1983, ranchers cleared much of those forests to graze cattle.

In the midst of this deforestation, during the 1970s, Costa Rica's leaders decided to protect some of the forests by establishing a system of nature reserves and national parks (Figure 9.39) using the buffer zone approach. The plan was successful and is now helping to sustain about 80% of the country's rich biodiversity. Within these zones, local people engage in sustainable agriculture, logging, hunting, and fishing.

Costa Rica now devotes a larger proportion of its land to conservation than does any other country. For this reason, it is one of the most biodiverse countries in the world. In just one park in Costa Rica, the number of known bird species is larger than the 925 known bird species in Canada and the United States. Costa Rica is home to an estimated 500,000 plant and animal species.

To reduce deforestation, Costa Rica's government has eliminated subsidies that reward people for converting forest to rangeland and, instead, pays landowners to maintain or restore tree stands and plants and animals in forested areas. Also, by 2008, the government had planted nearly 14 million trees. As a result, Costa Rica has more than doubled its area of forested land since 1980 when it had one of the world's highest deforestation rates.

CONSIDER this Between 1986 and 2008, the percentage of land covered by forest in Costa Rica increased from 21% to 51%.

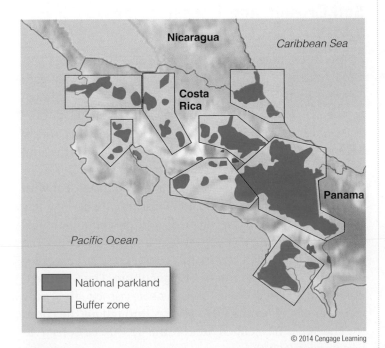

FIGURE 9.39 The green-shaded areas are protected reserves while areas shaded in yellow are buffer zones that humans can use for sustainable activities.

Let's REVIEW

- How much of the earth's land is truly protected from destructive human activities? How much of the earth's land do conservation experts say should be so protected?
- Explain the buffer zone concept, biosphere reserves, and wilderness protection.
- Define *biodiversity hotspots* and explain why some biologists believe this to be the best conservation strategy. Summarize the story of Costa Rica's successful efforts to protect its land.

Dealing with Desertification

Certain human activities, including overgrazing of rangelands, deforestation, and destructive forms of mining, have contributed to desertification. In order to deal with this growing problem, we will need to find more sustainable ways to get the benefits of these activities. We can also reduce the threat of climate change—which is projected to make drought far worse in some areas of the world—by cutting our emissions of climate-changing greenhouse gases and by sharply reducing deforestation.

For example, one of the biggest direct causes of desertification in many areas of the world is overgrazing. Rangelands can be managed much more sustainably by focusing on the number of animals grazing in a particular area and the amount of time that they are allowed to graze. This can be done through a method called *rotational grazing*, in which livestock graze in one area for a period of a few days or weeks, and then are moved to a new location.

FIGURE 9.40 On this stream bank along the San Pedro River in the state of Arizona (a), overgrazing had degraded the vegetation and soil. After grazing was banned, within 10 years, the area was restored through natural regeneration (b).

Overgrazing can be especially severe in *riparian zones*, areas of vegetation lying next to streams, rivers, and ponds where livestock tend to gather to drink water (Figure 9.40a). Protecting riparian zones by fencing them off allows for the possible restoration of these areas by secondary ecological succession (Figure 9.40b). To provide water for livestock, ranchers often locate water holes and tanks away from the riparian zones.

Preventing human activities that contribute to desertification is the best option for reducing it. However, on land that has experienced desertification, it is possible to restore ecosystems by planting trees (see *Making a Difference*, p. 251) and grasses, which anchor topsoil and help the soil to hold water. This is an expensive and difficult option, compared to the prevention approach.

Reducing Soil Erosion

One of the most important priorities of more sustainable agriculture is **soil conservation**. It involves using a variety of methods for reducing soil erosion and restoring soil fertility. These methods include *terracing*, *contour planting*, and *strip cropping*, which involve working with land contours to minimize water runoff (Figure 9.41a and b); keeping the soil covered with vegetation (Figure 9.41b, c, and d); reducing wind erosion by planting *windbreaks* (Figure 9.41c); and *conservation tillage*, which makes use of special planting machines that insert seeds through crop residues directly into the soil without having to plow up the soil (Figure 9.41d).

Another way to conserve the earth's topsoil is to retire some cropland and allow it to be reclaimed by ecological succession, as is being tried in the United States (see the following *For Instance*). The strategy is to identify croplands that are highly vulnerable to erosion, discontinue farming in these areas, and plant them with grasses or trees until their soils have been renewed.

FOR INSTANCE...

Soil Erosion and Conservation in the United States

In the 1930s, much of the Midwestern United States suffered under a prolonged drought. Combined with poor farming practices, this resulted in what became known as the "Dust Bowl" where much of the topsoil in several dry Great Plains states was lost to erosion by winds. Because of this, many Americans lost their farms and their livelihoods.

The U.S. government, in 1935, stepped in to deal with this disaster by passing the *Soil Erosion Act*. The new law set up programs to give farmers and ranchers technical assistance in using soil conservation practices, under the guidance of the Soil Conservation Service (SCS), an agency within the U.S. Department of Agriculture. The SCS identified and focused on *soil conservation districts*, the areas where erosion was the most severe, throughout the country. The SCS is now called the Natural Resources Conservation Service (NRCS).

Since the harsh lesson from the Dust Bowl years, and despite the resulting soil conservation programs, an

FIGURE 9.41 Soil conservation methods include (a) terracing, (b) contour planting and strip cropping, (c) use of windbreaks, and (d) conservation tillage.

estimated one-third of the original topsoil in the United States has now eroded away. Soil scientists estimate that in Iowa, which has some of the world's richest farmland, half of the topsoil is gone.

CONSIDER this According to the NRCS, about 90% of American farmland is, on average, losing topsoil 17 times faster than new topsoil is being formed.

However, the United States is sharply reducing its overall soil losses, partly through conservation-tillage farming (Figure 9.41d). In addition, the 1985 Food Security Act set up the Conservation Reserve Program in which farmers receive subsidy payments for taking highly erodible land out of production and planting it with grasses or trees for 10–15 years. Since 1985, these efforts have cut soil losses on U.S. cropland by 40%.

Yet between 2005 and 2012, some U.S. farmers were tempted by another offer of generous government subsidies for planting corn to make ethanol for use as a motor vehicle fuel. Many then took their land out of the conservation reserve to plant corn. Some politicians and agribusiness corporations have also begun a campaign to abandon or sharply cut back the country's highly successful soil conservation reserve program.

NUMB3RS

40%

The percentage by which soil losses in the United States were reduced through the Conservation Reserve Program

Reducing Soil Salinization

The best way to prevent soil salinization is to use less water in irrigation. Most of the world's farmers simply pump water into unlined ditches in crop fields or spray it

FIGURE 9.44 *Curtis Prairie*, in the University of Wisconsin's arboretum, Madison, Wisconsin, is a partial restoration of prairie on abandoned farm fields.

University of Wisconsin Arboretum

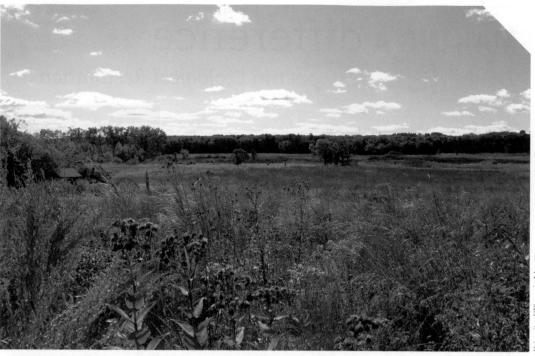

Curtis Prairie, now protected as part of the University of Wisconsin's arboretum, is one of the most successful of all prairie restoration efforts. Students from around the world regularly visit the site to study ecological restoration.

Another outstanding restoration story is that of Baobab Farm, a diverse natural area established on an abandoned limestone mining site outside of Mombasa in Kenya, Africa. By 1970, an open-pit mining operation run by the Bamburi Portland Cement Company had created several large pits. The company hired Dr. René Haller (Figure 9.45), a Swiss expert in horticulture, landscaping, and tropical agriculture, to oversee an attempt to restore the degraded land.

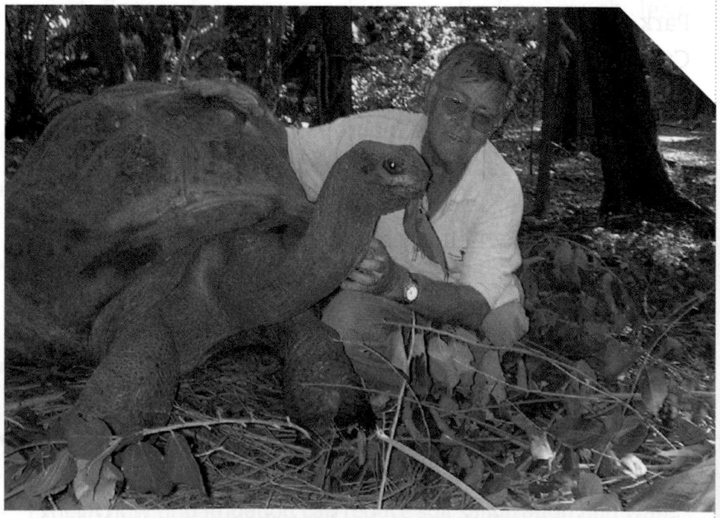

Lafarge Eco Systems Ltd.

FIGURE 9.45 Dr. René Haller sits with a giant tortoise on Baobab Farm, which he created on an abandoned limestone mining site in Kenya, Africa.

Haller started the project in 1971, first adding minerals to the soil to make it more able to support plants. Later, he oversaw the planting of 3,000 trees on the barren landscape. By 1977, Haller had transformed part of the mining site into a self-sustaining ecosystem containing a tropical forest area, wetlands, grassy pastures, lakes, and productive farms.

Baobab Farm, also known as Haller Park, now hosts towering trees, including more than 80 species of palms, along with more than 220 species of birds and clear pools containing healthy fish populations and crocodiles. Beyond the central area, cattle, sheep, and goats graze on grasslands that they share with antelopes, buffaloes, ostriches, and giraffes. The farm attracts about 100,000 tourists every year, and they enjoy walking and biking on trails throughout the farm site.

Baobab Farm also includes a vineyard where grapes are grown on steep hillsides, and groves of trees where honeybees are raised. The rapidly growing tree plantations provide a sustainable supply of firewood and building materials for local residents and vegetation for browsing livestock. Over the course of Haller's decades-old experiment, he has developed an array of sustainable ecological farming practices that he and his colleagues have shared with local farmers. Haller's work has allowed these farmers to better their own lives while helping to protect their fragile tropical environment.

While there have been plenty of success stories, scientists agree that there is much to learn about ecological restoration. It is an exciting and promising research frontier for environmental scientists and students.

Let's **REVIEW**

- Define *ecological restoration*. Summarize the four-step plan that has evolved for restoring degraded ecosystems.
- Describe the ecological restoration project taking place in Costa Rica's Guanacaste National Park.
- Explain how Curtis Prairie was partially restored.
- Summarize the story of the Baobab Farm restoration project.

What Would You Do?

Few of us have any control over land degradation processes such as the clear-cutting of trees or unsustainable mining practices. However, by making wise choices in buying and using forest and mineral resource products, we can send a message to those responsible for such forms of land degradation. We can let them know that we want their products to be produced as sustainably as possible and with minimal land degradation. We can also influence political processes that affect land resources. Here are some major ways in which many people are helping with solutions to the problems of land degradation:

Avoiding purchases that support land degradation.

- When buying lumber and other wood products, people can choose those that are certified as sustainably produced. For example, the paper used to produce this book came from trees that were sustainably grown.
- Many people make an effort to use recycled paper products as much as possible. More and more paper products, including this book, are being made partly from recycled fibers.

- A growing number of people are buying certified 100% organically produced vegetable and meat products, which are produced in ways that help to reduce soil erosion.

Taking steps to reduce personal impacts on land and to assist with the restoration of terrestrial ecosystems.

- Environmentally concerned consumers have *reduced* their use of paper, cardboard, paper bags, and other wood products. They also choose *reusable* products such as coffee mugs, nonplastic water bottles, and cloth shopping bags instead of disposable containers and bags. Many people are also *recycling* as much paper, cardboard, and other materials as possible and buying products made of these recycled materials.
- Some people are planting a tree or two, or joining in larger tree-planting projects.
- Some people volunteer or work for pay on ecological restoration projects, including abandoned lots in the towns and cities where they live.

Informing government and business leaders about their sustainability preferences.

- Some consumers are informing store managers, as well as suppliers of paper, building materials, and other wood products, of their desire to purchase sustainably produced products.
- Many citizens and interest groups are trying to convince their legislators, city officials, and other government representatives to protect more land from degradation.
- Similarly, citizens and groups are working with government representatives to end subsidies for land-degrading activities such as strip mining and clear-cutting of forests, and to replace them with subsidies favoring more sustainable uses of land.

Given these ideas and others that you might have gained from reading this module, what are three things you would do to help deal with the issue of land degradation?

KEYterms

THINKINGcritically

1. Suppose the continent of Africa was entirely defor- ested and was converted to desert. Describe three ways in which this might affect the environment in the area where you live. For each of these three effects, how might the change affect your life?

2. Why do you think that a tropical forest supports many different species of plants and animals, while a tree plantation supports far fewer species? Explain.

3. When vegetation grows back in a tropical forest within a few years after the forest has been cut, do you think the new vegetation could support the same variety of plants and animals that were present in the old for- est? Why or why not?

4. Only about 5% of the earth's land is truly protected from human activities that can degrade the land, and some scientists call for protecting four times as much land. Do you agree with them? Explain. If you disagree, what percentage of the earth's land do you think should be protected from harmful human activi- ties? Explain your reasoning.

5. Think of a natural land area, located near where you live or go to school, that is now degraded. It could be a wooded area, a field, an abandoned lot, or any other sort of land area that was once a healthy natural eco- system. Write a plan for how you would restore the area. (You could use the 4-point strategy listed on p. 257 as a basis for your plan.) Estimate the number of people you will need to help you. Estimate the cost of the project and how long it will take to restore the area to a functioning ecosystem.

6. A number of ecologists and economists argue that two of the best ways to reduce land degradation are to (a) include the economic value of the ecosystem ser- vices that forests (Figure 9.8) and other land ecosystems provide when evaluating how land should be used and (b) require those who use public lands for harvesting trees, grazing livestock, and mining minerals to pay for the environmental harm their activities cause by fund- ing ecological restoration projects. Explain why you are for or against each of these strategies. If you support these strategies, how would you implement them?

LEARNINGonline

Access an interactive eBook and module-specific interac- tive learning tools, including flashcards, quizzes, videos and more in your Environmental Science CourseMate, accessed through **CengageBrain.com**.

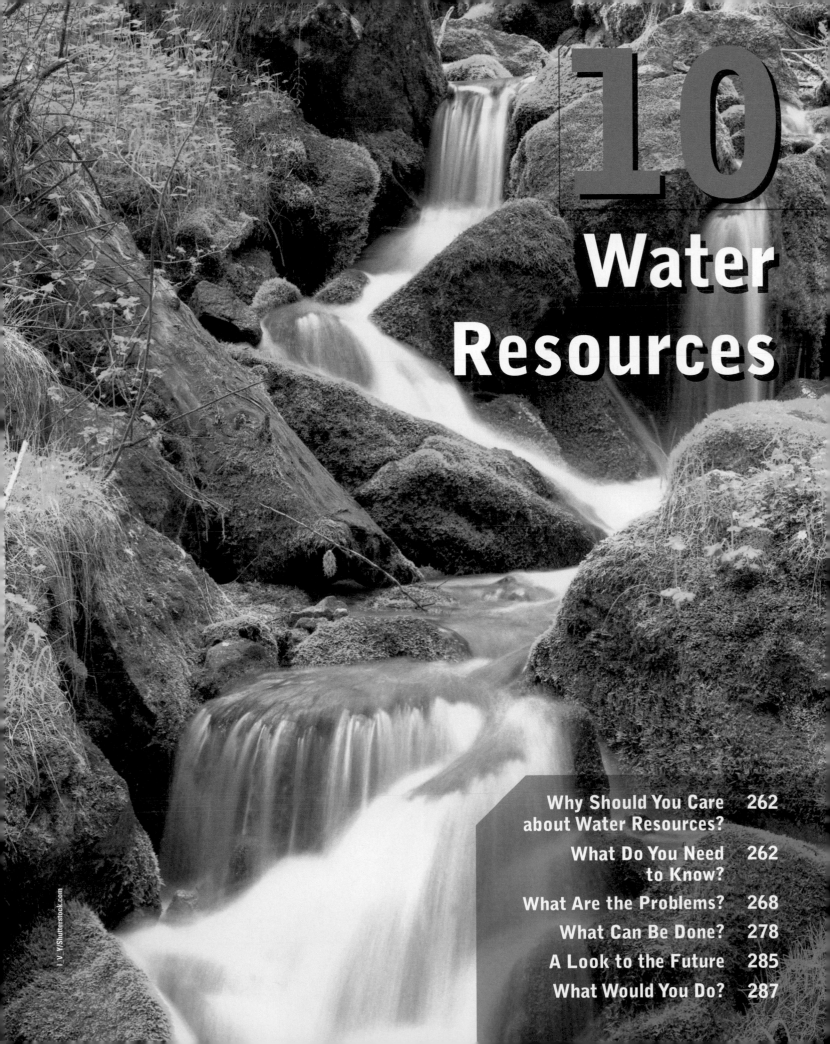

10 Water Resources

Why Should You Care about Water Resources?

Water is a wondrous and irreplaceable substance that we depend on for our survival. In fact, humans as well as most other life forms are made primarily of water. Without food, we can survive for several weeks, but without water, we can last only a few days.

We also need large amounts of water in order to produce the food and most of the other items we use in our daily lives. For example, it takes about 40 gallons of water, or about one full bathtub, to produce one cup of coffee. We also use water to dilute or remove a great deal of the wastes and pollutants we produce. In addition, water plays a key role in controlling the climate that makes it possible for us to thrive on this planet.

The earth has an abundant and renewable supply of water, but it is unevenly distributed. Some areas get too much (Figure 10.1a) and other areas get too little (Figure 10.1b). In water-rich areas, people do not generally fear water shortages. On the other hand, in some dry areas, people live with a shortage of water every day and spend as many as 3 hours per day collecting water from distant sources just to meet their most basic needs.

Despite water's importance, those of us who have it tend to use it freely without thinking about how our use of it affects the environment, and we often pollute commonly shared supplies of water. We also generally make it available at a very low cost to users, which is partly why this natural resource is often used unwisely and inefficiently.

As the human population grows and as the average rate of resource use per person rises, water shortages in some areas are becoming a major problem, mainly because of unequal distribution of the world's water resources. Many water resource experts warn that providing enough water for everyone is one of the most serious and poorly understood challenges that we will face during this century. This module examines the earth's water resources, the problems related to our use of this renewable but heavily used natural resource, and some possible solutions to these problems.

What Do You Need to Know?

Water's Unique Properties

The water molecule (H_2O) has a unique combination of properties (Figure 10.2) that make water one of our most important natural chemicals.

Water's most important property is the unusual forces of attraction between its molecules due to its unique molecular structure. Because of these forces of attraction, water has a high boiling point, which means that it takes a large amount of heat to separate molecules of liquid water and convert them to water vapor in the atmosphere. Without this property, the planet's oceans and all other bodies of water would have evaporated long ago.

FIGURE 10.1 Some areas, including this part of Dhaka, Bangladesh (a), get flooded regularly. Other areas get too little rain, including this drought-stricken area of India (b), where people must travel long distances every day just to get enough water to survive.

FIGURE 10.2 Water has unique properties that make it crucial for many life processes.

The heat that converts liquid water to water vapor is carried with the water vapor into the atmosphere where it is released when water vapor cools. At a certain point in this cooling process, the water molecules come back together again, or *condense*, to form liquid water. The ability of water to store and release heat in this way helps to distribute the sun's heat through the atmosphere from point to point around the globe. This is partly why much of the planet enjoys a moderate climate.

The forces of attraction between water molecules also allow liquid water to stick to a solid surface. This helps plants to move water through their hollow stems from their roots to their leaves. If water did not have this property, plants could not survive, and neither could we, because plants directly or indirectly provide us with all of our food.

Most liquids shrink when they freeze but liquid water expands when it becomes ice. Thus, the *density*, or mass per unit of volume, of ice is lower than that of liquid water, which is why ice floats on water. Without this unusual property, in cold climates, streams and lakes would freeze solid instead of just at their surfaces. In these bodies of water, fishes and other aquatic life as we know it would not exist.

Liquid water also has the ability to break apart, or *dissolve*, a number of certain compounds. This is why it is good for cleaning surfaces and for removing wastes. This property also allows water to transport dissolved nutrients into body tissues and to remove wastes from them. However, this ability of water to dissolve a variety of chemicals also means that bodies of water can be polluted by water-soluble wastes that we produce.

The Water Cycle

The earth's supply of water is fixed. It does not grow or shrink. But fortunately for us and other forms of life, the planet's fixed supply of water is constantly recycled and purified by the **water cycle**, or **hydrologic cycle** (see Module 1, Figure 1.20, p. 19).

The water cycle, driven by solar energy, involves three major processes: **evaporation**, the conversion of liquid water in the oceans, rivers, lakes, and soils into water vapor in the atmosphere; **transpiration**, the evaporation of water from plant leaves into the atmosphere; and **precipitation**, the return of water from the atmosphere to the planet's surface in the forms of dew, rain, sleet, snow, and hail.

Depending on the type of land surface, one-third or more of the precipitation that falls on land becomes **surface runoff**, freshwater from precipitation and snowmelt that flows over the ground. Most of this runoff remains on the earth's surface as **surface water**, or freshwater that is stored or that flows in streams, rivers, wetlands, and lakes. Some surface water evaporates into the atmosphere to begin the water cycle again. Some surface runoff seeps down into underground bodies of water stored in layers of sand, gravel, and rock called **aquifers**. This underground stored water is called **groundwater**. In the next section of this module, we consider some ways in which human activities are interfering with the earth's water cycle and its critical reserves of stored water.

The earth's water is always changing from liquid to vapor or to ice and back again as it moves around on the earth's surface and underground, through its atmosphere, and through living things on the planet (see Module 1, Figure 1.1, p. 2). Some of the water is stored temporarily as ice or snow in mountainous areas (Figure 10.3, p. 264). From there, it partially melts during warm weather, fills mountain lakes, and flows downward through mountain streams (see module-opening photo). These streams flow to rivers and lakes and many rivers empty into the oceans, where some of the water evaporates to continue the water cycle (Figure 10.4, p. 264).

Through this process, water gets purified in various ways. Most of the compounds dissolved in water are left on the earth's surface when the water is evaporated to the atmosphere. Some impurities are pulled out of water as it

Pichugin Dmitry/Shutterstock.com

flows through soil and rocky areas underground. Water is also filtered and partially purified by chemical and biological processes as it flows through soils, wetlands, streams, and lakes. This natural cycling and purification process works continually as long as it is not overloaded with harmful chemicals from one source or another.

KEY idea
Wherever it is not disrupted by human activities, the earth's fixed supply of water is continually recycled and purified by the hydrologic cycle.

Let's REVIEW

- Why should we care about the earth's water supply?
- What are five unique properties of water, and why are they important?
- What is the *water cycle (hydrologic cycle)* and why is it important?
- Define and distinguish among *evaporation*, *transpiration*, and *precipitation*.
- Define and distinguish among *surface runoff*, *surface water*, *aquifers*, and *groundwater*.
- Explain how water moves on and under the earth's surface and in its atmosphere and how this movement helps to purify water.

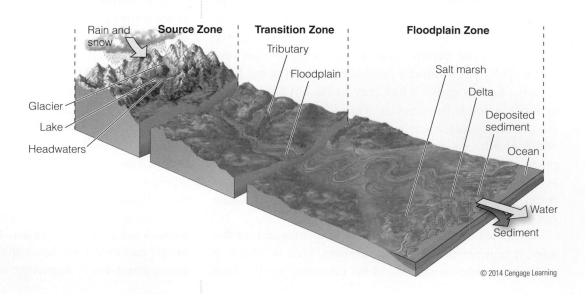

FIGURE 10.4 Water from melting snow and ice in the mountains flows downhill toward the oceans.

© 2014 Cengage Learning

The Earth's Water Supply

We live on a blue, watery planet. Satellite pictures show that liquid water covers almost three-fourths of the earth's surface (Figure 10.5), but only about 0.024% of this water is available to us as life-sustaining freshwater (Figure 10.6). The rest is too salty, lies too deep underground to be accessed with current technology, or is frozen in glaciers and polar ice caps. Yet this small percentage of all water amounts to a generous supply of freshwater that is continuously recycled, purified, and distributed in the water cycle (see Module 1, Figure 1.20, p. 19).

At least 30% of the world's available supply of freshwater is stored as groundwater in accessible aquifers. These groundwater supplies are usually recharged from above by precipitation sinking downward through exposed soil, or from the side by water seeping through the soil from streams. Because most aquifers are recharged very slowly, they can be depleted when we pump water from them faster than natural processes can recharge them. Some aquifers found deep underground get little, if any, recharge water. Large-scale withdrawal of water from these deep aquifers can deplete these ancient deposits of freshwater, which, on a human time scale, are nonrenewable.

CONSIDER this The U.S. Geological Survey estimates that the world's aquifers contain more than 100 times as much freshwater as is contained in all the world's streams, rivers, and lakes.

Nearly half of the world's people get their drinking water from aquifers. Groundwater supplies about 25% of the freshwater used in the United States in most years, according to the U.S. Geological Survey (USGS). This includes almost all of the drinking water in rural areas, one-fifth of that in cities, and around 40% of the water used to irrigate crops.

If the world's accessible freshwater were equally available to everyone on the planet, each of us would have plenty of water for healthy, sanitary living. However, water is not distributed evenly, and many millions of people go without enough water on most days. This is made worse in many areas by prolonged **drought**—an extended period of below-average precipitation and high evaporation. Projected changes in the global climate, which include more widespread drought, will further change the patterns of global water distribution during this century. This issue of distribution is one that we will explore further in the next major section of this module.

Ocean hemisphere   **Land–ocean hemisphere**

© 2014 Cengage Learning

FIGURE 10.5 Saltwater in the world's oceans covers about 71% of the planet's surface and roughly another 2% of its surface is covered by freshwater in wetlands, lakes, streams, and rivers.

Let's REVIEW

- How much of the earth's surface is covered by oceans?
- What percentage of earth's water supply is available to us as useable freshwater? Where is most of this available freshwater stored?
- Explain how aquifers are recharged and why they can be depleted.
- What is *drought*?

How We Use Water

The water cycle collects, purifies, and distributes water around the globe. However, as the human population has grown, our use of water has increasingly interfered with

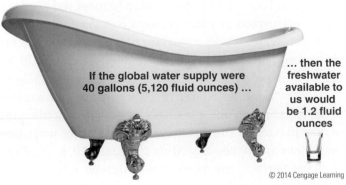

If the global water supply were 40 gallons (5,120 fluid ounces) … … then the freshwater available to us would be 1.2 fluid ounces

© 2014 Cengage Learning

FIGURE 10.6 Only about 0.024% of the world's water is available to us as usable freshwater. If we use a typical bathtub filled with water to represent all the water in the world, then the freshwater available to us would not quite fill a 1.5-ounce shot glass.

Photo credits: Left. Baloncici/Shutterstock.com. Right. Evgeny Karandaev/Shutterstock.com.

FIGURE 10.7 Irrigation is necessary for growing crops in dry areas, including this field of romaine lettuce in Yuma, Arizona.

Jeff Vanuga/USDA, Natural Resources Conservation Service

and even altered this cycle in various places on the earth. In order to understand these effects, we need to know how people use water.

Of all the freshwater withdrawn from surface and groundwater sources in the United States, about 40% is used to produce electricity, primarily for the cooling of the power plants, according to the USGS. Another 37% is used to irrigate crops (Figure 10.7), according to the USGS. However, because power plant cooling water is usually returned to its source, irrigation takes the largest share of freshwater that is withdrawn and not returned to its source. Public water supplies account for 13% of all freshwater withdrawals. The remaining 10% is used by industry, agriculture, and people who supply their own drinking water. Water is used in similar proportions for these purposes in most of the other more-developed countries of the world.

Our use of freshwater is increasing rapidly because of a combination of population growth and rising levels of water consumption per person. In addition, most communities and countries in the world do not put a high priority on **water conservation**—a variety of strategies for getting the most use out any amount of water that we withdraw from the water cycle. Globally, about two-thirds of the water we withdraw does not get used in this way, partly because it is available cheaply for people in areas where water is plentiful. For example, because of inefficient irrigation practices, irrigated crops in some areas receive only a very small fraction of the large amounts of water that are applied to them.

A concept that helps us to describe how we use water is the concept of **virtual water**, a measure of the water used to produce and deliver food and consumer products. For example, it requires about 160 gallons of water to produce the average loaf of bread, including the water used to raise and harvest the grains, to process them, and to bake and deliver the bread to a consumer (Figure 10.8). You can also think of this water as *hidden water* represented by the finished product.

Another concept that will be useful for understanding this module is that of a **water footprint**—an estimate of how much water it takes each year to meet the needs of a person, a population, or a country. Water footprints can be calculated for individuals, families, companies, communities, or nations. The **per capita water footprint** is a measure of how much water it takes to meet the needs of an average person in a country for a year. According to the American Water Works Association, the water we use for showering, gardening, and other home uses makes up 3% to 5% of this amount. The rest is virtual water (Figure 10.8).

Most of the water-related problems that we will explore in the next section occur because we are altering the rates of flow in parts of the water cycle in two major ways (see red arrows and boxes in Module 1, Figure 1.20, p. 19). First, in many parts of the world, we withdraw large quantities of freshwater from streams, lakes, and underground sources faster than natural processes can replenish them. We also transfer large volumes of water across land in order to irrigate crops (Figure 10.7) and to supply water for home and industrial uses.

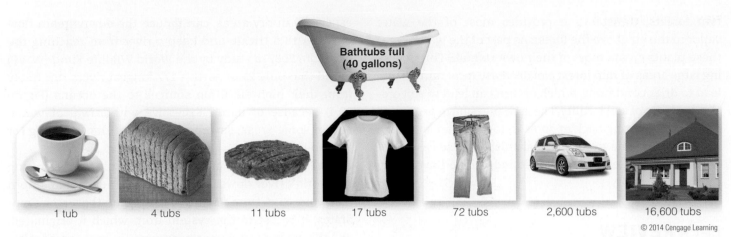

Bathtubs full
(40 gallons)

1 tub　　4 tubs　　11 tubs　　17 tubs　　72 tubs　　2,600 tubs　　16,600 tubs

© 2014 Cengage Learning

FIGURE 10.8 It takes large amounts of water to produce and deliver most consumer goods. (Data from UNESCO-IHE Institute for Water Education, UN Food and Agriculture Organization, World Water Council, and Coca-Cola Company)

Photo credits: Top row. Baloncici/Shutterstock.com. Bottom row, left to right. Aleksandra Nadeina/Shutterstock.com; Alexander Kalina/Shutterstock.com; Joe Belanger/Shutterstock.com; Wolfgang Amri/Shutterstock.com; Kateryna Larina/Shutterstock.com; Eky Chan/Shutterstock.com; Rafal Olechowski/Shutterstock.com.

One way to capture water for human uses is to build a **dam**, or a barrier across a river that blocks its flow. The result is a **reservoir**, or artificial lake behind the dam. Such a dam-and-reservoir system captures and stores surface runoff and releases it as needed to supply water for growing crops and for industries and cities. We also use dams to generate electricity (*hydroelectricity*) by allowing water to flow through turbines in the dam structure. Reservoirs are also used for recreational activities such as boating, fishing, and swimming.

For example, Lake Mead is a reservoir behind Hoover Dam that stores water flowing down the Colorado River on the border between the states of Nevada and Arizona (Figure 10.9). The reservoir is the source for 95% of the water used by the almost 2 million people living in the nearby Las Vegas metro area. Water in the reservoir is also used for irrigating crops in this dry area. Because of increasing population growth, rising rates of water use in this area, and a prolonged drought in the southwestern United States, Lake Mead's water level has been falling. According to a 2008 study by researchers Tim Barnett and David Pierce at Scripps Institution of Oceanography, there is a 50% chance that Lake Mead could run dry by 2021.

The second way in which humans have altered the water cycle is by clearing trees and other vegetation, and filling in wetlands for activities such as farming, road building, and urban development, and then covering much of the land with buildings and pavement. These artificial surfaces block the *infiltration*, the trickle of water down through the soil, that normally recharges groundwater supplies, and they increase surface runoff that can cause

flooding. Also, by destroying wetlands that have acted as natural sponges in absorbing excess rainwater, we have again contributed to flooding.

Clearing vegetation can also change the water cycle within certain ecosystems. For example, in dense tropical

FIGURE 10.9 Hoover Dam, completed in 1936, was built across the Colorado River on the border between the states of Nevada and Arizona. The reservoir behind it was named Lake Mead.

AndyZ./Shutterstock.com

rain forests, transpiration provides most of the water vapor in the air above the forest. As part of the water cycle, these plants create most of their own rainfall. Thus, clearing large areas of rain forest can decrease local rainfall and lead to drier conditions, which in turn can lead to soil erosion and loss of soil fertility. These changes have prevented trees from growing back in some deforested areas, which have then become tropical grasslands. In these cases, not only has the water cycle been altered, but so have entire ecosystems.

Let's REVIEW

- How is the freshwater that is used in the United States divided among agricultural, industrial, and home uses?

- What are three factors in our increasing use of freshwater? Define *water conservation*.

- Define *virtual water* and give an example of it. Define *water footprint* and *per capita water footprint*.

- Define *dam* and *reservoir*.

- What are the two major ways in which human activities have altered the hydrologic cycle?

What Are the Problems?

Surface Water Depletion

The portion of the planet's total annual surface runoff from precipitation that is available for human use is called the **reliable surface runoff**. In 2012, it accounted for about one-third of all surface runoff, and it varies from year to year and from one area to another.

According to a 2006 study by University of Michigan scientists, we are using a little over half of this reliable runoff for irrigation, drinking water, and other uses. In the arid American Southwest, up to 70% of the reliable runoff is withdrawn for human purposes, mostly for irrigation (Figure 10.7). Some water experts project that by 2025, we may be withdrawing as much as 90% of the world's reliable runoff to satisfy a larger population with greater irrigation and manufacturing demands for water, assuming we do not conserve water at much greater rates than we do today.

One of the major ways in which we make use of surface water is through dam-and-reservoir systems (Figure 10.9). While they provide great benefits to communities, these systems also create some problems. Damming the water in a river and withdrawing excessive amounts of it,

especially in dry areas, can reduce the downstream flow of water to a trickle and keep a river from reaching the ocean. In 2007, a study by the World Wildlife Fund (WWF) found that only 21 of earth's 177 longest rivers run freely from their high elevation sources to the oceans (Figure 10.4) because of dams, excessive withdrawal of water, and in some areas, severe drought (see the following *For Instance*).

There are several problems that result from reducing the world's river flows. First, more water evaporates from reservoirs than from flowing rivers. When the water evaporates, it is lost to the system from which it originated, and this makes water shortages worse in some areas. Also, largely because water in some river systems has dwindled, so have wildlife habitats. This is part of the reason that about one-fifth of the world's freshwater plant and fish species have become endangered or extinct.

Yet another problem associated with damming a river is that land is flooded to create the large reservoirs behind dams. The global total area of land flooded for reservoirs is roughly equal to that of the state of California, and millions of people have been displaced by these dam-and-reservoir systems. The world's largest dam-and-reservoir system, the Three Gorges Dam in China, flooded an area more than 370 miles long that included more than 1,500 villages and cities. About 1.4 million people were displaced.

Also, within several decades, most reservoirs fill up with mud, sediment, and silt moved downstream by their rivers. This makes a dam-and-reservoir system useless for generating electricity and for other purposes.

CONSIDER this
Some 40 to 80 million people have been displaced from their homes by dam-and-reservoir projects.

The depletion of reliable surface water supplies from river systems, largely due to damming, is creating conflicts among nations that share water resources in many of the world's river basins. These conflicts are likely to intensify as populations grow and water resource demands increase. For example, in the Middle East, many countries face rising tensions over water shortages because they must share water resources from just a few major river basins in this dry region, including the Jordan, Nile, and Tigris-Euphrates basins.

Lakes provide another source of surface water for millions of people around the world. As with many of the world's river systems, numerous lakes have been stressed to the point of collapse (see the following *For Instance*). This, too, is a growing problem in some parts of the world.

FOR INSTANCE...

Let's Compare Two Stressed Surface Water Supplies

The Colorado River Basin

The Colorado River, a major river system in the dry southwestern United States, flows 1,400 miles through seven states to the Gulf of California (Figure 10.10a). Since the early 1900s, 14 major dams and reservoirs have been constructed along this once free-flowing river to provide water, electricity, and flood control for ranchers, farmers, and cities. Two of these reservoirs—Lake Mead behind the Hoover Dam (Figure 10.9) and Lake Powell behind the Glen Canyon Dam (Figure 10.10b)—store about 80% of the water in the vast Colorado River basin.

The U.S. government built dams and reservoirs along the Colorado River to provide low-cost water and electricity for farmers and ranchers, and to encourage settlement in this desert region. Large and growing U.S. cities getting water from this river system include Phoenix, Arizona; Las Vegas, Nevada; and Los Angeles and San Diego in California. According to a Pacific Institute report, the number of people served by the Colorado River system in 2011 was almost 35 million. Water is withdrawn from this once-mighty, but now-tamed river to grow crops and raise livestock that account for about 15% of U.S. food production. Some of the water is also used to keep hundreds of desert golf courses green.

The Colorado River system faces several serious problems, including a modest flow of water (compared to other rivers its size), excessive water withdrawal to produce food and support large cities, and prolonged drought. As a result, the amount of water flowing to the mouth of the river has dropped dramatically since 1905 (Figure 10.11, p. 270) and since 1960, the river has rarely made it to the Gulf of California. Other problems are evaporation from reservoirs and leakage of reservoir water into the ground below, representing a major loss of water in this already water-short region.

Another problem is that as the flow of the river slows in the large reservoirs, it drops great amounts of suspended silt. Without the dams, this silt would feed the river's

FIGURE 10.10 The Colorado River basin (a) occupies a large area of land in the southwestern United States. The Glen Canyon Dam (b), built in 1963, created the huge reservoir called Lake Powell.

© 2014 Cengage Learning

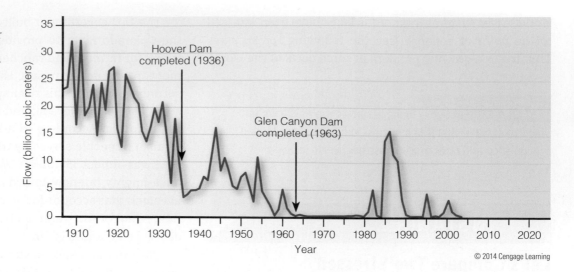

FIGURE 10.11 Since 1905, the flow of the Colorado River has dropped sharply because of multiple dams, water withdrawals for agriculture and urban areas, and prolonged drought. (Data from U.S. Geological Survey)

coastal delta where wetlands provide wildlife habitat and other valuable ecological services such as water storage and water purification. Thus, the damming has resulted in the destruction of these important wetlands and the loss of their important services.

Also, sometime during this century, these reservoirs may be too full of silt to store enough water for generating electricity or for controlling downstream flooding. Some analysts warn that these problems could eventually lead to intense legal battles over this dwindling water supply, a sharp drop in agricultural production, and the migration of many people from some of the major cities in this region.

CONSIDER this Every day, the amount of silt deposited on the bottoms of Lake Mead (Figure 10.9) and Lake Powell (Figure 10.10b) is equivalent to 20,000 dump truck loads.

What caused these problems? The primary factor is that this river system is in a very dry region that is projected to have serious drought conditions throughout all or most of this century. Another factor is that the government-built dams and reservoirs amount to a huge *government subsidy*—a beneficial form of assistance that has helped to increase the production of crops and cattle and to develop the big southwestern cities. However, this boost from the government has also had the effect of keeping water prices low, thus encouraging the inefficient use of irrigation water to grow thirsty crops such as cotton, rice, and alfalfa in this water-short area.

Some researchers argue that to avoid economic and ecological collapse in the Colorado's basin, the governments

of the seven affected states will have to pass and enforce much stricter water conservation measures, partly to remind people that they are living in a desert area and cannot expect to have green lawns and golf courses and to grow thirsty crops. These researchers also call for the federal and state governments to sharply increase the price of water withdrawn from the Colorado River in order to encourage water conservation and more sustainable use of the water resource.

The Draining of Lake Chad

Africa's Lake Chad used to be one of the world's largest lakes. But since the 1960s, its size has shrunk by 96% (Figure 10.12). This has happened because of three interacting factors:

- *Rapid population growth* in the four countries—Chad, Cameroon, Nigeria, and Niger—that surround the lake. Today, more than 20 million people in these countries get their drinking water from the lake.
- *Climate change*, as the region has become drier and subject to prolonged drought, which evaporates water from the lake faster than it is replaced.
- *Inefficient irrigation* has wasted water. Since 1963, a fourfold increase in use of mostly inefficient irrigation methods to feed the area's rapidly growing population has further drained the lake and the two rivers that feed it.

The shrinkage of the lake has devastated its fish populations that once were a major source of food for local people. This accelerated the need for irrigation water from the lake to grow more food, which further reduced the water level in the lake. Also, the overgrazing of livestock on the land surrounding the lake degraded the lush vegetation

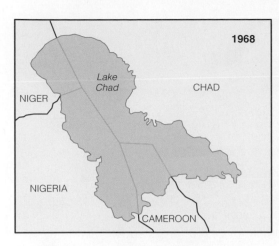
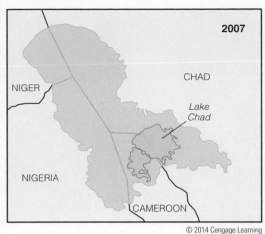

FIGURE 10.12 Africa's Lake Chad, in 1968, was 20 times as large as it was in 2007. (Data from NASA's Goddard Space Flight Center)

© 2014 Cengage Learning

that, through transpiration, had fed much moisture back into the atmosphere. This helped to reduce precipitation, making the drought worse.

This domino effect has greatly disrupted the area's ecosystems and ecological services. As a result, crops have failed, livestock have died, fish populations have dwindled, and the lake continues to shrink. Scientists warn that, as population and irrigation demands continue to increase, the entire aquatic system of Lake Chad is likely to suffer an ecological collapse.

Let's **REVIEW**

- What is *reliable surface runoff,* and about how much of it do humans use for drinking water and other purposes?

- List four problems that have resulted largely from the damming of rivers around the world.

- Summarize the story of how and why the Colorado River dams and reservoirs were built. What are four problems caused by the damming of the Colorado River? What are two major factors that contributed to these problems?

- Summarize the story of the draining of Lake Chad, giving three examples of how human activities contributed to the degradation of this lake's ecosystem.

Groundwater Depletion

Most aquifers are naturally renewed by precipitation and melting snow, unless the groundwater they contain is removed faster than these processes can recharge the aquifers. However, in many areas of the world, people are pumping water from aquifers faster than they can be recharged.

Most of this pumped groundwater is being used to irrigate grain crops, especially in the United States, China, and India. The U.S. Geological Survey estimates that groundwater is being withdrawn in the United States, on average, four times faster than it is being recharged. Figure 10.13 shows the areas of greatest aquifer depletion in the continental United States. One of the largest areas of concern is the vast Ogallala aquifer, which shows up as the large red area on this map (see *For Instance,* p. 273).

Engineers have learned how to tap into deeper aquifers as some more shallow ones have become depleted. However, because these aquifers are recharged extremely slowly, if at all, they can be depleted, essentially forever, by excessive pumping. Using pumps to bring water up from deep underground also takes lots of energy and is costly. For example, the arid desert nation of Saudi Arabia gets most of its water at a very high cost from deep aquifers

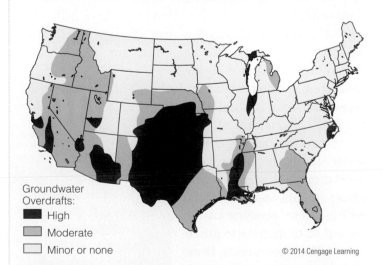

Groundwater Overdrafts:
- High
- Moderate
- Minor or none

© 2014 Cengage Learning

FIGURE 10.13 The shaded areas on this map show regions of the continental United States in which aquifers are becoming depleted because water is being removed much faster than it can be replenished. (Data from U.S. Water Resources Council and U.S. Geological Survey)

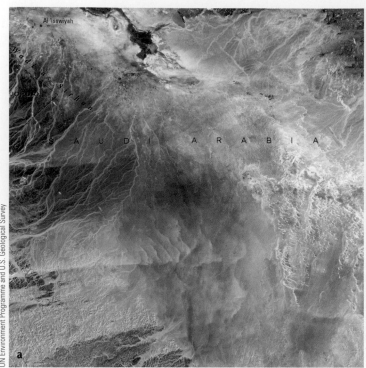

FIGURE 10.14 These satellite photos show crops irrigated by groundwater pumped from a nonrenewable aquifer in Saudi Arabia between 1986 before crops were planted **(a)** and 2004 **(b)**. Irrigated areas show up as green dots; brown dots indicate areas that have returned to desert as wells have gone dry.

and uses much of it to irrigate crops on desert land (Figure 10.14). This has allowed Saudi Arabia to become more self-sufficient, but water experts project that most irrigated agriculture in Saudi Arabia will disappear in 10 to 20 years as these nonrenewable aquifers are depleted of their water.

Overpumping of aquifers can disrupt urban and agricultural systems that depend on groundwater, and it can also disrupt natural ecosystems. Many wetlands, lakes, and river systems are fed partly by aquifers and as these aquifers are depleted, water levels in these systems drop. This pumping can cause plants and animals within these ecosystems to die out or to migrate to other areas. The ecosystems thus go into a state of decline and other connected ecosystems can also suffer.

NUMB3RS

500

The number of people in the world in millions who depend on grain produced through the unsustainable pumping of groundwater

Withdrawing large quantities of groundwater sometimes causes the porous rock in aquifers to collapse. When this happens, the land above the aquifer sinks. This *land subsidence* can compress an aquifer so that it cannot be recharged. As the land subsides, it can also damage building foundations and roads, and break water and sewer lines. Areas suffering from severe subsidence include Mexico City, Mexico; Beijing, China; Phoenix, Arizona; and California's San Joaquin Valley (Figure 10.15).

CONSIDER this Excessive withdrawal of groundwater in some sections of Mexico City, Mexico, has created areas of subsidence deep enough to swallow a four-story building.

In coastal areas, where many of the world's largest cities are located, the overpumping of groundwater can lead to the contamination of groundwater supplies. This occurs when salty ocean water seeps into spaces in aquifers where freshwater has been withdrawn. The resulting salty groundwater cannot be used for irrigation or for drinking and the problem can persist for many years.

Let's look more closely at the overpumping of one of the world's major aquifers.

Dick Ireland/U.S. Geological Survey, 1977

FIGURE 10.15 This pole shows subsidence from the over-pumping of groundwater for irrigation in California's San Joaquin Central Valley between 1925 and 1977. In 1925, the land in this area was at the top of this pole. This subsidence has continued since 1977.

FOR INSTANCE...

Overpumping of the Ogallala Aquifer

North America's largest aquifer is called the Ogallala. This precious deposit of water lies under eight states in the Great Plains—extending northward from Texas to South Dakota (Figure 10.16, p. 274). A large proportion of U.S. food crops are irrigated by water from the Ogallala, and it supplies drinking water for about 82% of the people living above it.

The Ogallala gets recharged very slowly, which makes it a nonrenewable resource, at least on a human time scale. Yet, the rapid pumping of this aquifer continues, for purposes of providing drinking water and growing crops that are used to feed people and livestock. In addition, an increasingly large portion of the corn that is irrigated by the Ogallala is used to make ethanol fuel for motor vehicles.

As a result, water is being removed from the Ogallala faster than natural processes can replenish it. In some areas, especially in the lower half of the Ogallala beneath Texas and Oklahoma, water is being withdrawn 10 to 40 times faster than the aquifer is being recharged (Figure 10.16 and much of the red area in the center of Figure 10.13). Consequently, crops in parts of this region are no longer irrigated because the sharp drop in the water table has greatly increased the cost of pumping water from the aquifer.

As in the case of the Colorado River, depletion of the Ogallala has been promoted by government subsidies that encourage farmers to increase their production of irrigated crops. Several water supply analysts have urged that these subsidies be discontinued and perhaps replaced with subsidies that encourage the more sustainable use of water in the Midwestern United States.

KEY idea Aquifers are renewable sources of freshwater only as long as water is not removed from them faster than they can be recharged by natural processes.

Problems Resulting from Water Transfers

In several areas of the world, water has been moved from water-rich areas to water-poor areas by means of canals and pipelines. This benefits the areas that receive the water, but water transfers can be harmful to the ecosystems and communities in the areas from which the water is taken.

A prime example of such a system and its mix of benefits and drawbacks is the California State Water Project, a vast water-transfer system that moves water from Northern California to the much drier southern part of the state and to Southern Arizona (Figure 10.17, p. 275). Since the 1960s, it has supplied irrigation water to central and Southern

FIGURE 10.16 This map shows where water levels in the Ogallala aquifer have dropped. The aquifer is at risk of being depleted, especially at its southern end beneath parts of Kansas, Oklahoma, Texas, and New Mexico. (Data from U.S. Geological Survey, 2008, *Ground Water Atlas of the United States*; and McGuire, V.L., 2011, *Changes in water levels and storage in the High-Plains Aquifer, pre-development to 2009*, USGS Fact Sheet 2011–3069)

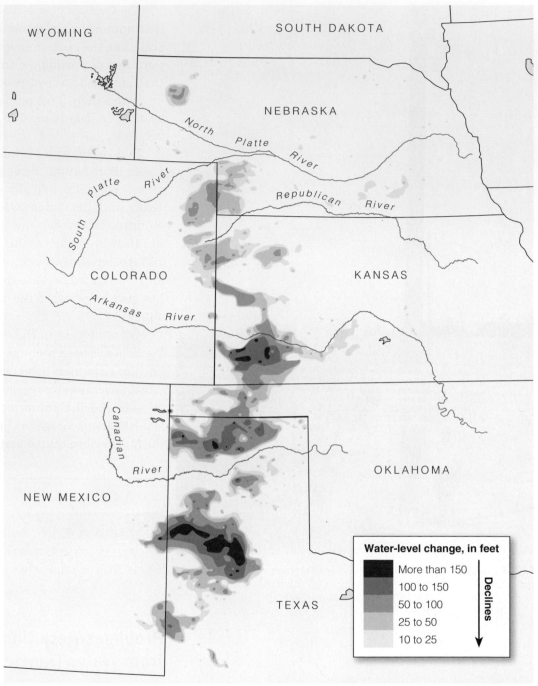

Water-level change, in feet

More than 150
100 to 150
50 to 100
25 to 50
10 to 25

Declines

© 2014 Cengage Learning

California. This system has enabled central California's dry San Joaquin Valley to become one of the nation's largest producers of fruits and vegetables.

At the same time, this project has generated considerable conflict for several decades. Three-fourths of the transferred water is used to irrigate fields in areas with desertlike climates to grow crops such as alfalfa and rice that are normally grown in water-rich areas. This means that a great deal of water is used and much of it evaporates and is therefore lost to the system.

Critics of this system also say that much of this water loss is due to inefficient irrigation methods. Studies indicate that just a 10% improvement in irrigation efficiency would yield enough water savings to meet the agricultural and urban needs of Southern California. However, because this water is subsidized for growers, the water prices are artificially low and farmers have little financial incentive to invest in more efficient irrigation methods.

While Northern California is water-rich, the water-transfer program has had serious environmental impacts

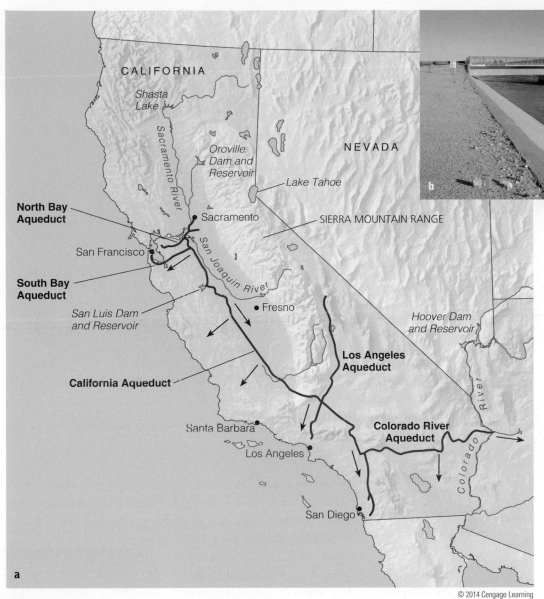

FIGURE 10.17 The California State Water Project (a) transfers large amounts of water from water-rich to water-poor areas. The black arrows show the general direction of water flow. This system makes use of lined canals (b) called *aqueducts*.

there. The withdrawal of large amounts of water has degraded the Sacramento River, threatened its fisheries, and reduced the river's ability to flush pollutants from San Francisco Bay. Consequently, the bay has suffered from pollution and often is short of the flowing freshwater that its coastal marshes and other ecosystems require. Wildlife species that depend on these ecosystems have declined, as have some ecological services.

Other systems have suffered similarly. In Mono Lake, on the eastern slope of the High Sierra to the east of San Francisco Bay, the water level dropped 35 feet after 1941 when water from its feeder streams was diverted as part of a water transfer project. The lake's ecosystem was in jeopardy, but the diversions were eventually halted and the lake is now being replenished to its natural levels.

Let's **REVIEW**

- What are three harmful effects of overpumping aquifers?
- Summarize the benefits and problems resulting from the pumping of water from the Ogallala aquifer.
- Describe the California State Water Project and summarize its benefits and drawbacks.

Growing Water Shortages

Some areas of the world suffer periodic shortages of freshwater that in many cases are getting worse, as shown in Figure 10.18 (p. 276). According to the World Water Council and the Water Resources Group, about 1.7 billion people—about one-fourth of the world's population—live in

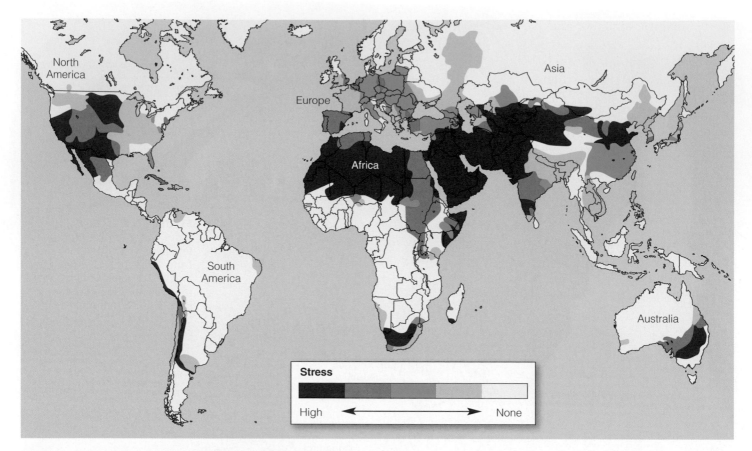

FIGURE 10.18 Shown here are the varying levels of stress that result from overuse of surface water in the world's major river basins, estimated by comparing the amount of water available with the amount used by humans. (Data from *The World Water Vision for the 21st Century*, World Water Commission, 2000; *Comprehensive Assessment of Water Management in Agriculture*, International Water Management Institute, 2007; and *Charting Our Water Future*, Water Resources Group, 2010)

such areas. Water shortages arise from a variety of factors, including dry climate conditions, prolonged drought, a growing population, higher levels of water use (especially in China, India, and South Africa), and inefficient management of water resources.

Currently, some 32 countries—most of them in the Middle East and Africa—face water scarcity (Figure 10.18). For example, according to the Chinese government, two-thirds of China's cities have faced water shortages since 2006. The United Nations projects that 2.7 billion people living in 48 countries, many of them in Asia, will face water scarcity by 2025, if water consumption continues at current rates.

It is mostly poor people who bear the brunt of water shortages (Figure 10.1b). According to the United Nations, about 1 billion people—one of every seven in the world—do not have regular access to enough clean water for drinking, cooking, and bathing. The World Health Organization

(WHO) estimates that about 2 million people—more than 80% of them children younger than age 5—die each year from diseases caused by drinking contaminated water. It is encouraging that since 1990, 2.4 billion people have gained access to safe drinking water supplies. However, some experts warn that projected climate disruptions in many parts of the world could reverse these positive trends.

Another reason why water shortages are so threatening to poor people is that water shortages hinder food production. For example, until recently, Australia was a major exporter of rice, but since 2002, Australia has experienced the worst drought in its recorded history. Since 2008, the Australian government has allocated no irrigation water for the dry southern agricultural areas that had produced most of the country's food. The resulting collapse of the country's rice harvest contributed to a doubling of the price of rice in the spring of 2008. This led to food riots in countries such as Egypt,

NUMB3RS

187

The average number of children under age 5 who die from waterborne diseases every hour

the Philippines, and Indonesia that had depended on Australian rice.

Another factor contributing to water shortages in many poor countries and affluent countries alike is drought, which is caused by shortages of precipitation and high evaporation rates, but can be made worse by overuse of water supplies. Severe drought can reduce stream flows, dry out soils (Figure 10.19), and cause a sharp drop in crop yields.

The United Nations estimates that the total area of the earth's land experiencing severe drought more than tripled between 1979 and 2007, representing almost a third of the planet's land area. By 2059, up to 45% of the world's land area could experience extreme drought, according to a 2007 study by climate researchers at NASA's Goddard Institute for Space Studies. In Syria, 160 villages had to be abandoned because of a 2008 drought.

In the United States, a 2007 U.S. Geological Survey study projected water shortages beginning in 2013 for at least 36 states, mostly in the western half of the country, resulting from a combination of drought, rising temperatures, population growth, and inefficient use of water. The U.S. Department of the Interior has identified *water hotspots* in 17 western states where there is intense and growing competition for scarce water supplies (Figure 10.20).

A growing number of analysts are sounding the alarm over water shortages that increasingly threaten human

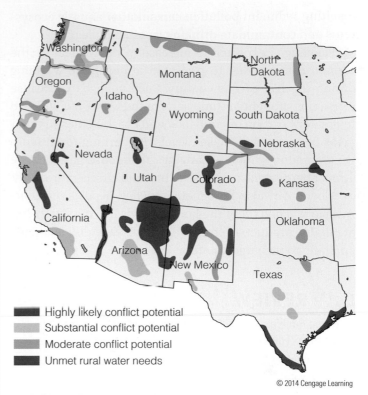

■ Highly likely conflict potential
□ Substantial conflict potential
▨ Moderate conflict potential
■ Unmet rural water needs

© 2014 Cengage Learning

FIGURE 10.20 By 2025, these water *hotspots* are likely to experience serious shortages of water. (Data from U.S. Department of the Interior, 2003)

health and food production. Some consider it to be one of the most serious problems of this century. In 2009, water expert Peter Gleick said, "Water is one of our most critical resources—even more important than oil." He and other analysts point out that there are alternatives to oil, but that for growing food and ensuring good human health, there is no substitute for water.

Flooding

Some areas get too much water. They suffer from natural flooding when streams overflow as a result of heavy rains or rapidly melting snow. This precipitation comes to some areas, such as Bangladesh (Figure 10.1a), in the form of heavy seasonal rains, called *monsoons*. Other areas are flooded more irregularly by hurricanes and other severe storms.

Floods kill thousands of people every year, and can cause severe damage to crops and human settlements. Most floods are natural events, but some are made worse by human activities that result in the removal of trees or the destruction of wetlands. Forests and wetlands both absorb excessive rainwater and provide natural flood protection. Deforestation also exposes forest soils to erosion, and flooding washes these soils into bodies of water. The

Anton Prado PHOTO/Shutterstock.com

FIGURE 10.19 This baked earth on land where crops once grew is the result of a long drought.

resulting sediment pollution can smother aquatic ecosystems and contaminate drinking water supplies.

Flooding is expected to increase in many areas as the climate changes due to projected atmospheric warming during the rest of this century. Such changes will cause worsening drought in some parts of the world, but will lead to greater flooding of croplands and coastal cities in other areas because of rising sea levels. By burning fossil fuels, we are adding climate-changing greenhouse gases to the atmosphere, which makes such harmful effects more likely.

Floods can also be beneficial. They deposit nutrient-rich silt washed off of the land and this has helped to create highly productive cropland. Floods also refill wetlands and help to recharge aquifers.

Let's **REVIEW**

- Name three areas of the world that are heavily stressed by overuse of surface water.
- How many of the world's people do not have access to enough clean water?
- How much of the earth's land area is now experiencing severe drought?
- Explain why flooding is a problem and how it can be worsened by certain human activities.

What Can Be Done?

Withdrawing More Water

Solutions to our water resource problems generally involve two major approaches: one, trying to increase water supplies by withdrawing or moving more water, and the other, working to conserve water and to make our supplies go farther. Attempts to find more water will have limited success, simply because there is a finite supply of water on the planet, and the bulk of it is unavailable for various reasons that we discussed earlier. However, we can enlarge the available supply somewhat.

Building more dams is one way to enlarge the water supply. Since the Egyptians built the first known dam about 5,000 years ago, we have increased the annual reliable runoff available for human use by nearly one-third. China is building more large dams to help meet its growing needs for water, electricity, and flood control (see the following *For Instance*). However, as we have discussed above, use of large dam-and-reservoir systems has its drawbacks. Figure 10.21 lists the major pros and cons of relying on these systems.

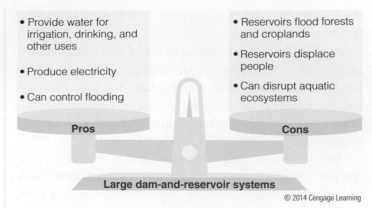

FIGURE 10.21 *Weighing the pros and cons* of large dam-and-reservoir systems.

CONSIDER this The reservoirs behind all the world's dams hold 3 to 6 times more water than do the flows of all the world's natural rivers combined.

Some water resource experts say we can rely more on groundwater to meet growing needs in areas where water shortages and drought are increasing. But other experts argue that it is too easy to overpump groundwater, as is happening in many areas of the world.

Deep aquifers could provide more water, although using this resource would be costly and these aquifers are nonrenewable on a human time scale. Scientific research indicates that some of these aquifers might have enough high-quality water to support billions of people for hundreds of years. But some experts urge caution in tapping deep aquifers until we know more about them. For example, we do not know how pumping out large volumes of this water might affect the aquifers and the land above them (Figure 10.15). Figure 10.22 summarizes the major pros and cons of relying more on groundwater.

FIGURE 10.22 *Weighing the pros and cons* of pumping more groundwater.

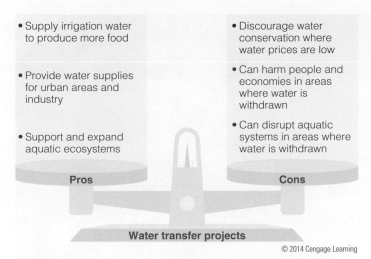

Pros	Cons
• Supply irrigation water to produce more food	• Discourage water conservation where water prices are low
• Provide water supplies for urban areas and industry	• Can harm people and economies in areas where water is withdrawn
• Support and expand aquatic ecosystems	• Can disrupt aquatic systems in areas where water is withdrawn

Water transfer projects

© 2014 Cengage Learning

FIGURE 10.23 *Weighing the pros and cons of water-transfer projects.*

We can also try to transfer more water from water-rich areas to drier areas. For example, China recently began the biggest water-transfer project in history. It will move a massive volume of water from its water-rich south to its water-short north at a cost of about $62 billion. As discussed above, this and other water-transfer systems have a number of drawbacks to weigh against their benefits.

Figure 10.23 lists the major pros and cons of transferring water.

FOR INSTANCE...

China's Three Gorges Dam

China has built the world's largest hydroelectric dam and reservoir, the Three Gorges Dam, on the Yangtze River (Figure 10.24). This massive dam is more than a mile long and the reservoir behind it stretches for more than 370 miles (roughly the distance between the cities of San Francisco and Los Angeles, California).

The reservoir will allow large cargo ships to travel deep into China's interior, which will help to boost trade in poor areas of China. The dam will also help contain the Yangtze River's floodwaters, which during the last 100 years have killed more than 500,000 people. This system will produce the same amount of electricity as 22 large coal-burning power plants, which will help China to reduce its dependence on coal for generating electricity. It could also help China to greatly reduce its air pollution and emissions of climate-changing carbon dioxide—both serious problems affecting China and the surrounding region.

However, as with all large dams, the Three Gorges Dam project has its drawbacks. For one, its vast reservoir flooded one of China's most beautiful areas, which included more than 1,500 cities, towns, and villages. It displaced about 1.4 million people from their homes.

Another problem is that the slower flow of water in the reservoir is depositing massive amounts of sediment that will fill the reservoir and shorten its projected lifespan. In addition, Chinese officials reported in 2010 that dense 2-foot-deep islands of garbage were threatening the flow of water over the dam. In some areas, they are compacted enough for people to walk on them. As it decays, this garbage is polluting the water.

Another concern is that the dam prevents much of the river's nutrient-rich sediment from reaching farms downstream as it has, historically, every year during spring flooding. These farms will become less productive unless they use large amounts of commercial fertilizer, which will likely raise farmers' costs and lead to more water pollution.

Some geologists point out that the area where the dam is built is prone to earthquakes. They warn that a major earthquake might collapse the dam and cause a flood that could kill millions of people. Engineers claim that the dam is designed to withstand any earthquake, but many small cracks have already appeared in the dam's wall.

Yet another problem will result from flooded trees and other vegetation rotting underwater in the reservoir. This will produce large emissions of methane gas—a potent greenhouse gas that will contribute to projected global climate change—erasing some of the benefit of using hydropower instead of coal to generate electricity in China.

FIGURE 10.24 The Three Gorges Dam on China's Yangtze River is the largest dam-and-reservoir system in the world.

According to some critics, in the long run, the dam's problems will outweigh its benefits. They argue that it would have been much cheaper and less harmful to build a series of smaller dams. The Chinese government, on the other hand, judged that the benefits to millions of people will outweigh its drawbacks. Time will tell which side is right.

Let's REVIEW

- Summarize the major pros and cons of building more large dam-and-reservoir systems.
- Summarize the major pros and cons of relying more on groundwater to expand water supplies.
- Summarize the major pros and cons of transferring water from water-rich areas to drier areas.
- Explain the major benefits and drawbacks of the Three Gorges Dam project.

Desalinating Seawater

It is tempting to think of the ocean as a potential water resource because it would provide a seemingly endless supply. But we would have to get the salt out of it to make it useful. **Desalination** is the set of processes that are used to remove salt from salty water.

In one such method, called *distillation*, saltwater is heated until it evaporates and leaves behind salts in solid form or in a salt-laden liquid. The water vapor is then cooled and condensed as freshwater. Another method is *reverse osmosis*, also called *microfiltration*. In this method, salt is removed by using high pressure to force saltwater through a membrane filter.

Currently, desalination meets less than 1% of the world's demand and only about 0.4% of the U.S. demand for freshwater, according to a National Academy of Sciences report. The growth of desalination around the world has been limited by three major problems. First is the high cost of building desalination plants and operating them, largely because it takes a great deal of energy to desalinate water. Second, pumping large volumes of saltwater and using chemicals to sterilize that water can harm certain species of organisms that live in the saltwater environments where the water is removed. Third, desalination produces huge quantities of salty wastewater. Dumping this concentrated salty water, even in the deep ocean, can threaten aquatic life, and dumping it on land can contaminate both surface waters and groundwater. Both of these disposal options are also costly.

FIGURE 10.25 This desalination plant lies on the Persian Gulf in Dubai, the major city of the United Arab Emirates.

Given the high costs of desalination, most countries cannot yet afford to desalinate and pump water overland to enlarge their freshwater supplies. Desalination tends to work best for water-short coastal countries and cities that can afford its high cost—countries such as Saudi Arabia and the United Arab Emirates (Figure 10.25). There are now more than 14,500 desalination plants operating around the world. Most of them are located in the arid nations of the Middle East, the Mediterranean Sea, and the Caribbean Sea.

Let's REVIEW

- Name and describe the two major types of desalination technology.
- What are three major problems associated with desalination?

Conserving Water

Many experts argue that water shortages will continue to grow worse in some areas of the world unless people begin to conserve water by using it more efficiently.

According to Mohamed El-Ashry of the World Resources Institute, because of leaks and other water losses that could be prevented, as much as 66% of the water used throughout the world and about 50% of the water used in the United States is lost through evaporation and other causes. El-Ashry estimates that we could meet most of the world's projected future water needs by cutting such water losses to 15%. This would slow the depletion of aquifers and surface water resources, and lessen the need for building large and expensive dams and water-transfer projects that can harm people and ecosystems.

There are countless ways to conserve water. Let's look at a few of them.

Raise water prices and shift water subsidies. According to water resource experts, the most effective first step toward serious improvements in water conservation would be to make it more expensive. When it is inexpensive, as it is in most places in the world, there is little incentive for people to conserve water.

For example, governments often provide farmers with subsidies that help to make irrigation water and the fuel for pumping it available at below-market prices. These subsidies stimulate local economies, encourage crop production, and help keep the prices of food and other goods low for the benefit of consumers. But they also have the effect of encouraging farmers to pump water from rivers and aquifers without thinking about the overuse and depletion of these resources and the likely results, including reduced crop yields and higher food prices.

Raising water prices can encourage water conservation, but it can also make it difficult for low-income farmers and city dwellers to buy enough water to meet their basic needs. The South African government recognized this danger and when it raised water prices, it provided a two-tiered set of rates. Each household gets enough free or low-priced water to meet its basic needs, and those who exceed this basic level of water use pay higher prices.

Some water experts argue for replacing government subsidies that encourage water use with equal subsidies that would encourage water conservation, especially in agricultural and water-short areas. Proponents of such a *subsidy shift* argue that, while the same amount of taxpayer dollars would be spent, a great deal of water could be saved. This would be especially effective if it were applied to agriculture, which accounts for about 70% of the world's freshwater use.

Increase irrigation efficiency. The United Nations estimates that reducing the global use of water for irrigation by only 10% would save enough water to meet the world's estimated additional water needs of farms, cities, and industries through 2025.

On average, about 60% of the water used for irrigation does not reach the targeted crops. Much of this water is simply poured or sprayed onto fields where it flows through unlined ditches (Figures 10.7 and 10.26, left). Typically, about 40% of this water is lost through runoff and evaporation.

However, there are crop irrigation technologies that greatly reduce water losses. Center-pivot systems (Figure 10.26, center) can get 80% to 95% of the water they use onto the crops using pumps to spray water lightly and at close range. Because the pivot systems rotate in a circle, lands irrigated by this method appear as circular fields (Figure 10.27, p. 282). Drip irrigation systems (Figure 10.26, right) get 90% to 95% of the water to the crops using a network of perforated plastic tubes at or below the ground level. Small pinholes in the tubes allow water to drip slowly close to the roots of the crop plants. Farmers can increase their crop yields by 20–90% by switching from gravity flow systems to one of these more efficient systems.

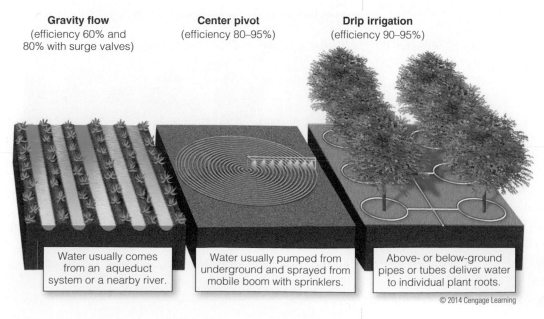

Gravity flow
(efficiency 60% and 80% with surge valves)

Center pivot
(efficiency 80–95%)

Drip irrigation
(efficiency 90–95%)

Water usually comes from an aqueduct system or a nearby river.

Water usually pumped from underground and sprayed from mobile boom with sprinklers.

Above- or below-ground pipes or tubes deliver water to individual plant roots.

© 2014 Cengage Learning

FIGURE 10.26 The traditional gravity-flow irrigation method (left) is far less efficient than newer methods such as center-pivot irrigation (center) and drip irrigation (right).

FIGURE 10.27 Cropland irrigated by center pivot irrigation appears as green circles.

In the United States, drip irrigation accounts for only 7% of crop growth. The main drawback of drip irrigation systems is their high costs. However, if governments provided subsidies and other assistance for using these systems, they would likely be more widely used and a great deal of water could be saved, especially in drier agricultural areas. Also, there are efforts underway to develop inexpensive, highly efficient irrigation systems that poor farmers can use to save water (see *Making a Difference*, below).

Other ways to increase efficiency in crop irrigation are:

- Irrigating at night to reduce evaporation.
- Growing several crops on each plot of land to hold and use more water per acre.
- Growing water-thirsty crops only in areas with ample water.
- Importing foods that require a lot of water to grow (see *A Look to the Future*, p. 285).
- Using treated urban wastewater for irrigation.

Since 1950, Israel has improved its irrigation efficiency by 84% and irrigated 44% more land by using many of these techniques. It treats and reuses 30% of its municipal sewage water for crop production and plans to treat and reuse 80% of this water by 2025. The Israeli government

MAKING A difference

Paul Polak: Providing Affordable Drip Irrigation

Paul Polak is a psychiatrist who has dedicated himself to helping people escape poverty. In 1981, he founded International Development Enterprises (IDE), a nonprofit corporation devoted to the development and distribution of affordable, efficient irrigation technologies for use by the world's poor farmers.

A conventional drip irrigation system (Figure 10.26, right) can cost as much as $6,000 per acre of irrigated land. Polak saw that such systems are far too costly for poor farmers who typically struggle to survive on an income of a little more than a dollar per day, according to United Nations estimates. He directed IDE to develop a drip irrigation kit that can cost as little as $37 per acre and can easily be scaled up or down for different size plots.

IDE's irrigation systems have helped many poor farmers to grow fruits and vegetables, and to increase their annual incomes. By 2009, IDE had distributed more than 100,000 of these systems to small-scale farmers in India. Polak now has a goal to provide 295,000 Indian farmers with these systems, and he has received financial help from private foundations.

IDE has also developed a leg-powered treadle pump (Figure 10.A) for distributing irrigation water to crops. The pump costs only about $25 to purchase and install, and uses 60–70% less water than a conventional gravity-flow system uses to irrigate the same amount of cropland. More than 1.5 million of these pumps are now being used in India and Bangladesh. This simple tool has paid for itself in most cases, increasing farmers' net profits by 100% to 400%.

FIGURE 10.A This inexpensive foot-powered treadle pump was developed by IDE.

also phased out most water subsidies and sharply raised the price of irrigation water. The country imports most of its wheat and meat, thus saving the water that would be used to produce these foods domestically, and it uses drip irrigation to grow vegetables, fruits, and flowers that need less water.

FIGURE 10.28 A great deal of water is lost through leaking valves on water pipes. Most of these leaks are smaller than this one, but altogether, they add up to very large losses.

> KEY idea Farmers can use a mix of traditional and high-tech irrigation methods to save as much as 60% of the water they use.

Let's REVIEW

- How can government subsidies discourage water conservation?
- What is a subsidy shift and how could it encourage water conservation?
- What is a gravity-flow irrigation system? How do center-pivot and drip irrigation systems save water?
- What are five other ways to save water in irrigation?

Conserve water in industry. It takes large amounts of water to produce most consumer products (Figure 10.8). Some manufacturers have found ways to reduce their water use and to collect and recycle much of the water they use. For example, in the United States more than 95% of the water used to make steel is recycled. However, most industries do not come close to this level of water recycling.

Most industrial processes could be redesigned to cut water use and increase water-use efficiency. Raising water prices would spur research and development of these processes. Currently, many water management authorities charge a flat rate for water use, and some even give a price break to those who use more, which encourages water consumption and discourages water conservation. About one of every five U.S. public water systems, for example, do not even use water meters, and they charge a single low rate for almost unlimited use of high-quality water by industries—a guaranteed way to keep water-use rates high.

Industries and cities can also use water more efficiently by fixing leaking pipes, valves (Figure 10.28), and water tanks, as well as by metering water use. Factories, businesses, and communities could also join forces in capturing, purifying, and recycling water. In Singapore, for example, all sewage water is treated at reclamation plants for reuse by industry.

Conserve water in homes. According to a 2006 study by research scientists A.Y. Hoekstra and A.K. Chapagain, the average American lifestyle requires about 1,800 gallons of water on a typical day—enough water to fill about 45 bathtubs. That makes the U.S. per capita water footprint about 657,000 gallons, or more than 16,000 bathtubs full per year. Figure 10.29 shows per capita water footprints for several other countries.

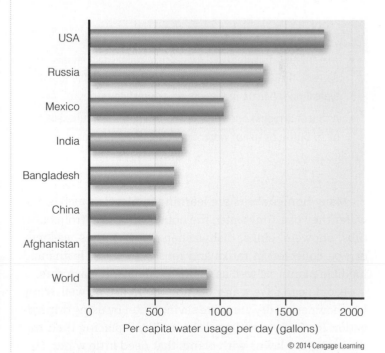

© 2014 Cengage Learning

FIGURE 10.29 This bar graph show the per capita water footprints of selected nations, expressed in gallons of water per day. (Data from Hoekstra, A.Y. and A.K. Chapagain, "The Water Footprints of Nations," *Water Resources Management Journal*, 2006)

Individuals do not directly consume all of this water. Recall that these numbers include the virtual water used to produce and ship the food and other products used by a typical person (see *A Look to the Future*, at right). Water resource experts say there are plenty of ways in which people can conserve water by making some changes within a typical home.

Toilet flushing is the single largest use of domestic water in the United States. Government standards require that all new toilets be made to use no more than 1.6 gallons of water per flush. But there are toilets on the market that use far less water and work well. For example, Niagara Conservation began selling a toilet in 2010 with a powerful 0.8 gallons per flush for $90. There are also composting toilets that use no water and produce a dry and odorless material that can be used as a soil conditioner.

Another way in which water is lost is through leaky pipes and faucets, which could be checked regularly and repaired. The United Nations estimates that 40–60% of all water used in major cities in less-developed countries is lost from leaks in water pumps, valves (Figure 10.28), mains, and pipes. These losses average 10–30% in more-developed countries such as the United States. Water experts say that, especially in water-short areas, fixing these leaks should be an urgent priority that would cost far less than building dams or importing water.

CONSIDER this

- The amount of water used for a single toilet flush equals or exceeds the average daily amount of water available to a typical poor person in the world's poorest countries.
- A faucet dripping once per second can drip the equivalent of 75 bathtubs full of water down the drain in a year.

Many homeowners are learning to recycle much of the water they use. *Gray water*, the household water from bathtubs, showers, sinks, dishwashers, and clothes washers, can be collected in tanks and reused to water lawns and inedible plants, as well as to flush toilets and wash cars.

People can save water outside the house as well. Many homeowners in dry areas are saving water by using drip irrigation for flowers and gardens, and by replacing their traditional grass lawns with plants that need little water. This water-thrifty landscaping reduces water use by 30–85%.

Use less water to remove wastes and recycle sewer water. We use huge amounts of clean water to flush away household, animal, and industrial wastes. Most sewage treatment plants dump much of this nutrient-rich treated wastewater—and untreated waste in times of overflows due to storms—into oceans, rivers, and lakes. Some of these wastes include potentially valuable plant nutrients that could be used as fertilizer, but instead, conventional sewage treatment systems end up transferring them to bodies of water.

We could make use of these nutrients by converting some of the treated waste to fertilizer. However, the sewage stream also contains toxic chemicals from some household and industrial wastes, which makes it difficult to recycle wastewater from conventional sewage treatment plants. However, we could make greater use of waterless composting toilets that convert human fecal matter into a dry and odorless, nutrient-rich material that can be used to fertilize soil and save large amounts of water and energy in the process.

Another alternative that would save a great deal of freshwater is to use wetlands-based sewage treatment systems in place of conventional treatment plants. These innovative systems mimic natural wetland ecosystems in purifying water. Another system that mimics nature involves a series of tanks containing progressively more diverse communities of aquatic plants and other organisms that filter wastes from the water. More than 800 municipalities around the world are using such systems and saving a great deal of freshwater.

Let's REVIEW

- What are two ways to save water in industry?
- What are four ways to save water in homes?
- How can we reduce the amount of freshwater that we use to move wastes? How does a wetlands-based sewage treatment system work?

Reducing Flooding

One of the most important ways to reduce flooding is to preserve forests and wetlands, because they absorb excess water and release it slowly, thereby providing natural flood control. Degraded forests and wetlands can be restored in order to regain this valuable ecological service that has been lost in so many areas of the world.

We can also use dams (Figure 10.9) to help control flooding. But large dam-and-reservoir systems have serious drawbacks as well as benefits (Figure 10.21). Other engineering approaches are to straighten and deepen streams and rivers to channel flood waters, a method called channelization, and to build levees or floodwalls along the banks of rivers. Both methods can reduce upstream flooding, but they speed the flow of flood waters, which often results in downstream flooding. They can also eliminate aquatic habitats within a river's floodplain and reduce the recharging of aquifers lying beneath the floodplain.

We can reduce our contribution to the coastal flooding that is projected to increase as the global climate warms during this century, leading to rising sea levels. To do this, we would have to cut our emissions of carbon dioxide and other greenhouse gases dramatically and put a stop to massive deforestation.

Finally, we could relieve some water shortages by managing flood waters. This would include applying the ancient strategy of collecting and storing rainwater in cisterns and tanks, as people in some arid regions have done on a small scale for centuries. This could be done on a much larger scale in major cities in dry areas of the world. The occasional heavy rains that sometimes flood parts of these cities could be channeled into reservoirs for use during dry times. This would relieve the stresses that such cities put on groundwater and surface water resources.

Let's **REVIEW**

- How do forests and wetlands help to reduce flooding?
- Summarize the problems of using engineering approaches to control flooding.
- How can we try to reduce projected coastal flooding?
- How can we use floods to relieve water shortages?

A Look to the Future

Reducing Water Footprints by Measuring Virtual Water

With water shortages growing in number and severity, many people and countries in the world will have no choice in the future but to reduce their water footprints. Those of us who live in water-short areas can help to accomplish this by making more efficient use of freshwater. However, because our water footprints are made up mostly of virtual water—our indirect, or hidden, uses of water—we will also have to find ways to monitor and reduce our use and waste of virtual water.

How much water do we use? According to the American Water Works Association, the average American directly

uses about 69 gallons of water a day—enough water to fill about 1.7 typical bathtubs full of water. Most of this water is used for flushing toilets (27%), washing clothes (22%), taking showers (17%), and running faucets (16%). Another 14%, on average, is lost to our use through leaking pipes, valves, and faucets. Compared to these daily personal uses, though, much larger amounts of water—roughly 25 times as much—are used to produce and deliver the food and consumer products that we each use. This virtual water, plus the water that we use directly every day make up our water footprints (Figure 10.29).

KEY idea Our water footprints are made up mostly of virtual water contained in the products we use every day.

When we look at a loaf of bread, a hamburger, or a cotton T-shirt, we have no way of knowing how much water was used to produce and deliver these items to us (Figure 10.8). Estimating virtual water amounts is not easy. Depending on the assumptions used and on how much of the production and delivery process is included, it can lead to a range of values. But these estimates give us at least a rough idea of the sizes of our water footprints.

There are many ways to use this information in order to monitor and perhaps reduce a water footprint. One way is to think about the food we eat. For example, according to the World Water Council, nearly 38% of the world's virtual water goes into producing meat, while about 8% goes into vegetable production. Considering the amount of water used to deliver and prepare these foods, a vegetarian diet uses about half as much virtual water as the typical meat-based diet. Similarly, a diet including less than the average amount of meat reduces per capita water use.

It is also important to consider where our food comes from because a large water footprint can do much more environmental damage in dry areas of the world than in areas that have plenty of water. For example, irrigation water is used to grow crops in a desert area not far from Phoenix, Arizona (Figure 10.30, p. 286). Much of the water used to grow these crops comes from the Colorado River system (Figure 10.10a). This river system is already heavily stressed (see *For Instance*, p. 269). Thus the virtual water in produce from the Arizona desert comes at a much greater environmental cost than would the virtual water in produce from water-rich areas.

One of the reasons we do not know about these costs is that they are not reflected in food and product prices.

FIGURE 10.30 Partially subsidized irrigation water from the Colorado River is transferred to grow crops in this desert area near Phoenix, Arizona, often using inefficient irrigation systems.

For example, much of the water that is transferred to Arizona for irrigating crops (Figure 10.30) is lost to evaporation. This loss is not accounted for in the produce prices because subsidies from the federal government and the state of Arizona make the water inexpensive, which enables the growers to raise crops in a desertlike climate using inefficient irrigation technology.

If such subsidies were removed, the cost of the virtual water in highly irrigated produce would have to be paid by producers who would then pass the cost along to consumers through higher prices. Such *full-cost pricing* (see Module 1, Figure 1.28, p. 28) would give people more incentive to reduce their water footprints by choosing produce carefully, according to where it was grown. Most water experts argue that as long as water is underpriced, we will continue to use it inefficiently.

Excessive pumping of groundwater for the irrigation and production of food and other consumer products contributes to water shortages in some areas of the world. In these areas, some countries are conserving water by importing virtual water through food and other imports. For example, arid Egypt imports more than 310,000 gallons of virtual water—more than 7,700 bathtubs full of water—for every ton of wheat that it imports instead of producing it domestically.

In fact, this has led to an international virtual water trading system with water-rich countries exporting virtual water to countries that do not have enough water to feed their people. However, this system is now being strained because there are too many countries that want to import virtual water and not enough exporters to meet the rising demand for it. At the same time, in two of the world's largest food-exporting countries—Australia and the United States—certain food-producing areas have been hit by prolonged droughts. The exporting of virtual water is adding to the strains on water supplies in these areas, and in other areas of the world. This shows that on a global scale, a country can more than offset its conservation of direct water with its exporting of virtual water.

A growing number of environmental scientists and economists argue that during this century of projected rapid climate change, water will become, for many countries, the kind of precious resource that oil became in the 20th century. The degree to which this will happen depends on how fast the human population grows and on how rapidly the climate changes. These analysts also argue that, especially in water-short areas, measuring water footprints and keeping track of virtual water use will become increasingly important.

Let's REVIEW

- What are the two major components of a water footprint, and what is the larger of these two?
- What are two ways in which an individual can use information about virtual water to shrink his or her water footprint?
- How are some nations in dry areas of the world using virtual water to extend their real water supplies?
- What is a problem that has resulted from the growing use of this strategy?

What Would You Do?

This module began by asking why you should care about water resources. You have learned that the answer is, water is unique and critical to our survival, as well as to the survival of other living things and the earth's ecosystems. You also learned how water is being mismanaged and used inefficiently at a time when there are growing water shortages in many parts of the world.

We also considered the fact that our water footprints are made up mostly of virtual water and that today's average consumer has a large and far-reaching water footprint. Now, a growing number of people are recognizing this as well as the threat of worsening water shortages around the world, and they are trying several ways to reduce their water footprints. Here are three key strategies:

Using water wisely.
- Showering instead of taking a bath can cut bathing water use by two-thirds.
- Washing only full loads of dishes in dishwashers and clothing in clothes washers cuts the number of loads and the amount of water required.
- Using drip irrigation for gardens, and watering gardens and lawns at night or in the early morning can greatly reduce water losses from evaporation.

Cutting daily water losses.
- Checking for and fixing all leaks in home faucets, sinks, toilets, tubs, and outdoor faucets can lead to considerable water savings.

- Using water-saving showerheads, faucet aerators, and new, more efficient toilets can cut water use by half or more in a typical home.
- Turning sink faucets on and off as needed while washing dishes or brushing teeth, instead of leaving the water running, saves a lot of water.

Thinking globally: considering how daily choices affect water footprints.
- When buying rice, beef, cotton, leather, and other products that require large amounts of water to produce, many people are trying to buy only those products that came from water-rich areas.
- A growing number of people are replacing their lawns with plants that are native to the areas where they live and that do not require a lot of watering.
- By washing the car by hand instead of using a commercial car wash that does not recycle its water, one can save hundreds of gallons with each washing.

Every time we use water, no matter where we live, we have some impact on the environment. When all of our individual environmental impacts are added up, regardless of how small each single one is, the total impact is huge. All of us working together can lessen this total impact by reducing our water footprints.

KEYterms

THINKINGcritically

1. Why does it matter that excessive withdrawal of water from aquatic systems for human uses can reduce river flows, shrink lakes, dry up wetlands, and deplete aquifers? Explain your reasoning.

2. What do you think would happen if the water were to stop cycling on and beneath the earth's surface and in its atmosphere? List three immediate effects that this would have on your life.

3. What is the source of drinking water used in your community? If it is groundwater, identify the aquifer, do some research, and write a description of its size and other characteristics. If it is surface water, identify the body or bodies of water involved, do some research, and describe the characteristics of this source. Is this source of water being stressed? Explain.

4. Imagine a drought-stricken area where city dwellers and farmers compete for water resources. Representatives of each side claim that their needs should get the higher priority in decisions about allocating water.

Would you side with the farmers or the city dwellers in such a dispute? Explain your thinking.

5. Explain why you would support or oppose **(a)** raising the price of water while providing low rates for minimal use to help poorer families and individuals, and **(b)** providing government subsidies to encourage farmers, households, and businesses to increase their water-use efficiency.

6. Do some research online or in a library to estimate your own water footprint. (There are several websites available for this purpose.) As part of this exercise, list everything you do throughout the day that involves your direct use of water. Also, analyze the virtual water you use, by considering at least five common items you use daily and five kinds of food that you regularly eat, and listing the amount of virtual water that is used to produce each of these things. Finally, consider whether there are ways for you to use the water more efficiently, and if there are, describe them.

LEARNINGonline

Access an interactive eBook and module-specific interactive learning tools, including flashcards, quizzes, videos and more in your Environmental Science CourseMate, accessed through **CengageBrain.com**.

11

Water Pollution

Rechitan Sorin/Shutterstock.com

Why Should You Care about Water Pollution?

Look in the mirror. What you see is about 60% water, and your life processes depend on it. You could survive for weeks without food, but without water, you would last for only a few days.

The earth has an abundant supply of this key ingredient of life, but that supply is all we have. It is *finite*. However, water is continually recycled and purified in the global *water cycle* (see Module 1, Figure 1.20, p. 19) that provides several of the earth's vital ecosystem services. Liquid water continually flows through *aquatic*, or water-based, ecosystems—oceans (Figure 11.1a), rivers (Figure 11.1b), lakes (Figure 11.1c), and wetlands—and through underground layers of gravel and sand as *groundwater*. Much water is also temporarily stored in large masses of ice called *glaciers* (Figure 11.1d).

Water pollution is any change in water quality that causes harm to humans and other living organisms using the water, or that generally makes water unsuitable for human uses such as swimming, fishing, and boating. It is a major health threat in most of the world's less-developed countries. In 2011, about 2 million people, 80% of them children younger than age 5, died from infectious diseases transmitted through polluted water.

Water pollution can be a health problem in more-developed countries as well. Every year, thousands of people are sickened by drinking water from polluted wells in rural areas and from swimming at beaches or in lakes and rivers where the water contains disease-causing organisms.

Wildlife species are also threatened by water pollution, especially aquatic life such as fish, shellfish, aquatic birds,

FIGURE 11.1 Oceans (a) cover nearly three-quarters of the earth's surface and contain saltwater. Freshwater flows in rivers and streams (b) and is stored in lakes such as this one in the Swiss Alps (c), and in glaciers such as this one in Patagonia, Argentina (d).

Galyna Andrushko/Shutterstock.com

Tischenko Irina/Shutterstock.com

Lazar Mihai-Bogdan/Shutterstock.com

Ramunas Bruzas/Shutterstock.com

and mammals that live in or near aquatic systems. When water pollution harms one species, it can threaten other species in an ecosystem that depend on the injured species for their food or other needs. In some cases, water pollution has caused whole ecosystems to collapse.

So the answer to why we should care about water pollution is quite simple: polluted water can make us sick or kill us, and it can disrupt aquatic ecosystems. In this module, we explore the different types of water pollution, as well as its various causes and its harmful effects. We also look at ways to deal with water pollution, especially ways to prevent it.

What Do You Need to Know?

The Water Cycle

Water has a unique combination of properties that make it essential for the earth's life. One of water's important properties is that it is a good *solvent*, which means it has the ability to *dissolve*, or break apart, many different compounds. This makes it important to life because it can dissolve and carry *nutrients*, or the chemicals necessary for life's processes, into the tissues of living organisms. It also flushes unneeded chemicals out of the body's tissues, just as it flushes away wastes in our homes and communities. However, water's ability to dissolve substances also means that water-soluble wastes can pollute it fairly quickly. For example, the opening photo of this module shows water pollution from a copper mine.

The earth's overall supply of water in its solid, liquid, and gaseous forms does not grow or shrink. However, fortunately for us and other forms of life, the planet's fixed supply of water is constantly recycled and purified by the **water cycle**, or **hydrologic cycle**—the continual movement of water in various forms on and under the earth's surface and in its atmosphere (see Module 1, Figure 1.20, p. 19).

The water cycle, driven by solar energy, involves three major processes: **evaporation**, the conversion of liquid water in the oceans, rivers, lakes, and soils into water vapor in the atmosphere; **transpiration**, the evaporation of water from plant leaves into the atmosphere; and **precipitation**, the return of water from the atmosphere to the planet's surface in the forms of dew, rain, sleet, snow, and hail. In this cycle, the planet's water is always changing from liquid to vapor or to ice and back again as it moves through the four parts of earth's life-support system (see Module 1, Figure 1.1, p. 2).

Depending on the type of land surface, one-third or more of the precipitation that falls on land becomes **surface runoff**, freshwater from precipitation and snowmelt that flows over the ground. Most of this runoff remains on the earth's surface as **surface water**, or freshwater that is stored or that flows in streams, rivers, wetlands, and lakes (Figure 11.1). Some surface water evaporates into the atmosphere to begin the cycle again. Some surface water seeps down into underground bodies of water, called **aquifers**, that are stored in layers of sand, gravel, and rock. This water stored underground is called **groundwater**.

Through this process, water gets purified in various ways. Most of the compounds dissolved in water are left on the earth's surface when the water is evaporated to the atmosphere. Some impurities are pulled out of water as it flows through soil and sand and gravel underground. Water is also filtered and partially purified by chemical and biological processes as it flows through wetlands, streams, lakes, and soils. This natural cycling and purification process works continually as long as it is not overloaded with harmful chemicals from one source or another.

> ## KEY idea
> Wherever it is not disrupted by human activities, the earth's fixed supply of water is continually recycled and purified by the hydrologic cycle.

Let's REVIEW

- What is *water pollution* and why should we care about it?
- Why is it important that water has the ability to dissolve a variety of chemicals?
- What is the *water cycle*, or *hydrologic cycle*, and why is it important? Define and distinguish among *evaporation*, *transpiration*, and *precipitation*.
- Define and distinguish among *surface runoff*, *surface water*, *aquifers*, and *groundwater*.

Carbon, Nitrogen, and Phosphorus Chemical Cycles

Another vital chemical cycle is the *carbon cycle*, which endlessly circulates life-sustaining carbon (C) through the four major parts of the earth's life-support system (see Module 1, Figure 1.21, p. 20).

A key chemical in the carbon cycle is carbon dioxide (CO_2) gas, which makes up 0.039% of the volume of air in the earth's atmosphere. It is a vital component of the atmosphere's temperature control system. If the carbon cycle removes CO_2 from the atmosphere faster than it is replenished, the atmosphere cools, and if CO_2 is added to the atmosphere faster than it can be removed, the atmosphere warms. Global and regional climates are determined largely by changes in the average atmospheric temperature over at least 30 years and, thus, CO_2 also plays a key role in the earth's climate conditions.

The carbon and water cycles interact. If excess CO_2 builds up in the atmosphere, the planet's average atmospheric temperature rises. This affects the water cycle by evaporating more water into the atmosphere from the soil and from bodies of water, especially the oceans. Increased evaporation and the resulting increase of atmospheric moisture content can cause heavier rains and excessive flooding in some areas. This, in turn, can add to water pollution because floods can wash harmful chemicals into streams, lakes, and ocean waters.

Carbon dioxide is also dissolved in the earth's water, primarily in the oceans. Changes in atmospheric levels of CO_2 affect the levels of dissolved CO_2 in the oceans' waters. This helps to determine the level of *acidity* of these waters, which in turn helps to determine what life forms can live there. Increased ocean acidity can threaten the existence of coral reefs and various ocean organisms that have calcium carbonate shells. This is because more acidic water makes it harder for these organisms to form their shells, and when the water gets to a certain level of acidity, it can actually dissolve calcium carbonate. Because increased ocean acidity from higher levels of dissolved CO_2 can harm ocean organisms, it is another form of water pollution.

Two other important nutrients are nitrogen (N) and phosphorus (P). Similar to carbon, they continually move through the earth's life-support system in the **nitrogen cycle** (Figure 11.2) and in the **phosphorus cycle** (Figure 11.3). Nitrogen and phosphorus are widely used as *nitrates* (NO_3^-) and *phosphates* (PO_4^{3-}) to make fertilizers. When they are applied improperly, these chemicals can flow off crop fields, lawns, and golf courses into streams, lakes, and ocean waters, where they stimulate the growth of

FIGURE 11.2 This diagram illustrates the nitrogen cycle (blue arrows) and human impacts on this cycle (red arrows). *See an animation based on this figure at* **www.cengagebrain.com**.

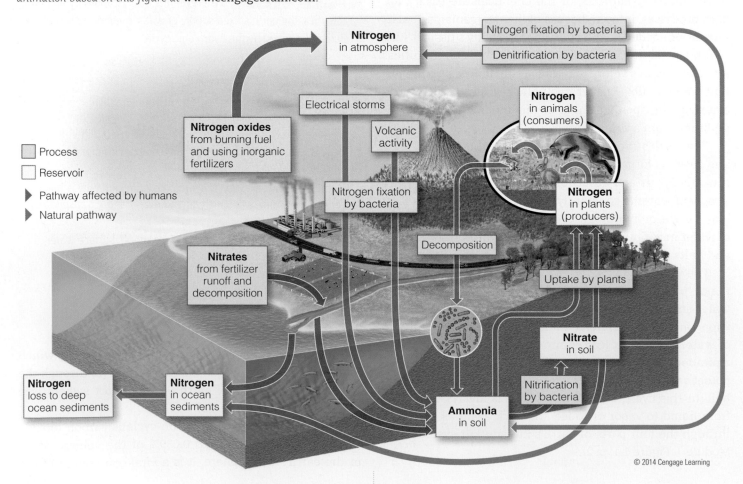

© 2014 Cengage Learning

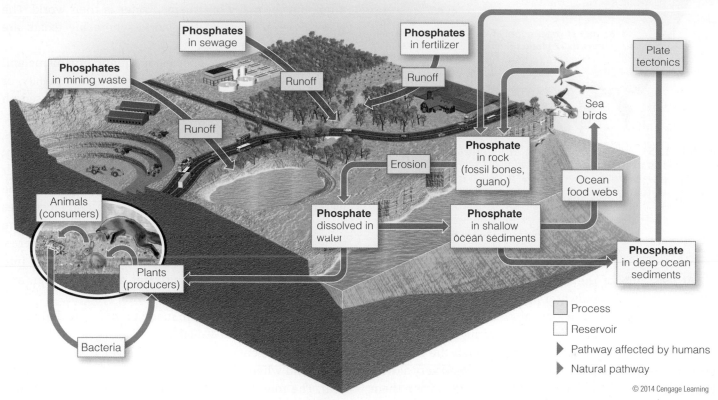

FIGURE 11.3 This diagram illustrates the phosphorus cycle (blue arrows) and human impacts on this cycle (red arrows). *See an animation based on this figure at* **www.cengagebrain.com.**

producers such as algae and various aquatic plants. Thus, the release of nitrogen and phosphorus in large amounts into the environment contributes to water pollution that can disrupt aquatic ecosystems.

We explore how changes in nutrient cycles affect bodies of water and how human activities are affecting these natural cycles in the next major section of this module.

Let's **REVIEW**

- Define *carbon cycle.* Why is carbon dioxide a key component of the carbon cycle? How is this cycle related to the water cycle?

- Define *nitrogen cycle* and *phosphorus cycle,* and explain how each is related to water pollution.

The Earth's Water Supply

Satellite pictures show that liquid water covers about 73% of the earth's surface (Figure 11.4), but only about 0.024% of this water is available to us as life-sustaining

freshwater (Figure 11.5, p. 294). The rest is too salty, lies too deep underground to be accessed with current technology, or is frozen in glaciers and polar ice caps. Yet this small percentage of all water amounts to a generous supply of freshwater that is continually recycled, purified, and distributed in the water cycle (See Module 1, Figure 1.20, p. 19).

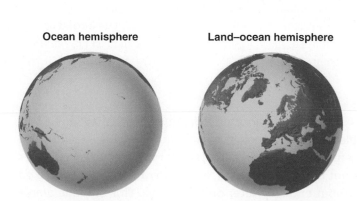

Ocean hemisphere   **Land–ocean hemisphere**

© 2014 Cengage Learning

FIGURE 11.4 Saltwater in the world's oceans covers about 71% of the planet's surface and roughly another 2% of its surface consists of freshwater in wetlands, lakes, streams, and rivers.

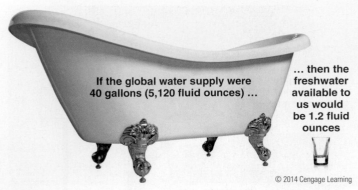

If the global water supply were 40 gallons (5,120 fluid ounces) …

… then the freshwater available to us would be 1.2 fluid ounces

© 2014 Cengage Learning

FIGURE 11.5 Only about 0.024% of the world's water is available to us as usable freshwater. If we use a typical bathtub filled with water to represent all the water in the world, then the freshwater available to us would not quite fill a 1.5-ounce shot glass.

Photo credits: Left. Baloncici/Shutterstock.com. Right. Evgeny Karandaev/Shutterstock.com.

At least 30% of the world's available freshwater supply is stored as groundwater in accessible aquifers. These groundwater supplies are usually recharged from above by precipitation sinking downward through exposed soil, or from the side by water seeping through the soil from streams. Because most aquifers are recharged very slowly, they can be depleted when we withdraw water from them faster than natural processes can recharge them.

Nearly half of the world's people get their drinking water from aquifers. Groundwater also supplies about 25% of the freshwater used in the United States in most years, according to the U.S. Geological Survey (USGS). This includes almost all of the drinking water in rural areas, one-fifth of that in cities, and around 40% of the water used to irrigate crops.

CONSIDER this   The U.S. Geological Survey estimates that the world's aquifers contain more than 100 times as much freshwater as is contained in all the world's streams, rivers, and lakes.

If the world's accessible freshwater were equally available to everyone on the planet, each of us would have plenty of water for healthy, sanitary living. However, water is not distributed evenly, and many millions of people go without enough water on most days. This is made worse by the fact that human activities have polluted much of the available water supply in many areas of the world, making it unusable, at least until natural processes can be allowed to purify it.

Water as Habitat

We might think of water as something that comes out of a tap, but for many organisms, water is their world. The water-covered areas of the world where life exists are called **aquatic life zones**.

The great variety of species of aquatic organisms and where they live are determined largely by the *salinity*, or salt content, of the water that surrounds them. Based on this, aquatic life zones are divided into two major categories. Waters in the oceans and coastal wetland areas, as well as in some inland seas and lakes are very salty, and the aquatic life zones in these waters are *saltwater* or *marine zones*, also called **marine aquatic systems**. The waters in rivers, streams, inland wetlands, and most lakes have a much lower salt content, and their aquatic life zones are *freshwater zones*, or **freshwater aquatic systems**. Some systems such as *estuaries*, where rivers empty into oceans, contain a mix of saltwater and freshwater, and they are generally classified as marine systems.

Most bodies of water are made up of layers defined by *water temperature* and *availability of light*. Lakes and oceans, for example, generally have a top layer that light penetrates and where the water is warmer, at least part of the year. The middle layer receives some light and is colder, while the bottom layer of water is dark and colder than the layers above. Another key factor that helps to determine the types and numbers of organisms found in these layers is the amount of oxygen contained in a given volume of the water, or *dissolved oxygen content*.

As on land, the organisms in aquatic life zones each play one of three major roles. One such role is that of a *producer*—an organism that makes the *nutrients*, the chemicals it needs in order to survive, from the chemicals and energy that it gets from its environment (Figure 11.6a). In aquatic systems, producers are mostly shore plants, plant-like organisms called *algae*, and tiny floating plants called *phytoplankton*.

A second major role is that of the *consumer*—any organism that gets its nutrients by eating producers and other consumers (Figure 11.6b, c, and d). The third major role is that of a *decomposer*—any organism that feeds on or otherwise breaks down the wastes and remains of plants and animals. As they do this, aquatic decomposers release nutrients from these wastes and remains into the water to be recycled by producers. As in undisturbed land ecosystems, there is little or no waste in aquatic life zones, because the nutrients that aquatic creatures need are constantly recycled among these three types of organisms.

FIGURE 11.6 Aquatic life includes producer organisms such as tiny drifting plants, including blue-green algae that can cover the surface of a pond (a). Three types of consumer organisms are drifting animals, ranging from tiny to very large, called *zooplankton*, such as this jellyfish (b); strongly swimming animals called *nekton*, such as turtles (c), fishes, and whales; and bottom-dwelling animals called *benthos* such as this Caribbean spiny lobster (d).

Let's **REVIEW**

- What percentage of the earth's surface is covered by water? What percentage of this water is available to us as useable freshwater?

- How much of the available freshwater is stored underground and how many of the world's people use this groundwater?

- What is an *aquatic life zone*?

- Distinguish between saltwater, or marine, and freshwater zones.

- Define and distinguish among the producer, consumer, and decomposer roles played by aquatic organisms and give an example of each.

- Why is there little or no waste of nutrients in undisturbed aquatic ecosystems?

Now we look more closely at the different kinds of aquatic systems and at some of the important ecosystem services they provide.

Marine Aquatic Systems

The largest aquatic system is, of course, the *global ocean* (Figure 11.4)—a single and continuous body of water that geographers have divided into four large areas separated by the continents. The largest of these is the Pacific Ocean, which contains more than half of the earth's water and covers one-third of its surface. The others are the Atlantic, the Arctic, and the Indian Oceans.

The oceans contain three major types of marine aquatic systems, or marine zones: the coastal zone, the open sea, and the ocean bottom (Figure 11.7, p. 296). Each of these zones hosts a different variety of life forms.

The **coastal zone** is the relatively shallow water that extends from the shore to the edge of the *continental shelf* (the part of each continent that extends out under the

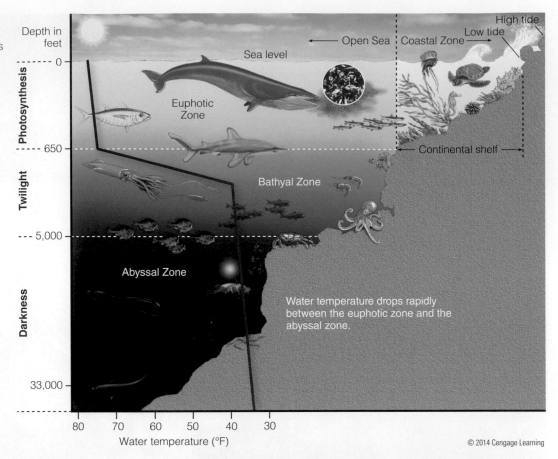

FIGURE 11.7 This diagram shows the major marine aquatic life zones. Actual zones vary greatly in terms of depth, temperature (red curve), and other factors.

Depth in feet

Photosynthesis

Twilight

Darkness

0

650

5,000

33,000

Sea level

Open Sea ← | Coastal Zone →

High tide

Low tide

Euphotic Zone

Continental shelf

Bathyal Zone

Abyssal Zone

Water temperature drops rapidly between the euphotic zone and the abyssal zone.

80   70   60   50   40   30

Water temperature (°F)

© 2014 Cengage Learning

water). The coastal zone is the warmest of ocean waters and while it makes up less than 10% of the world's ocean area, it contains 90% of all marine species, including most of the seafood species that people eat.

The coastal zone hosts a number of important types of aquatic systems. One is the *estuary*, or the partially enclosed body of water where a river meets the sea and saltwater mixes with freshwater (Figure 11.8). It contains a rich mix of nutrients from streams, rivers, and runoff from the land, which can pollute estuaries in most developed coastal areas.

Another important coastal system that is usually found within or near estuaries is a *coastal wetland*—a coastal land area covered with water all or part of the year. Coastal wetlands include *coastal marshes* and *mangrove forests* (Figure 11.9a and b). Similar to estuaries, these systems receive inputs of nutrients from rivers and runoff from nearby land, and they get ample sunlight, so they host a variety of producer and consumer species.

Also located in the shallow waters along many coastlines are *sea-grass beds* (Figure 11.10), another type of coastal ecosystem that supports a rich variety of sea life. A typical sea-grass bed contains at least 60 species of underwater plants that support fishes and other organisms.

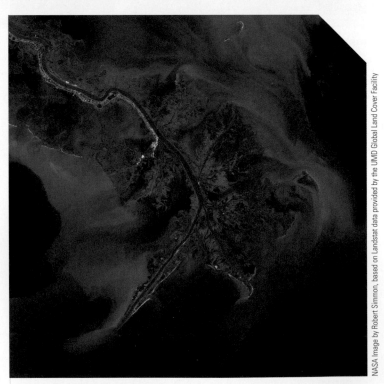

NASA Image by Robert Simmon, based on Landstat data provided by the UMD Global Land Cover Facility

FIGURE 11.8 This satellite photo shows the Mississippi River estuary where the river empties into the Gulf of Mexico. Nutrient-rich sediment flowing into the gulf has created a massive bloom of algae.

Diane Uhley/Shutterstock.com

Nickolay Stanev/Shutterstock.com

FIGURE 11.9 Two examples of coastal wetlands are **(a)** coastal marshes, and **(b)** mangrove forests.

Sea-grass beds also help to stabilize shorelines that might otherwise be eroded by wave action.

Another type of coastal system, crucial to marine biodiversity in tropical areas of the world, is a *coral reef* (Figure 11.11, p. 298)—a structure made of calcium carbonate formed by tiny animals called *polyps*. Each polyp builds a protective crust around its body, and these crusts accumulate to make a reef. Living in the tissues of the polyps' bodies are tiny, single-celled algae that provide the polyps with food and oxygen, and help them to produce calcium carbonate. These algae give the reefs their spectacular coloring. Over time, a reef grows into a complex structure that serves as habitat for an astonishing variety of marine animals. Coral reefs are among the world's most diverse and productive ecosystems.

Coastal aquatic systems support marine biodiversity and provide important services to human communities. For example, coral reefs are home to one of every ten fish that humans catch and eat. Mangrove forests supply building material and firewood to coastal communities.

Rich Carey/Shutterstock.com

FIGURE 11.10 Sea-grass beds provide habitats for a variety of marine species.

iStockphoto/Martin Strmko

FIGURE 11.11 This colorful coral reef is located off the coast of the Red Sea.

Mangrove forests, coastal wetlands, and coral reefs reduce storm damage, flooding, and coastal erosion by absorbing waves and storm surges. Coral reefs also support a large part of the tourism industry, which is important to many poorer coastal nations.

The second major marine life zone is the **open sea**, the vast volume of the ocean that lies beyond the edges of the continental shelves where the ocean floor drops off sharply into deeper waters. The open sea is divided into three *vertical zones* (Figure 11.7), which vary in the amount of sunlight they receive and in their water temperatures. The deepest zone in the open ocean is also the third major marine life zone, which biologists refer to as the **ocean bottom** life zone. It contains many species that get their food from the steady supply of dead and decaying organisms that drift down from the upper zones.

The open sea and ocean bottom life zones contain rich marine biodiversity and provide humans with much of their food supply. These are the zones where industrial fishing fleets have netted large quantities of fish and other marine creatures every day for hundreds of years. The open sea also helps to moderate the earth's climate by absorbing carbon dioxide, a *greenhouse gas* that is helping to warm the atmosphere and change the earth's climate.

NUMB3RS

90%

The estimated percentage of all marine species that live in coastal aquatic systems

Let's REVIEW

- What are the three major types of marine aquatic systems contained within the oceans?

- Define and describe the *coastal zone*. Distinguish between an estuary and a coastal wetland. What are two types of coastal wetlands and why are they important?

- What are sea-grass beds and what benefits do they provide? What are coral reefs and why are they important?

- Define and describe the *open sea* and *ocean bottom* life zones. What ecosystem services do they provide?

Freshwater Aquatic Systems

Freshwater systems include *standing* bodies of freshwater, such as lakes (Figure 11.1c), ponds, and inland wetlands, and *flowing* systems, such as rivers and streams (Figure 11.1b).

A *lake* is a large body of standing freshwater fed by rain and snow, surface runoff, streams, and groundwater flowing upward through springs on its bottom or near its shores. Many lake basins were gouged out of the earth by retreating glaciers about 10,000 years ago. Earthquakes and volcanic activity have also created some of the world's lakes.

Ecologists classify lakes according to the amount of plant nutrients—including chemicals such as phosphates (PO_4^{3-}) and nitrates (NO_3^-)—contained in their waters. Nutrients get into lakes naturally through sediments, plant and animal wastes and remains, and other substances that wash off surrounding land—a process called **eutrophication**. People add nutrients to lakes through chemicals such as lawn and farm fertilizers that wash off the land, and through chemicals emitted into the air that fall into the lakes. This acceleration of natural eutrophication by humans is called **cultural eutrophication**.

Lakes that have a low level of these nutrients are called *oligotrophic lakes* (Figure 11.12a). These lakes tend to have very clear water, reflecting blue in the sunlight, and many of them are fed by ice-cold mountain streams. Because of their low levels of nutrients, they usually contain only small populations of phytoplankton and fishes. A lake that has a high level of plant nutrients is called a *eutrophic lake* (Figure 11.12b). These lakes typically have murky brown or green water. Because of their high levels of nutrients, they are rich with algae and other producers in their water and with plant life on their shores. Most lakes fall somewhere between these two extremes.

Another important freshwater aquatic system is a *stream*, or a narrow body of water flowing along a defined

FIGURE 11.12 Crater Lake in the state of Oregon (a) is an example of an *oligotrophic lake* with a low level of plant nutrients. This *eutrophic lake* (b) has such a high level of plant nutrients that it is covered by mats of algae.

course from a higher elevation to a lower elevation. Streams are formed when surface water moves down sloping land—often on the sides of a valley—to a lower point where it collects and forms a flowing body of water. The land area off which this water flows is the stream's *watershed*, or *drainage basin*.

One stream can flow into another and form a larger stream. Large streams that are fed by other smaller streams and that flow into yet larger streams, lakes, or oceans are called *rivers*. Water flowing downhill moves through three zones (Figure 11.13): the *source zone* containing headwaters such as springs or melting snow; the *transition zone* containing wider, lower-elevation streams; and the *floodplain zone* containing rivers.

A third type of freshwater aquatic system is an *inland wetland*, which is land that is located away from ocean coasts and is saturated or covered with freshwater all or part of the year. There are several types of inland wetlands, including *marshes* (dominated by grasses and reeds, Figure 11.14a, p. 300) and *swamps* (dominated by trees and shrubs, Figure 11.14b).

Inland wetlands are important because they usually host great varieties of insects, amphibians, reptiles, birds, and mammals. They also trap some toxic chemicals and excessive plant nutrients running off farm fields or other developed lands. These chemicals would otherwise often flow directly into streams or lakes. Wetlands also reduce flooding and erosion by absorbing excess rainwater, melting snow, and stream overflows and then releasing this water slowly.

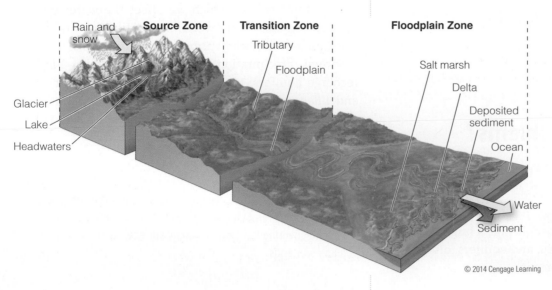

© 2014 Cengage Learning

FIGURE 11.13 As it moves to the sea, the downhill flow of water in a river system passes through three zones.

FIGURE 11.14 This inland marsh (a) in the Florida Everglades is home to many species including this great white egret. This algae-covered swamp (b) is located in South Carolina.

Inland wetlands, streams, and lakes all provide routes for water flowing down into aquifers and thus they help to maintain groundwater supplies. These freshwater systems also supply people with valuable food products such as cranberries and wild rice, and they provide recreation for birdwatchers, nature photographers, boaters, anglers, and waterfowl hunters.

Let's REVIEW

- Define and distinguish between *eutrophication* and *cultural eutrophication*.
- Distinguish between *oligotrophic* and *eutrophic* lakes.
- Explain how streams and rivers form. What is a watershed, or drainage basin?
- Describe the three zones through which the water in a river system flows.
- What is an inland wetland and what are two types of such systems? Why are inland wetlands important?

What Are the Problems?

Human Impacts on Natural Cycles

Human activities affect the water cycle in three major ways (see red arrows and boxes in Module 1, Figure 1.20, p. 19). First, on a global scale, people take a huge volume of freshwater from streams, lakes, and aquifers, often faster than the water cycle can replace it.

Second, we clear the land of its trees and other plants in order to farm it, to dig mines, to build roads, and to expand cities. We also cover much of the land with pavement and buildings. Precipitation runs off these artificial surfaces and does not trickle down through the earth beneath them. Thus, aquifers beneath paved-over areas are recharged much more slowly or not at all.

Another effect of these artificial surfaces is that much more runoff goes into streams than would be the case if the land were absorbing it, and the streams then flood more easily. Farm fields also cause increased runoff, and together, urban and rural runoff pollutes streams, lakes, and wetlands with excessive nitrates and phosphates. This transfer of important plant nutrients from land to aquatic systems often leads to excessive growth of algae and other plants in aquatic systems. This form of water pollution results from disrupting the normal flows in the nitrogen and phosphorus chemical cycles.

The third way in which we affect the water cycle is by draining and filling wetlands for farming and urban development. Left undisturbed, wetlands act like sponges, absorbing precipitation and holding it, which helps to recharge aquifers. However, when wetlands are drained and filled, runoff increases along with the pollution of streams. The risk of flooding goes up, while the rate of aquifer recharge goes down.

KEY idea Human activities interfere with the water cycle by removing water faster than it is replaced, clearing and covering land with artificial surfaces, and draining and filling wetlands.

Human activities also affect the carbon, nitrogen, and phosphorus cycles in several ways. We are altering the carbon cycle (see Module 1, Figure 1.21, p. 20) by adding large amounts of CO_2 to the atmosphere when we burn carbon-containing fossil fuels. We also clear vegetation from forests that, if left in place, would remove some CO_2 from the atmosphere. Thus, we are disrupting the balance between the addition and the removal of CO_2 that was established by natural systems within the carbon cycle over millions of years.

This imbalance is altering the climate in ways that are leading to more precipitation and flooding in some areas and prolonged, drier conditions in other areas. Increased flooding often pollutes aquatic systems with harmful chemicals and excess plant nutrients. Extended drought can concentrate harmful chemicals in streams, rivers, and lakes by reducing the level and flow of their waters. Marine systems are also being disrupted by the rising acidity of ocean waters resulting from their absorption of large amounts of atmospheric CO_2 due to this growing imbalance in the carbon cycle.

Human activities also affect the nitrogen cycle (see red arrows in Figure 11.2) and the phosphorus cycle (see red arrows in Figure 11.3). Nitrogen in the form of nitrates is released from fertilizers. Phosphorus in the form of phosphates is also contained in fertilizers as well as in many detergents. These nutrients flow from farm fields, urban lawns, golf courses, and sewage systems into streams, ponds, and lakes, and they can overfertilize these aquatic systems. This leads to explosive growths of algae (Figures 11.6a and 11.12b) and various plants such as water

hyacinths, which in turn can disrupt aquatic ecosystems. Some of these excess nutrients end up in rivers that flow into coastal waters, where they overfertilize and upset coastal aquatic systems.

According to scientists working on the 2005 Millennium Ecosystem Assessment, since 1950, human activities have more than doubled the annual release of nitrogen from the land into the rest of the environment. They also project that this annual release of nitrogen from human activities will double again by 2050.

Water Pollution

Human activities can change water in many ways. Chemicals that we dump into water can reduce its usefulness for drinking, fishing, and swimming, as well as for irrigating crops. When we use water as a coolant in coal-burning and nuclear power plants, the heated water that is released back to its source can change the nature of that stream or body of water. These changes fit under the umbrella of water pollution.

Any substance or any heat energy that causes water pollution is called a **pollutant**. Pollutants can enter water from **point sources**, or single, easily identifiable points such as the end of a drainpipe (Figure 11.15a) or a sewer line. Examples of point sources are factories, sewage treatment plants that fail to remove all pollutants from wastewater, underground mines, and ships.

Pollutants can also enter water from **nonpoint sources**, which are areas, rather than points, from which pollutants flow into water. Examples include cropland, livestock feedlots, logged forests, and golf courses. Sediment eroded from farmlands is the largest form of nonpoint-source water pollution (Figure 11.15b). Other major agricultural

FIGURE 11.15 This drain pipe (a) is a point source of water pollution. Nonpoint-source sediment eroded from farmland (b) flows into and disrupts streams; sometimes it can even change a stream's course or dam it up.

vasakkohaline/Shutterstock.com

Tim McCabe/Natural Resources Conservation Service

pollutants are fertilizers and pesticides, bacteria from live-stock and food-processing wastes, and salts that wash out of the soil on irrigated cropland.

Another significant category of sources is industrial activities. Most of these are point sources such as factories that release a variety of harmful chemicals into waterways (Figure 11.15a). But this category also includes construction sites, which are nonpoint sources, and power plants that emit heated water into natural bodies of water (a form of pollution called *thermal pollution*).

The third biggest source of water pollution is mining activities. Surface mining involves digging up vast areas of land and leaving large piles of rubble (Figure 11.16). These *spoils* piles are a major source of eroded sediments and toxic chemicals that can seep into groundwater or run off into bodies of surface water near the mining sites. For example, the opening photo of this module shows water pollution from an open-pit copper mine. Underground mining also releases chemicals into groundwater and waterways, usually to a far lesser degree than surface mining does.

There are numerous different types of water pollutants. Table 11.1 lists the major types along with a brief description of their harmful effects, examples of each, and typical sources.

FIGURE 11.16 Rock, sand, and soil removed from an open-pit surface mine can pollute the water and air.

Let's **REVIEW**

- List three ways in which human activities interfere with the water cycle. How do human activities affect the carbon, nitrogen, and phosphorus cycles? What are some usual results in terms of water pollution?

- Define *pollutant*. Define and distinguish between *point sources* and *nonpoint sources* of pollutants and give an example of each.

- What are three major sources of water pollutants? What is the largest of these sources?

- Name five major water pollutants and, for each one, list an example of the pollutant, some of its effects, and its major sources.

TABLE 11.1 Major Types of Water Pollutants and Their Effects and Sources

Type/*Effects*	Examples	Major Sources
Organisms that infect other organisms *Cause diseases*	Bacteria, viruses, parasites	Human and animal wastes
Wastes that use up oxygen dissolved in water (oxygen-demanding wastes) *Deplete dissolved oxygen needed by aquatic species*	Animal wastes and plant debris	Sewage treatment plants, feedlots, food-processing facilities, pulp mills
Plant nutrients *Cause excessive growth of algae and various aquatic plant species*	Nitrogen and phosphorus compounds such as nitrates (NO_3^-) and phosphates (PO_4^{3-})	Sewage, animal wastes, inorganic fertilizers
Organic chemicals (chemicals that contain carbon) *Add toxins to aquatic systems*	Oil, gasoline, plastics, pesticides, cleaning solvents	Industry, farms, households, construction sites, mining sites
Inorganic (non-carbon-containing) chemicals *Add toxins to aquatic systems*	Acids, bases, salts, metal compounds	Industry, households, surface runoff, mining sites
Sediments *Disrupt aquatic plants and the animals that feed on them, and other ecological processes*	Eroded soil, silt	Farm fields, construction sites, mining sites
Chemical elements (heavy metals) *Cause cancer, disrupt immune and endocrine systems*	Lead, mercury, arsenic	Certain landfills, household chemicals, mining waste, industrial activities
Thermal *Makes some species vulnerable to disease*	Heat	Coal-burning and nuclear power plants and industrial plants

Pollution of Rivers and Lakes

According to the World Commission on Water for the 21st Century, half of the world's 500 rivers are heavily polluted, and most of those polluted streams run through one or more of the world's less-developed countries. A large proportion of the pollutants in these streams come from discharges of untreated sewage and industrial wastes. Many of these countries cannot afford to build waste treatment plants and most of them do not have, or do not enforce, laws for controlling water pollution.

For example, industrial wastes and sewage pollute more than two-thirds of China's rivers and streams (Figure 11.17). In addition, one-third of China's rivers are too polluted for use as sources of irrigation water. The water in some of these rivers is too toxic to touch, let alone to drink. In Latin America and Africa, most streams in urban and industrial areas are also severely polluted.

Zhao Weiming/UNEP/Peter Arnold, Inc.

FIGURE 11.17 This highly polluted river in China receives point-source pollutants from thousands of factories.

Surface-water pollution is not as big a problem in most of the world's more-developed countries where water pollution laws have helped to control it. However, problems still occur in these countries because of accidental or deliberate releases of toxic inorganic and organic chemicals from industrial and mining sites and from overflowing sewage treatment plants, and because of lax enforcement of the laws in some cases. Fertilizer and pesticide runoff from cropland and animal feedlots is also a problem in many areas of more-developed countries.

Urban growth in many of the world's countries is making some water pollution problems worse. For example, as urban areas grow, their wastewater treatment systems, which usually contain both storm water and sewage, can be strained. Heavy rains can overwhelm them, causing untreated sewage and storm water to flow into streams and other waterways or into treated water discharged from sewage treatment plants.

Fortunately, polluted streams can cleanse themselves of moderate levels of oxygen-demanding wastes if they are not overloaded with these pollutants. This cleansing happens for two reasons. First, as freshwater flows from upstream, it dilutes and washes away pollutants that have built up in streams. Second, bacteria in the water break down some pollutants and this helps with the dilution process. Those pollutants that can be broken down, or degraded, in this way are called **biodegradable pollutants**.

However, this natural cleansing function is hindered when a river's flow is reduced by dams, heavy water withdrawals, or drought. It also does not happen when streams are overloaded with biodegradable pollutants or with **nondegradable pollutants**—those that cannot be broken down naturally—such as toxic lead and arsenic.

Lakes can also be polluted by fertilizer and pesticide runoff, and by other chemicals from surrounding land. Point-source discharges of oil and other toxic chemicals affect many lakes. Some pollutants, including nondegradable, toxic heavy metals (such as lead and mercury, see Table 11.1) enter lakes from the air after they are discharged by engines burning leaded gasoline, coal-burning power plants, and other sources. These pollutants can kill lake-bottom life and fish as well as the birds and other animals that feed on those aquatic organisms. They can also harm humans who eat fish that contain dangerous levels of these toxic chemicals.

Lakes in urban or agricultural areas often undergo cultural eutrophication as various human activities result in high inputs of plant nutrients. The common sources of these nutrients are runoff from fertilized farm fields, suburban lawns, and golf courses. Other nutrients originate in animal feedlots and municipal sewage treatment plants. Nutrients may flow directly into lakes that border these sites or they may be present in streams that flow into the lakes.

During hot weather, high levels of nutrients in a lake often cause rapid growths of algae—called *algal blooms* (Figures 11.6a and 11.12b)—and thick growths of water hyacinth and other aquatic plants. These dense colonies of plant life often block sunlight that is needed by phytoplankton. As a result, these plankton cannot produce nutrients that support fish populations and many fish can die, sometimes all at once in *fish kills*. When dense growths of algae die, they are decomposed by bacteria, which take up the dissolved oxygen in a lake's water. This, too, can kill fish and other aquatic animals that need oxygen to survive.

According to the U.S. Environmental Protection Agency (EPA), about one-third of the 100,000 medium-to-large lakes and 85% of the large lakes near major U.S. cities have some degree of cultural eutrophication. The International Water Association also estimates that more than half of the lakes in China suffer from this sort of pollution. Because most lakes have little or no flow and only seasonal mixing of their waters, they are much less able to recover from high levels of pollution than are streams. Large estuaries can suffer some of the same water pollution effects as are seen in lakes (Figure 11.18).

Lakes and rivers can also be polluted by trash (Figure 11.19), which reduces the usefulness of these bodies of water for recreation. Some of this debris also pollutes the

FIGURE 11.19 Plastics and other forms of trash pollute bodies of water such as this river.

water by releasing harmful chemicals that can kill aquatic organisms. Also, some aquatic animals die or become ill after ingesting pieces of plastic, mistaking them for food.

Let's REVIEW

- About how many of the world's rivers are heavily polluted?
- What are two major sources of stream pollution?
- How can wastewater treatment systems pollute surface waters?
- What are *biodegradable wastes*? How can streams cleanse themselves? When does this natural cleansing process fail to work?
- What are algal blooms, what causes them, and how can they affect lakes?

Ocean Pollution

About 4 of every 10 people in the world, and more than half of all Americans, live on or near a seacoast, and coastal population numbers are steadily rising. Consequently, human activities greatly affect ocean waters. An enormous amount of pollutants flow into coastal waters around the world and this has an impact on all coastal aquatic systems, especially estuaries, coastal wetlands, coral reefs, and mangrove swamps (Figure 11.20).

For example, according to a 2006 UN Environment Programme report, 80–90% of the sewage from cities in less-developed countries located on ocean coasts was dumped into oceans untreated. The volume of the global ocean is huge, and in deeper waters, such oxygen-demanding

FIGURE 11.18 This dense bloom of blue-green algae grew in the St. Johns River in Jacksonville, Florida.

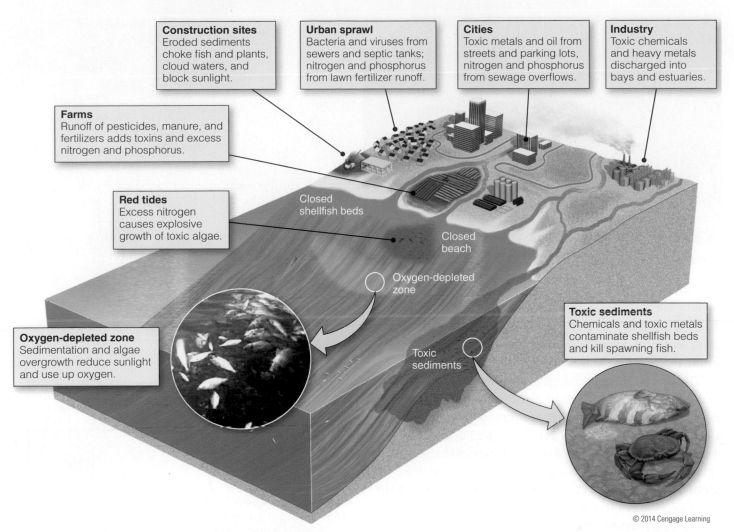

Construction sites
Eroded sediments choke fish and plants, cloud waters, and block sunlight.

Urban sprawl
Bacteria and viruses from sewers and septic tanks; nitrogen and phosphorus from lawn fertilizer runoff.

Cities
Toxic metals and oil from streets and parking lots, nitrogen and phosphorus from sewage overflows.

Industry
Toxic chemicals and heavy metals discharged into bays and estuaries.

Farms
Runoff of pesticides, manure, and fertilizers adds toxins and excess nitrogen and phosphorus.

Red tides
Excess nitrogen causes explosive growth of toxic algae.

Closed shellfish beds

Closed beach

Oxygen-depleted zone

Toxic sediments
Chemicals and toxic metals contaminate shellfish beds and kill spawning fish.

Oxygen-depleted zone
Sedimentation and algae overgrowth reduce sunlight and use up oxygen.

Toxic sediments

© 2014 Cengage Learning

FIGURE 11.20 Pollutants and wastes produced by factories, farms, coastal cities, and residential areas end up in coastal aquatic zones—especially estuaries, coastal wetlands, coral reefs, and mangrove swamps.

wastes can be diluted and degraded by bacteria. However, in more shallow coastal waters, pollutants such as untreated sewage can overwhelm these natural purifying processes.

Even in more-developed countries where sewage is treated, high volumes of waste and heavy rains can overwhelm sewage treatment systems. Studies of some U.S. coastal waters found colonies of viruses thriving in raw sewage and in the effluents of sewage treatment plants. (These plants do not remove viruses from treated water.) In suburban and rural coastal areas, most homeowners use on-site sewage treatment systems called *septic tanks*, and these systems can also leak or overflow.

Runoff of sewage and farm wastes into coastal waters usually includes large quantities of nitrate and phosphate nutrients, which can cause algae populations to explode, and some algae can be harmful. These *harmful algal blooms*,

sometimes called brown, green, or red toxic tides (Figure 11.21, p. 306), release toxins that poison fish and kill some fish-eating birds and other animals, including pet dogs that accidentally swallow the water while swimming.

Harmful algal blooms often result in the closing of recreational beaches, which reduces tourism in beach areas. Each year, on average, harmful algal blooms lead to the poisoning of about 60,000 Americans who eat shellfish contaminated by the algae. A 2008 study by scientists at China's Dalian Maritime University found that water pollutants such as nitrates and phosphates that cause algal blooms seriously contaminated about half of China's shallow coastal waters.

Another problem arising from algal blooms is the formation of *oxygen-depleted zones* in some coastal areas. As oxygen-using bacteria decompose massive numbers of algae, they can deplete the water of dissolved oxygen. Oxygen-consuming fish and bottom-dwelling organisms die off or abandon these areas, and they are therefore sometimes

FIGURE 11.21   This is a toxic red-tide algal bloom.

called *dead zones* (even though they still host bacteria and other organisms). In 2011, the World Resources Institute mapped these zones and found 415 of them around the world, including 50 in U.S. waters.

The formation of an oxygen-depleted zone is a process that starts with high levels of plant nutrients in the water that feed the algal blooms. These nutrients include nitrates and phosphates from fertilizer runoff and animal wastes. A classic case of this type of situation occurs in the Gulf of Mexico where the Mississippi River empties into the gulf (Figures 11.8 and 11.22). The river's water contains high levels of plant nutrients from thousands of farms, factories, municipalities, and sewage treatment plants in the vast Mississippi River basin (see top map in Figure 11.22).

The gulf is the site of one of the world's largest oxygen-depleted zones. It forms every year during the spring and summer, often covering an area larger than the state of Massachusetts. The largest oxygen-depleted zone in the world lies in the Baltic Sea of northern Europe.

Many other pollutants (Table 11.1) flow into coastal systems, including lead, mercury, arsenic and other toxic metals from industrial operations, oil from city streets and parking lots, eroded soils from construction sites, and runoff of sediments, pesticides, and fertilizers from farm fields. These pollutants cause eutrophication of estuaries (Figure 11.18), which can disrupt their ecosystems (see *For Instance*, p. 308).

The type of coastal system that is perhaps the most threatened by pollution is coral reefs, because they grow slowly and can be easily disrupted. Corals thrive only in clear and fairly shallow water. Sediment and other pollutants can cloud the waters over coral reefs or settle on the reefs themselves, blocking sunlight and killing off the producer organisms that live there.

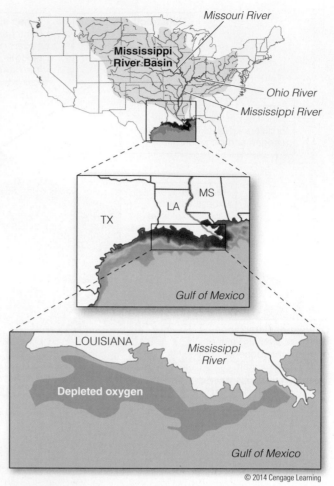

FIGURE 11.22   A vast oxygen-depleted zone forms in the Gulf of Mexico every spring (bottom map). The center map, based on a satellite image, shows the inputs of nutrients from the Mississippi River watershed (top map) into the gulf during the summer of 2008. The reds and greens represent high concentrations of phytoplankton and river sediment. (Data from NASA)

CONSIDER this   Since the 1950s, coastal development, pollution, and overfishing have damaged or destroyed at least 45% of the world's coral reefs. Another 25–33% of all coral reefs could be lost by 2050.

Oil pollution in coastal waters is an especially difficult problem. Oil, gasoline, and other petroleum products reach ocean waters from many different sources, including natural seepage, oil tanker spills and leaks, and ruptures from offshore drilling rigs. The worst-ever rupture of this type happened in the Gulf of Mexico during the spring and summer of 2010 (Figure 11.23). However, studies have shown that the largest overall source of ocean oil pollution is the runoff from city streets and industrial

FIGURE 11.23 On April 20, 2010, natural gas leaking from a ruptured oil well borehole caused an explosion on the British Petroleum (BP) *Deepwater Horizon* drilling platform in the Gulf of Mexico that killed 11 people and released massive amounts of crude oil into the gulf's waters.

sites combined with leaks in coastal area pipelines and oil-handling facilities around the world.

Chemicals in oil called *volatile organic hydrocarbons* are deadly to many aquatic organisms, poisoning and killing them very quickly. Other chemicals in oil form tarlike globs in the water that coat the feathers of swimming and diving birds as well as the fur of marine mammals. This oil

FIGURE 11.24 Plastic items, including discarded fishing nets (a) are polluting ocean waters. The Great Pacific Garbage Patch (b) is made up of two huge, swirling masses of trash and tiny particles of broken down plastics that are floating or suspended slightly underwater.

coating destroys their natural heat insulation and buoyancy, causing many of them to drown or die from loss of body heat. Certain components of oil sink to the ocean floor and smother bottom-dwellers such as crabs, oysters, mussels, and clams. Some oil spills have killed coral reefs.

In the open ocean away from coastal areas, pollution is a growing problem. One source of pollutants is ships of all kinds, including cruise ships, some of which carry more than 6,000 passengers and 2,000 crewmembers. These ships generate large amounts of waste that can include toxic substances such as paints and used oil. Several nations, including the United States, have enacted laws that prohibit the dumping of such waste, but this is not the case worldwide, and it is difficult to enforce such laws.

A growing pollution problem in the oceans is plastic items dumped from ships and left as litter on beaches. Scientists estimate that plastics kill up to 1 million seabirds and 100,000 marine mammals and sea turtles every year. Plastic garbage also threatens the lives of countless fish that ingest bits of plastic or become entangled in larger pieces of plastic such as plastic bags and discarded or lost fishing nets (Figure 11.24a).

In the northern Pacific Ocean near the Hawaiian Islands, researchers have been studying two huge swirling masses of solid wastes that have accumulated in those waters, carried by intersecting ocean currents. They are known collectively as the Great Pacific Garbage Patch (Figure 11.24b). The patch is primarily made up of plastic items, most of them washed or blown off beaches or dumped into coastal cities' storm drains that empty into the Pacific or into rivers that flow to the ocean.

a

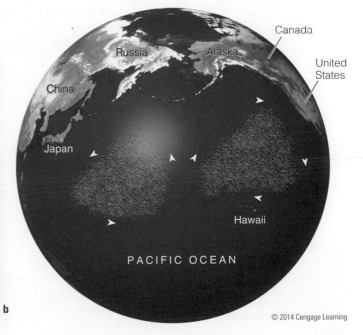

b

Canada
Russia
Alaska
United States
China
Japan
Hawaii
PACIFIC OCEAN

Much of the plastic in this garbage patch, as well as in others that have been discovered in each of the world's major oceans, has been partially decomposed into particles about the size of rice grains. Research shows that these particles can clog the digestive systems of some marine mammals, seabirds, and fishes that swallow them. These marine species can then no longer get the nutrients they need, and as a result, they essentially starve to death. Researchers found one dead sea turtle in Hawaii that had more than a thousand pieces of plastic in its stomach and intestines.

The particles in garbage patches can also contain PCBs, DDT, and other long-lasting toxic chemicals. As larger aquatic animals eat small fish that have ingested these chemicals, the toxins can move up the food chain, building to high concentrations in the larger predators. These chemicals can then end up in our own bodies if we eat marine species that have ingested them.

FOR INSTANCE...

Pollution in the Chesapeake Bay

The Chesapeake Bay (Figure 11.25) is the largest estuary in the United States. It is bordered by six states and fed by 9 large rivers and 141 smaller streams and creeks.

As the U.S. population has grown, so has the number of people living on the Chesapeake's shores. Between 1940 and 2010, that number grew from 3.7 million to almost 17 million.

With this population growth, the amount of wastes entering the estuary from human sources grew dramatically. Phosphate and nitrate levels rose sharply in many parts of the bay, causing algal blooms. Most of the phosphates come from sewage treatment plants and industrial plants in the hundreds of towns and cities scattered throughout the Chesapeake's huge watershed. Most of the nitrates come from nonpoint sources—primarily suburban and agricultural lands from which fertilizers and animal wastes run off into streams that feed the estuary. Some air pollutants also fall into the bay.

Since 2009, some parts of the Chesapeake have shown lower levels of phosphates, nitrates, and sediment. However the overall concentrations of these pollutants are still too high to be judged safe and healthy for the bay, according to a 2011 Chesapeake Bay Foundation statement.

The results of this pollution include oxygen-depleted waters that threaten aquatic life. Since 1980, a once-thriving fishing industry that harvested tons of oysters, blue crabs (Figure 11.26), and fishes every year has declined sharply, largely because of pollution and diseases, but partly because of overfishing. Birds and mammals that

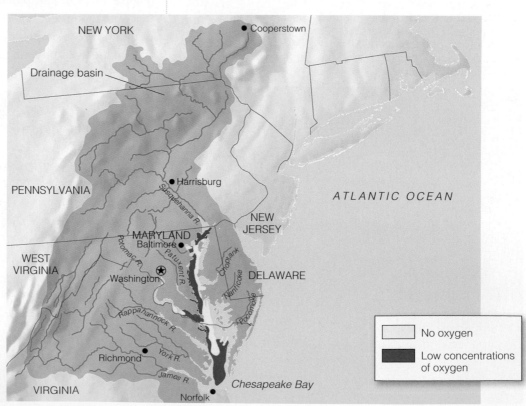

FIGURE 11.25 The Chesapeake Bay, the largest estuary in the United States, is severely degraded as a result of water pollution from point and nonpoint sources in the six states that surround it.

© 2014 Cengage Learning

FIGURE 11.26 The harvests of Chesapeake Bay's famous blue crabs have fallen sharply since 1980, largely because of water pollution in the bay.

feed on fish and shellfish are also in decline in the Chesapeake. Beach closings are common every summer.

Chesapeake Bay is so shallow that people can wade through much of it. Because of the extremely low rate of flow from the bay to the Atlantic Ocean, only about 1% of the wastes entering the bay are flushed out into the ocean. According to the Chesapeake Bay Foundation, despite some progress in cleaning up this estuary, the bay in 2011 was "still dangerously polluted."

Let's **REVIEW**

- What proportion of the world's people live in coastal areas?
- List three sources of coastal water pollution.
- What are harmful algal blooms? Summarize their causes and effects.
- What are oxygen-depleted zones? How do they form and what are their effects?
- Why are coral reefs perhaps the most threatened of all coastal aquatic systems?
- Summarize the problems of oil pollution in oceans. What is the largest source of oil pollution?
- Why are plastic wastes a growing pollution problem? What is the Great Pacific Garbage Patch?
- Summarize the story of pollution in the Chesapeake Bay region.

Groundwater Pollution

Groundwater can also become polluted and this has become a serious problem for the billions of people in the world who depend on groundwater supplies for their drinking and irrigation water. It can result in serious health problems.

Pollutants can seep into groundwater from many different sources (Figure 11.27, p. 310). Common groundwater pollutants are fertilizers, pesticides, gasoline, motor oil, paint thinners, and other potentially toxic chemicals. Individuals can add to this problem without realizing it. For example, someone living in a rural or suburban home that is served by a septic system might dump used paint thinner down a drain. Septic systems have been known to leak into groundwater that is used as a source of drinking water in these rural and suburban areas. Thus, people who dump harmful chemicals into their septic systems may unknowingly be polluting their own drinking water supplies.

Most groundwater flows very slowly and it can eventually cleanse itself of pollutants similar to what a stream does. However, because it flows so slowly, groundwater can take decades to thousands of years to cleanse itself of biodegradable pollutants. When the pollutants are nondegradable, groundwater pollution is a permanent and dangerous problem (see Table 11.1, p. 302).

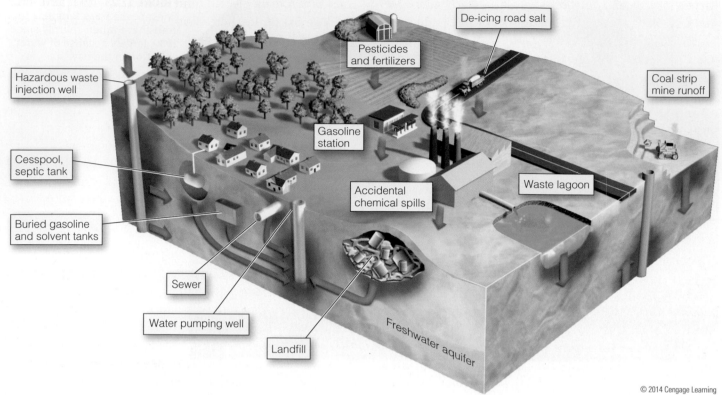

© 2014 Cengage Learning

FIGURE 11.27 This is a simplified illustration of the major sources of groundwater pollution.

Labels in figure:
- Hazardous waste injection well
- Cesspool, septic tank
- Buried gasoline and solvent tanks
- Sewer
- Water pumping well
- Landfill
- Pesticides and fertilizers
- Gasoline station
- Accidental chemical spills
- De-icing road salt
- Coal strip mine runoff
- Waste lagoon
- Freshwater aquifer

It is difficult and expensive to detect and monitor groundwater pollution, and many countries cannot afford to do it. However, scientists who study the problem have found it to be a serious and growing global concern. For example, in China, where 7 of every 10 people depend on groundwater for their drinking water, the Chinese government reported in 2006 that aquifers in 9 of every 10 Chinese cities were polluted or overpumped and could take hundreds of years to recover.

In the United States, the EPA has studied several sources of groundwater pollution, including 26,000 industrial waste ponds, one-third of which have no liners to prevent toxic liquids from seeping into aquifers. Also, there are more than 200 liquid hazardous waste disposal sites where these wastes are injected into deep wells, some of which leak toxic substances into aquifers used for drinking water.

Another problem is the more than 500,000 underground tanks that have leaked gasoline, diesel fuel, home heating oil, or toxic solvents into groundwater. By 2011, the EPA had overseen the costly cleanup of about 413,000 of these tanks. However, every year, the number of leaking underground tanks increases, and this concerns some scientists.

Let's **REVIEW**

- List five common groundwater pollutants and five sources of groundwater pollution.
- Why is it more difficult for groundwater to cleanse itself than for streams to do so?

What Can Be Done?

Cleanup versus Prevention

There are two ways to deal with pollution of any kind—to try to clean it up and to prevent it from occurring. When an oil tanker or an oil well ruptures and spills oil into coastal waters, people spend a lot of money and thousands of hours of volunteer time cleaning up the oil so that it will not ruin beaches and kill wildlife. However, scientists estimate that current cleanup methods can recover no more than 15% of the oil from a major spill, and significant damage is usually done to ecosystems and wildlife before the oil can be cleaned up.

Many analysts argue that the money and energy used to clean up this pollution could go a lot farther if it were invested in preventing oil spills from happening. One way to do this is to have stronger safety regulations and more inspections of oil wells, drilling platforms (Figure 11.23), and oil refineries, as well as other land-based industries.

Similarly, there are ways to clean up lakes that are showing the harmful effects of cultural eutrophication. We can mechanically remove excess weeds by using machines that dredge them up. We can try to control the growth of algae and weeds by spraying herbicides into ponds and lakes. We can pump air through lakes to prevent their waters from becoming oxygen-depleted. But all of these methods are expensive, energy-intensive, and potentially threatening to some aquatic organisms, compared to preventing the flow of plant nutrients into the lakes in the first place, a measure we will explore further in the next sections.

Another area where prevention is the best policy is that of drinking water supplies. We can purify polluted water in water purification plants, but this is a cleanup approach. In some cities where drinking water is taken from reservoirs or lakes, officials have found that the prevention approach works better. For example, New York City pays upstate landowners to protect the watersheds around reservoirs and lakes used to supply the city's water. Land in these watersheds is left undisturbed and this helps to prevent the release of pollutants into the reservoirs. The city found this approach to be far less expensive than building water purification plants.

Similarly, there are ways to clean up groundwater pollution and ways to prevent it. However, cleaning up polluted groundwater is very difficult and expensive. As with other forms of water pollution, it is far less expensive to take measures to prevent this pollution. These strategies fall under the general categories of *legal measures*, *technical solutions*, and *political and economic tools*. These categories often overlap and they are interrelated in almost every application, but they give us a good framework for examining methods that can be used to help prevent water pollution.

> ## KEY idea
> It is far more effective, less expensive, and often much easier to deal with pollution by preventing it rather than by cleaning it up.

Legal Measures

In the 1960s, water pollution had become so extreme in some of the more-developed countries like the United States that it became a major political problem. Voters put pressure on elected officials to deal with it and, since 1970, most of the more-developed countries have passed laws and regulations that have sharply reduced water pollution.

For example, the U.S. Clean Water Act spells out permissible levels of major water pollutants and requires polluters to get permits limiting the amounts of pollutants that they can discharge into waterways. According to the EPA, this law led to numerous improvements in water quality. Between 1992 and 2009, the proportion of the U.S. population served by sewage treatment plants increased from 33% to 75%. The percentage of streams found to be safe for swimming and fishing rose from 36% to 60% of those tested.

Global water pollution problems can also be reduced by both national and international laws and agreements. For example, oil pollution in the oceans has been reduced by measures that require or encourage the use of double hulls (basically, a hull within a hull) on oil tankers. There are still a large number of single-hulled tankers in use, but since 1990, the number of tanker oil spills has dropped in almost every year, even though oil tanker traffic has increased, overall.

Pollution laws and regulations can prevent pollution directly by making it illegal, and they can also have the effect of encouraging researchers and engineers to come up with long-term solutions to pollution's underlying problems. For example, pollution control regulations on U.S. livestock operations have motivated scientists to find ways to keep animal wastes out of nearby waterways. Consequently, some farmers are now converting these wastes into natural gas in devices called *digesters*. Undigested wastes that are a by-product of these devices can also be recycled into the soil, or they can be converted to biodiesel fuel for cars.

Let's REVIEW

- What are two major ways to deal with water pollution?
- Give two examples of how it can be less expensive to prevent water pollution than to clean it up.
- What are two effects that laws and regulations can have in the area of water pollution problems?

Technological Solutions

One of the biggest and most difficult pollution problems is nonpoint-source pollution. It is relatively easy to pass and enforce legal measures against point-source pollution. We can usually identify where the pollution is coming from—the end of a pipe (Figure 11.15a), for example—and usually the owner of the pipe can be found and required to stop the pollution. However, it is often harder to find and deal with the party or parties responsible for pollutants that are flowing from a broad area of land into a river or lake.

Legal issues aside, there are a number of technological solutions to the problems of nonpoint-source pollution. The single largest water pollution problem is sediment from soil erosion (Figure 11.15b), and farmers have found several ways to reduce it. One of the oldest methods is *terracing*, in which a farmer creates a series of broad, nearly level terraces that run across the land's contours (Figure 11.28a). Each terrace retains water for its crops, slowly releasing it to the next level down, and this controls runoff and reduces soil erosion on steep slopes.

Another way to control runoff and soil erosion on sloping land is to use a combination of *contour planting* and *strip cropping* (Figure 11.28b). Farmers work across the slope of the land rather than up and down, as they plow and plant their rows of crops. Like a small terrace, each row holds some water for its plants and slowly releases the excess water downslope to other plants. Crops are planted in alternating strips, and farmers harvest one crop at a time, with the remaining alternating crop strips left to catch runoff and reduce soil erosion.

On flat land, wind erosion can be a problem that adds to water pollution, and it can be reduced through *alley cropping* (Figure 11.28c). With this method, farmers plant one or more crops together in strips, or alleys, between rows of trees or shrubs. These tree or shrub rows block some wind and also provide shade, which reduces water loss by evaporation, and this further helps to reduce erosion by keeping the soil as moist as possible. Similarly, some farmers plant rows of trees, called *windbreaks*, or *shelterbelts* (Figure 11.28d), around the edges of crop fields to block wind and reduce erosion.

For centuries, farmers have regularly plowed and *tilled*, or turned over and loosened, the topsoil in their fields before planting seeds. This opens the topsoil to wind and rain and increases the likelihood of soil erosion. Today, many farmers practice *minimum-tillage farming*, also called *conservation-tillage farming*, by using special tillers and planting machines that drill seeds directly into unplowed soil. This method helps greatly to reduce topsoil erosion and the resulting water pollution from sediment and fertilizer runoff.

Still another way to reduce water pollution near farm fields is to plant strips of vegetation between cultivated

(a) Terracing

(b) Contour planting and strip cropping

(c) Alley cropping

(d) Windbreaks

© 2014 Cengage Learning

FIGURE 11.28 These are four planting methods that farmers have used for centuries to help control soil erosion.

fields and streams or lakes (Figure 11.29) to block the flow of sediments and nutrients. Farmers can also locate their feedlots and animal waste sites away from sloping land and floodplains.

In addition, farmers can reduce the flow of plant nutrients into streams and ponds near farmland by fertilizing the fields with organic manure instead of manufactured

Lynn Betts/USDA, Natural Resources Conservation Service

FIGURE 11.29 This natural vegetation was planted between farm fields and a creek in Story County, Iowa.

commercial fertilizers. Manure tends to cling more closely to soil, while some commercial fertilizers are applied as granules, which can more easily wash into streams. Farmers can also reduce the amount of fertilizer that runs off into surface waters by using slow-release fertilizer and by using no fertilizer on steeply sloping land.

In urban areas, some of the same techniques can be used to prevent or reduce the runoff of chemicals and other pollutants from lawns, golf courses, streets, and parking lots into streams and lakes. The problem is that such pollutants usually flow from land owned by several different parties. If all but one of them agrees to work on controlling runoff, the pollution and cultural eutrophication will still occur. However, once the flow of plant nutrients into a lake or stream is stopped, these waterways can usually recover from cultural eutrophication.

Dealing with pollution in bigger bodies of water, such as the Gulf of Mexico (Figures 11.22 and 11.23), Chesapeake Bay (Figure 11.25), and other heavily populated coastal zones around the world, is vastly more difficult, but not impossible. Farmers working upstream from these bodies of water can use the preventive measures described here. Another helpful measure advocated by many environmental scientists is to restore and reestablish inland and coastal wetlands that have been filled or degraded in most of these populated areas. These wetlands act as natural pollution filters and they provide the ecosystem service of sharply reducing the flow of plant nutrients into inland and coastal waters.

One of the most important ways to prevent water pollution is through sewage treatment. In urban areas of more-developed countries, most wastewater from homes and businesses, and sometimes the runoff from city streets and parking lots, flows through sewer pipes to sewage treatment plants. There, it typically passes through two levels of treatment. The first, called *primary sewage treatment*, is a physical process that removes most solids and sediments (Figure 11.30, left, p. 314). A second level, called *secondary sewage treatment*, is a *biological* process in which bacteria remove as much as 90% of degradable, oxygen-demanding organic wastes (Figure 11.30, right).

A sewage treatment system that includes primary and secondary treatment methods typically removes most but not all solids and oxygen-demanding organic wastes, toxic metal compounds, and synthetic organic chemicals. The system does remove about 70% of the phosphorus and 50% of the nitrogen in waters flowing through it.

Most disease-causing organisms survive the primary and secondary treatment processes, so many treatment systems include a final disinfection process, which uses chlorine to kill bacteria. This process does not, however, kill viruses. Chlorine has been known to react with organic materials in water to form small amounts of compounds that might increase the risks of cancer, miscarriages, and damage to human nervous, immune, and endocrine systems. Through research and testing, public health scientists have established levels of chlorine in drinking water that they consider to be safe.

In most rural and suburban areas of more-developed countries, homeowners have on-site septic systems. In these systems, household sewage and wastewater is pumped into a settling tank where bacteria decompose the solid wastes. The resulting partially treated wastewater seeps from the tank into a broad drainage field made of gravel or crushed stone lying just below the topsoil. From there, the wastewater percolates down through the soil beneath the gravel, a process that filters out most potential groundwater pollutants, while soil bacteria decompose other degradable materials. About one-fourth of all U.S. homes make use of these on-site systems.

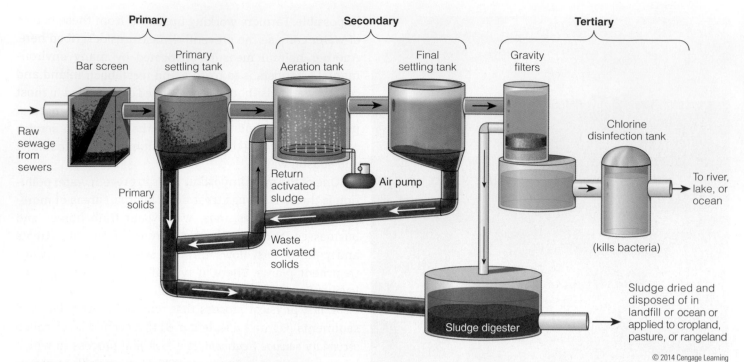

Primary · Secondary · Tertiary

Bar screen · Primary settling tank · Aeration tank · Final settling tank · Gravity filters

Raw sewage from sewers

Primary solids

Return activated sludge · Air pump

Waste activated solids

Chlorine disinfection tank

To river, lake, or ocean

(kills bacteria)

Sludge digester

Sludge dried and disposed of in landfill or ocean or applied to cropland, pasture, or rangeland

© 2014 Cengage Learning

FIGURE 11.30 This is a simplified model of a sewage treatment system employing primary and secondary sewage treatment processes.

Some scientists argue for a greater focus on preventing toxic and hazardous chemicals from reaching sewage treatment plants where they can be discharged as effluents into waterways. One way to do this would be to require industries and businesses to remove toxic and hazardous wastes from the water they send to municipal sewage treatment plants. Another way would be to encourage industries to reduce or eliminate their use and waste of toxic chemicals, and this goal could be met with proven technological solutions.

For example, industries can find nontoxic substitutes for some toxic chemicals, and they can find ways to recycle some toxic chemicals. Both approaches have been used by several companies, including 3M, a manufacturer of thousands of consumer products. Since it began focusing on pollution prevention in 1975, 3M has prevented more than 1.5 million tons of pollutants from reaching the environment.

Many scientists have argued that sewage treatment systems can be improved with a greater focus on the prevention of water pollution. For example, several communities have built unconventional, but highly effective, *wetland-based sewage treatment systems*. Another approach is to pass sewage through a series of tanks that use living organisms to treat the sewage (see *Making a Difference*, at right). Such systems use a more ecological approach to treating sewage and industrial wastewater.

Let's REVIEW

- What are six methods that farmers use to reduce sediment pollution from soil erosion?

- What are two ways in which we can prevent or reduce runoff from lawns, golf courses, streets, and parking lots lying near bodies of water?

- Why is it a good idea to restore wetlands near bodies of surface water?

- Distinguish between primary sewage treatment and secondary sewage treatment, and summarize how a conventional sewage treatment system works. What is a septic tank system? Describe how it works.

Economic and Political Solutions

Technological solutions and legal measures often go hand in hand, as legal requirements frequently spur technological inventions or refinements to existing pollution control technology. Economic tools and political processes can enhance both legal and technological approaches.

For example, the EPA is experimenting with one economic approach called a *discharge trading policy*. Under this program, factories and other sources of pollution each receive permits to release a certain amount of pollutants into waterways. Permit holders who pollute below their allowed levels can sell their unused permits to those who want to pollute at levels above what their issued permits allow.

MAKING A difference

John Todd: Working with Nature to Treat Wastewater

Biologist John Todd has developed an ecological approach to treating sewage. Todd calls his system a *living machine* (Figure 11.A). More than 800 cities and towns around the world and 150 in the United States are using this type of system to treat their sewage.

The first step in the living machine's purification process is to pass sewage through rows of large open tanks exposed to sunlight, each row containing a more complex set of plants and other organisms than the one before it. The first set of tanks contains algae and a variety of microorganisms that decompose organic wastes. When the sun shines, it speeds up this process by providing more energy for it. The wastewater then flows into tanks containing aquatic plants such as water hyacinths, cattails, and bulrushes that use the wastewater's nutrients for their own growth.

After flowing though several of these tanks, the water passes through a human-made wetland containing sand, gravel, and bulrushes and other plant species that filter out the remaining organic wastes. Some of the plants also absorb toxic metals such as lead and mercury, and some contain natural antibiotic compounds that kill disease-causing bacteria. Next, the wastewater flows into aquarium tanks, where snails and zooplankton consume most of the remaining harmful bacteria.

After 10 days in this system, wastewater flows into a second artificial wetland for final filtering and cleansing. The water can be purified by using ultraviolet light or a device called an ozone generator to kill any remaining harmful bacteria. (This system, like conventional treatment systems, does not eliminate or kill viruses.) The living machine produces clean water that can be released to the environment or used for irrigation, toilet flushing, industrial processes, cleaning equipment and animal areas, and filling fish ponds. It also yields a nutrient-rich sludge that can be sold as compost. Operating costs for these systems are about the same as those for a conventional sewage treatment plant.

Ocean Arks International

FIGURE 11.A Biologist and inventor John Todd demonstrates a *living machine* at a sewage treatment plant in Providence, Rhode Island.

John Todd argues that conventional sewage treatment is a process of transferring potential plant nutrients from homes and industries to rivers, lakes, and oceans where they overload these aquatic systems. By contrast, Todd's living machine uses sunlight and a variety of living organisms to remove these nutrients from wastewater. They are then returned to the land to support plant growth instead of being dumped into waterways to disrupt aquatic systems.

This economic approach essentially creates a market in which people buy and sell rights to pollute waters. The total amount of pollution allowed under all permits is supposed to be low enough to avoid overloading natural aquatic systems in a given area with pollutants. It can work if the cap on total pollution levels in any given area is set low enough and if the caps are lowered every few years to encourage pollution prevention and improvements in pollution control technologies.

Critics of this system argue that discharge trading could allow pollutants to build up to dangerous levels in places where a large number of credits are bought. They also argue that political processes could result in caps being set too high, which would allow for too much pollution overall. In addition, the caps are often not lowered.

The pros and cons of a discharge trading policy are summarized in Figure 11.31 (p. 316).

Political processes can be used to weaken pollution controls, but they can also be used to strengthen efforts to prevent and control pollution. One example is the emerging use of a strategy called **integrated coastal management**. This is a community-based effort to develop and use coastal resources more sustainably, and to prevent pollutants from entering coastal aquatic systems. More than 100 coastal communities around the world are using this strategy.

In this grassroots political process, citizens, business owners, commercial fishermen, developers, scientists, and lawmakers meet to identify problems and to set goals for solving pollution problems and for using their shared

- Puts an upper limit on water pollution discharges in a given area

- Encourages polluters to cut pollution and then make money by selling unused permits

- Makes polluters aware of their discharges by putting a price on them

Pros

- Allows wealthy polluters to continue polluting at a cost that is affordable to them

- Allows for pollution hotspots in aquatic systems where trading results in a high density of permit holders

- Relies largely on self-reporting by polluters

Cons

Discharge trading policy

© 2014 Cengage Learning

FIGURE 11.31 *Weighing the pros and cons of a discharge trading policy.*

marine resources in the most sustainable ways possible. In order for the process to work, all participants must be willing to make trade-offs that can lead to the greatest number of ecological and economic benefits for all. For example, a shrimp farmer might have to install equipment that will prevent the shrimp farm wastes from entering coastal waters. This would improve the health of the coastal ecosystem and have the effect of making the coastal area more attractive to tourists, which might improve the shrimp farmer's business in the long run.

In 1983, the Chesapeake Bay Program was established to make use of integrated coastal management techniques (see *For Instance*, p. 308). Since then, citizens' groups, communities, state legislatures, and the federal government have worked together to reduce the flow of pollutants into the bay and to restore wetlands around its shores and sea grasses on its floor.

As a result, the Chesapeake's phosphorus and nitrogen levels have dropped considerably, in spite of the fact that the population in the bay's watershed has continued to grow. While some environmental scientists note that there is still a lot of room for improvement in the bay, many analysts see the Chesapeake Bay Program as a good example of a grassroots political process that has worked for the benefit of both people and the environment.

Many environmental scientists and analysts argue that the integrated coastal management model can be applied to other water pollution problems. Cultural eutrophication, stream pollution, and groundwater pollution are all problems that typically involve a large and diverse number of people, organizations, and interests. The energy and resources of these various and often competing parties can be harnessed, as it was in the Chesapeake Bay Program, to tackle these difficult environmental problems.

Let's **REVIEW**

- Describe how a discharge trading policy works. Summarize its pros and cons.
- Define *integrated coastal management* and give an example of how it works. What are three other pollution problems that might be served by applying the integrated coastal management model?

A Look to the Future

Ocean Acidification

Because of our large-scale burning of carbon-containing fossil fuels—coal, natural gas, oil, and gasoline and diesel fuel made from oil—we have been dumping huge amounts of carbon dioxide (CO_2) into the atmosphere. There is a great deal of evidence that this increase in atmospheric CO_2 is warming the atmosphere and that this warming will contribute to rapid climate change during this century.

If climate change takes place as projected, it will also affect the world's oceans. Sea levels are projected to rise, and hurricanes, which form over the oceans, are expected to become more intense. However, we hear little about what could prove to be one of the biggest and longest-lasting environmental threats of all: **ocean acidification**— the process in which ocean waters absorb CO_2, which reacts with the water to form a weak acid called carbonic acid, which increases the acidity of ocean water.

This absorption of CO_2 by ocean water is part of the natural carbon cycle. However, when the atmosphere is overloaded with CO_2, the rate of absorption increases to the point where ocean acidity begins to have harmful effects. For example, a high level of acidity slows the rate at which certain marine creatures—notably corals and shellfish such as clams, oysters, and lobsters—can build their calcium carbonate shells and exoskeletons. If ocean acidity increases as projected during this century, the formation of corals could be reduced by well over half, causing already-threatened reef structures to age without being renewed and, thus, to weaken. In turn, they will likely suffer more damage and erosion from storms and large waves.

Scientific studies indicate that increasing acidity is already causing problems. Pacific oysters, for example, have not successfully reproduced in the wild since 2004. In addition, since 1990, there has been a 14% decline in the formation of corals on Australia's Great Barrier Reef. Computer models project that if we do not significantly slow the rate at which we are burning fossil fuels, average ocean acidity will double by the end of this century. In short, projected sharp increases in acidity throughout the oceans will likely have serious consequences for many marine ecosystems and for the organisms they support, including many commercial fish species.

Assuming the problem is as severe as many scientists think, what can we do about it? First, we could quickly step up research on the still poorly understood biological and ecological effects of rising ocean acidity. Second, we could try to persuade politicians, educators, and business leaders to take the issue far more seriously. And third, we can all take steps as quickly as possible to sharply reduce our emissions of CO_2—the root of this problem.

According to scientists on the InterAcademy Panel on International Issues, the projected increase in ocean acidity potentially ranks as one of the largest and most critical of all the known changes to the earth's chemistry, and it could take many centuries to reverse such a change.

Let's REVIEW

- Define *ocean acidification* and describe the effects it is having.
- What are three ways to deal with this problem?

What Would You Do?

The ultimate source of most water pollution can be traced to the individual level and to the choices we all make. This does not make the problem any easier to solve. None of us can control what a factory in another country does to pollute the waters near it. But each of us can make personal choices that affect the surface waters and groundwater where we live. Also, by making careful purchasing decisions, we can at least try to affect the actions of factory owners and other businesses with which we interact, either by supporting them or by denying them our business. We can also try to influence decision makers at all levels of governments. Here are some ways in which people are working to reduce or prevent water pollution, both locally and globally:

Being mindful of how their daily activities affect water supplies and aquatic systems.
- Using only as much water as they need.
- Not dumping chemicals or unused medicines down home or municipal drains or into aquatic systems.
- Preventing their yard wastes from entering the storm water systems in their communities.

Making personal choices carefully.
- Trying to buy only products that were made by companies that do their best to prevent or reduce their own water pollution.
- Choosing recreational activities that minimize pollution of aquatic systems, such as canoeing and swimming, and avoiding motorized water activities.
- Fertilizing their yards or gardens with manure or compost instead of with commercial fertilizers.

Being socially and politically active in helping to prevent pollution.
- Talking about water pollution problems and solutions with their friends and families.
- Supporting politicians who give pollution control a high priority and calling or e-mailing them often to express concerns about water pollution.
- In talking with other people about water pollution, emphasizing the importance of preventing pollution so that it does not have to be cleaned up.

KEYterms

aquatic life zones, p. 294
aquifer, p. 291
biodegradable pollutants, p. 303
coastal zone, p. 295
cultural eutrophication, p. 298
eutrophication, p. 298
evaporation, p. 291

freshwater aquatic systems, p. 294
groundwater, p. 291
hydrologic cycle, p. 291
integrated coastal management, p. 315
marine aquatic systems, p. 294
nitrogen cycle, p. 292

nondegradable pollutants, p. 303
nonpoint sources, p. 301
ocean acidification, p. 316
ocean bottom, p. 298
open sea, p. 298
phosphorus cycle, p. 292
point sources, p. 301

pollutant, p. 301
precipitation, p. 291
surface runoff, p. 291
surface water, p. 291
transpiration, p. 291
water cycle, p. 291
water pollution, p. 290

THINKINGcritically

1. How are the carbon and nitrogen cycles connected to the water cycle? What are two examples of human activities that affect the carbon cycle and two examples of human activities that affect the nitrogen cycle? How does each of these activities affect the water cycle?

2. If you were in charge of controlling pollution in a certain lake or stream, explain the strategy you would use for dealing with pollution from (a) a pipe discharging chemicals from a factory; (b) a parking lot next to a shopping center; (c) a farm field; and (d) the lawns around the homes located next to the stream or lake.

3. What difference do you think it makes if a lake goes from being oligotrophic to being eutrophic, primarily because of the nutrients added to the lake by human activities? Explain the affects that this might have on aquatic life in and around the lake. Explain how this might affect people who use the lake for recreation. List three ways to reduce the nutrient inputs from human activities that create a eutrophic lake.

4. Explain the connection between farms located upstream from an oxygen-depleted zone of the ocean, such as that in the Gulf of Mexico, and the process that causes the oxygen depletion. What are three specific steps that could be taken by the farmers upstream to help break this connection?

5. Would you be willing to pay higher taxes in order to fund measures that would prevent the pollution of the groundwater that serves as your drinking water supply? If so, about how much of a hike in your taxes would you be willing to pay, stated in terms of a percentage increase?

6. When you flush your toilet, where does the wastewater go? Find out where your community's sewage treatment plant is (if there is one) and learn where the wastewater goes when it leaves the treatment plant. Try to visit a local sewage treatment plant to see what it does with wastewater. Compare the processes it uses with those shown in Figure 11.30. What happens to the sludge produced by this plant? What improvements, if any, would you suggest for this plant?

LEARNINGonline

Access an interactive eBook and module-specific interactive learning tools, including flashcards, quizzes, videos and more in your Environmental Science CourseMate, accessed through **CengageBrain.com**.

12

Air Pollution

Why Should You Care about Air Pollution?

Air pollution is the presence of chemicals in the air at levels high enough to harm humans, other organisms, ecosystems, or human-made materials or objects. With every breath you take, you inhale not only life-giving oxygen, but also small amounts of other gases along with tiny droplets and small solid particles of various chemicals, some of which can threaten your health.

Air pollution, at one time or another, becomes a threat to just about everyone on the planet, as well as to other forms of life, and that is reason enough to care about it. We all face possible harm from **air pollutants**—agents that cause air pollution (Figure 12.1a). Pollutants can also harm trees (Figure 12.1b), damage metals and stone statues, cause large economic losses, and contribute to long-term, highly disruptive climate change.

Some air pollutants come from natural sources such as volcanic eruptions and dust storms. Others come from human activities, including the burning of gasoline and diesel fuel in cars, trucks, and ships; the use of wood and charcoal for cooking and heating; and the use of coal, oil, and natural gas for fueling furnaces, factories, and electric power plants.

Outdoor air pollution is a serious problem, but even worse in some areas is the problem of *indoor air pollution*. The World Health Organization (WHO) estimates that, every year, about 2.4 million people die from illnesses related to exposure to air pollutants. That is an average of nearly 274 people every hour. About 1.5 million of these deaths (171 per hour) are caused by indoor air pollution. The Environmental Protection Agency (EPA) estimates that in the United States, 150,000 to 350,000 people die from such illnesses every year. These estimates do not include the many millions of people who suffer from asthma attacks and other respiratory disorders as a result of exposure to outdoor and indoor air pollutants.

Because we have to breathe, we can hardly escape from air pollutants, no matter where we live. But we can find ways to reduce air pollution levels and to prevent certain chemicals from building up to harmful levels in the air that we must inhale. In fact, our health and well-being depend on our finding ways to do so. This module examines the problems of outdoor and indoor air pollution and possible solutions to these problems.

What Do You Need to Know?

The Nature of the Atmosphere

The **atmosphere** is the mixture of gases that surround the earth. It is made up of layers, the innermost two of which contain most of these gases (Figure 12.2).

The **troposphere**, the atmospheric layer nearest the earth's surface, contains most of the planet's air—the mixture of gases that we depend on for staying alive. About 78% of this mixture is nitrogen gas (N_2), and oxygen gas (O_2) makes up another 21%. The remaining 1% of the air we breathe is composed of water vapor (H_2O), argon (Ar) gas,

FIGURE 12.1 This policeman in Bangkok, Thailand (a), wears a smog mask every day to try to protect himself from the harmful health effects of inhaling air pollutants, mostly from vehicle exhausts. Trees such as these in Poland (b) have been killed by prolonged exposure to rain and fog that contains acidic air pollutants.

Mark Edwards/Peter Arnold, Inc.

Bildermann/Dreamstime.com

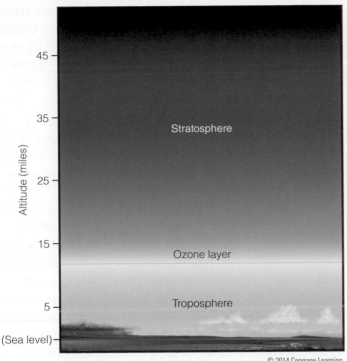

FIGURE 12.2 Most of the air we breathe is found in the atmosphere's innermost layer called the *troposphere*. The next layer up is called the *stratosphere*, which includes the *ozone layer*. We focus on these layers in this module. *See an animation based on this figure at* www.cengagebrain.com.

carbon dioxide (CO_2), and other gases, plus very small, or *trace*, amounts of dust, droplets of various chemicals, and particles of *soot*, or unburned carbon, sometimes called *black carbon particles*.

CONSIDER this If the earth were the size of a large tomato, the troposphere would be no thicker than the tomato's skin.

The **stratosphere** is the earth's second layer of air. It contains much of the atmosphere's small amount of ozone gas (O_3), which lies in a sub-layer of the lower stratosphere known as the **ozone layer**. This stratospheric ozone gas serves as a sunscreen for the planet by filtering out about 95% of the sun's harmful ultraviolet (UV) radiation before it can reach the earth's surface.

KEY idea The troposphere provides us with life-sustaining oxygen, and the stratosphere contains ozone that protects us from most of the sun's deadly ultraviolet radiation.

Natural Atmospheric Warming

The earth would be too cold for most life as we know it, were it not for the **greenhouse effect**—the ability of the atmosphere to temporarily store some of the sun's energy received by the earth as heat. This effect is somewhat like that of a glass greenhouse that collects solar energy and temporarily holds it as heat.

This natural, life-sustaining atmospheric warming occurs primarily because of the presence of small concentrations of four natural gases in the atmosphere: water vapor (H_2O), carbon dioxide (CO_2), methane (CH_4), and nitrous oxide (N_2O). They are the most important examples of **greenhouse gases**, or gases that interact with solar energy to have a warming effect on the earth's atmosphere.

Here is how the greenhouse effect works: A tiny fraction of the sun's total energy is intercepted by the earth, and an even smaller amount reaches the planet's surface (see Module 1, Figure 1.15, p. 14). As this high-quality solar energy interacts with the earth's atmosphere, land forms, oceans, and life forms, it is degraded into lower-quality energy, as required by the second law of thermodynamics (see Module 1, p. 12). When some of this lower-quality energy, or heat, reflects off the planet's surface back through the troposphere toward space, it interacts with molecules of the greenhouse gases and is degraded to an even lower-quality energy that we experience as heat. Some of this heat flows back into space, while some temporarily remains in the lower troposphere, warming the air and the earth's surface. Eventually, this heat, too, flows back into space as more high-quality solar energy arrives to undergo the same process.

Let's REVIEW

- What is *air pollution* and why should you care about it? Define *air pollutant*.
- What is the *atmosphere*? Define and distinguish among the *troposphere*, the *stratosphere*, and the *ozone layer*, and explain the importance of each to life on the earth.
- What is the *greenhouse effect* and why is it important to life on the earth? Define *greenhouse gases* and name the four most important ones in the atmosphere.

The Carbon Cycle

The **carbon cycle** is the continual circulation of various compounds of life-sustaining carbon (C) through the air, water, soil, and living organisms (see Module 1, Figure 1.21, p. 20).

A key compound in the carbon cycle is carbon dioxide (CO_2) gas, which makes up about 0.039% of the volume of air in the troposphere. As an important greenhouse gas, it is a key component of the atmosphere's temperature control system. If the carbon cycle removes CO_2 from the atmosphere faster than it is replenished, the atmosphere cools, and if CO_2 is added to the atmosphere faster than the carbon cycle can remove it, the atmosphere warms up. Global and regional climates are determined largely by changes in the average atmospheric temperature over at least 30 years. Thus, any change up or down in the average amount of CO_2 in the atmosphere over three decades or longer affects the earth's climate.

Another part of the carbon cycle is the earth's life forms, all of which contain carbon. Natural processes such as erosion bury some of the remains of dead plants, animals, and microbes, and over millions of years, these materials become compressed between layers of sediment under high pressure and heat. This eventually converts them to what we call **fossil fuels**—ancient deposits of carbon-containing substances, including coal, oil, and natural gas, that can be burned. When these fuels are extracted and burned completely, the carbon they contain returns to the atmosphere as CO_2. Currently, fossil fuels supply about 85% of the world's energy and thus they are by far the world's largest human source of certain air pollutants and CO_2.

Extensive scientific evidence indicates that our addition of large amounts of CO_2 to the atmosphere has played a role in warming the atmosphere faster than it would have warmed if we had not burned huge amounts of coal, oil, and natural gas. This faster warming is partly due to the very rapid rate at which we have added CO_2 to the atmosphere over the last 200 years, and especially since 1950, and this rate is still increasing.

According to most climate scientists, this increase in the average atmospheric temperature will likely alter global and regional climates during this century. Climate models indicate that these changes will very likely lead to rising sea levels, disruption of food production and water supplies, higher precipitation and heavy flooding in some areas, extreme and prolonged drought in other areas, and more intense weather events such as heat waves and hurricanes. The rising average atmospheric temperature is also likely to speed up chemical reactions that cause air pollution.

Because many of these changes will harm humans, other organisms, and life-sustaining ecosystems, the U.S. Environmental Protection Agency (EPA) has classified CO_2 as an air pollutant. You might ask how a natural chemical such as CO_2, which we emit into the atmosphere every time we exhale, can be an air pollutant. The answer is that any chemical, if it reaches high enough levels in the air, can cause harm and thus can be classified as an air pollutant.

The Nitrogen and Sulfur Cycles

Another important chemical cycle is the **nitrogen cycle** in which different chemical forms of nitrogen repeatedly cycle through air, water, soil, and living organisms (Figure 12.3).

Two major human activities have affected the nitrogen cycle and added nitrogen-containing pollutants to the atmosphere. One is the burning of fossil fuels, which generates high temperatures that convert some of the N_2 in the air to the gases nitric oxide (NO) and nitrogen dioxide (NO_2), which can act as air pollutants. Atmospheric levels of the greenhouse gas nitrous oxide (N_2O) also go up when fossil fuels are burned. The other major activity that affects the nitrogen cycle is the agricultural use of commercial nitrate fertilizers and animal manure, both of which release N_2O into the atmosphere.

Yet another important natural cycle is the **sulfur cycle**, in which various chemical forms of sulfur circulate through the earth's air, water, soil, and living organisms (Figure 12.4, p. 324). Sulfur is a component of several proteins and vitamins that are vital to life processes. One sulfur compound also plays a role in the formation of clouds which in turn affects climates.

Sulfur is commonly stored as ions, called sulfate (SO_4^{2-}) ions, within salts found in underground rocks and minerals and in sediments that build up on ocean bottoms. Particles of these salts enter the atmosphere through dust storms, forest fires, and sea sprays as part of the natural sulfur cycle. Sulfur also enters the atmosphere from volcanoes as hydrogen sulfide (H_2S)—a poisonous gas that is released from rotting vegetation in swamps, tidal flats, and bogs as well. Volcanoes also release sulfur dioxide (SO_2), another harmful air pollutant.

The natural sulfur cycle is affected by human activities that release large amounts of sulfur dioxide (SO_2) into the atmosphere (as shown by the red arrows in Figure 12.4). These activities include the burning of sulfur-containing coal and oil to produce electricity and to make steel, cement, and other materials. Another such activity is the conversion of sulfur-containing oil into gasoline, heating oil, diesel fuel and other useful products. Still another is smelting—the processing of sulfur-containing metallic minerals to obtain important metals such as copper and lead.

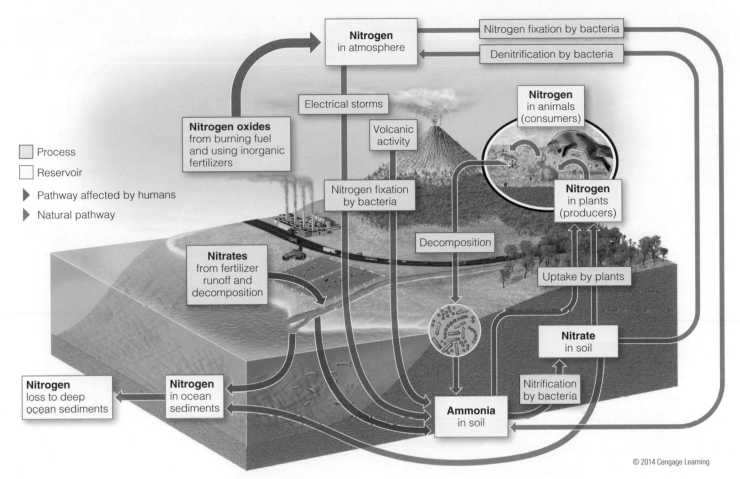

Process
Reservoir
▶ Pathway affected by humans
▶ Natural pathway

Nitrogen in atmosphere

Nitrogen fixation by bacteria

Denitrification by bacteria

Electrical storms

Nitrogen oxides from burning fuel and using inorganic fertilizers

Volcanic activity

Nitrogen in animals (consumers)

Nitrogen fixation by bacteria

Nitrogen in plants (producers)

Nitrates from fertilizer runoff and decomposition

Decomposition

Uptake by plants

Nitrate in soil

Nitrogen loss to deep ocean sediments

Nitrogen in ocean sediments

Nitrification by bacteria

Ammonia in soil

© 2014 Cengage Learning

FIGURE 12.3 The nitrogen cycle is shown here by the blue arrows, and human impacts on this cycle are shown by the red arrows. *See an animation based on this figure at* www.cengagebrain.com.

Let's REVIEW

- Define *carbon cycle.* What is the role of carbon dioxide (CO_2) in the carbon cycle? What are *fossil fuels* and how does the use of these fuels affect the carbon cycle?

- Explain how CO_2 can affect the earth's climate and why CO_2 has been classified as an air pollutant.

- What is the *nitrogen cycle* and why is it important to life on the earth? Describe two human activities that have affected the natural nitrogen cycle in ways that increase outdoor air pollution.

- What is the *sulfur cycle* and why is it important to life on the earth? List three human activities that have affected the natural sulfur cycle in ways that increase outdoor air pollution.

KEY idea Certain human activities add air pollutants to the atmosphere that alter the earth's life-sustaining carbon, nitrogen, and sulfur cycles.

Major Air Pollutants

The carbon, nitrogen, and sulfur cycles are vital to almost all life forms. However, they also produce air pollutants that can interfere with life processes, especially when these cycles are disrupted by human activities such as smelting and the burning of fossil fuels. Let's take a closer look at some of these pollutants.

Sulfur dioxide. *Sulfur dioxide* (SO_2) is a colorless gas that smells like rotten eggs. About a third of the SO_2 in the atmosphere comes from natural sources such as volcanic activity, while the rest comes mostly from the human activities listed in the previous section. Sulfur dioxide can intensify some people's breathing problems. It also contributes to a haze in the air that reduces visibility, and it can damage trees, crops, soils, and aquatic life in lakes.

Sulfuric acid and sulfate salts. Sulfur dioxide in the atmosphere is often converted into tiny suspended droplets of *sulfuric acid* (H_2SO_4) and particles of *sulfate salts* (SO_4^{2-}). These suspended acidic droplets and particles eventually fall to the earth's surface or are washed out of the air by rain or snow as *acid rain* or *acid deposition*. These sulfuric

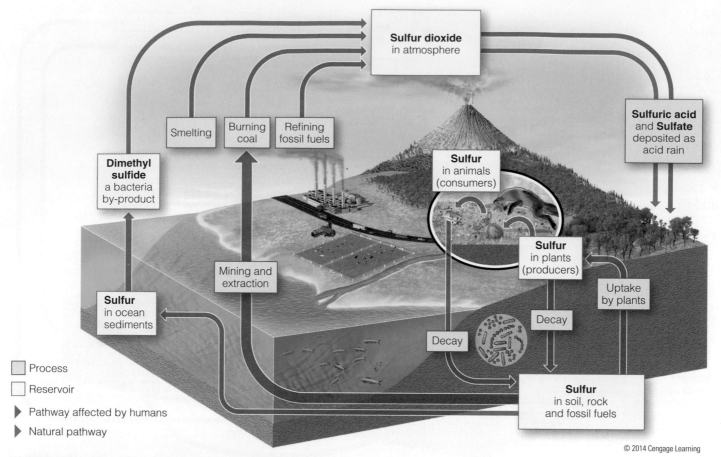

Process (box)

Reservoir (box)

▶ Pathway affected by humans

▶ Natural pathway

© 2014 Cengage Learning

FIGURE 12.4 The sulfur cycle is shown here by the blue arrows and human impacts on this cycle are shown by red arrows. *See an animation based on this figure at* www.cengagebrain.com.

pollutants have the same ill effects as sulfur dioxide, and they can also harm trees (Figure 12.1b), corrode metals, and damage materials such as paint and the stone used in statues and on buildings.

Nitrogen compounds. Two types of *nitrogen oxides* (both commonly referred to as NO_x) act as air pollutants. The first, *nitric oxide* (NO), is a colorless gas produced as part of the nitrogen cycle (Figure 12.3) by lightning and by various bacteria in soil and water. It also forms when the nitrogen (N_2) and oxygen (O_2) gases in air react with the high temperatures in automobile engines and coal-burning power and industrial plants (see module-opening photo). The second NO_x is *nitrogen dioxide* (NO_2), a reddish-brown gas. NO and NO_2 can irritate the throat, nose, and eyes, and aggravate asthma, bronchitis, and other lung ailments.

In the atmosphere, NO_2 can react with water vapor to form *nitric acid* (HNO_3) and *nitrate salts* (NO_3^-), which are components of acid deposition. A fifth air pollutant that also contains nitrogen is *nitrous oxide* (N_2O). It is produced when fossil fuels are burned in motor vehicles and it is emitted into the air from nitrate-containing fertilizers and animal wastes. Some tropical soils and the oceans are natural sources of this gas. Nitrous oxide is a greenhouse gas that can contribute to atmospheric warming. It also plays a role in depleting life-sustaining ozone in the stratosphere (Figure 12.2), a topic that we cover later in this section.

Carbon oxides. *Carbon monoxide* (CO) is a colorless, odorless, highly toxic gas that forms when carbon-containing fuels and other materials are incompletely burned. Major sources of this pollutant are burning forests and grasslands, people smoking tobacco, and inefficient stoves and open fires used for cooking indoors.

When we inhale CO, it reduces the ability of our blood to transport oxygen to body cells and tissues. Long-term exposure to CO can worsen lung diseases such as asthma and emphysema and trigger heart attacks. At high levels, CO can cause drowsiness, confusion, headache, nausea, and, ultimately, death.

Another carbon oxide is *carbon dioxide* (CO_2), a colorless, odorless greenhouse gas. About 93% of the CO_2 in the troposphere comes from the natural carbon cycle (see Module 1, Figure 1.21, p. 20). As noted above, certain human activities have added large amounts of CO_2 to the atmosphere, and because these rising levels of CO_2 are contributing to climate change and causing harm to humans and other organisms, CO_2 is classified as a pollutant.

Ozone. *Ozone* (O_3) is highly reactive gas that is a major ingredient of *photochemical smog*, a form of air pollution common in many urban areas. Ozone in the stratosphere helps sustain life, but ozone in the troposphere can act as an air pollutant. Considerable scientific evidence indicates that some human activities are raising the levels of harmful ozone in the air near the ground and reducing the amount of beneficial ozone in the ozone layer of the lower stratosphere (Figure 12.2). We discuss these effects later in this module.

Ozone can aggravate lung and heart diseases, lessen people's resistance to colds and pneumonia, and irritate the throat, nose, and eyes. It also damages plants, paints, fabrics, and the rubber in tires.

Volatile organic compounds (VOCs). Organic compounds that exist as gases in the atmosphere are called *volatile organic compounds* (VOCs). They include hydrocarbons emitted by certain plants, and methane (CH_4), which is also emitted by natural sources such as certain plants and rotting vegetation in wetlands. However, about two-thirds of all emissions of methane come from human sources, including landfills, rice paddies, dairy- and beef-cattle operations, and leaks from oil and natural gas wells and pipelines. Another important VOC is benzene, which is used to make dry-cleaning fluids, industrial solvents, and many other products, including gasoline, certain drugs, synthetic rubber, and plastics.

Methane is a greenhouse gas that plays a role in determining global and regional climates. Excessive levels of

this gas in the atmosphere due to human activities will likely contribute to projected climate change during this century. Thus, at a certain atmospheric level, methane becomes an air pollutant. Many VOCs contribute to the formation of a type of outdoor air pollution known as photochemical smog and some, such as benzene, can damage bone marrow, harm a developing fetus, and cause cancer (especially leukemia) in humans.

Particulates. *Suspended particulate matter* (SPM) is a broad term that includes a variety of solid particles and liquid droplets small enough to be suspended in air. The smallest of these particles and droplets can be smaller than one one-hundredth of the diameter of a hair on your head. Most SPM in outdoor air comes from natural sources such as wildfires and sea spray. However, more than a third of it comes from human sources, especially coal-burning power and industrial plants, diesel engines, plowed fields, and construction sites. Indoor SPM comes from tobacco smoke, inefficient heating and cooking stoves, and open cooking fires.

SPM can irritate your nose and throat. It can also damage your health or even shorten your life by damaging lungs, causing asthma attacks, and aggravating bronchitis and other lung ailments. It also contributes to haze in the air, and it corrodes metals and discolors paints and clothes.

Major categories. To understand where some air pollutants come from and how we can deal with them, it is helpful to group them into two categories (Figure 12.5). The first category, called **primary air pollutants**, includes all chemicals that are emitted directly into the air from natural processes and human activities and that build up to harmful levels (see *The Big Picture*, pp. 326–327). The second category is **secondary air pollutants**—the harmful

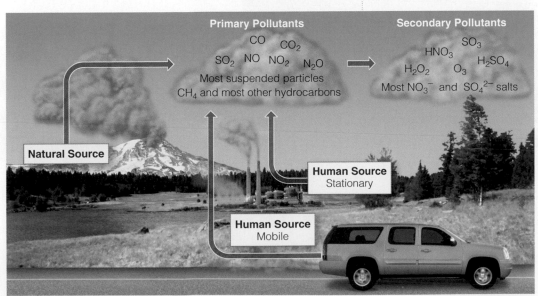

FIGURE 12.5 Primary air pollutants come from natural and human sources. Secondary air pollutants form in the atmosphere.

© 2014 Cengage Learning

Sources of Primary Outdoor Air Pollutants

Air pollutants come from natural and human sources. Examples of natural sources of primary air pollutants are wildfires, volcanic eruptions, and some plants, such as certain oak and poplar species, that release volatile organic chemicals. Most of these air pollutants settle out of the air fairly quickly, are removed by precipitation and chemical cycles, or are spread out over large areas and diluted to harmless levels.

Wildfire caused by lightning

Volcanic eruption

326

Most human inputs of primary air pollutants come from three major sources: the smokestacks of coal-burning power and industrial plants with inadequate air-pollution-control technologies; forests that are deliberately burned to provide land for growing crops and grazing cattle (especially in tropical areas); and the engines of cars, trucks, trains, aircraft, and ships that burn gasoline, diesel fuel, or other fossil fuels.

Angela Waye/Shutterstock.com

Coal-burning industrial plant

Herbert Giradet/Peter Arnold, Inc.

Digital Vision/Thinkstock

Deliberate burning of forests

Motor vehicles

chemicals that form in the air by reacting with primary pollutants or with chemicals naturally found in the air.

To study and deal with major outdoor air pollutants, scientists also find it helpful to group the major sources of these pollutants. *Stationary sources* are those sources that stay in one place, such as a burning field or forest, or a factory. *Mobile sources* are those that move, including motor vehicles, aircraft, and ships. Within those broad categories are several more specific sources (see *The Big Picture*, pp. 326–327).

Let's REVIEW

- What are the major sources and effects of the following air pollutants: **(a)** sulfur dioxide, **(b)** sulfuric acid and sulfate salts, **(c)** nitrogen compounds, **(d)** carbon oxides, **(e)** ozone, **(f)** volatile organic compounds, and **(g)** particulates?

- Distinguish between *primary* and *secondary air pollutants*, and give two examples of each type.

- What are two major natural sources of primary outdoor air pollutants? What are the three major human sources of primary outdoor air pollutants?

What Are the Problems?

Indoor Air Pollution

According to the World Health Organization (WHO) and the World Bank, the world's most serious air pollution problem, in terms of human health, is indoor air pollution, especially in less-developed countries. Worldwide, it harms and kills many more people than outdoor pollution does.

In poorer areas of the world, most indoor air pollution comes from the burning of fuels in open fires for heating and cooking (Figure 12.6), or in unvented or poorly vented stoves. These fuels include wood, charcoal, dung, crop residues, and coal. Indoor fires result in dangerous levels of indoor particulate air pollution. Another source of indoor air pollution in many countries is tobacco smoke.

In more-developed and rapidly developing countries, the worst indoor air pollutants come from other sources, especially chemicals used in the manufacturing of building materials and other home products (Figure 12.7). For example, formaldehyde is a chemical that is used in the manufacture of furniture, wall paneling, foam insulation, and particleboard. It can aggravate asthma and other breathing problems, irritate skin and eyes, and cause dizziness and headaches.

Joerg Boethling/Peter Arnold, Inc.

FIGURE 12.6 This woman in India is cooking over an open fire in her home, exposing herself and others to carbon monoxide, tiny particles of soot, and other air pollutants.

According to the EPA and some public health agencies, the four most dangerous indoor air pollutants found in many U.S. homes are: *tobacco smoke*; *formaldehyde*; *radioactive radon-222 gas* that can seep into houses from natural underground rock deposits, and *very small particulates*. In most of the more-developed countries, especially in urban areas, the health risks from exposure to indoor air pollutants are usually greater than those of outdoor air pollutants because most people spend the bulk of their time inside buildings or in vehicles, where pollutant levels can also be high.

NUMB3RS

1 of every 5
The number of U.S. commercial buildings in which employees are subject to health risks from indoor air pollution

CONSIDER this

- Levels of 11 common air pollutants in U.S. commercial buildings and homes are typically 2 to 5 times and sometimes as much as 100 times higher than they are outdoors.

- Levels of air pollutants inside cars in traffic-clogged urban areas can be up to 18 times higher than outside levels.

KEY idea Indoor air pollution, primarily smoke and soot from wood and coal fires and chemicals used in building materials and products, kills and harms more people than outdoor air pollution does.

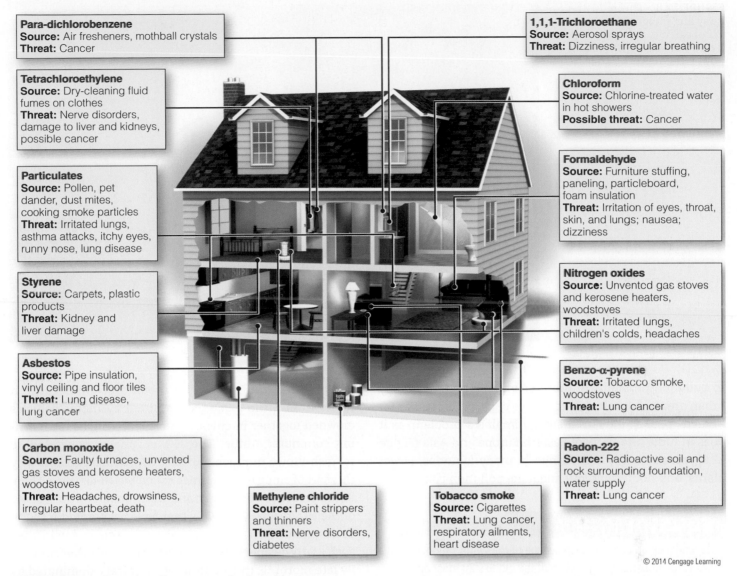

Para-dichlorobenzene
Source: Air fresheners, mothball crystals
Threat: Cancer

Tetrachloroethylene
Source: Dry-cleaning fluid
fumes on clothes
Threat: Nerve disorders,
damage to liver and kidneys,
possible cancer

Particulates
Source: Pollen, pet
dander, dust mites,
cooking smoke particles
Threat: Irritated lungs,
asthma attacks, itchy eyes,
runny nose, lung disease

Styrene
Source: Carpets, plastic
products
Threat: Kidney and
liver damage

Asbestos
Source: Pipe insulation,
vinyl ceiling and floor tiles
Threat: Lung disease,
lung cancer

Carbon monoxide
Source: Faulty furnaces, unvented
gas stoves and kerosene heaters,
woodstoves
Threat: Headaches, drowsiness,
irregular heartbeat, death

Methylene chloride
Source: Paint strippers
and thinners
Threat: Nerve disorders,
diabetes

1,1,1-Trichloroethane
Source: Aerosol sprays
Threat: Dizziness, irregular breathing

Chloroform
Source: Chlorine-treated water
in hot showers
Possible threat: Cancer

Formaldehyde
Source: Furniture stuffing,
paneling, particleboard,
foam insulation
Threat: Irritation of eyes, throat,
skin, and lungs; nausea;
dizziness

Nitrogen oxides
Source: Unvented gas stoves
and kerosene heaters,
woodstoves
Threat: Irritated lungs,
children's colds, headaches

Benzo-α-pyrene
Source: Tobacco smoke,
woodstoves
Threat: Lung cancer

Radon-222
Source: Radioactive soil and
rock surrounding foundation,
water supply
Threat: Lung cancer

Tobacco smoke
Source: Cigarettes
Threat: Lung cancer,
respiratory ailments,
heart disease

© 2014 Cengage Learning

FIGURE 12.7 Indoor air pollutants can be found in many areas of a typical home in a more-developed or developing country. (Data from U.S. Environmental Protection Agency)

Let's REVIEW

- What is the world's most serious air pollution problem and in what kinds of countries is it most serious?
- What are the biggest sources of indoor air pollution in most less-developed countries? What are the biggest sources in more-developed countries?
- List the four most dangerous air pollutants found in many U.S. homes and buildings.

Industrial Smog

As late as the 1950s, people in hundreds of cities and towns around the world burned coal in furnaces and stoves to heat their homes and cook their food. Coal was also burned in power plants and factories without the air pollution–control technologies that are available today.

This dependence on highly polluting coal exposed people in cities, especially during winter, to **industrial smog**: a mixture of sulfur dioxide, various suspended solid particles, and suspended droplets of sulfuric acid. This deadly mixture is sometimes called *gray-air smog* because it typically has a gray color. The term *smog* is a combination of the words *smoke* and *fog*.

Currently, most of the urban areas of more-developed countries seldom see industrial smog. This is because a majority of coal-burning industrial and power plants in these countries are now equipped with reasonably good air pollution–control technologies. Some plants also have tall smokestacks that transfer many of the pollutants downwind where they are dispersed, although this can lead

FIGURE 12.8 This industrial (gray air) smog over an Eastern European city comes primarily from a large coal-burning power plant.

Antonin Vodak/Shutterstock.com

to pollution problems in those downwind areas. In many urban areas of developing countries such as China (see *For Instance*, p. 332), industrial smog remains a problem, as it does in India and parts of Eastern Europe and Asia (Figure 12.8), where coal is still burned in many factories, power plants, and homes without adequate pollution controls.

CONSIDER this Because of its heavy dependence on coal, China has some of the world's highest levels of industrial smog and 16 of the world's 20 most polluted cities.

We could greatly reduce industrial smog worldwide, as has been demonstrated since the 1970s in Western Europe and the United States where this form of pollution was sharply reduced quite rapidly. This was accomplished through methods that we discuss later in this module.

Photochemical Smog

Another form of smog that is very much a problem in many countries is **photochemical smog**—a mixture of secondary pollutants generated from primary pollutants such as volatile organic compounds (VOCs) and nitrogen oxides (NO and NO_2) that have been exposed to sunlight. These secondary pollutants include ozone (O_3) and various organic compounds.

Photochemical smog is an urban problem because it forms from nitric oxide (NO) emitted by motor vehicles crowded together in cities, especially during heavy morning commuter traffic. The NO is quickly converted to reddish-brown nitrogen dioxide (NO_2), which explains why this sort of smog is sometimes called *brown-air smog*. Then, in the presence of sunlight and warm temperatures, some of the NO_2 reacts with VOCs that have evaporated into the atmosphere from gasoline, solvents used in businesses such as dry cleaners and some factories, and other sources. By late morning, this mixture of chemicals, dominated by highly reactive ground-level ozone, typically reaches peak levels. It is at this point when people are most subject to the harmful effects of photochemical smog, including irritation of the eyes and the respiratory tract.

Cities that have a lot of motor vehicles along with a sunny, warm, and dry climate suffer most from photochemical smog. Examples of such cities are Denver, Colorado; Salt Lake City, Utah; Los Angeles, California (Figure 12.9a); Mexico City, Mexico; São Paulo, Brazil; Buenos Aires, Argentina; Santiago, Chile (Figure 12.9b); Rome, Italy; Shanghai, China; Bangkok, Thailand; Tokyo, Japan; Jakarta, Indonesia; and Sydney, Australia.

Air pollutants that make up industrial and photochemical smog are not just urban problems because many of these chemicals can travel long distances in the atmosphere (see *For Instance*, p. 332). Some long-lived air pollutants, such as particles of toxic mercury and lead, return to earth's surface in areas far downwind of the cities where

FIGURE 12.9 Photochemical smog is common in Los Angeles, California (a), and in Santiago, Chile (b).

they were generated. Once they fall onto land or into water, some of them can get into plant and animal tissues and from there into our food (Figure 12.10).

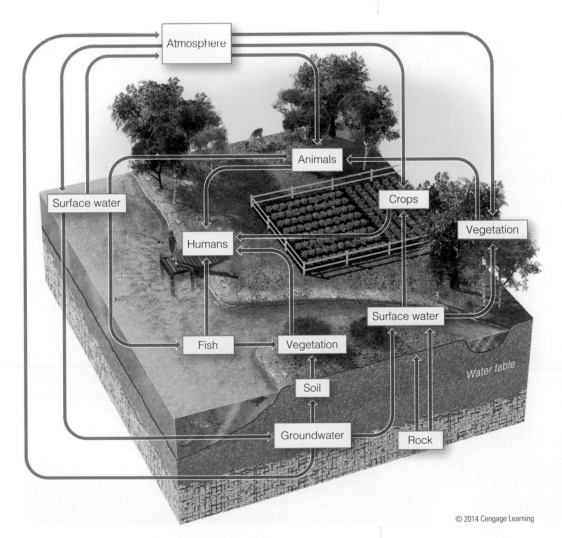

Atmosphere

Animals

Surface water

Crops

Vegetation

Humans

Fish

Vegetation

Surface water

Water table

Soil

Groundwater

Rock

© 2014 Cengage Learning

FIGURE 12.10 Some long-lived harmful chemicals emitted into the atmosphere can circulate through the biosphere and end up in the air we breathe, the water we drink, and the food we eat.

South Asia's Massive Pollution Cloud

In some areas of the world, pollution reaches levels at which huge, long-lived clouds full of pollutants form. Driven by high winds, these clouds can cover vast areas and move long distances.

Since 2002, the UN Environment Programme (UNEP) has worked with scientists such as V. Ramanathan, director of the Center for Atmospheric Sciences at the Scripps Institution of Oceanography in La Jolla, California, to track and study huge, dark brown clouds of air pollutants that regularly form over much of South Asia. Satellite images show that these clouds typically cover much of India, Bangladesh, and China (Figure 12.11a), and nearby areas of the Indian Ocean and the western Pacific Ocean. The clouds pollute the air of many major cities such as Shanghai, China (Figure 12.11b).

Analysis shows that these clouds contain a mix of soot, acidic compounds, hundreds of organic compounds, and particles of toxic metals such as lead and mercury. These pollutants are emitted into the atmosphere primarily by the burning of coal, other fossil fuels, and wood in factories and homes in South Asia's densely populated cities. The clouds also contain smoke and ash particles, produced by the clearing and burning of forests for planting crops, and by dust blowing out of drought-stricken regions.

CONSIDER this The pollution clouds over South Asia are typically about 2 miles thick and cover an area about the size of the continental United States.

During much of each year, most of the people living in the densely populated Asian cities under these clouds

FIGURE 12.11 This cloud of pollution lies over eastern China (a). Under the cloud, the air in cities such as Shanghai, China (b), is heavily polluted.

The SeaWiFS Project, NASA/Goddard Space Flight Center, and Orbimage

iStockphoto/George Corbin

see brown or gray polluted skies (Figure 12.11b). The thick clouds also slow the growth of crops and other plants by blocking some of the solar energy they need for photosynthesis. In addition, acidic particles and droplets falling from the clouds onto the earth's surface damage crops, trees, and aquatic life in lakes, streams, and coastal waters.

The clouds also affect the health and life expectancy of hundreds of millions of people in South Asia. UNEP scientists estimate that air pollutants in these clouds contribute to the deaths of at least 700,000 people every year.

These clouds affect more than just the areas in which they are generated. Satellites have tracked long-lived air pollutants from these clouds as they moved from South Asia across the Pacific Ocean to the U.S. West Coast. According to EPA estimates, on some days in the skies above Los Angeles, California, about two-thirds of the black carbon particles, one-third of toxic mercury particles, and one-fourth of all particulates have blown over from South Asia. This shows how the atmosphere connects all regions of the world.

Let's REVIEW

- What is *industrial smog*, how does it form, and what are its major sources?
- What is *photochemical smog*, how does it form, and what are its major sources?
- Describe South Asia's massive pollution cloud. How does it affect people in much of Asia, and how is it affecting the United States?

Acid Deposition

What can local communities do about air pollutants emitted by coal-burning power and industrial plants? One strategy that is widely used in more-developed countries is to require coal-burning facilities to build tall smokestacks that emit the pollutants high into the atmosphere where the wind can dilute and disperse them.

This strategy has succeeded in reducing pollution in the immediate areas around coal-burning plants, but it has contributed to a regional air pollution problem in downwind areas. That is the problem of **acid deposition** (Figure 12.12), often called *acid rain*—acidic particles and

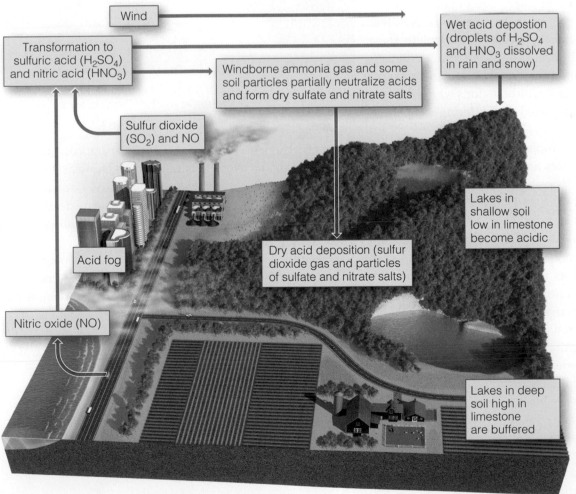

FIGURE 12.12 Acid deposition can harm crops and trees, deplete soil nutrients, and upset aquatic life in lakes. *See an animation based on this figure at* **www.cengagebrain.com**.

© 2014 Cengage Learning

droplets falling from the air or washed out of the air by precipitation onto land and into aquatic systems.

When primary pollutants such as sulfur dioxide and nitrogen oxides are emitted into the atmosphere by coal-burning facilities and motor vehicles, prevailing winds can transport them as far away as 600 miles. While in the atmosphere, some of these primary pollutants are converted into secondary pollutants (Figure 12.5), including tiny droplets of sulfuric acid, nitric acid vapor, and particles of acid-forming sulfate and nitrate salts.

Depending on prevailing winds, precipitation, and other weather factors, these acidic chemicals remain in the atmosphere for two days to two weeks. Gravity and precipitation eventually return the suspended particles and droplets to the earth's surface in two forms. One is *wet deposition*, which consists of acidic rain, snow, fog, and cloud vapor. The other is the *dry deposition* made up of acidic particles.

Acid deposition is a problem that affects regions that are downwind from coal-burning power and industrial plants and from urban areas with large numbers of cars. Some regions can tolerate moderate inputs of acids because their soils contain basic compounds that will react with and neutralize, or *buffer*, the acids. However, repeated exposure to acid deposition can deplete these neutralizing compounds. Areas that have thin, acidic soils without buffering compounds are especially sensitive to acid deposition. The same is true of regions where decades of acid deposition have depleted the buffering compounds in their soils.

Acidic particles and droplets in the air contribute to human respiratory diseases and to irritations of the eyes, nose, and throat. Acid deposition can also damage metal and stone building materials and statues (Figure 12.13).

A combination of acid deposition and other air pollutants such as ozone can also harm forests. These chemicals can remove calcium, magnesium, and other essential plant nutrients from soils and release ions of lead, cadmium, mercury, and aluminum that are toxic to trees (Figure 12.14). These changes in soil chemicals rarely kill trees directly. But they can stress and weaken trees, making them more susceptible to diseases, insect attacks, drought, and other stresses. High-elevation forests (Figure 12.1b) are especially vulnerable to damage from acid deposition because they are regularly exposed to highly acidic clouds and fog. Also, many of these forests have thin soils that cannot buffer the acids.

Another effect of acid rain is that it can remove toxic metals such as lead and mercury from soils and rocks and transfer them into nearby streams and lakes where fish

Lorpic99/Dreamstime.com

FIGURE 12.13 This sculpture has been eroded by long-term exposure to acid deposition and other forms of air pollution.

can pick them up. These toxins can then build up in the tissues of fish and eventually kill them. There are thousands of acidified lakes in Sweden, Norway, and Canada where the fish have all died off. Fish populations are declining for this reason in thousands of other lakes, including several hundred in the northeastern United States.

Acidified lakes present a problem to many people who depend largely on fish for their food supply. Consuming fish that are contaminated with mercury compounds can cause serious health problems. Most U.S. state governments have advised people—especially pregnant women and small children—not to eat fish caught from waters that are contaminated with these toxic compounds.

KEY idea Acid deposition, primarily a regional problem caused mostly by the burning of coal in power and industrial plants and by the use of motor vehicles, can threaten human health, forests, and aquatic life in lakes.

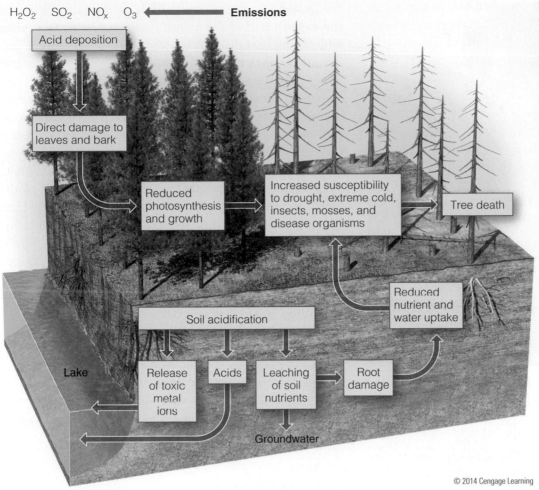

H_2O_2 SO_2 NO_x O_3 ← **Emissions**

Acid deposition

Direct damage to leaves and bark

Reduced photosynthesis and growth

Increased susceptibility to drought, extreme cold, insects, mosses, and disease organisms

Tree death

Reduced nutrient and water uptake

Soil acidification

Lake

Release of toxic metal ions

Acids

Leaching of soil nutrients

Root damage

Groundwater

© 2014 Cengage Learning

FIGURE 12.14 Prolonged exposure to acid deposition can damage and weaken trees, harm soils, and pollute groundwater and surface water in forests. In general, it can greatly weaken forest ecosystems. *See an animation based on this figure at* **www.cengagebrain.com**.

Let's **REVIEW**

- What is *acid deposition*, how does it form, and what are its major sources? Why is it considered a regional, rather than a local, problem?
- Describe the harmful effects of acid deposition on human health and on metal and stone sculptures and building materials. How does it affect forests and lakes?

Health Effects of Air Pollution

Air pollution kills at least 2.4 million people a year, according to studies by the WHO and the World Bank. Most of these deaths occur in Asia, including more than 650,000 deaths annually in China. About 6 of every 10 of these deaths result from breathing indoor air pollutants in homes, polluted factories, and other workplaces. In the United States, according to the EPA, between 150,000 and 350,000 deaths per year are related to air pollution.

Air pollution is especially dangerous to children under age 5, the elderly, pregnant women, smokers, factory workers, and people who have heart or respiratory problems.

Among air pollutants, one of the major culprits is very small particulates. These particles can bypass the human body's defenses and become lodged deep in the lungs. There, they contribute to heart attacks and strokes, as well as to asthma attacks, which can cause wheezing, coughing, and difficulty breathing (Figure 12.15, p. 336).

NUMB3RS

274

The number of people dying every hour (worldwide average) from prolonged exposure to various air pollutants

CONSIDER this   Asthma, often made worse by prolonged or even brief exposure to various air pollutants, kills about 250,000 people every year, worldwide—an average of 28 deaths every hour. This includes 4,000 Americans per year.

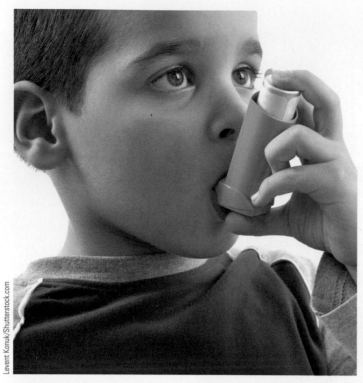

FIGURE 12.15 This boy is using an inhaler to help reduce the severity of an asthma attack, which can be triggered by air pollution.

According to EPA estimates, prolonged exposure to very small particulates is responsible for about 22,000 to 52,000 deaths per year in the United States (Figure 12.16). Much of this particulate pollution is emitted by coal-burning power plants, most of which are located in the central part of

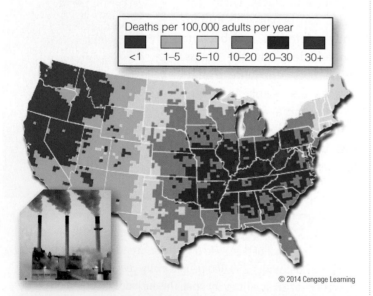

Deaths per 100,000 adults per year

| <1 | 1–5 | 5–10 | 10–20 | 20–30 | 30+ |

© 2014 Cengage Learning

FIGURE 12.16 This map shows how the death rates from air pollution vary across the United States. (Data from U.S. Environmental Protection Agency)

the United States. Emissions from these plants blow into the eastern half of the country and this helps to explain why the highest death rates from very small particle pollutants occur in the eastern half of the United States (red and purple areas of map, Figure 12.16).

Years of breathing polluted air can lead to other lung ailments such as chronic bronchitis, which produces excess mucus that causes a chronic cough and breathing difficulties, similar to those of many smokers. Another respiratory problem is emphysema (Figure 12.17), which leads to acute shortness of breath by causing a stiffening of the lungs' expandable air sacs. Anyone can get emphysema from prolonged smoking or exposure to air pollutants.

Let's **REVIEW**

- About how many of the world's people die every year from prolonged exposure to air pollution? About how many of these deaths occur in China and how many occur in the United States?

- Who are the people most susceptible to illness or death from breathing air pollutants? Which type of pollutant is a major culprit? What is one of the major sources of this pollutant?

- List three types of illness that can be made worse by exposure to air pollutants.

Climate Change

Scientists have collected a great deal of strong evidence indicating that the earth's atmosphere has been warming, especially since 1975. This evidence indicates that this warming is very rapid, compared to long-term historic changes in the average atmospheric temperature, and that it is very likely to alter the global climate during this century. There is equally strong evidence that human activities, especially the burning of fossil fuels and the clearing of many of the world's forests, are responsible for much of this warming effect.

Since 1975, the earth's average atmospheric temperature has risen by about 1 F°, and the first decade in this century (2000–2009) was the warmest decade since 1881. According to most of the world's climate scientists, a major factor in this atmospheric warming has been the dramatic rise in atmospheric levels of CO_2 and other greenhouse gases. Such higher levels are caused mainly by human activities that have been adding CO_2 to the atmosphere faster than the carbon cycle (see Module 1, Figure 1.21, p. 20) can remove it.

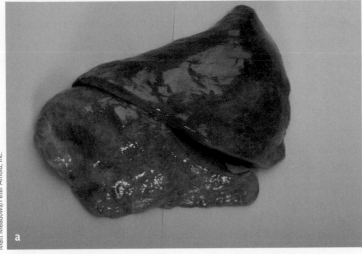

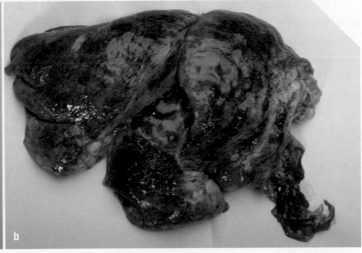

FIGURE 12.17 These are photos of a normal lung (a) and the lung of a person who suffered from emphysema (b).

In addition to historical temperature data, there is plenty of other evidence that the atmosphere is warming quickly. For example, many of the world's mountain glaciers (Figure 12.18a) are melting at higher than historic rates during summer months, as is the floating sea ice in the Arctic and some of the land-based ice in Greenland (Figure 12.18b). According to climate scientists, while this melting has been slow, it appears to be accelerating. The world's average sea level will rise if the net loss of Greenland's land-based ice continues over a number of decades.

Other evidence of atmospheric warming comes from mathematical models of the earth's climate system developed by climate scientists. When these scientists run the models on some of the world's most powerful computers, they project that the earth's average atmospheric temperature is likely to rise by 3.6 F° to 8.1 F° by the end of this century, depending on the rate of greenhouse gas emissions—a huge increase in a short time.

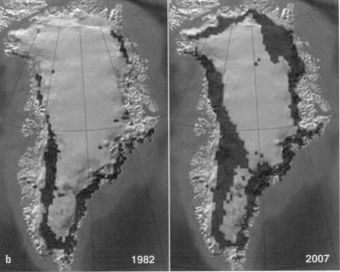

© 2014 Cengage Learning

FIGURE 12.18 The Athabasca Glacier in the Canadian province of Alberta (a) retreated considerably between 1992 and 2009, and continues to do so. Between 1982 and 2007, the summer melting of glacial ice (shown in red) in parts of the nearly ice-covered country of Greenland (b) increased dramatically. In 2010, scientists reported that this melting had continued to expand. (Data from Konrad Steffen and Russell Huff, University of Colorado, Boulder)

The U.S. National Academy of Sciences (NAS) has calculated that if the atmosphere warms as projected, abrupt climate change affecting ecosystems around the globe could happen during this century. Such a change would likely be irreversible for thousands of years, and it would be the end of the overall favorable climate that has allowed the human species to thrive during the past 10,000 years, since agriculture began near the end of the last ice age.

According to NAS and various other scientific studies, if such warming occurs, the many changes that are likely to occur during this century would include:

- the collapse of some economically and ecologically important ecosystems such as tropical rain forests and coral reefs;
- partial flooding of many low-lying coastal cities (Figure 12.19) due to rising sea levels caused by the expansion of seawater as it warms and by the melting of much of the earth's land-based glacial ice (Figure 12.18);
- drought-stricken, dried-out forests consumed in vast wildfires;
- grasslands, dried out by prolonged drought, turning into dust bowls that can no longer be used to grow crops;
- the extinction of at least a third and possibly half of the world's species; and
- increased photochemical smog (Figure 12.9).

This is a worst-case estimate. However, if just a few of the NAS projections are accurate, human economies and

political systems will be severely strained and jeopardized, as will many of the ecosystems, natural services, and wild species on which human systems depend.

Let's REVIEW

- List three types of evidence that point to rapid atmospheric warming that could disrupt the earth's climate.
- How have humans contributed to atmospheric warming?
- Describe five of the likely major effects of this projected change in the earth's climate.

Stratospheric Ozone Thinning

Recall that the ozone layer in the stratosphere contains much of the earth's ozone and protects us and a number of other species from harmful ultraviolet (UV) radiation. Imagine the dismay of the scientists who discovered several decades ago that part of this protective planetary sunscreen was disappearing.

Using measurements made over several decades from satellites, aircraft, and weather balloons, these researchers found considerable seasonal thinning of the ozone layer over Antarctica during September and October. This annual loss of ozone is commonly called an ozone hole, but the more accurate term is *ozone thinning*. Based on these measurements as well as on mathematical and chemical models, the researchers warned that stratospheric ozone thinning posed a serious threat to humans, to other animals, and to some primary producers (mostly plants) that use sunlight to support the earth's food webs.

The origin of this serious environmental problem was the discovery of the first chlorofluorocarbon (CFC) compound in 1930. Chemists then developed a number of similar CFC compounds, many known by their trade name of Freons. These useful chemicals were nontoxic, noncorrosive, and inexpensive to make. They were used as coolants in refrigerators and air conditioners, as gases used to make foam insulation and packaging, as cleaning agents for electronic parts, and as propellants in aerosol spray cans.

In 1974, the use of these chemicals was questioned because of research carried out by chemists F. Sherwood Rowland and Mario Molina (see *Making a Difference*, at right). They and other scientists had discovered that these long-lived chemicals were rising into the stratosphere. There, under the influence of high-energy UV radiation, the CFC molecules were breaking down and releasing chlorine, fluorine, and bromine atoms that reacted with and destroyed ozone faster than it was forming in some parts of the stratosphere.

© 2014 Cengage Learning

FIGURE 12.19 If the average sea level rises by more than 3 feet during this century, as various models of the earth's climate system project it will, the areas shown here in red in the state of Florida will be flooded. (Data from Jonathan Overpeck and Jeremy Weiss based on U.S. Geological Survey Data)

MAKING A difference

F. Sherwood Rowland and Mario Molina Sounded an Alarm

In 1974, two chemists at the University of California–Irvine issued a report that would send shockwaves through the scientific and business communities. Professors F. Sherwood Rowland (1927–2012) and Mario Molina (Figure 12.A) had calculated that UV-absorbing ozone in the stratosphere was being destroyed by chlorofluorocarbons (CFCs). The research report caused panic in the $28-billion-per-year CFC industry.

Rowland and Molina decided to go public with the results of their research because it indicated a serious threat to the ozone layer that helps to sustain most of the life on earth. They called for an immediate ban on the use of CFCs as propellants in spray cans, noting that other, safer propellants were available.

With lots of profits and jobs at stake, the politically and economically powerful CFC industry, led by the DuPont Company, mounted a vigorous attack against the scientists' calculations and conclusions. But the two researchers defended their work, expanded their research, and explained the results of their investigations to other scientists, to the media, and to elected officials.

In 1988, after 14 years of delaying tactics, DuPont officials agreed that CFCs were depleting the ozone layer. They also agreed to stop making CFCs and to begin selling chemicals that they had developed as substitutes for CFCs.

In 1995, Rowland and Molina were awarded the Nobel Prize in Chemistry for their work on the effect of CFCs on the ozone layer. In presenting the prize, the Royal Swedish Academy of Sciences said that the two researchers had contributed to "our salvation from a global environmental problem that could have had catastrophic consequences."

FIGURE 12.A The research of the late F. Sherwood Rowland (a) and Mario J. Molina (b) revealed how CFCs and certain other chemicals destroy stratospheric ozone.

Thus arose the term **ozone depletion**, the process by which ozone thinning occurred as CFCs destroyed some of the ozone molecules in the stratosphere. Researchers eventually found that CFC molecules could be especially long-lived and potent ozone destroyers. In fact, scientists estimate that some types of CFC molecules can last in the stratosphere and keep on destroying ozone for up to 385 years.

Satellite data and other measurements in combination with computer models indicate that since 1976, 75–85% of the observed losses of stratospheric ozone were caused by CFCs and other *ozone-depleting chemicals* released by humans into the atmosphere. Whenever a spray can with CFC propellant was used, whenever CFC coolants leaked from air conditioners and refrigerators, and whenever CFC solvents were used to clean electronic parts, these chemicals rose into the troposphere and eventually reached the stratosphere where they reacted with and destroyed ozone.

In 1984, researchers refined their estimates, finding that at least 40–50% of the ozone in the lower stratosphere

over Antarctica was disappearing every year during September, as it still is today. Figure 12.20 (p. 340) shows a colorized satellite image of ozone thinning over Antarctica during September of 2010.

The thinning ozone layer is a concern especially in parts of Australia, New Zealand, South America, and South Africa. Every year for a few weeks in these areas of the Southern Hemisphere, large masses of ozone-depleted air flow northward from Antarctica. As a result, levels of biologically damaging UV radiation near the ground increase by 3–20% in these areas, and people there are more subject to skin cancers, eye cataracts, and severe sunburns, while many other organisms besides people are also harmed.

KEY idea Widespread emissions of chlorofluorocarbons have led to seasonal decreases in levels of ozone in the stratosphere, which has allowed more harmful ultraviolet radiation to reach parts of the earth's surface.

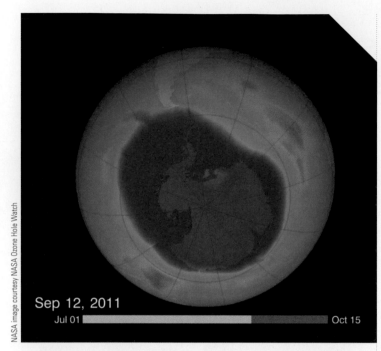

Sep 12, 2011

Jul 01 ▮▮▮▮▮▮▮▮▮▮ Oct 15

FIGURE 12.20 This colorized satellite image shows massive ozone thinning over Antarctica (blue and purple areas) during September of 2011. The concentration of ozone has decreased by 50% or more in the large area in the center of this image.

Let's REVIEW

- Describe the thinning of ozone in the stratosphere and define *ozone depletion*. Why are CFCs such potent ozone destroyers?

- Why should we be concerned about ozone depletion?

- Describe the roles of chemists F. Sherwood Rowland and Mario Molina in alerting the world to this serious environmental problem.

What Can Be Done?

Two Ways to Deal with Air Pollution

Any coal-burning power or industrial plant emits large quantities of carbon dioxide and certain outdoor air pollutants. A cooking fire inside a dwelling (Figure 12.6) can also expose occupants to dangerous levels of indoor pollutants. We can try to deal with these problems by asking two quite different questions. One question is "how can we clean up the emissions?" The other is "how can we avoid producing the emissions in the first place?"

The answers to these questions lead to two different ways of dealing with air pollution. One approach is called **pollution cleanup**, any method that involves collecting and disposing of air pollutants after they have been

produced. For example, there are several ways to capture pollutants as they enter smokestacks and then convert them to a form that can be disposed of somewhere, usually ash that is buried in landfills or an ash slurry that is stored in settling ponds.

The other approach, called **pollution prevention**, focuses on eliminating or sharply reducing the production of air pollutants (see *A Look to the Future*, p. 344). For example, we can switch from high-pollution fuels such as coal for generating electricity to cleaner-burning fuels such as natural gas or to very low-polluting alternatives such as solar cells and wind turbines.

Another way to reduce the production of air pollutants is to cut the unnecessary waste of energy. For example, almost half of the energy used in the United States is unnecessarily wasted. Sharply reducing this waste would greatly reduce air pollution and would also save money. Let's take a closer look at these options.

Reducing Indoor Air Pollution

In terms of the number of people affected worldwide, the two largest health threats from indoor air pollution are tobacco smoke and the smoke and soot from indoor fires (Figure 12.6) and leaky, inefficient cooking and heating stoves.

Some countries have banned smoking in all public buildings and are mounting programs to educate people about the dangers of smoking. Indoor air pollution in less-developed countries could be greatly reduced if leaky stoves and indoor fires could be replaced by inexpensive, energy-efficient stoves that are now available (Figure 12.21). These stoves vent their exhausts to the outside of a home, sharply cutting indoor air pollution. Mounting an international program to provide poor people with these stoves would go a long way toward improving the health of the poor.

In areas with adequate sunshine, people can use stoves powered by solar energy to cook food (Figure 12.22). In many parts of the world, these simple, low-pollution technologies could help people avoid cooking indoors with wood, charcoal, and dung. Because people in less-developed countries often must find firewood anywhere they can, giving them more efficient fuel-burning stoves or solar stoves and ovens would also reduce the unsustainable cutting of trees from many areas that are threatened by deforestation.

In more-developed countries, where the chemicals used in building materials and home products are more often a threat, there are several steps that could be taken to deal with the indoor air pollution they cause. For example, we could find less-hazardous substitutes for building materials, furniture, carpets, and other indoor products that contain formaldehyde and other harmful chemicals. We can also use devices called heat exchangers to bring fresh outside air into a building without losing heat to the outside or bringing in more heat than is desired. In addition, there are technologies available for preventing infiltration of radioactive radon gas into a building.

Let's **REVIEW**

- What are two ways for people in less-developed countries to reduce indoor air pollution caused by the use of open fires and inefficient stoves?
- List three ways to prevent indoor air pollution caused by the chemicals in some building materials and home products.

Reducing Outdoor Air Pollution

As with indoor air pollution, there are ways to clean up outdoor air pollution, most of which involve technological means. For example, new cars that burn gasoline or diesel fuel are now fitted with devices that collect or treat the pollutants before they enter the atmosphere. Similarly, many pollutants can be removed from factory and power plant smokestacks. However, such pollution removal can be costly.

Several scientists and analysts argue that a better approach is to avoid creating the pollutants in the first place (see *A Look to the Future*, p. 344). Short of preventing pollution altogether, there are also many ways to sharply reduce it. These approaches generally fall under the categories of technological, legal, and market-based solutions.

Scientists and engineers are busy designing and improving technologies that can help us to reduce pollution. These technologies include superinsulation for buildings, energy-efficient lighting and appliances, highly fuel-efficient vehicles, hybrid/electric cars that burn little or no fossil fuels (Figure 12.23, p. 342), and power plants that burn low-sulfur coal. In all of these examples, pollutants are still created—in the burning of fuel to make electricity for charging the electric car batteries, for example, and in the burning of low-sulfur coal.

However, some of these technologies reduce the use of energy and the resulting emissions of various pollutants. For example, the use of plug-in hybrid electric vehicles can reduce air pollution if the electricity used to recharge their batteries is generated by low-polluting sources such as wind and solar power, rather than by coal-burning power plants.

FIGURE 12.21 This energy-efficient Turbo Stove™ being used by an Indian family can greatly reduce indoor air pollution and the illness and deaths it causes.

FIGURE 12.22 This solar cooker uses a curved mirror surface that concentrates direct solar energy on a pot of food or water.

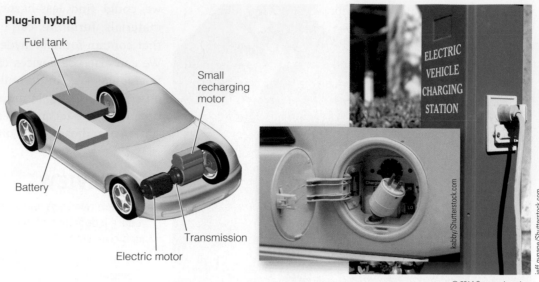

FIGURE 12.23 A plug-in hybrid electric vehicle is an electric car that uses a small gasoline engine to recharge its batteries. Drivers can also recharge the batteries by plugging them into charging stations (inset photos) or outlets at their homes or businesses.

Plug-in hybrid

Fuel tank

Small recharging motor

Battery

Transmission

Electric motor

ELECTRIC VEHICLE CHARGING STATION

kabby/Shutterstock.com

jeff gynane/Shutterstock.com

© 2014 Cengage Learning

Legal approaches to reducing air pollution have worked well in many of the world's more-developed countries as well as in some less-developed countries. For example, the U.S. Congress passed the Clean Air Acts in 1970, 1977, and 1990. These federal laws led to regulations that limit emissions of six key air pollutants: carbon monoxide (CO), sulfur dioxide (SO_2), nitrogen dioxide (NO_2), suspended particulate matter (SPM), ozone (O_3), and lead (Pb).

This approach has worked well, with the combined emissions of the six key pollutants dropping by about 67% between 1980 and 2010, even though the U.S. population grew by 36%. Also during that time, total vehicle miles driven by all Americans almost doubled, and energy consumption in general grew by almost half. Specifically, emissions of the key pollutants declined as follows between 1980 and 2010: lead dropped by 97%, suspended particulate matter by 83%, carbon monoxide by 71%, sulfur dioxide by 69%, volatile organic compounds by 63%, and nitrogen oxides by 52%. However, during this same period, CO_2 emissions increased by 16%.

The successful reduction of most of these key pollutants in the United States is an excellent example of democracy in action and of the benefits that can result from reducing outdoor air pollutants. By 1970, Americans had become fed up with the poor quality of outdoor air, and they pressured lawmakers to pass and enforce the air quality laws over strong objections from the oil and coal industries, utilities, and car companies. The laws eventually saved the country a lot of money and lives. There has also been a growing grassroots movement in the United States and Europe to phase out coal—by far the

dirtiest fuel to burn—and to replace it with much cleaner wind and solar power coupled with energy-efficiency improvements.

The United States has also used economic, or market-based, tools to help reduce some forms of outdoor air pollution. Since 1990, producers of outdoor air pollutants in the United States have been able to buy and sell permits that give them the right to emit specified levels of certain air pollutants. This *emissions trading* or *cap-and-trade program* began with the Clean Air Act of 1990, with the goal of reducing SO_2 emissions.

This act allows owners of coal-burning power plants to buy and sell rights to emit a certain amount of SO_2 each year. If a particular power plant emits less than its legal allotment of SO_2, it has a surplus of pollution credits. The utility can then use these credits to keep polluting at a higher level in one or more of its other plants. It can also sell the credits to other power companies that are exceeding their emission limits or to private citizens or groups.

Supporters of the cap-and-trade approach argue that it is more effective than having the government dictate how to control air pollution. Critics of this approach contend that it allows utilities with older, dirtier power plants to continue polluting by buying the rights to pollute from companies that have done a good job of reducing their SO_2 emissions. This can create air pollution hotspots.

How successful an emissions trading program is in lowering overall emissions of SO_2 or any other air pollutant depends on two factors: first, setting the initial cap low enough so that polluters have a strong incentive to lower their emissions and, second, lowering the cap every few years to promote continuing improvements in the reduction and prevention of air pollution emissions.

The emissions trading program for reducing the release of pollutants from U.S. electric power plants has been quite successful. According to EPA estimates, SO_2 emissions dropped by 68% between 1990 and 2010. Also, while utilities and other SO_2-emitting industries had predicted very high costs, the actual cost of this successful air pollution reduction program was less than one-tenth of what the industries had projected.

The best approach to most air pollution problems, according to many experts, would be a mix of technological, political, market-based, and lifestyle solutions that would sharply reduce and even prevent the creation of pollutants. There are many examples, a few of which are listed in Figure 12.24. Many individuals and several colleges, universities, local governments, and private companies (see the following *For Instance*) have seen the wisdom

Preventing Emissions of Air Pollutants

Use solar energy, wind, geothermal and other renewable energy sources instead of coal to generate electricity.

Use electric vehicles powered by electricity generated by renewables.

Give people incentives to get old, inefficient vehicles off the road.

Sharply improve fuel efficiency in all new motor vehicles.

Walk, bike, and use mass transit to get around.

Buy locally grown food to avoid pollution caused by long-distance shipping.

Cut energy use and save money by insulating heavily, plugging air leaks, and using energy-efficient windows in homes and buildings.

Use only energy-efficient lighting and appliances.

© 2014 Cengage Learning

FIGURE 12.24 These are some of the many ways to prevent emissions of air pollutants.

Photo credits (top to bottom): Photodisc/Getty Images; Jim Henderson/www.wikimedia.org; artwork by Patrick Lane; arrtwork by Patrick Lane.

of such an approach. They are applying pollution prevention strategies and saving a lot of money as well as helping to keep the earth's air and water cleaner.

FOR INSTANCE...

Pollution Prevention by a Major U.S. Company

The 3M Company, based in Minnesota, makes more than 60,000 different products in 100 manufacturing plants. In 1975, it began a pioneering Pollution Prevention Pays (3P) program.

As part of its 3P program, 3M focused on redesigning its manufacturing processes and equipment in order to cut its use of hazardous raw materials, and in turn, to reduce its output of air and water pollutants and toxic wastes. It also recycled or reused as many of its waste materials as possible. The company found that it could sell some of its recycled waste materials to other companies as raw materials.

The Pollution Prevention Pays program was a remarkable success that has inspired other companies to undertake similar programs. Between 1975 and 2010, the 3P program prevented more than 3 billion pounds of pollution and wastes from reaching the environment. As a result, the company saved more than $1.4 billion in costs of raw materials, waste disposal, and compliance with U.S. pollution laws and regulations, making the 3P program a superb example of how pollution prevention pays. Since 1990, a growing number of companies have adopted similar pollution and waste prevention programs.

Let's **REVIEW**

- What is the cleanup approach to air pollution and what is its major drawback?
- Give an example of a technological approach to reducing or preventing air pollution.
- Explain how the United States has used laws and regulations to reduce outdoor air pollution, and summarize the results of these efforts.
- Explain how a market-based emissions trading, or cap-and-trade, program can help reduce outdoor air pollution. How successful has the SO_2 emissions trading program been in the United States?
- List six ways to prevent emissions of air pollutants.
- Summarize the story of the 3M Company's pollution prevention program.

Protecting the Ozone Layer

Scientific models used to study the ozone depletion problem indicate that, even if we could immediately halt all production of ozone-depleting chemicals, it would still take about 60 years for the ozone in the stratosphere to return to its 1980 level and about 100 years to recover to pre-1950 levels.

After researchers had identified this serious problem (see *Making a Difference*, p. 339), several nations met in 1987 to develop the *Montreal Protocol* and again in 1992 to develop the *Copenhagen Amendment*. These international agreements, signed by 196 of the world's countries as of 2012, established a plan to phase out key chemicals that have been depleting the protective ozone in the stratosphere.

These agreements are important examples of how nations can cooperate to help reduce the severity of a very serious global environmental problem, using a prevention approach. One of the reasons for this unusually high level of international cooperation was that scientists had made a clear case, using solid scientific evidence, that ozone depletion was a serious threat to the well-being of people and the environment. As a result of this strong action by governments, chemical companies were spurred to find safer alternatives to CFCs, and some of these substitute chemicals proved to be more profitable than CFCs were for the companies.

Scientists have been measuring the results of rollbacks in the use of CFCs due to the Montreal Protocol and its amendments. Computer models project that if the cooperating nations continue to abide by these historic agreements, stratospheric ozone levels should return to 1980 levels by 2068 (18 years later than originally projected) and to 1950 levels by 2108.

Figure 12.25 lists some ways in which you can protect yourself from harmful UV radiation, as we try to prevent further damage to the stratospheric ozone layer.

Let's REVIEW

- Explain what has been done to try to slow ozone depletion in the stratosphere.
- Why is the Montreal Protocol an important example of international cooperation?
- List six ways in which you can reduce your exposure to harmful UV radiation.

Reducing Exposure to UV Radiation

Stay out of the sun, especially between 10 A.M. and 3 P.M.

Do not use tanning parlors or sunlamps.

When in the sun, wear protective clothing and sunglasses that protect against UV-A and UV-B radiation.

Be aware that overcast skies do not protect you.

Do not expose yourself to the sun if you are taking antibiotics or birth control pills.

When in the sun, use a sunscreen with a protection factor of at least 15.

Examine your skin and scalp at least once a month for moles or warts that change in size, shape, or color and sores that do not heal. If you observe any of these signs, consult a doctor immediately.

© 2014 Cengage Learning

FIGURE 12.25 There are a number of ways to reduce your exposure to higher levels of harmful UV radiation.

Photo credits (top to bottom): www.photos-public-domain.com; Vidux/Shutterstock.com; artwork by Patrick Lane.

A Look to the Future

Pollution Prevention

Several of the world's more-developed countries have managed to reduce their outdoor air pollution levels during the past 3 to 4 decades. This has been done mostly with an emphasis on controlling air pollution, either by directing emissions to where they will be diluted (such as through tall smokestacks that vent pollutants high into the troposphere) or by collecting pollutants before they enter the atmosphere (by filtering them and removing them from tailpipes and smokestacks).

Many environmental scientists point to two problems with any cleanup approach that tries to deal with air pollutants after they have been produced. First, it is only a temporary solution as long as levels of population and consumption grow without continuing improvements in pollution-control technology. For example, we have reduced emissions of some air pollutants by using catalytic converters in car exhaust systems in most industrialized countries (Figure 12.26). These devices have sharply reduced emissions of carbon monoxide (CO), volatile organic compounds

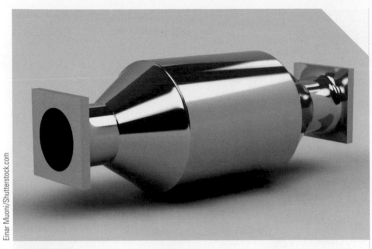

FIGURE 12.26 This is a catalytic converter used in most cars produced since 1975. It contains chemicals called *catalysts* that reduce the production of various harmful air pollutants within a vehicle's exhaust.

(VOCs), and nitrogen oxides (NO and NO₂) from vehicle exhaust pipes. However, the effectiveness of this approach has declined as both the number of cars and the average distance traveled per car has increased.

Second, when we remove a pollutant from one part of the environment, it often ends up as a pollutant in another part of the environment. For example, we can use various technologies to remove harmful chemicals from the smokestack emissions of coal-burning power plants. But this leaves us with a toxic ash that can pollute surface water or groundwater unless it is safely stored in a lined landfill or some other secure storage facility.

To environmental and health scientists, the better option is to shift to *preventing* air pollution. With this approach, scientists, engineers, and business executives would focus their creativity and efforts on answering the question, "how can we avoid producing outdoor and indoor air pollutants in the first place?" Accomplishing a major cultural shift to air pollution prevention would require revising existing air pollution laws and regulations, and also developing new air pollution laws that encourage air pollution prevention. Figure 12.24 lists some specific ways to bring about some important changes in the way we think about and deal with air pollution.

There are many important reasons for making this shift. One is that prevention works better than cleanup. For example, atmospheric levels of lead in the United States dropped by 97% between 1980 and 2010—much more than the drops in levels of other major outdoor air pollutants. This was because the use of lead in gasoline in the United States was legally phased out (Figure 12.27), with the result that lead as an air pollutant was virtually eliminated.

Another good reason for shifting to pollution prevention is that, in the long run, prevention is far less expensive than control and cleanup. Several companies have saved billions of dollars by preventing the production of pollutants instead of trying to collect and dispose of them (see *For Instance*, p. 343).

FIGURE 12.27 Use of leaded gasoline was phased out in the United States by 1995, and by 2011, had been phased out in most of the world's countries. In addition to its toxic effects on humans, especially children, lead also prevents catalytic converters (Figure 12.26) from working properly.

Yet another reason for making the shift is that pollution prevention is good for the health of humans and all other species. By preventing the release of harmful chemicals into the air, water, and soil, we help to prevent these chemicals from entering our bodies through the air we breathe, the water we drink, and the food we eat.

Some countries are making progress toward implementing pollution prevention as a major policy. In 2000, delegates from 122 countries developed a global treaty known as the Stockholm Convention on Persistent Organic Pollutants (POPs), which was strengthened in 2009. This treaty regulates the use of 21 widely used persistent organic pollutants (such as DDT and PCBs) that can accumulate in the fatty tissues of humans and other meat-eating organisms. Most environmental and health scientists view the POPs treaty as an important advance in international environmental law and pollution prevention. It has now been ratified or approved by 168 countries, but not by the United States.

FIGURE 12.28 A growing number of people in the United States, like these people in New Orleans, Louisiana, as well as in European and other countries, are protesting the construction of new coal-burning power plants to try to prevent new sources of dangerous air pollution and reduce the threat of climate change.

CONSIDER this

- Using blood tests and statistical sampling, researchers at New York City's Mount Sinai School of Medicine concluded that nearly everyone on the earth is likely to have detectable levels of POPs in their bodies.

- We know very little about the likely long-term health effects of this involuntary chemical experiment on humankind.

Also in 2000, the Swedish Parliament enacted a law that, by 2020, will ban any chemical that is persistent in the environment and that can build up in living tissue. Under this pollution prevention law, industries must show that any chemical they use is safe for humans, other species, and the environment. In other words, chemicals are assumed to be guilty until proven innocent—the reverse of the current policy used in the United States and most other countries, where chemicals must be proven to be toxic or otherwise dangerous before they can be banned or strictly regulated. There is strong opposition in the United States to such laws, mostly from the industries that produce and use potentially harmful chemicals.

In the 1970s, there was similar opposition from industry to the passage of the first set of air and water pollution laws that resulted in much cleaner air in the decades to follow. But those laws were passed because individual citizens and groups put strong political pressure on elected

officials as well as economic pressure on companies through their purchasing decisions. Analysts argue that with similar pressure from citizens and consumers (Figure 12.28), we can make the shift to prevention of outdoor and indoor air pollution in the decades to come.

Let's REVIEW

- What are two problems with depending on control and cleanup methods to deal with air pollution?

- What are three good reasons to make the shift from pollution control to pollution prevention?

- Describe the efforts of some countries to prevent pollution from persistent organic pollutants (POPS) and the efforts of Sweden to prevent air pollution.

What Would You Do?

This module began with the question of why you should care about air pollution. The obvious answer is that we all must breathe to stay alive, and breathing polluted air is hazardous to our health.

There are some sources of air pollution that we as individuals cannot directly control, such as factories and power plants. However, we can often have an indirect effect on these sources by exerting pressure on elected officials and on company managers to control or prevent pollution from these sources. Also, there are many ways to directly reduce or prevent air pollution through our daily activities and choices. Here are some ways in which people are working to reduce or prevent air pollution:

Making their homes as energy efficient and pollution-free as possible.

- Many are choosing only home-use products that do not contain formaldehyde or other pollutants, and they are testing their homes for radioactive radon, asbestos, and other pollution hazards.
- People are replacing inefficient, leaky stoves and furnaces with more energy-efficient stoves (Figure 12.21) and furnaces. Indoor cooking fires can be replaced by new stoves that vent pollutants to the outside; such stoves can be donated to poor families through various charity organizations.
- Homeowners are heavily insulating ceilings and walls, sealing air leaks in their homes, and installing energy-efficient windows, appliances, and lighting.

Using energy-efficient transportation options.

- Instead of driving, more people are walking, using a bicycle, or taking a bus or subway whenever possible.
- Many are choosing to drive only energy-efficient vehicles—those that get at least 40 miles per gallon.
- Hybrid-electric and all-electric vehicles also help to cut pollution if they are recharged with electricity generated from solar, wind, or geothermal power plants and not from coal-burning or nuclear power plants.

Making all purchases with environmental and energy impacts in mind.

- By eating less meat or no meat, people can cut air pollution by reducing the demand for industrialized livestock farming, which takes much more energy than does fruit and vegetable farming, and thus results in more air pollution.
- By buying only products that they can reuse or recycle, and by trying to minimize packaging in all purchases, people can reduce the demand for nonreusable and nonrecyclable products and packaging and thus, they can help to reduce pollution of all kinds.
- Many people are careful to buy primarily from companies that are trying to reduce their environmental impacts by cutting energy waste, reducing their dependence on fossil fuels, and preventing pollution.

KEYterms

THINKINGcritically

1. Emissions from motor vehicles play a major role in the formation of the photochemical smog found in most of the world's cities (Figure 12.9). List three ways in which you could reduce your contribution to photochemical smog.

2. Check with your utility company to find out whether the electricity you use comes from a coal-burning power plant, a nuclear power plant, or some other source. Find out whether you can get some or all of your electricity from a renewable energy resource such as solar, wind, or hydroelectric power.

3. Tall smokestacks can reduce local air pollution but lead to greater air pollution and acid deposition in downwind areas. Would you favor banning the use of tall smokestacks as a way to promote greater emphasis on preventing acid deposition and other forms of air pollution? Explain. If your answer is yes, would you still favor this action if it raised your electricity bill by 10%? How about by 20% or 50%?

4. If you live in the United States, list three important ways in which your life would be different today if actions by U.S. citizens between the 1970s and 1990s had not led to the Clean Air Acts of 1970, 1977, and 1990, because of intense political opposition by the affected industries.

5. Should we phase out the use of nonrenewable fossil fuels such as coal, oil, and natural gas over the next 50 years because of their major contribution to air pollution and projected climate change? Why or why not?

6. Think of three activities you took part in yesterday in which you might have used energy inefficiently, thus increasing the amount of air pollutants added to the atmosphere. How would you improve the energy efficiency of these activities?

LEARNINGonline

Access an interactive eBook and module-specific interactive learning tools, including flashcards, quizzes, videos and more in your Environmental Science CourseMate, accessed through **CengageBrain.com**.

13

Climate Change

Why Should You Care about Climate Change?

Since humans discovered agriculture about 10,000 years ago, we have benefited from living during a period with a fairly stable climate. This began to change in 1980 when average atmospheric temperatures started rising. Now, after more than three decades of atmospheric warming, we are seeing global **climate change**—long-term changes in the earth's average temperature, precipitation, and other environmental factors.

In 2007, more than 2,500 of the world's top climate scientists analyzed over 29,000 sets of data and published a report projecting likely changes in the earth's average temperature and climate during this century. According to these and many other climate experts, what the world faces today is not just climate change but *rapid climate change* during this century, with most of it caused by human activities, primarily the burning of fossil fuels and the clearing of forests, or *deforestation*.

According to these widely accepted scientific conclusions, climate change is taking place now and will very likely accelerate as atmospheric temperatures rise. If this occurs, it will change life as we know it by causing some areas of the world to get too little water (Figure 13.1a) and causing other areas to get too much (Figure 13.1b). Extreme weather events such as severe droughts, heat waves, storms, and floods are likely to increase in frequency and intensity in various parts of the world.

There will also likely be changes in where food can be grown and where people and many of the world's plants and animals can live and thrive. For example, production of corn and other crops (Figure 13.1a) may drop in some areas such as the U.S. Midwest because of a hotter and drier climate. Polar bears (see module-opening photo), which survive by hunting seals on floating Arctic ice, may become extinct as warmer temperatures continue to melt that ice.

Projected climate change has the potential to cause considerable harm to our economies and our societies, as well as to hundreds of generations of humans and many other species. This is why most climate scientists see it as one of the most important and urgent environmental threats that humanity faces—and something about which we ought to care deeply.

In this module, we examine the projected major effects of climate change during this century and the possibilities for dealing with the serious threat of rapid climate change. We start by considering a common point of confusion concerning the difference between weather and climate, and we look at how this helps to fuel controversy surrounding the threat of climate change.

What Do You Need to Know?

Weather and Climate

Many people, including some scientists, confuse the terms *weather* and *climate*. In order to learn about and evaluate the nature and effects of climate change, it is very important to know the difference.

Weather consists of short-term changes in atmospheric variables such as temperature and precipitation

FIGURE 13.1 Two possible consequences of projected climate change during this century are **(a)** crop losses from excessive heat and lack of water in some areas and **(b)** prolonged and heavy flooding in other regions such as this area of Thailand that suffered heavy rains in 2011.

in a given area over a period of hours or days. Weather can change quickly and tends to fluctuate from day to day and from year to year between warmer and cooler and between wetter and drier periods. **Climate**, by contrast, is determined by the *average* weather conditions, especially temperature and precipitation, in a particular area over periods ranging from at least three decades to thousands of years. These definitions have been adopted by the World Meteorological Association.

Has the climate where you live become warmer or cooler over the past 30 years? To answer this question you would need to do two things. First, find data for the average annual temperature for your area over at least the past 30 years. Second, plot these average annual temperatures on a graph and note whether they have generally risen, dropped, or stayed about the same during the entire time plotted. We have to use this same process to draw any meaningful conclusions about changes in the global climate.

We all know a lot about weather because we experience it every day, but only a small group of specialized climate scientists and researchers in related fields know much about climate. This leads to confusion and fuels controversy over whether climate change is really occurring. For example, people incorrectly conclude that a year or even a decade or two of warmer or cooler weather can confirm or refute statements about climate change. You can save yourself a lot of confusion and frustration by ignoring any conclusions about global warming (or global cooling) and climate change that are not based on at least 30 years of weather data.

The Earth's Major Climate Zones

Scientists have divided the earth's climate into major zones based mostly on their differences in average temperature and average precipitation over long periods of time (Figure 13.2). In terms of average temperature, we can

FIGURE 13.2 This map shows the earth's major climate zones along with major warm and cold ocean currents that distribute some of the earth's moisture and heat.

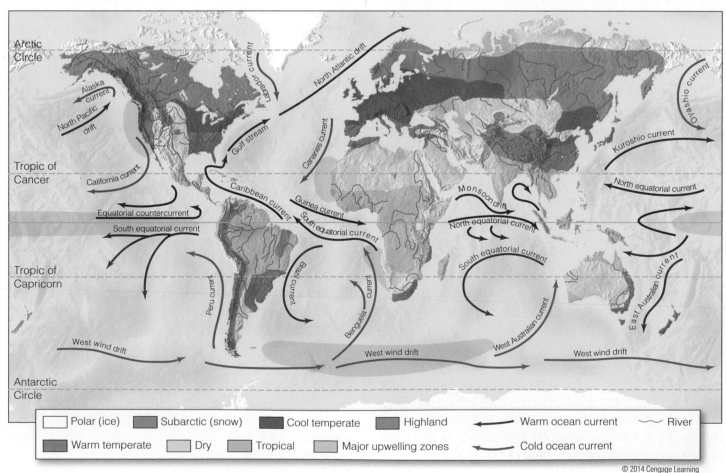

Polar (ice)	Subarctic (snow)	Cool temperate
Warm temperate	Dry	Tropical

Highland — Warm ocean current — River
Major upwelling zones — Cold ocean current

© 2014 Cengage Learning

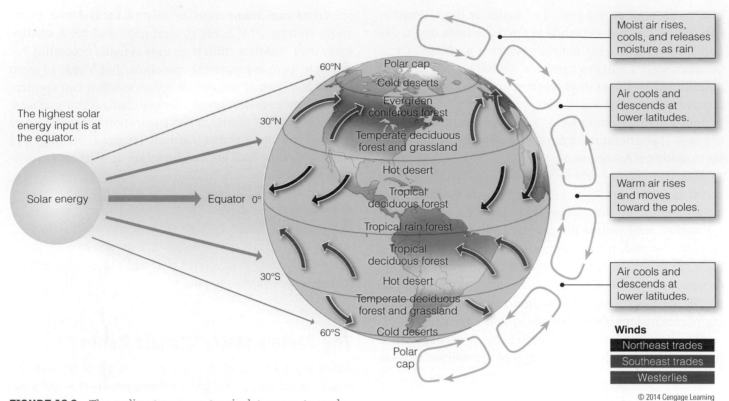

The highest solar energy input is at the equator.

Solar energy

Equator 0°

60°N
Polar cap
Cold deserts
Evergreen coniferous forest

30°N
Temperate deciduous forest and grassland

Hot desert
Tropical deciduous forest
Tropical rain forest
Tropical deciduous forest

30°S
Hot desert
Temperate deciduous forest and grassland

60°S
Cold deserts
Polar cap

Moist air rises, cools, and releases moisture as rain

Air cools and descends at lower latitudes.

Warm air rises and moves toward the poles.

Air cools and descends at lower latitudes.

Winds
Northeast trades
Southeast trades
Westerlies

© 2014 Cengage Learning

FIGURE 13.3 Three climate zones—tropical, temperate, and polar—are created by uneven solar heating of the planet, the rotation of the earth on its axis, and the resulting prevailing winds and ocean currents that distribute heat and moisture unevenly around the globe.

divide the earth's climate into three major zones: *tropical* (warm temperatures), *temperate* (moderate temperatures), and *polar* (cold temperatures).

These zones exist because heat and moisture are distributed unevenly around the planet (Figure 13.3). The equator receives the largest input of solar energy. The intensely heated air at the equator expands, rises, and moves north or south toward one of the earth's poles. The earth's rotation then deflects the movement of air over different parts of the planet, creating *prevailing winds* that blow almost continually. Note in Figure 13.3 that these winds are either *westerly* (blowing primarily from the west) or *easterly* (blowing primarily from the east), depending on how far north or south we go from the equator. The easterly winds are also referred to as the *northeast trade winds* and the *southeast trade winds*.

Prevailing winds drive the massive movements of ocean water known as *ocean currents*, which, like the prevailing winds, flow in fairly predictable patterns (Figure 13.2). Together, these winds and currents help to distribute heat and moisture in the atmosphere to different parts of the planet. In this way, they play a major role in determining various regional climates and climate zones.

Let's REVIEW

- Define *climate change*. Why should you care about climate change?
- Define and compare *weather* and *climate*. What are two major factors that help to determine climate?
- What are earth's three major climate zones? Explain how these zones were created.

Atmospheric Warming and Cooling

Life on planet Earth depends on the **greenhouse effect**— the process by which the lower atmosphere temporarily stores some of the energy received from the sun as heat. This warming occurs primarily because of the presence in the lower atmosphere of **greenhouse gases**—gases that have the effect of temporarily holding some of the sun's energy as heat. As some of the sun's energy is reflected from the planet's surface back toward space, greenhouse gases interact with this energy and release heat into the earth's lower atmosphere (see Module 1, Figure 1.15, p. 14). Four naturally occurring and abundant greenhouse gases are *water vapor* (H_2O), *carbon dioxide* (CO_2), *methane* (CH_4), and *nitrous oxide* (N_2O).

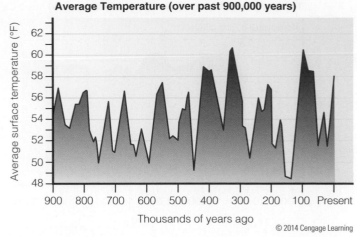

Average Temperature (over past 900,000 years)

© 2014 Cengage Learning

FIGURE 13.4 This graph shows the estimated global average temperatures of the atmosphere near the earth's surface over 900,000 years. (Data from NASA's Goddard Institute for Space Studies, Intergovernmental Panel on Climate Change, National Academy of Sciences, National Aeronautics and Space Agency, National Center for Atmospheric Research, and National Oceanic and Atmospheric Administration)

Throughout earth's long history, its average atmospheric temperature, precipitation levels, and other factors that determine its climates have changed many times in many ways. Factors affecting the planet's changing climate have included multiple large volcanic eruptions, changes in the sun's output of energy, the gradual growth and shrinkage of oceans as land-based ice melted or expanded, and devastating impacts by large meteors and asteroids. Thus, over the past 900,000 years, the earth's atmosphere has experienced prolonged periods of both cooling (which led to ice ages) and warming (which melted much of the ice) (Figure 13.4).

During the last ice age, much of the Northern Hemisphere was covered with ice in the form of glaciers (Figure 13.5, left). Near the end of this ice age, the atmosphere began to warm and over the next several thousand years, most of the ice melted (Figure 13.5, right). This drastically raised the sea level and changed the distribution of water on the earth's surface and in its atmosphere, making conditions favorable for a growing variety of organisms and ecosystems.

As a result, for roughly the past 10,000 to 12,000 years, we and millions of other species have lived with a fairly warm and stable climate over most of the earth's surface

(Figure 13.6, p. 354). This has allowed the human population to flourish, to develop agriculture, and to build societies and cities. It also helped to support dramatic and accelerating human population growth.

Let's REVIEW

- What is the *greenhouse effect* and why is it important to life on the earth? What are *greenhouse gases*? List four naturally occurring greenhouse gases.

- How has the earth's average atmospheric temperature changed **(a)** over the last 900,000 years and **(b)** during the 10,000 to 12,000 years since the last ice age ended?

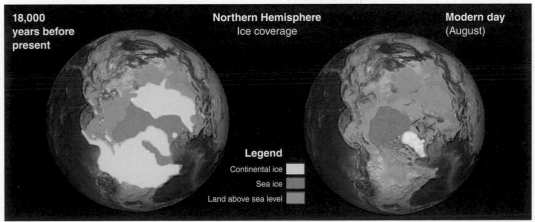

FIGURE 13.5 These maps show the changes in ice coverage in the Northern Hemisphere during the past 18,000 years. (Data from the National Oceanic and Atmospheric Administration)

© 2014 Cengage Learning

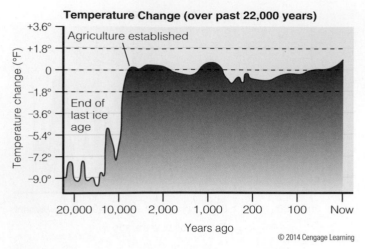

FIGURE 13.6 The estimated changes in the global average temperature of the atmosphere near the earth's surface over the past 22,000 years are shown in this graph. (Data from NASA's Goddard Institute for Space Studies, Intergovernmental Panel on Climate Change, National Academy of Sciences, National Aeronautics and Space Agency, National Center for Atmospheric Research, and National Oceanic and Atmospheric Administration)

Carbon Dioxide and Climate

Carbon dioxide (CO_2) plays an important role in regulating the average temperature of the earth's atmosphere, which together with precipitation, helps to determine the earth's overall climate. When the carbon cycle (see Module 1, Figure 1.21, p. 20) removes CO_2 from the atmosphere faster than it is added, all other things being equal, the average global atmospheric temperature drops because there is less CO_2 available to absorb energy and release heat. Similarly, when CO_2 is added to the atmosphere faster than the carbon cycle can remove it, the earth's atmospheric temperature rises. Carbon dioxide levels have long-term consequences because CO_2 molecules, on average, remain in the atmosphere for at least 100 years.

In 1894, Swedish chemist Svante Arrhenius (1859–1927) hypothesized that certain human activities, especially the greatly increased burning of coal during the 1800s, were adding CO_2 to the atmosphere faster than the carbon cycle could remove it, and that if they were kept up, these activities would lead to a warmer atmosphere. Since then, countless measurements and experiments have supported Arrhenius' hypothesis (Figure 13.7).

Carbon dioxide plays the key role in atmospheric warming, but there are other factors, including changes in the output of energy from the sun. Another important factor is the concentrations of other greenhouse gases such as methane (CH_4) and nitrous oxide (N_2O), both of which have also been rising and are much more potent as greenhouse gases than CO_2.

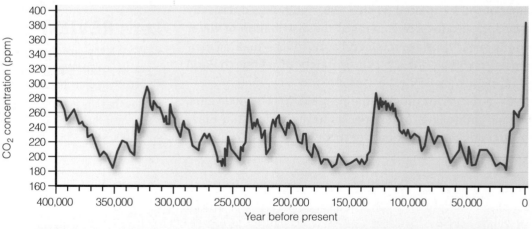

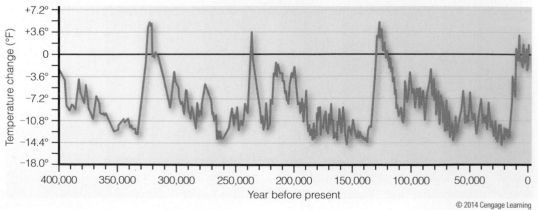

FIGURE 13.7 These graphs compare changes in the estimated average atmospheric temperatures and CO_2 levels during the last 400,000 years. (Data from Intergovernmental Panel on Climate Change, the U.S. National Center for Atmospheric Research, National Oceanic and Atmospheric Administration, National Aeronautics and Space Agency, and F. Vimeux, et al. *Earth and Planetary Science Letters*, 2002, vol. 203: 829–843)

U.S. Geological Survey/National Ice Core Laboratory

U.S. Geological Survey/National Ice Core Laboratory

FIGURE 13.8 Scientists study ice cores, such as this one (a) obtained by drilling a deep hole in an Antarctic glacier. Thousands of such cores from around the world are stored at (b) the U.S. National Ice Core Center in Denver, Colorado.

> **KEY idea** All other things being equal, the atmosphere warms if carbon dioxide is added faster than the carbon cycle can remove it, and the atmosphere cools if the carbon cycle removes carbon dioxide faster than it is added.

How Scientists Study Climate

Scientists have a number of ways to estimate historic atmospheric temperature changes such as those shown in Figures 13.4, 13.6, and 13.7, and the likely effects of such changes on the earth's climate. They have studied tree rings that reflect the annual growth of trees, and they have examined ancient fossils, marine sediments, and tiny bubbles of ancient air found in ice cores drilled from glaciers (Figure 13.8). Scientists also use these and other techniques to estimate precipitation levels over long periods of time, and they collect data to estimate and measure changes in average sea levels and ocean temperatures.

Since 1880, atmospheric temperatures have been measured at various locations and altitudes around the world. Since 1982, satellites with infrared sensors have also been measuring the temperature of various parts of the earth's surface on a daily basis.

These historic estimates and measurements show an overall increase in the global atmospheric temperature since 1880, with a sharp trend upward since 1980. The ups and downs in annual atmospheric temperatures shown in Figure 13.9 are the results of various short-term weather events that have temporarily heated or cooled the

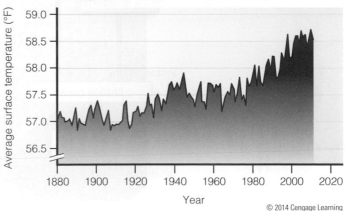

© 2014 Cengage Learning

FIGURE 13.9 This graph shows the global average temperatures of the atmosphere at the earth's surface between 1880 and 2011. (Data from NASA's Goddard Institute for Space Studies, Intergovernmental Panel on Climate Change, National Academy of Sciences, National Aeronautics and Space Agency, National Center for Atmospheric Research, and National Oceanic and Atmospheric Administration)

What Are the Problems?

Rapid Atmospheric Warming

In the early 1980s, climate scientists became increasingly concerned about how rising average annual atmospheric temperatures (Figure 13.9) and CO_2 levels (Figure 13.10) might affect the global climate. In 1988, in response to these concerns, the United Nations and the World Meteorological Organization established the Intergovernmental Panel on Climate Change (IPCC) to study historical temperature and climate changes, and to project future changes in the earth's climate.

The IPCC network now includes more than 2,500 climate experts from more than 130 countries who volunteer their time to help us learn about past and future climate change. The IPCC issued four reports summarizing its findings in 1990, 1995, 2001, and 2007, and a fifth report is due in 2014. The fourth and most comprehensive report in 2007 was based on more than 29,000 sets of data, collected mostly between 2002 and 2005. As a result of this massive analysis of climate data, the large majority of IPCC and other climate scientists generally agreed on three points:

1. Since 1980, the lower atmosphere on average has warmed at an accelerating rate by about 1.4 F° (Figure 13.9) and CO_2 levels have been increasing since 1958 (Figure 13.10).
2. Computer models and other data indicate that most of the temperature increase since 1980 is due to human activities (Figure 13.12), especially the burning of fossil fuels, which has raised atmospheric CO_2 levels, and the rapid clearing of forests and other vegetation that, if left in place, would take up much of the excess CO_2 from the atmosphere.
3. The climate is beginning to change, and if human activities that promote such change continue at their current or higher rates, we will likely experience further rapid atmospheric warming and long-term climate disruption during this century.

Does such general agreement mean that climate scientists have absolutely proven that atmospheric warming and climate change are real and are caused largely by human activities? No, because as discussed in Module 1 (p. 6), scientists can never prove anything absolutely. Instead, they strive to establish a high degree of certainty in their results.

Here are three of the thousands of pieces of evidence that IPCC and other climate scientists have used to support their major conclusions.

- The 25 warmest years on record have occurred since 1980, and the 10 warmest years have occurred since 1998.
- In some parts of the world, glaciers have been melting and floating sea ice has been shrinking for several decades.
- Rainfall patterns are changing. Extreme and prolonged drought is getting worse in various parts of the world due to rising temperatures and too little rainfall (Figure 13.1a). Other areas are experiencing extreme and more frequent flooding from too much rainfall (Figure 13.1b).

A large and growing amount of scientific evidence supports the three general conclusions of the 2007 IPCC report as well as more recent projections. According to a 2009 study by the American Meteorological Society, the atmosphere may warm by twice as much as projected by the IPCC in 2007. Other newer data indicate that sea ice and many glaciers are melting more rapidly than many scientists projected earlier. In 2010, the U.S. National Academy of Sciences—the nation's leading scientific body—said that: "a strong and credible body of evidence shows that climate change is occurring, is caused largely by human activities, and poses significant risks for a broad range of human and natural systems." Increasingly, the IPCC is being criticized as being too conservative in its projections.

Figure 13.13 shows the projected range of the global average atmospheric temperature for the remainder of this century, based on the 2007 IPCC study and more recent improved measurements and climate model projections. Note that the projected range of future temperature increases is 2 F° to 11 F°. More current research indicates that the upper range of temperature changes during this century could reach 13 F° if the world doesn't sharply reduce its use of fossil fuels and deforestation. Climate scientists are continually gathering more data and trying to improve their models to reduce the degree of uncertainty.

According to the IPCC reports and other research findings, the real threat is not simply climate change but the speed at which climate change is projected to occur. Most of the major changes in the earth's climate have taken place over thousands to tens of thousands of years (Figure 13.4). However, we are very likely to experience significant climate change in less than a hundred years with major long-term harmful effects—a process that some scientists are referring to as *climate disruption*.

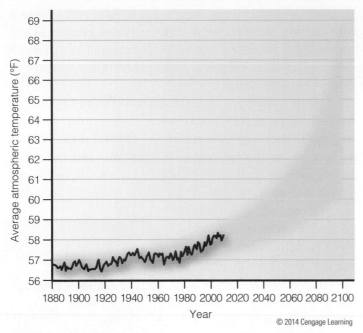

FIGURE 13.13 The red curve shows the combined estimated and measured changes in the average temperature of the earth's atmosphere between 1880 and 2011, and the yellow area shows the projected range of temperature changes during the remainder of this century, based on climate models.

The likely effects of climate disruption are numerous. For example, some areas will get hotter and some will get colder; some will get drier and some will get wetter. Some forests are likely to die back and their ecosystems will change. More frequent and far-reaching forest fires are likely to rage in areas that become drier and this could lead to an era of megafires. They will make matters worse by adding more climate-changing CO_2 and soot to the atmosphere and by reducing vegetation that removes CO_2 from the atmosphere. Experts warn that such effects could be irreversible for hundreds to thousands of years. This would make the planet a much less hospitable place for many future generations of humans and other species. We look at the effects of climate disruption in more detail later in this module.

Some scientists disagree with the IPCC conclusions. This is to be expected because scientific knowledge advances through vigorous debate and constant pressure to improve evidence and models. In fact, the criticisms of those who disagree with some of the more detailed findings of the IPCC have led to better climate data and climate models that further support the IPCC findings.

When evaluating the opposing arguments in this very important environmental issue, there are two major factors to consider. First, in light of the measurements and climate models, a large majority of the world's climatologists

and other climate experts generally agree with the IPCC conclusions. In fact, because the IPCC findings involve inputs from so many people, and because scientists tend to be conservative in making conclusions, they may be underestimating the rate and effects of climate change.

Second, most of the scientists, economists, politicians, and talk-show hosts who disagree with the IPCC conclusions are not experts in the extremely complex field of climate science. Many of their statements and conclusions are about weather, not climate, and some widely repeated statements are issued by individuals and groups that have a financial interest in our continuing to use fossil fuels at current or higher rates.

Now, let's take a closer look at the possible causes and effects of rapid climate change.

CONSIDER this

- Changes of just a few degrees in the global average atmospheric temperature can set into motion climate changes that will likely alter the face of the planet and its life for hundreds to thousands of years.

- Major climate models project that the average temperature of earth's atmosphere is likely to rise by 2 F° to 11 F° in less than 100 years—an extremely high rate compared to atmospheric warming periods of the past, each of which took place over many thousands of years (Figure 13.4).

Let's REVIEW

- What is the IPCC?
- What are the major conclusions of the 2007 IPCC report and the 2010 report by the U.S. National Academy of Sciences?
- List three pieces of evidence that IPCC scientists have used to support their conclusions.
- Summarize the likely effects of rapid climate change during this century.
- Explain why disagreement among scientists is to be expected. What are two important factors to keep in mind in evaluating opposing arguments?

Human Activities and Climate Change

Scientists measure CO_2 levels in the atmosphere in *parts per million*, or *ppm*—one part CO_2 for every million parts air. Data from the National Oceanic and Atmospheric Administration (NOAA) show that between 1850, in the midst of the Industrial Revolution, and 2011, the average atmospheric concentration of CO_2 rose from 285 ppm to 393 ppm. Most of this rise has occurred since 1960 (Figure 13.10) when industrialization was spreading around the globe. In a 2007 study, climate scientists Christopher Field and Gregg Marland estimated that unless CO_2 emissions are reduced by 80% over the next two decades, atmospheric CO_2 levels are likely to rise to 560 ppm by 2050 and to 1,390 ppm by 2100. Other studies have supported this estimate.

The problem is that many scientists believe that even the current CO_2 level of 393 ppm is too high for us to avoid some degree of climate disruption. NASA climate scientist James Hansen (see *Making a Difference*, p. 362) and other climate scientists estimate that we need to prevent CO_2 levels from rising above 450 ppm. Climate models indicate that this level is likely to be an irreversible *climate change tipping point* that would lead to large-scale climate disruptions that would last for hundreds to thousands of years.

In 2011, the largest CO_2 emitters were, in order, China, the United States, the European Union (with 27 countries), Indonesia, Russia, Japan, and India. The United States alone has contributed about 25% of the excess CO_2 that has accumulated in the atmosphere since 1850, compared to about 5% contributed by China. Although China's recent CO_2 emissions are high and growing rapidly, the United States emits about five times more CO_2 per person than China emits and almost 200 times more CO_2 per person than the poorest countries emit.

Another major contributor to atmospheric warming, second only to carbon dioxide, is dark-colored soot particles (known as black carbon), which absorb heat and release it into the atmosphere. While soot particles remain in the atmosphere for only a few weeks, compared to 100 years or more for CO_2 molecules, soot is playing a major role in atmospheric warming.

Soot is produced when a carbon-containing fuel is burned incompletely and emits black smoke containing tiny particles of carbon into the air. The main sources of soot are inefficient coal-burning power and industrial plants without up-to-date air pollution controls, inefficient stoves and open fires used by billions of people for heating and cooking, and millions of motor vehicles, locomotives, and ships that burn diesel fuel inefficiently. Another source is the smoke from forest fires, many of which are deliberately set to clear land, especially in tropical regions.

According to the IPCC and other researchers, two major human activities—the burning of fossil fuels and the clearing of vast tracts of forestland, especially in tropical areas—are largely responsible for the rise in atmospheric levels of soot and CO_2 and, thus, for the observed atmospheric warming since 1980 (Figure 13.9). These activities have interfered with the carbon cycle (see Module 1, Figure 1.21, p. 20) in two ways. First, we have emitted huge amounts of CO_2 into the atmosphere by burning fossil fuels (Figure 13.14) and by burning forests (Figure 13.15) to make way for plantations, cattle grazing, and human

NUMB3RS

19

The average number of tons of carbon dioxide that each American added to the atmosphere in 2009

FIGURE 13.14 By (a) burning coal in power and industrial plants and (b) burning gasoline and diesel fuel in motor vehicles, we emit large quantities of CO_2 into the atmosphere.

Stockbyte/Thinkstock

FIGURE 13.15 In Brazil's Amazon basin, large areas of tropical forest are burned every year to clear land for plantation crops and livestock grazing. These fires emit huge quantities of CO_2 into the atmosphere.

settlements. Studies have shown that these inputs of CO_2 can remain in the atmosphere for at least 100 years.

The second major way in which we have interfered with the carbon cycle also has to do with the clearing of forests, especially tropical forests (Figure 13.16). Because we have cleared massive amounts of trees and other vegetation, the carbon cycle now removes far less CO_2 than it would if the forests had not been cleared.

KEY idea

There is considerable and growing evidence that some human activities are disrupting the carbon cycle, which helps regulate the average temperature of the atmosphere and thus greatly influences the earth's global climate.

We now turn to a more detailed examination of the possible effects of climate disruption. In the next major section of this module, we consider what can be done about this growing problem.

Let's REVIEW

- What is the current level of CO_2 in the atmosphere in parts per million? What is the level that some scientists think we should not exceed?

- How is soot produced and what role does it play in atmospheric warming?

- What two human activities are primarily responsible for rising levels of CO_2 in the atmosphere and how do they interfere with the carbon cycle?

S. Channarrith-UNEP/Peter Arnold, Inc.

FIGURE 13.16 In many tropical areas, forests are being cleared faster than they can be replenished. The result of clearing a rain forest in Chiang Mai, Thailand, is shown here.

MAKING A difference

James Hansen: Climate Change Sentinel

As head of NASA's Goddard Institute for Space Studies since 1981, James Hansen (Figure 13.A) has made major contributions to the science of climate change. He is one of the world's most respected climate research scientists and has developed cutting-edge computer models of the earth's climate.

Hansen published his first climate change projections in 1981. He and his colleagues steadily improved their climate models, and in 1988, he testified to the U.S. Senate that global temperatures were higher than ever before measured. He also testified that most of the recently recorded atmospheric warming was likely due to greenhouse gas emissions from human activities, and that extreme weather events such as severe droughts and extensive flooding were likely to become more frequent.

His pioneering projections seem ordinary now, but his testimony released a storm of controversy driven by coal, oil, and utility companies. These and other industries opposed to Hansen's testimony have a vested interest in making sure that emissions of carbon dioxide and other greenhouse gases are not regulated or taxed. Hansen's climate models and temperature and climate change projections have been confirmed by reports of the Intergovernmental Panel on Climate Change (IPCC) and by more recent research.

Based on what their models project, Hansen and other climate researchers warn that a rise in the global average atmospheric temperature of more than 5 F° could eventually push us

FIGURE 13A James C. Hansen

past three major climate change tipping points—the disappearance of Arctic summer sea-ice, the melting of most of Greenland's ice sheet, and the melting of most Himalayan glaciers. If this happens, it would have long-lasting, disastrous effects on human civilization and on many ecosystems.

In 2004, Hansen reported that he was told by high-ranking government officials not to talk or write about the role of human activities in climate change. However, he refused to have his freedom to publish the results of his research suppressed for political reasons, and he has continued to sound the sentinel alarm. In 2006, he received the American Association for the Advancement of Science Award for Scientific Freedom and Responsibility. In 2009, the American Meteorological Society awarded him its highest award for his "outstanding contributions to climate modeling, understanding climate change, and for clear communication of climate science in the public arena."

Melting Ice

Data gathered over many decades show that atmospheric warming has been more severe in the world's polar (Arctic and Antarctic) regions than in most other areas of the earth. Since the 1950s, the Arctic has warmed by 5 to 7 F°—twice as fast as in other parts of the world, on average. Scientists believe this is because of a number of factors, including the tendency of ocean currents to transfer heat toward the planet's poles. Also, as more Arctic and Antarctic snow and ice melts, it reflects less sunlight, meaning that more solar energy is absorbed in polar areas, which speeds up the warming effect.

As a result of this faster warming, satellite data show that the average area of summer ice floating in the Arctic Sea (Figure 13.17) dropped by about 39% between 1979 and 2010. The floating ice is also getting thinner. If this trend continues, floating Arctic summer sea ice is likely to disappear by 2030 to 2040. This will also sharply reduce the Arctic's polar bear population (see module-opening photo and Figure 13.17) because polar bears depend on the ice for hunting seals—their main food supply. Other animals, such as some species of seals, also depend on this ice for finding their food.

This loss of Arctic sea ice could also affect weather and climate in other areas of the globe, because this ice plays a large role in how moisture is picked up and distributed through the atmosphere. In 2010, NOAA researcher James Overland projected that shrinking areas of Arctic sea ice would likely lead to less rainfall and snowfall in the already dry American West and more precipitation, flooding, and

FIGURE 13.17 Floating ice in the Arctic Sea is disappearing because sharp increases in the Arctic atmospheric temperature are speeding summertime melting of this ice.

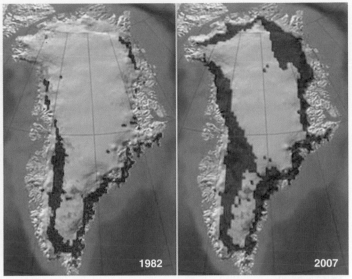

1982   2007

FIGURE 13.18 Partial melting at the edges of the land-based ice sheet in Greenland increased dramatically between 1982 and 2007. (Data from Konrad Steffen and Russell Huff, University of Colorado, Boulder)

colder, snowier winters in western and southern Europe, eastern Asia, and eastern North America.

The ice sheet that covers about 80% of Greenland— the world's largest island—has been melting around the edges during the summer at a slow but increasing rate (Figures 13.18 and 13.19). Complete melting of Greenland's land-based glaciers would add enough water to the oceans to raise the global sea level by as much as 23 feet. This is highly unlikely, but continued partial summer melting of Greenland's ice could raise the global average sea level by several feet.

Other major masses of land-based ice are found in mountain glaciers. In mountain ranges around the world, these ice masses act as reservoirs by storing water as ice during cold seasons and releasing it slowly to streams and major rivers during warmer seasons. Over the past 30 years, many of the world's mountain glaciers have been melting and shrinking at accelerating rates (Figure 13.20, p. 364) as the earth's atmosphere has warmed. For example, in 1850, Glacier National Park in the United States had 150 named glaciers. By 2010 only 25 were left, all of them shrinking. If most of the world's mountain glaciers shrink

FIGURE 13.19 During summer months, water from melting ice pours off Greenland's vast ice sheet in several locations such as this one, photographed in 2008. Since then, the melting has accelerated.

FIGURE 13.20 Much of the Muir Glacier in Alaska's Glacier Bay National Park and Preserve melted between 1948 and 2004, and it continues to melt today.

at similar rates in the future, as is projected, hundreds of millions of people who rely on glacial meltwater for annual irrigation and hydropower could face severe water, power, and food shortages.

Melting Permafrost

Another greenhouse gas that is causing great concern among scientists is methane (CH_4), which is 23 times more effective per molecule at holding heat in the atmosphere

FIGURE 13.21 Ice forms in underground layers of frozen soil called *permafrost* (a), in polar regions such as this area of Arctic tundra (b). These areas have held stores of methane and carbon dioxide for thousands of years.

than is CO_2. However, it typically remains in the atmosphere for 12 years compared to at least 100 years for carbon dioxide. Methane is emitted in huge quantities by cattle all around the world, as they digest their food, and by leaks from natural gas wells and pipelines and facilities that use natural gas, which is mostly methane. It also comes naturally from bodies of water such as wetlands where large amounts of vegetation are decaying.

Large amounts of methane (along with CO_2) are also found in normally frozen underground layers of soil, called **permafrost** (Figure 13.21a) that lie beneath polar grasslands such as Arctic tundra (Figure 13.21b). Methane is also found in peat bogs and under some lakes. Because polar regions are warming much faster than other parts of the planet, some of this permafrost is beginning to melt

and to release some of the methane and CO_2 that have been stored in it for thousands of years. Some scientists refer to these permafrost deposits as climate-change time bombs, which could eventually release amounts of methane and CO_2 that would be many times the total amounts now in the atmosphere, resulting in rapid and catastrophic atmospheric warming and climate change.

Let's REVIEW

- Describe the trends and likely long-term effects of the melting of (a) floating sea ice in the Arctic, (b) land-based glaciers in Greenland, and (c) many of the world's mountain glaciers.
- How does methane compare with CO_2 as a greenhouse gas?
- Define *permafrost*, and describe the possible effects of the melting of permafrost in Arctic tundra soils on atmospheric warming and climate change.

Rising Sea Levels

Because Arctic sea ice floats, its melting does not affect the world's average sea level, just as a floating ice cube that melts does not raise the level of water in a glass. However, as more ice melts from land-based glaciers, water is added to the oceans, causing sea levels to rise. Also, as ocean waters warm up, they expand into a larger volume, which also contributes to the rise in sea levels.

According to a 2008 U.S. Geological Survey report, the world's average sea level is most likely to rise 3–6.5 feet during this century and to keep rising for centuries. This is an example of how the 2007 IPCC report may have underestimated the effects of atmospheric warming. The rise in sea level is very likely to be even greater if Greenland's ice sheet (Figures 13.18 and 13.19) continues to melt at current or higher rates.

Even if the world were to make drastic reductions in greenhouse gas emissions starting now, scientists project that the world's average sea level is likely to rise at least 3 feet by 2100, simply because of the built-in effects of the current atmospheric warming. According to the IPCC, this projected change could threaten at least one-third of the world's coastal ecosystems, especially coral reefs that contain much of the world's aquatic biodiversity and coastal wetlands that produce much of the world's seafood and rice. Even this minimal rise will also likely disrupt many coastal fisheries.

Such a rise in sea level would flood several low-lying island nations (Figure 13.22) as well as many barrier islands and coastal areas. In doing so, it would contaminate freshwater aquifers lying near these coastlines. Figure 13.23 shows areas of the state of Florida that would be flooded with an average sea level rise of 3 feet. Other low-lying coastal areas and cities throughout the world would also be flooded by this projected rise in sea level.

Malbert/Dreamstime.com

FIGURE 13.22 Projected climate change could submerge Male, the capital city of the Maldives, along with the 1,200 other small islands that make up this nation in the Indian Ocean.

© 2014 Cengage Learning

FIGURE 13.23 Areas of Florida, shown here in red, will likely be flooded if the average sea level rises by 3 feet. (Data from Jonathan Overpeck and Jeremy Weiss based on U.S. Geological Survey data)

Ocean Warming and Acidification

As the global average atmospheric temperature has risen, so has the global average temperature of the ocean's waters. Just as with the atmosphere, some areas of the oceans have warmed faster than others. Warmer tropical waters in several areas are already leading to the bleaching and destruction of temperature-sensitive coral reefs that serve as centers of marine biodiversity (Figure 13.24).

Pawel Borówka/Shutterstock.com

FIGURE 13.24 Healthy coral reefs serve as habitats and food sources for an amazing variety of marine species.

Our massive inputs of CO_2, while they have led to the warming of the atmosphere and ocean waters, are also increasing the acidity of the world's ocean water in a process called **ocean acidification**, which could lead to further disruption of some ocean ecosystems and the human economies that depend on them. Jane Lubchenco, marine ecologist and head of the National Oceanic and Atmospheric Administration (NOAA), calls ocean acidification the "equally evil twin" of atmospheric warming.

As part of the carbon cycle (see Module 1, Figure 1.21, p. 20), the oceans, which cover almost three-quarters of the planet's surface, have played a key role in slowing atmospheric warming by removing about 30% of the excess CO_2 that we have put into the atmosphere since around 1800. However, most of the CO_2 that the oceans absorb from the atmosphere is converted to carbonic acid (H_2CO_3) in the seawater. As a result, the acidity of ocean surface water has increased by about 30% since 1850, according to a 2005 study by the U.K. Royal Society. Other studies have confirmed these findings. At current rates of CO_2 absorption, by 2100 the world's oceans will be 150% more acidic than they were in 1800 and more acidic than they have been during any of the past 20 million years.

When the acidity of ocean water rises, the level of carbonate ions (CO_3^{2-}) in the water decreases. These ions are vital to marine organisms that build shells and to those that build coral reefs (Figure 13.24). Thus, more acidic ocean water not only dissolves or weakens shells and structures such as reefs, but it makes it harder for marine organisms to replace them.

In 2009, American coral reef biologist Nancy Knowlton projected that "coral reefs will cease to exist as physical structures by 2100, perhaps 2050." Because shell-building organisms are such a critical part of the ocean food web, and because a huge variety of marine species inhabit or depend on coral reefs, ocean acidification is a major threat to all life in the seas and to humans and other species that depend on it.

Let's REVIEW

- What effect is the melting of mountain glaciers around the world having on the oceans? What is the other way in which atmospheric warming is raising sea levels?
- What will be the major harmful effects of rising sea levels caused by atmospheric warming?
- How are warmer ocean waters affecting coral reefs? What is *ocean acidification* and what are its likely harmful effects?

Extreme Weather Events and Worsening Drought

According to IPCC and other climate scientists, extreme weather events such as intense and longer heat waves, damaging storms, and extreme cold and blizzard conditions will probably become more frequent as the atmosphere continues to warm. Research also indicates that warmer ocean waters will likely help to increase the intensity and perhaps the frequency of tropical storms and hurricanes, resulting in more deaths and destruction.

Because a warmer atmosphere can hold more moisture, some areas of the world will experience more flooding (Figure 13.1b) from much heavier rains and more snowfall. Other areas will see much drier weather and, in the worst cases, **drought**, an extended period of dry weather resulting from a combination of higher-than-normal temperatures and lower-than-normal precipitation. One of the major harmful effects of drought is that it dries up soil, which can lead to crop losses (Figure 13.1a). In recent decades, drought has affected a growing area of the earth's land and, in some areas, is becoming more severe and long-lasting.

In 2010, the areas suffering from drought totaled more than 30% of the earth's land (excluding Antarctica)—an area about 5 times larger than the continental United States. For example, southern Australia, parts of Thailand (Figure 13.16), northern China, and much of the western United States are all experiencing severe, long-term drought.

Some of this drought is related to natural short-term changes in weather patterns. However, long-term climate change will likely contribute to prolonged and severe drought. According to a 2007 study by climate researchers at the U.S. National Aeronautics and Space Administration (NASA), by 2059, up to 45% of the world's land area could be suffering from extreme drought. In addition, some scientists estimate that by the end of this century, droughts could be 10 times more severe than they are today.

To make matters worse, drought reduces the amount of vegetation that removes CO_2 from the atmosphere. In addition, as the land becomes drier, more forests and grassland burn, which adds CO_2 to the atmosphere. More CO_2 means more atmospheric warming, which could lead to more severe and prolonged drought in a runaway cycle of change.

NUMB3RS

45%

The area of the earth's land projected to be suffering from prolonged drought by 2059

Changes to Food Production

According to the IPCC, moderate climate change could improve farming in parts of Canada, Russia, and Ukraine, but food production there will be limited by relatively poor soils. If warming goes beyond moderate levels, agriculture in most areas of the world is very likely to decline because of drier soils, scorching temperatures (Figure 13.1a), more severe and more frequent extreme weather events, and rising populations of many crop-eating insects and fast-growing weeds. This will likely lead to wide swings in annual food production yields and prices.

IPCC scientists note that food could be plentiful for a few decades in a warmer world because of the longer growing season in northern regions. However, by 2050, because of sharply declining food production in tropical areas, some 200–600 million poor people could face starvation and malnutrition due to the effects of climate change. Experts fear that international conflicts over access to food and water will grow, and that there will be millions of *environmental refugees*—people fleeing floods, drought, and excessive heat in search of food and water and new places to live.

Let's REVIEW

- What are three extreme weather events that might increase in frequency with more atmospheric warming?
- What is *drought* and how will climate change affect it globally?
- How might climate change affect food production?

Threats to Human Health

Scientists and public health experts have warned that projected climate change could have several harmful effects on human health. For example, according to IPCC researchers and other experts, although there will be fewer deaths due to extreme cold weather in a warmer world, the number of deaths and illnesses due to more frequent, intense, and persistent heat waves in some areas is likely to rise. Already, in 2010, an intense heat wave and fires from dry conditions in Russia claimed 56,000 lives, and 19 nations set all-time high-temperature records.

Climate change might also lead to a rise in the incidence of infectious diseases. Warmer temperatures and higher levels of CO_2 are very likely to favor rapidly multiplying insects, microbes, toxic molds, and fungi that can transmit such diseases. Microbes that cause tropical infectious diseases such as dengue fever, yellow fever, and malaria are likely to spread from tropical to temperate areas.

On average, about 2,700 people—90% of them under age 5—die each day from malaria. Public health experts have estimated that this death toll could double as the atmosphere warms and the mosquito species that transmit the malaria parasite spread from tropical to temperate areas.

Also, plants producing pollens that cause allergies and asthma attacks are likely to expand their ranges in a warmer, CO_2-rich environment. Higher atmospheric temperatures will also increase the rate of chemical reactions that produce photochemical smog, a form of urban air pollution that plagues many people with respiratory problems such as asthma.

Another effect on human health could be increasing hunger and malnutrition due to declines in food production in some areas of the world. One area that is projected to be hardest hit by climate change is sub-Saharan Africa where crop losses due to drought are already high and where hunger and malnutrition are now major problems.

Declining Biodiversity

Declining biodiversity is a major environmental problem even without climate change (see Module 1, p. 15). *Extinction*, or the disappearance of whole species, is now occurring at the highest rate in recorded human history.

FIGURE 13.25 Because of warmer winters, populations of mountain pine beetles have exploded and munched their way through large areas of (a) whitebark pine forests (orange-colored trees that are dead or dying) in Yellowstone National Park in the United States, and (b) lodgepole pine forests in the Canadian province of British Columbia (orange area).

According to the 2007 IPCC report, as well as more recent studies, atmospheric warming and the resulting climate disruption will accelerate extinctions. The main problem for many threatened species is that changes in climate can destroy or fragment their habitat, as is the case with polar bears in the Arctic that depend on sea ice that is melting at increasing rates (see module-opening photo). Plants and animals in areas undergoing such environmental changes must adapt to these changes, move to other areas, or go extinct.

Many of the earth's plants and animals do not need satellite surveys, thermometers, or climate models to sense that temperatures are rising and that their habitats are disappearing. Studies show that during the past 40 years, many of the world's animals (especially certain species of birds, bees, ants, and fishes) have been expanding their ranges north or south out of tropical regions toward the earth's poles. Some plants and animals are moving up mountain slopes to find cooler conditions as their habitats change in warmer, lower altitudes.

Scientists report that some extinctions related to climate change are happening now on every continent. According to the IPCC, nearly one of every three land-based plant and animal species could disappear if the average global temperature changes by 2.7–4.5 F°. This number could grow to 7 of every 10 species if the temperature change exceeds 6.3 F°.

For those species that can adapt quickly enough to climate change, atmospheric warming could expand their ranges and populations. But some of these organisms—such as weeds, fire ants, disease-carrying organisms, and beetles that kill trees (Figure 13.25)—will likely threaten many other species as they expand. According to some

biologists, the net effect will likely be a loss of biodiversity, with rare and specialized species declining and more adaptable generalist species thriving.

Let's REVIEW

- What are three threats to human health that will likely be worsened by climate change?
- What is the main problem for many threatened species resulting from climate change?
- What are the IPCC projections about higher rates of extinction due to climate change?

The Range of Projected Effects

Figure 13.26 summarizes several climate-model projections for the major effects of rising global atmospheric carbon dioxide levels and average temperatures. According to IPCC and other climate scientists, some degree of climate change is now unavoidable, mostly because we have delayed taking any significant action on this urgent problem for almost four decades.

It is not just the average atmospheric temperature increase, but also the rate at which that temperature will rise that will determine which of the effects listed in Figure 13.26 are most likely to occur. Some scientists say that if we can reduce the human contribution to atmospheric warming by about 80% as quickly as possible, we might avoid the worst of these projected effects. (We consider this challenge further in *A Look to the Future*, near the end of this module.)

KEY idea Some amount of climate change is unavoidable because of the massive amounts of greenhouse gases that we have released into the atmosphere, especially since 1980. Climate scientists project that the level of severity of change depends on how much we decrease or increase our greenhouse gas emissions from now on.

FIGURE 13.26 These are some of the projected impacts of climate change based on average global CO_2 levels and atmospheric temperatures. (Data from 2007 Intergovernmental Panel on Climate Change Report and Nicolas Stern, *The Economics of Climate Change: The Stern Report*, Cambridge University Press, 2006)

3.6°F Warming with 450 ppm CO_2 (now unavoidable effects)	5.4°F Warming with 550 ppm CO_2 (potentially avoidable effects)	7.2°F Warming with 650 ppm CO_2 (potentially avoidable effects)
Forest fires worsen	Forest fires get much worse	Forest fires and drought increase sharply
Prolonged droughts intensify	Prolonged droughts get much worse	Water shortages affect almost all people
Deserts spread	Deserts spread more	Crop yields fall sharply in all regions and crops die out in some regions
Major heat waves more common	Major heat waves and deaths from heat increase	Tropical diseases spread even faster and further
Fewer winter deaths in higher latitudes	Irrigation and hydropower decline	Water wars, environmental refugees, terrorism, and economic collapse become widespread
Conflicts over water supplies increase	1.4 billion people suffer water shortages	Methane emissions from melting permafrost accelerate
Modest increases in crop production in temperate regions	Water wars, environmental refugees, and terrorism increase	Ecosystems such as coral reefs, tropical forests, alpine and Arctic tundra, polar seas, coastal wetlands, and high-elevation mountaintops begin collapsing
Crop yields fall by 5–10% in tropical Africa	Malaria and several other tropical diseases spread faster and further	Glaciers and ice sheets melt faster
Coral reefs affected by bleaching	Crop pests multiply and spread	Sea levels rise faster and flood many low-lying cities and agricultural areas
Many glaciers melt faster and threaten water supplies for up to 100 million people	Crop yields fall sharply in many areas, especially Africa	At least half of plant and animal species face premature extinction
Sea levels rise enough to flood low-lying coastal areas such as Bangladesh	Coral reefs severely threatened	
More people exposed to malaria	Amazon rainforest may begin collapsing	
High risk of extinction for Arctic species such as the polar bear	Up to half of Arctic tundra melts	
	Sea levels continue to rise	
	20–30% of plant and animal species face premature extinction	

What Can Be Done?

Two General Strategies

According to most climate scientists, our best way to avoid climate disruption and its most serious, long-lasting harmful effects is to sharply reduce greenhouse gas emissions as quickly as possible. By doing so we would be following the adage: "When you find you have dug yourself into a hole, the first thing to do is to stop digging."

With this strategy, the urgent goal would be to keep CO_2 levels and atmospheric temperatures from exceeding the levels shown in the two right-hand columns of Figure 13.26. A second option would be to accept some degree of warming as unavoidable and to find ways to soften the harmful effects of modest climate change. Most analysts call for pursuing both of these strategies because even if we can adapt to modest climate change, our efforts will be overwhelmed by severe climate disruption if we do not sharply cut our greenhouse gas emissions.

According to the latest climate models, we must cut greenhouse gas emissions by at least 40% by 2020 and 80% before 2050 to achieve the first goal. Some experts, such as James Hansen (see *Making a Difference*, p. 362) and Lester R. Brown, say that we need to achieve an 80% reduction in emissions by 2020 to prevent serious climate disruption. We now examine some ways to sharply reduce greenhouse gas emissions.

Cutting Energy Waste

According to the U.S. Department of Energy, about 43% of the commercial energy used in the United States is unnecessarily wasted and even more is wasted in many other countries. This is an enormous source of energy that we can tap into more quickly, at a lower cost, and with a lower environmental impact than any other energy resource. Reducing this waste of energy would also reduce CO_2 emissions caused by our use of fossil fuels.

Another name for reducing the unnecessary use and waste of energy is **energy conservation**. It involves doing things such as turning the thermostat setting down at night (or programming a thermostat to do this automatically and to raise the setting

NUMB3RS

43%

The percentage of all commercial energy used in the United States that is unnecessarily wasted

in the morning) and adding insulation and plugging leaks in houses. The best way to cut energy use and waste is to improve **energy efficiency**: the measure of how much work we get from each unit of energy we use. Most of our energy waste and a good deal of our climate-changing CO_2 emissions are due to our use of inefficient lighting, appliances, and other devices (see the following *For Instance*).

FOR INSTANCE...

Shifting Away from Energy-Wasting Technologies

All over the world, people use a wide variety of devices every day that waste large amounts of energy, thus contributing to atmospheric warming and projected climate change. Start with the light switch on the wall. Are you using any incandescent lightbulbs where you live? These bulbs use just 5 to 10% of the electricity they burn to create light. The rest—90 to 95% of the electricity they use—generates heat that goes into the environment as wasted energy.

Because coal is burned to produce most of the world's electricity, the use of incandescent lightbulbs adds large amounts of climate-changing CO_2 to the atmosphere, along with a number of other harmful air pollutants. As a result, these bulbs are being phased out in the United States and in many other parts of the world and are being replaced with more energy efficient compact fluorescent bulbs and LED bulbs.

Where does the electricity that you use for your lighting come from? If it is a nuclear power plant, then some 82% of the energy used to feed electricity to your home is wasted. That is because the long, complex process of building the nuclear plant, making it safe to operate, providing its fuel, and safely storing its highly radioactive waste takes a lot of energy and reduces the net amount of energy provided by the use of nuclear power.

Another energy-wasting device is the typical internal combustion engine that powers most cars and trucks. It wastes about 80% of the energy it uses. So only about 20% of the money we spend on gasoline goes to moving us from one place to another.

A coal-burning power plant, if we include the mining and delivery of coal in our calculations, wastes 75% of the energy it uses. In leaky and poorly insulated buildings, the waste factor for heating them is 50–70%. In addition, most industrial electric motors, used in factories to manufacture a huge volume of goods every day, waste fully half of the energy they use.

Many scientists and engineers urge us to phase out our use of outdated, energy-wasting, and climate-changing energy technologies. They call for replacing them with more energy-efficient, low-carbon, and cleaner energy technologies for producing electricity, powering motor vehicles, and heating and cooling buildings. We already have such technologies, and they could be phased in over the next two decades.

Let's REVIEW

- What are two general strategies for dealing with climate change?
- About how much of the commercial energy used in the United States is unnecessarily wasted?
- Define *energy conservation* and *energy efficiency*.
- List five commonly used devices that waste large amounts of energy.

KEY idea
One of the best ways to cut energy waste, tap into a large source of unused energy, and help slow atmospheric warming is to improve *energy efficiency*—the measure of how much work we can get from each unit of energy we use.

Shifting to Low-Carbon Sources of Energy

Coal-burning power plants provide 44% of the electricity used in the United States and 80% of that used in China. When it is burned, coal emits more CO_2 and more air pollutants per unit of energy than any other fossil fuel. Thus, reducing coal use is one of the best ways to reduce emissions of CO_2 as well as a variety of harmful air pollutants that sicken and kill large numbers of people.

According to many economists, one reason coal is so widely used despite these problems is that the numerous harmful environmental effects of using coal are not included in the market price of coal-fired electricity. As a result, consumers are unaware of these harmful effects. Economists say that this is unlikely to change unless coal is taxed heavily for each unit of energy produced or for each unit of CO_2 emitted into the atmosphere. So far, however, coal and electric utility companies in the United States and in other coal-rich countries have had enough economic and political power to prevent such taxation.

They have also been able to acquire generous government *subsidies* (payments intended to help businesses get established) and tax breaks that help them to keep their prices low, which in turn helps to hide coal's harmful environmental effects.

A growing number of businesses and consumers would like to shift to a mix of low-carbon, cleaner renewable energy resources for generating electricity, such as wind power, solar energy, and heat stored in the earth's crust (geothermal energy). Numerous scientists, economists, and business leaders call for shifting subsidies and tax breaks away from coal and toward such a mix of renewable resources.

Some analysts call for relying more on natural gas to help make the transition away from coal, because burning it produces less CO_2 and fewer other air pollutants per unit of energy. However, recent research indicates that wells used to extract natural gas and pipelines used to transport the gas together contain enough leaks to account for a large amount of climate-changing methane emissions.

There is also a growing interest in expanding the use of nuclear energy to generate electricity, in order to replace coal-burning power plants and cut CO_2 emissions. However, many proponents of this idea do not realize that, when we include all phases of nuclear power production—in what is called the *nuclear fuel cycle*—nuclear power is far from carbon-free. While operating, the plants produce no CO_2, but large amounts of CO_2 are generated in mining the fuel, building the plants, and storing the growing amounts of highly radioactive wastes.

Even though the nuclear fuel cycle produces less CO_2 per unit of electricity than coal-fired power plants produce, use of nuclear power is extremely costly and can lead to accidents that emit long-lived radioactive material into the air and water, as happened in Japan in 2011 (Figure 13.27, p. 372). Guaranteeing public safety in and around nuclear power plants takes a great deal of time, energy, and money. Use of the nuclear fuel cycle produces dangerous radioactive wastes that must be safely stored for thousands of years, and it spreads the knowledge and technology that can be used to develop nuclear weapons.

Another major source of greenhouse gases is the transportation sector, with millions of vehicles around the world emitting these gases every day. Scientists and engineers are developing transportation options that rely more on renewable energy resources. For example, plug-in hybrid electric vehicles and electric cars and buses could some day replace most of the gasoline- and diesel-powered vehicles now in use. These energy-efficient vehicles would run mostly on electricity, and their batteries could

FIGURE 13.27 This is one of the three reactor buildings at Japan's Fukushima Daiichi nuclear power plant that were heavily damaged by an earthquake and flooding in 2011. The accident released large amounts of dangerous radioactive material into the atmosphere and into coastal waters.

be recharged by wind-generated electricity or by solar power or other renewable energy sources that involve low carbon emissions.

Not all renewable energy sources are low- or no-carbon options. For example, the amount of energy that motorists get from ethanol fuel made from corn is little more than the amount of energy it takes to produce this fuel. The latter includes energy used to fuel tractors and to make fertilizers, most of which comes from fossil fuels, which release CO_2 into the atmosphere when burned.

Let's **REVIEW**

- Why is it that reducing use of coal is one of the best ways to reduce emissions of climate-changing CO_2? Why is coal so widely used, according to some economists?
- Explain why nuclear power is not carbon-free.
- How could we lower carbon emissions within the transportation sector?
- Why is corn-based ethanol not a low-carbon option?

Engineering the Earth's Climate

By adding excessive amounts of carbon dioxide and other greenhouse gases to the earth's atmosphere and oceans for over 200 years, we unknowingly have been carrying out a climate-engineering experiment that could have extremely harmful and long-lasting consequences. Now, some scientists have proposed that we use other engineering schemes, called *geoengineering*, to help us in dealing with climate change. Figure 13.28 shows some of these proposed schemes.

Geoengineering proposals can be attractive because they hold out hope that we can escape serious climate change without having to sharply cut our use of fossil fuels. However, these schemes involve carrying out new global experiments on our still poorly understood life-support systems with little or no information on how well the geoengineering devices will work or on what their side effects may be. Some scientists argue that these devices could make matters worse.

We can divide current geoengineering proposals into two broad categories: those that seek to reduce atmospheric warming by removing excess CO_2 from the atmosphere and from systems that emit it and then storing it somewhere—known as *carbon sequestration* strategies—and those that would reduce the amount of sunlight reaching the earth's surface in order to cool the atmosphere—known as *global dimming* technological fixes.

One carbon sequestration approach involves efforts to remove CO_2 from the atmosphere and store it in soils and plant tissues. This could be done by preserving and restoring natural forests and by planting billions of trees in plantations throughout the world. Research indicates that the first of these options, storing CO_2 in natural forests, should be the top priority, because natural forests store about 60% more CO_2 per unit of land area and are more resilient to climate change and other disturbances than are tree plantations.

Another carbon sequestration scheme would involve dropping massive amounts of iron powder into ocean waters to spur blooms of *phytoplankton*, tiny aquatic organisms that use iron as a nutrient. These organisms remove CO_2 from the atmosphere as they grow. When they die, they sink to the ocean floor, carrying the carbon with them.

The problem is that the long-term effects of adding large amounts of iron to ocean ecosystems are largely unknown. Oceanographer Sallie Chisholm and several other researchers warn that iron fertilization could increase emissions of greenhouse gases by encouraging the growth of marine organisms that deplete the oxygen

AP Photo/AIR PHOTO SERVICE, File

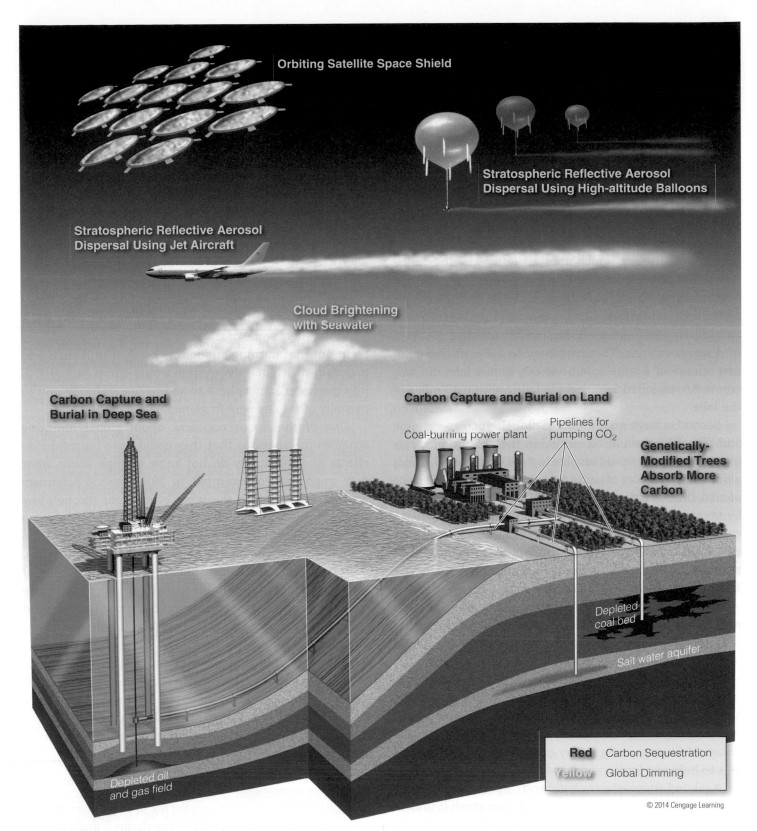

Orbiting Satellite Space Shield

Stratospheric Reflective Aerosol Dispersal Using High-altitude Balloons

Stratospheric Reflective Aerosol Dispersal Using Jet Aircraft

Cloud Brightening with Seawater

Carbon Capture and Burial in Deep Sea

Carbon Capture and Burial on Land

Coal-burning power plant

Pipelines for pumping CO_2

Genetically-Modified Trees Absorb More Carbon

Depleted coal bed

Salt water aquifer

Depleted oil and gas field

| **Red** | Carbon Sequestration |
| **Yellow** | Global Dimming |

© 2014 Cengage Learning

FIGURE 13.28 Various geoengineering schemes have been proposed in hopes of warding off climate change or offsetting its harmful impacts.

KEYterms

carbon capture and
   storage (CCS), p. 374
climate, p. 351
climate change, p. 350

drought, p. 367
energy conservation,
   p. 370

energy efficiency, p. 370
greenhouse effect, p. 352
greenhouse gases, p. 352

ocean acidification, p. 366
permafrost, p. 364
weather, p. 350

THINKINGcritically

1. In answer to the question of why we should care about climate change, some people might say we should not. They might argue that we have little control over something as huge and complex as the climate system, so we should just go on living as we have been without worrying about it. How would you respond to such an argument? Explain.

2. Imagine that you hear a debate between two scientists on the radio. One says that there is no really good evidence that climate change is occurring as projected or that atmospheric warming is mostly due to human activities. The other disagrees on both counts. List three questions that you would like to ask both of these scientists.

3. China relies on coal for two-thirds of its commercial energy supply and 80% of its electricity, partly because the country has a great deal of coal. Yet China's coal burning has made it the world's top CO_2 emitter and has probably sped up the process of atmospheric warming. Do you think China is justified in using coal intensively, as other countries, including the United States, have done and continue to do? Explain. What are China's alternatives?

4. One way to slow the rate of climate change is to reduce the clearing of forests that absorb CO_2 from the atmosphere—especially in tropical less-developed countries where intense deforestation is taking place. Should the United States and other more-developed countries pay less-developed countries to stop cutting their forests? Explain.

5. Do you believe that carbon emissions should be taxed in order to put a high price on carbon? Write a short speech that you would give to explain your position.

6. What are three of your daily habits that directly add greenhouse gases to the atmosphere? Which, if any, of these things would you be willing to give up to help slow projected climate change?

LEARNINGonline

Access an interactive eBook and module-specific interactive learning tools, including flashcards, quizzes, videos and more in your Environmental Science CourseMate, accessed through **CengageBrain.com**.

14

Wastes

Why Should You Care about Wastes?

Human societies create enormous amounts of solid and liquid wastes. Most of the solid wastes that people produce are deposited in open dumps, buried in sanitary landfills, or burned in incinerators. However, a lot of this waste is casually thrown away and gets scattered across the land as litter (see module-opening photo), much of which ends up in lakes, rivers, and oceans. In fact, so much waste material has flowed into the seas that gigantic swirling masses of plastic and other solid wastes are now found in all of the oceans, the most famous being the Great Pacific Garbage Patch (Figure 14.1), which some estimate is at least the size of the state of Texas.

As our population numbers and rate of resource use per person increase, our output of wastes also grows. Here are three important reasons for caring about the huge amount of wastes that we produce. First, much of this waste, even as we bury it or burn it, ends up polluting the land, soil, air, and water (Figure 14.1) on which we depend for maintaining our health and well-being, our lifestyles, and our economies. For example, waste incinerators can pollute the air unless they are equipped with expensive air pollution control devices.

Second, as our wastes pile up and pollute the environment, they harm or crowd out plant and animal communities, on land and in lakes, wetlands, rivers, and oceans. This wildlife that makes up the earth's *biodiversity*, is an important component of the planet's life-support systems (see Module 1, Figure 1.1, p. 2).

Third, most of the wastes we produce are wasted resources. According to the U.S. Environmental Protection Agency (EPA), up to 80% of the waste materials that we throw away could be reused or recycled. We could save energy and money, and put much less stress on the earth's life-support system, by reducing the amount of matter that we use and waste, and by putting much more emphasis on recycling and reusing items and materials that we typically throw away.

In this module, we look at the nature and origins of the wastes we produce, and at the resulting environmental and human health problems. We also consider several ways to reduce this serious environmental threat.

What Do You Need to Know?

Natural Capital, Chemical Cycling, and Energy Flow

We depend completely on the earth's **natural capital**—the resources and ecological services provided by nature that help us to survive and thrive, and that support human

FIGURE 14.1 The Great Pacific Garbage Patch **(a)** is the name given to two vast areas in the northern Pacific Ocean that are filled with swirling masses of trash **(b)** made up mostly of plastic items and partly decomposed plastic particles floating on or near the surface.

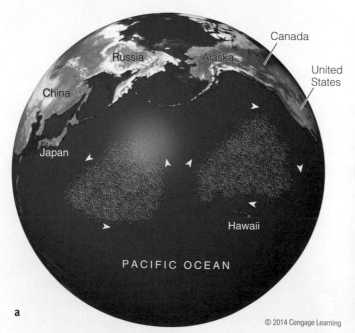

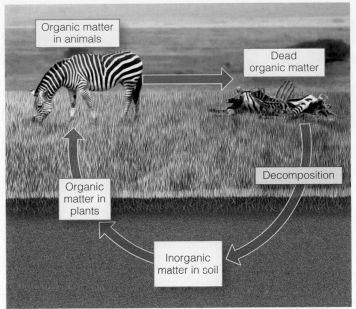

FIGURE 14.2 Chemical cycling is a vital natural service that endlessly recycles the earth's *nutrients*, or the chemicals that organisms depend on for survival.

© 2014 Cengage Learning

economies (see Module 1, Figure 1.13, p. 13). One of these vital ecological services is **chemical cycling**—the continual circulation of key chemicals from the environment (soil, water, and air) through organisms and back to the environment (Figure 14.2). For example, water is continually recycled by the *water cycle*, and the earth's finite supply of carbon moves continually through the *carbon cycle* (see Module 1, Figures 1.20, p. 19, and 1.21, p. 20). Other chemicals also flow in such cycles.

Chemical cycling allows the earth's organisms to obtain their *nutrients*, or the chemicals vital to their life processes. The material wastes left behind by any one organism become resources for other organisms as they take part in the recycling of these vital chemicals. This is the main reason that there is little or no material waste in nature.

Over billions of years, life on the earth has developed the processes of chemical cycling, driven by two factors. First, the planet receives no significant inputs of chemicals from outer space other than from meteorites. Second, no new supplies of chemicals can be created on earth because the *scientific law of conservation of matter* states that whenever matter undergoes a physical or chemical change, no atoms are created or destroyed. For these reasons, the earth's life forms, through trial-and-error, developed ways to recycle the chemicals that were available, including the wastes of other organisms.

The earth's life and our human economies also depend on the continuous, one-way flow of essentially inexhaustible, high-quality energy from the sun (see Module 1, Figure 1.15, p. 14). While nutrients can be recycled, high-quality energy cannot, because of the *second law of thermodynamics*. According to this scientific law, whenever energy is changed from one form to another in a physical or chemical change, we end up with lower-quality or less usable energy than we started with.

Unlike matter that flows within cycles, high-quality solar energy makes a one-way trip from the sun. It is converted to other forms of energy by organisms through their feeding interactions and other life processes, and much of it eventually flows into the environment as low-temperature heat. Solar energy cannot be recycled. Thus, life on the earth is sustained indefinitely by a combination of this one-way flow of energy and chemical cycling (see Module 1, Figure 1.19, p. 18).

KEY idea Our lives and economies depend on the one-way flow of energy from the sun, the earth's natural capital, and the recycling of the key chemicals needed by all forms of life.

Let's REVIEW

- Give three reasons why we should care about the wastes we produce.
- What are *natural capital* and *chemical cycling*, and why are they important? What are nutrients?
- Why is there very little material waste in nature?
- What two natural processes sustain life on the earth?

Economic Systems and Waste Production

In an **economic system**, goods and services are produced, distributed, and consumed to satisfy people's needs and wants. Three types of *capital*, or economic resources—natural capital, human capital, and manufactured capital—are

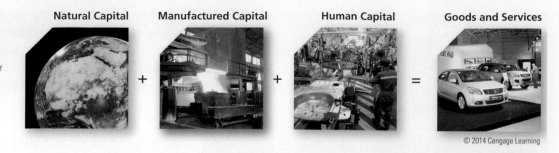

FIGURE 14.3 Most human economic systems use three types of resources, or *capital*, to produce goods and services.

Photo credits (left to right): NASA; United States Library of Congress's Prints and Photographs division/fsac.1a35062; Vasily Smirnov/Shutterstock.com; Mile Atanasov/Shutterstock.com.

Natural Capital Manufactured Capital Human Capital Goods and Services

© 2014 Cengage Learning

used to produce the goods and services that we buy (Figure 14.3).

Economic growth occurs when the world, a nation, or a locality increases its capacity to provide goods and services to its human population. Accomplishing such growth requires population growth (more people producing and consuming goods), more production and consumption of goods and services per person, or both.

Most of today's advanced industrialized countries have **high-consumption, high-waste economies**—economic systems that are devoted primarily to boosting economic growth by increasing the flow of matter and energy resources to produce more goods and services every year (see Module 1, Figure 1.22, p. 21). These economies convert much of the high-quality matter and energy resources from this flow into waste, pollution, and low-quality energy (mostly waste heat). *Ecological economists* distinguish themselves from other economists by arguing that all human economic systems are subsystems of the earth's life-support system (Figure 14.4). They point to a growing amount of evidence indicating that the outputs of high-waste economies can be harmful to the earth's

life-support system. We explore the resulting problems in the next section of this module.

Let's REVIEW

- What is an *economic system* and what three types of resources does such a system depend on?
- What is *economic growth*?
- Define *high-consumption, high-waste economies*. What are some of the outputs of such economies?
- What do ecological economists say about human economic systems?

What Are the Problems?

Harmful Costs of Economic Growth

Ecological economists contend that economic systems based on ever-increasing economic growth will eventually become unsustainable for two reasons. First, these systems will likely deplete or degrade much of the earth's

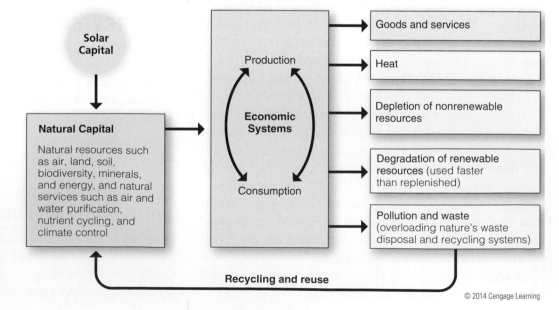

FIGURE 14.4 Many economists argue that all economies are human subsystems of the earth's life-support system because these systems depend on natural resources and services provided by the earth. *See an animation based on this figure at* www.cengagebrain.com.

© 2014 Cengage Learning

natural capital to the point where it will no longer support these systems. Opponents of this view argue that we can find technological solutions to the problems of depletion and degradation. But many economists point out that there are no substitutes for vital natural resources such as air, water, fertile topsoil, and biodiversity, or for ecosystem services such as climate control and chemical cycling (Figure 14.2).

The second reason that economic growth as we know it may become unsustainable is that high-waste economic systems are already exceeding the capacity of the environment to handle the growing amounts of wastes that these systems produce (Figure 14.5). This is presenting a number of environmental and health-related challenges that we will examine in this section.

Exclusion of Harmful Costs from Market Prices

In general, the market prices that we pay for most products do not include a number of *hidden costs* stemming from the harm done to the environment and to human health as a result of the production, use, and disposal of products. For example, the market price we pay for a computer includes the costs of raw materials, labor, marketing, and shipping. But it does not include various harmful environmental and health effects, such as the pollutants and wastes that result from the mining of raw materials to make the computer, the manufacturing process, and the ultimate disposal of that computer.

These hidden costs represent short- and long-term harmful effects on the earth's life-support systems, as well as on the health of people in current and future generations. Most people do not connect these harmful effects to owning a computer, partly because the costs of these effects are excluded from the computer's market price. However, sooner or later, the computer's owner and other people in a society pay these hidden costs in the forms of higher health-care premiums, poorer health, and higher taxes for pollution control and waste management.

Many economists believe that if these costs were included in product pricing, consumers would have a way to measure the harmful environmental impacts of the goods and services they use. They would, therefore, be less willing to buy products that have high environmental impacts. These economists also argue that as long as such

FIGURE 14.5 Growing volumes of waste are causing (a) water pollution and (b) land degradation, among several other problems.

costs are excluded, there will be little incentive for producers and consumers to try to reduce their contributions to the growing problems of pollution and waste.

Let's REVIEW

- What are two reasons given by ecological economists for their argument that high-consumption economies are not sustainable in the long run?
- What are hidden costs and what are the effects of excluding these costs from the market prices of most products and services?

Threats to Natural Capital

Environmental scientists point to a growing body of scientific evidence that high-waste economies are leading to globally widespread **natural capital degradation**—the depletion and degradation of the earth's natural resources and ecosystem services and the production of growing quantities of wastes and pollutants (Figure 14.6). The resulting stresses on the earth's life-support system—its land, air, water, and wildlife—are increasing because of the steady growth of the human population, the growing average amount of resources that each person uses every year, and the resulting pollution and wastes. In the long run, natural capital degradation threatens our lives and health, and the economies that support our lifestyles.

In 2005, the United Nations released its *Millennium Ecosystem Assessment*. According to this 4-year study by 1,360 experts from 95 countries, during the past 50 years, human activities have degraded about 60% of the earth's natural services. According to this study, "human activity is putting such a strain on the natural functions of earth that the ability of the planet's ecosystems to sustain future generations can no longer be taken for granted." In other words, humanity is not living sustainably.

One rough measure of the stresses that we are putting on natural systems is our **ecological footprint**: the area of land and water needed to supply a given number of people with the renewable resources they use and to absorb the pollution and wastes they produce (see Module 1, Figure 1.23, p. 24). We can estimate the ecological footprint of one person, of a population such as that of a city or country, and of the entire human population.

According to the World Wildlife Fund (WWF), in order to maintain our current global rate of resource use with a long-term supply of renewable resources, and to deal with the resulting wastes and pollution indefinitely, we would

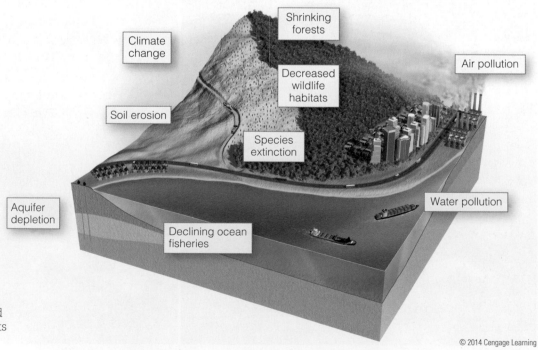

FIGURE 14.6 Growing numbers of people striving to consume more resources are degrading renewable natural resources and ecosystem services in many parts of the world.

© 2014 Cengage Learning

need roughly the equivalent of 1.3 earths. By about 2035, according to WWF projections, if both the world population and the average rate of renewable resource use per person continue to grow as they are now, we will need the equivalent of two earths to sustain us indefinitely.

In a typical year, the average American consumes about 30 times more resources than the average Indian and about 100 times more than the average person in the world's poorest countries. This explains why the United States is also the world's largest producer of wastes and pollutants per person (see *The Big Picture*, pp. 388–389). According to William Rees and Mathis Wackernagel, developers of the ecological footprint concept, if everyone in the world consumed as many renewable resources as the average American does today, the earth could support only about 1.3 billion people indefinitely—about one-fifth of the current population of 7 billion.

On the other hand, in the United States and in most other affluent countries, the air is clearer, drinking water is purer, and most rivers and lakes are cleaner than they were in the 1970s. In addition, people on average are living longer and food supplies are more abundant and safer.

Three factors played key roles in these improvements in environmental quality. One was that some of the wealth in these affluent countries was used to promote scientific research and technological advances that improved human health and reduced these countries' outputs of wastes and pollutants. The second was that large numbers of citizens of these countries insisted that businesses and governments work toward improving environmental quality. The third factor was that some of the industries that have produced pollution and wastes moved from wealthier countries to poorer ones to avoid labor costs and environmental regulations. Some analysts argue that in this way, wealthier countries have exported much of their pollution and wastes.

NUMB3RS

5

The number of planet Earths we would need to sustain the world's current population if everyone consumed resources at the rate of the average American

KEY idea

As our ecological footprints grow, we are depleting and degrading the earth's natural capital.

Let's REVIEW

- Define and give two examples of *natural capital degradation*.
- Define *ecological footprint*.
- How many earths would it take to supply, indefinitely, the amount of the renewable resources currently used around the world and to absorb the wastes and pollution resulting from this level of resource use?
- Compare the ecological footprint of a typical person living in the United States with that of a typical person living in India.
- How many planet Earths would we need if everyone in the world were to have an ecological footprint as big as that of a typical American?
- What are three factors that played roles in the improvements to environmental quality in the more-developed countries?

Solid Waste

One major category of the wastes we produce is **solid waste**, which consists of all discarded materials that are not liquid or gaseous (see *The Big Picture*, pp. 388–389). There are two major types of solid waste. One is **industrial solid waste** produced by mining and industries in order to provide people with goods and services. The other is **municipal solid waste (MSW)**, often called *garbage* or *trash*, which consists of solid waste produced by homes and workplaces, and by people who drop litter.

In many countries, the fastest-growing category of MSW is **electronic waste**, or **e-waste**—discarded TV sets, computers, cell phones, and other electronic devices that represent a loss of valuable resources that could be recovered. Most of this e-waste contains toxic metals that can end up in the air, water, and soil, and eventually in our bodies as well as in those of other animals.

Most MSW breaks down very slowly, if at all. Toxic lead and mercury never break down because they are chemical elements. An aluminum can takes 500 years, a plastic six-pack holder, 100 years, and plastic bags, 10–20 years to break down.

CONSIDER this

Based on EPA estimates, every year in the United States,

- the number of tires thrown away, if lined up, would encircle the earth three times;
- people throw away a total of 22 billion nonreturnable plastic bottles; and
- the office paper thrown away could be used to build an 11-foot-high wall from coast to coast across the country.

Solid Waste in the United States

The United States leads the world in total solid waste production and in the average amount of solid waste produced per person. The country has only 4.6% of the world's population but produces about 25% of the planet's solid waste. According to the U.S. Environmental Protection Agency (EPA), about 98.5% of the solid waste produced in the United States is industrial solid waste from mining (76%), agriculture (13%), and industry (9.5%).

Mining wastes

Industrial wastes

Just 1.5% of all U.S. solid waste is MSW. Still, the amount of MSW produced each year in the United States would fill about 24 million garbage trucks, which if they were lined up bumper-to-bumper, would reach almost halfway to the moon. Each American, on average, throws away about 7.1 pounds of trash per day, according to 2010 estimates by Columbia University scientists. This amounts to about 102 tons of trash in a typical lifetime.

The fastest-growing category of MSW is electronic waste. Every year, Americans discard 160 million or more cell phones, more than 48 million personal computers, and millions of television sets, printers, MP3 players, and other electronic devices.

Bakalusha/Shutterstock.com

Electronic wastes (e-wastes)

Péter Gudella/Shutterstock.com

Household and workplace wastes

About 69% of the MSW in the United States is buried in landfills, and 7% is incinerated. The remaining 24% of this MSW gets recycled in various ways, according to Columbia University scientists. Between 1960 and 1990, both the total amount of MSW in the United States and the amount of MSW per person almost doubled. However, since then, both have leveled off, mostly because more waste is being recycled.

Let's REVIEW

- Define *solid waste*, *industrial solid waste*, and *municipal solid waste (MSW)* and give an example of each.
- Define *electronic waste*, or *e-waste*, and list three examples of e-waste.

Hazardous Waste

Another major type of waste is **hazardous**, or **toxic**, **waste**, which is any waste that threatens human health or the environment because it is poisonous, dangerously chemically reactive, flammable, or corrosive. Examples include industrial solvents, car and household batteries (containing acids as well as toxic lead and mercury), household pesticides, hospital medical waste, and toxic ash from incinerators and coal-burning power plants. E-waste is also a fast-growing source of hazardous wastes (see the following *For Instance*). Most homes contain a variety of harmful chemicals such as those shown in Figure 14.7.

There are also numerous industrial sites containing barrels that are leaking hazardous wastes (Figure 14.8), although certain laws have led to the cleanup of the worst of these sites. Other industrial sites have large waste

FIGURE 14.8 These barrels are leaking hazardous wastes.

storage ponds that can present hazards. In 2010, a storage pond near an aluminum plant in Hungary ruptured, sending a wall of toxic red sludge across the nearby countryside. The chemicals burned many people, killing 9 and injuring 120, and devastated homes, cars, crops, livestock, and wildlife.

CONSIDER this Extracting and processing the materials used to build a medium-size television set produces an amount of hazardous wastes roughly equal to the weight of a large pickup truck.

Another form of hazardous waste is the highly radioactive waste produced by nuclear power plants and nuclear weapons facilities. The fuel for a reactor in a nuclear power plant is a form of the chemical element uranium (U) that produces energy when its nuclei split apart and release energy that is used to boil water and produce electricity. After 3 or 4 years in a reactor, this highly radioactive fuel is no longer effective and must be removed and replaced with new fuel.

The waste generated by used-up fuel, called *spent fuel*, is usually stored temporarily in deep pools of water (Figure 14.9a). At some nuclear power plants, after about 5 years of cooling, the still highly radioactive spent fuel is then transferred and stored temporarily in sealed dry-storage casks made of heat-resistant metals and concrete (Figure 14.9b). These wastes remain dangerously radioactive for at least 10,000 years and possibly much longer. To guarantee public safety, they will eventually have to be stored

What Harmful Chemicals Are in Your Home?

General
Dry-cell batteries (mercury and cadmium)
Glues and cements

Cleaning
Disinfectants
Drain, toilet, and window cleaners
Spot removers
Septic tank cleaners

Gardening
Pesticides
Weed killers
Ant and rodent killers
Flea powders

Paint Products
Paints, stains, varnishes, and lacquers
Paint thinners, solvents, and strippers
Wood preservatives
Artist paints and inks

Automotive
Gasoline
Used motor oil
Antifreeze
Battery acid
Brake and transmission fluid

© 2014 Cengage Learning

FIGURE 14.7 Many commonly used products contain harmful chemicals.

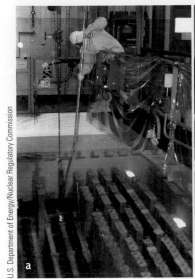

FIGURE 14.9 Highly radioactive spent fuel from a nuclear reactor is usually stored for cooling in a deep pool of water contained within a steel-lined concrete basin (a), and later in a dry-storage cask (b).

in essentially permanent facilities, but after six decades of research, scientists and governments have yet to find a scientifically safe and politically acceptable way to do this.

CONSIDER this

- According to a report by a Nevada state agency, after 10 years in storage, spent fuel from a nuclear reactor still emits enough radiation to kill a person standing 40 inches away in less than 3 minutes.

- If all of the world's nuclear power plants were shut down tomorrow, we would still have to protect ourselves, and many hundreds of future generations of people, from the intensely radioactive solid wastes currently on hand by using safe storage facilities that do not yet exist.

FOR INSTANCE...

The Growing Problem of Hazardous E-waste

Electronic waste, or e-waste, is the fastest-growing solid and hazardous waste problem in the United States and in the world (see *The Big Picture*, p. 389). Scientists are now analyzing this problem, looking for solutions to it.

Although an estimated 80% of the materials in electronic devices can be recycled or reused, most e-waste goes to incinerators and landfills. These discarded resources include high-quality plastics and valuable metals such as silver, gold, platinum, copper, and aluminum. E-waste is also a source of toxic and hazardous pollutants, including lead, mercury and polyvinyl chloride (PVC) plastic. These and other chemicals can contaminate the air, surface water, groundwater, and soil, and cause serious health problems and even early death for e-waste workers.

Much of the e-waste in the United States that is not buried or incinerated is shipped for recycling to Asia (mostly to China, India, and Bangladesh) and to poor African nations where environmental regulations are weak and labor is cheap. Workers in these countries dismantle, burn, and treat the e-waste with acids to recover valuable metals and reusable parts. This exposes these workers—many of them children—to toxic metals and chemicals. Scrap left over is often dumped into waterways and fields or burned in open fires, which exposes many people to toxic dioxins in the resulting air pollution.

Some 179 countries, not including the United States, have signed and ratified an international agreement, called the Basel Convention, which bans exports of hazardous waste. However, much hazardous e-waste is smuggled among countries and the United States still exports it.

Let's REVIEW

- Define and give two examples of *hazardous*, or *toxic*, *waste*.
- Why should we be concerned about the radioactive waste produced by nuclear power plants? How are these wastes now stored?
- Why does e-waste represent wasted resources? Explain why e-waste is a growing problem.

What Can Be Done?

Waste Management and Waste Reduction

In considering the problems of solid and hazardous waste, there are two questions we can ask. One is "what do we do with these wastes?" The other is "how can we avoid creating so much solid and hazardous waste in the first place?"

In the past, people have focused mostly on the first question, answering it by developing an approach called **waste management**, which involves a set of methods for handling waste with the goal of reducing its environmental impacts. These methods generally result in moving the wastes from one part of the environment to another, usually by burying them (which can pollute soil and groundwater), burning them (which can pollute air and soil), or shipping them to another location (which makes it someone else's problem).

Now, many environmental scientists, governments, and corporations are focusing more on the second question and taking a prevention approach called **waste reduction**, which involves producing and using goods and services with the goal of generating much less waste. We are also learning to mimic the way nature deals with wastes by treating them as potential resources to be reused, recycled, or composted.

A group of top scientists who were appointed by the U.S. National Academy of Sciences to study the problems of solid and hazardous waste have called for us to rely much more on waste reduction (Figure 14.10). So far, the United States and most industrialized countries are not following their advice. According to a study by Columbia University scientists, about 69% of the MSW produced in the United States in 2010 was buried in landfills, 7% was incinerated, and the remaining 24% was recycled or composted. The study also estimated that, on average, each American produces about 7.1 pounds of MSW per day. The output of MSW has become one of the most accurate indicators of economic growth by some measures. However, many environmental scientists consider the growing MSW output to be a glaring indicator of unsustainable living. And MSW is only 1.5% of the total amount of solid waste produced in the United States (see *The Big Picture*, pp. 388–389).

Waste management and waste reduction both have environmental and economic advantages and disadvantages. Many analysts call for making the most of the advantages of both by using **integrated waste management**—a strategy that involves a mix of waste-management and waste-reduction tools and methods (Figure 14.11). This strategy could be used to implement many of the scientists' recommendations listed in Figure 14.10.

According to some scientists and economists, we could cut the amount of solid wastes we produce by 75–90% if we followed the experts' recommendations (Figure 14.10) and made intensive use of integrated waste-management strategies (Figure 14.11). We now look more closely at these recommendations.

> **KEY idea** We could benefit from viewing solid wastes as potential resources that could be reused or recycled.

Let's REVIEW

- Distinguish between *waste management* and *waste reduction*.
- What are three priorities suggested by a panel of scientists for dealing with solid and hazardous wastes? For each strategy, what are four important steps that we can take?
- What is *integrated waste management*?

FIGURE 14.10 A panel of scientific experts chosen by the U.S. National Academy of Sciences recommended these priorities for dealing with solid waste. (Data from U.S. Environmental Protection Agency and U.S. National Academy of Sciences)

First Priority	Second Priority	Third Priority
Pollution and Waste Prevention by Producers	**Pollution and Waste Prevention by Consumers**	**Waste Management**
Change production processes to reduce use of harmful chemicals	Reduce use of harmful chemicals and products in general	Treat waste to make it less harmful
Find less harmful substitutes for harmful chemicals in production	Repair and reuse as much as possible	Incinerate solid wastes
Reduce materials used for production and packaging	Recycle nonreusable items and compost food waste	Bury wastes in sanitary landfills
Make products more durable, reusable, repairable, and recyclable	Buy and use recyclable and reusable products	Release treated waste into environment for dispersal or dilution

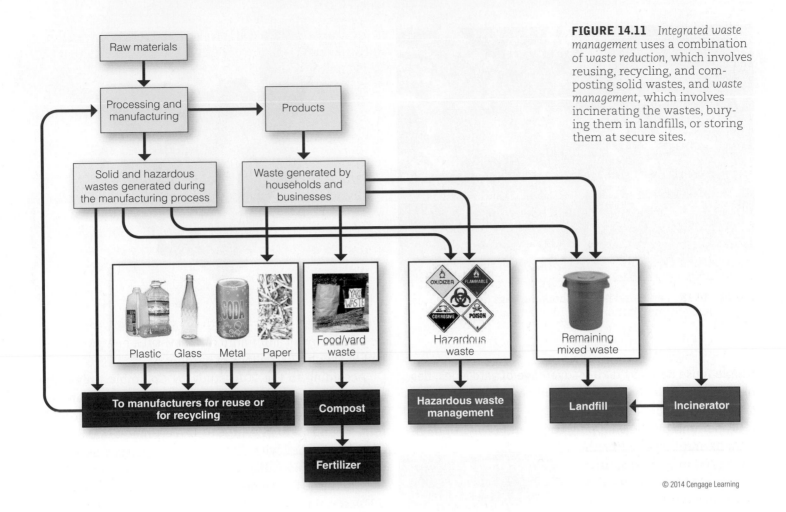

FIGURE 14.11 *Integrated waste management* uses a combination of *waste reduction*, which involves reusing, recycling, and composting solid wastes, and *waste management*, which involves incinerating the wastes, burying them in landfills, or storing them at secure sites.

© 2014 Cengage Learning

Reduce, Reuse, and Recycle

It's easy to talk about waste reduction, but making it happen involves some real challenges. We can handle some of these challenges by applying the 3 Rs of resource use: reduce, reuse, and recycle.

To **reduce** involves consuming less, which automatically reduces one's outputs of wastes and pollution. This ultimate waste-reduction strategy involves asking a very important question when deciding whether or not to purchase an item: Do I really need it, or can I get along without it? This is not always easy to decide and it is complicated by the fact that most product advertising aims to convince us that we need something.

Reuse involves using a product or material at least twice. For example, we can wash and reuse a durable travel mug (Figure 14.12a) every day instead of using paper or plastic coffee cups (Figure 14.12b) and then throwing them away. This saves us money and reduces our output of solid wastes and pollution. Most coffee shops and fast food restaurants give discounts to people who bring their own reusable cups.

Some governments are taking steps to encourage the use of refillable containers. Beverage containers that cannot be refilled and reused have been banned in Finland, in Denmark, and on Canada's Prince Edward Island. Many

a

b

FIGURE 14.12 A growing number of people are reusing their own stainless steel, plastic, or glass coffee mugs (a) at work, in coffee shops, and in school cafeterias and restaurants, instead of using throwaway paper (b) or plastic cups.

FIGURE 14.13 Paper and plastic shopping bags (a) can be much more environmentally harmful than reusable cloth bags (b).

people reuse items by taking advantage of yard sales, flea markets, secondhand stores, and online sites such as eBay and The Freecycle Network.

Paper and plastic shopping bags can be environmentally harmful. They litter the landscape in many countries (Figure 14.13a), as well as large areas of ocean waters. Plastic bags also block drains and clog sewage systems, and they can kill birds and other wild animals that try to eat them or that become entangled in them. Many people carry reusable cloth shopping bags (Figure 14.13b) instead of choosing either paper or plastic bags for carrying groceries and other purchased items home.

In Ireland, a tax of about 30¢ per plastic shopping bag has encouraged people to switch to reusable bags and has cut plastic bag litter by more than 90%. Several other countries have followed Ireland's example, reducing this major component of their high-waste economies.

To **recycle** involves collecting discarded solid materials and converting them into new materials. For example, we can crush and melt discarded aluminum cans to make new aluminum cans or other aluminum products (Figure 14.14). This costs much less and uses much less energy and raw materials than does making aluminum cans from mined aluminum. Consequently, it saves us money, makes aluminum ore resources last longer, and reduces our output of solid wastes and pollution. The EPA estimated that

NUMB3RS

45 **million**
The number of plastic bags used every hour in the United States

recycling 1 ton of aluminum cans saves the energy equivalent of about 1,700 gallons of gasoline.

According to the 2010 study by Columbia University scientists, between 1960 and 2010, the recycling rate for MSW produced in the United States increased four-fold, from around 6% to 24%. In 2010, EPA estimates of U.S. recycling rates varied with different materials such as aluminum drink cans (50%), wastepaper (49%, Figure 14.15a), steel scrap (34%, Figure 14.15b), glass bottles and jars (33%), certain plastic soft drink bottles (29%), and plastics (7%). Compared to making paper from wood pulp, recycling paper uses 64% less energy and produces 35% less water pollution and 74% less air pollution, and no trees are

FIGURE 14.14 Instead of discarding aluminum cans, we can recycle them.

FIGURE 14.15 In 2010, almost half of the wastepaper (a) and more than a third of the steel scrap (b) in the MSW produced in the United States were recycled.

cut down. Recycling 1 ton of mixed wastepaper can save the energy equivalent of about 165 gallons of gasoline.

About 95% of magazines and books produced in the United States are printed on *virgin paper* made totally from trees. In producing this textbook, the publisher used paper that was made of a mix of recycled fibers and wood from certified sustainably-grown forests. (See a description of the Cengage Learning sustainable publishing program on p. xxiv.)

Only about 7% by weight of the plastic wastes in the United States were recycled because there are many different types of plastic, some of which are difficult and expensive to separate from other plastic products and to recycle. However, progress is being made in the recycling of plastics (see *Making a Difference*, p. 397).

Studies show that the health, environmental, and economic benefits of recycling far outweigh the costs. For example, the revenues brought in by the U.S. recycling industry are much greater than those of the waste-management industry. Experts estimate that the United States could recycle 60–80% of its MSW with a combination of economic incentives (such as free pickup of recyclable and reusable items) and consumer education.

KEY idea By reusing and recycling MSW items, we can make supplies of nonrenewable resources last longer, save ourselves money, and reduce solid waste and pollution.

One form of recycling is **composting**, which involves mimicking nature's recycling of key chemicals by allowing decomposer bacteria to break down organic wastes to produce *compost* (Figure 14.16), a material that can be added to soil to supply it with plant nutrients. Compost also helps to slow soil erosion, retain water in soil, and improve crop yields.

Homeowners can compost yard trimmings, vegetative food scraps, and many other organic wastes in composting drums (Figure 14.17a, p. 396) and use the compost as a soil conditioner in gardens and flowerbeds. These wastes

FIGURE 14.16 This is a handful of compost, a soil conditioner produced by the biological decomposition of various organic materials.

FIGURE 14.17 Homeowners can make compost in a backyard composter drum (a). Compostable wastes are also recycled at community composting facilities (b).

make up about 27% of urban wastes in the United States. Some communities collect compostable organic wastes from homes and businesses and compost them on a large scale at open-air sites (Figure 14.17b) or in large closed buildings.

CONSIDER this   The city of San Francisco, California, reuses, recycles, or composts about 70% of its MSW and plans to increase this percentage to 75%. What efforts are your city or school making to reduce, reuse, and recycle materials?

Recycling is very popular, partly because it eases the consciences of people living in high-waste economies. Many people think they can help to care for the environment mainly by recycling their aluminum cans and newspapers. Recycling is an important way to help deal with the massive amount of solid wastes that we produce. However, we can cut our resource waste and our outputs of solid wastes and pollutants more effectively by reducing resource consumption and by reusing resources, because these approaches focus on preventing the production of wastes and pollution.

Why don't we do more reuse and recycling? There are three major reasons. First, most of the harmful environmental costs of producing, using, and discarding products are not included in their market prices, so the prices of disposable items are low, and throwing these things away seems cheaper to consumers. Second, reuse and recycling industries usually get fewer government *subsidies*, or payments designed to help them stay in business, than do *extractive industries*—those that extract resources from the earth to make products. Third, the prices of recycled materials are not predictable, mostly because buying goods that are made of recycled materials is not a priority for most governments, businesses, and consumers.

Some governments have encouraged reuse and recycling by giving subsidies and tax breaks to industries and individuals who reuse or recycle certain materials. They can fund these subsidies by reducing or discontinuing subsidies and tax breaks for extractive industries. Also, several city and state governments purchase products made of recycled materials to meet their own needs, thereby helping to raise the demand and lower the prices of these products. In addition, the European Union has a law that requires manufacturers of electronic products and appliances to take back and recycle or reuse these products when they are no longer useful to the consumers who bought them. The manufacturers factor the costs of these programs into the prices of their products.

Let's **REVIEW**

- What are the 3 Rs of resource use? Define and give an example of each. What are three important benefits of reusing and recycling MSW items?
- What is *composting*? Which two of the 3 Rs are the most important for reducing waste production?
- Give three reasons for why reuse and recycling are not more widespread practices.
- List three ways in which governments have encouraged the reuse and recycling of products and materials.

MAKING A difference

Mike Biddle Learned How to Recycle Plastics

Engineer Mike Biddle, while working at Dow Chemical, decided he wanted to do something about the growing problem of plastic waste. In 1994, he and a business partner, Trip Allen, founded MBA Polymers, Inc., with the goal of developing a profitable process for recycling high-value plastics.

For raw materials, MBA Polymers turned to the wastes generated in the manufacture of cars, appliances, computers, and other electronic products. The challenge was to develop an automated process for separating plastics from nonplastic waste items and then separating various types of plastic from each other. Biddle and Allen met that challenge and invented a new

recycling process that produces plastic pellets that can be used to make new products.

MBA Polymers pellets cost less than newly made plastics because the company's process uses 90% less energy than it would use to make a new plastic. Another cost saving results from the fact that they make use of a large stream of free or inexpensive raw materials in the form of discarded plastic items. Also, the greenhouse gas emissions from this process are much lower than are those produced in the process for making new plastics. With plants in California, China, and Austria, MBA Polymers is considered a world leader in plastics recycling.

Dumping or Burying Solid Waste

In many less-developed countries, solid wastes are simply dumped into pits or piled in large heaps on the land in what are called **open dumps** (Figure 14.18). These have caused a number of problems over the years, including pollution of surface waters, air, and groundwater.

In most of the world's more-developed countries, the problems of open dumps have largely been solved with the use of **sanitary landfills**—facilities in which solid wastes are compacted and covered, daily or weekly, with a fresh layer of clay or a layer of plastic foam (Figure 14.19, p. 398). This covering helps to keep the wastes dry and thus reduces leakage of contaminated water from a landfill. It also helps to reduce odors and lessens the risk of fire, a common problem in open dumps. The best of these landfills have strong double liners and systems for collecting *leachates*, the liquids that would otherwise seep out of the landfill, possibly contaminating groundwater or surface waters.

Michael Zysman/Shutterstock.com

FIGURE 14.18 Open dumps like this one are found in many locations around the world, especially near the large cities of less-developed countries.

FIGURE 14.19 This state-of-the-art *sanitary landfill* eliminates or sharply reduces environmental problems associated with older landfills and open dumps.

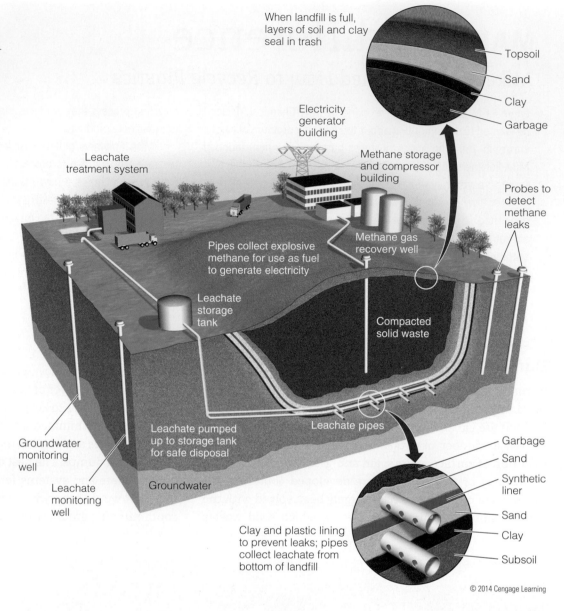

When landfill is full, layers of soil and clay seal in trash
- Topsoil
- Sand
- Clay
- Garbage

Electricity generator building

Methane storage and compressor building

Leachate treatment system

Probes to detect methane leaks

Pipes collect explosive methane for use as fuel to generate electricity

Methane gas recovery well

Leachate storage tank

Compacted solid waste

Groundwater monitoring well

Leachate monitoring well

Leachate pumped up to storage tank for safe disposal

Leachate pipes

Groundwater

Garbage
Sand
Synthetic liner
Sand
Clay
Subsoil

Clay and plastic lining to prevent leaks; pipes collect leachate from bottom of landfill

© 2014 Cengage Learning

By weight, about 80% of the MSW in Canada and 69% of that in the United States are buried in sanitary landfills. About 85% of China's rapidly growing amount of solid waste is disposed of primarily in rural areas in open dumps or in poorly designed and poorly regulated landfills that have no lining or only a thin single lining. Figure 14.20 lists the major pros and cons of using sanitary landfills to dispose of solid waste.

Burning Solid Waste

Around the world, more than 600 large **waste-to-energy incinerators** (89 in the United States) burn MSW and use the heat to boil water and make steam for generating electricity or for heating buildings. Figure 14.21 shows the major components of this type of incinerator.

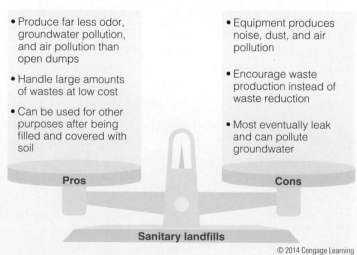

Pros
- Produce far less odor, groundwater pollution, and air pollution than open dumps
- Handle large amounts of wastes at low cost
- Can be used for other purposes after being filled and covered with soil

Cons
- Equipment produces noise, dust, and air pollution
- Encourage waste production instead of waste reduction
- Most eventually leak and can pollute groundwater

Sanitary landfills

© 2014 Cengage Learning

FIGURE 14.20 *Weighing the pros and cons* of using sanitary landfills.

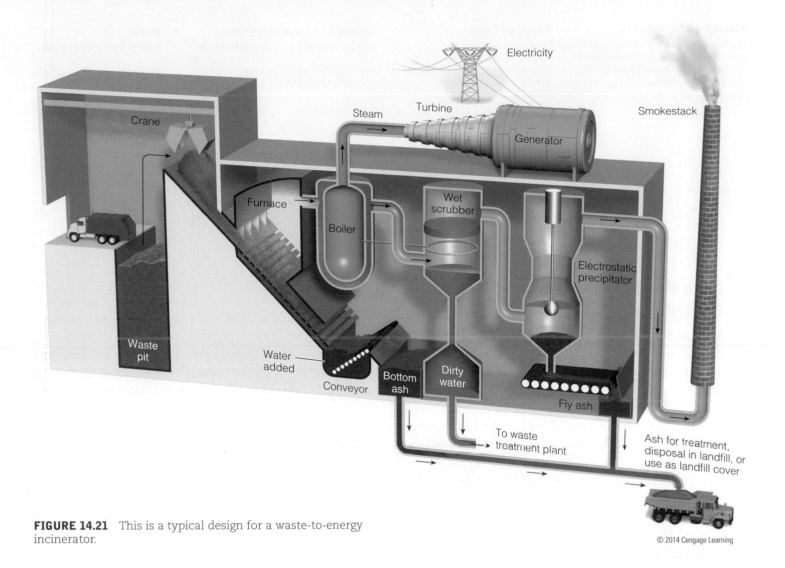

FIGURE 14.21 This is a typical design for a waste-to-energy incinerator.

© 2014 Cengage Learning

Countries vary greatly in how much they rely on this technology. For example, Germany incinerates about 34% of its MSW, while only about 7% of all MSW in the United States and 1% of that in Poland are burned. Incinerators do not turn a profit for their owners unless they are fed large volumes of trash every day. This encourages trash production and discourages waste reduction, reuse, and recycling. Figure 14.22 lists the major pros and cons of using incinerators to burn solid waste.

Let's **REVIEW**

- Distinguish between an *open dump* and a *sanitary landfill*.
- Summarize the major pros and cons of disposing of solid waste in sanitary landfills.
- What is a *waste-to-energy incinerator*?
- Summarize the major pros and cons of incinerating solid waste.

Pros
- Saves land that would have to be used for landfills
- Reduces the need to burn fossil fuels in power plants
- Sale of energy reduces cost

Cons
- Without strict pollution controls and monitoring, it releases harmful air pollutants
- Encourages waste production instead of waste reduction and recycling
- Usually difficult to build because of citizen opposition

Incinerating solid waste

© 2014 Cengage Learning

FIGURE 14.22 *Weighing the pros and cons* of incinerating solid waste.

FIGURE 14.23 The U.S. National Academy of Sciences has recommended these priorities for dealing with hazardous waste. (Data from U.S. National Academy of Sciences)

Produce Less Hazardous Waste	Convert to Less Hazardous or Nonhazardous Substances	Put in Long-Term Storage
Change industrial processes to reduce production of hazardous waste	Allow wastes to decompose naturally in temporary storage	Bury wastes in sanitary landfills
Find nonhazardous substitutes for hazardous materials in production	Incinerate wastes but prevent release of hazardous air pollutants	Inject liquid wastes into deep underground wells
Reuse or recycle hazardous materials	Treat wastes with thermal, chemical, physical, or biological processes	Pump liquid wastes into underground salt formations
Sell reusable hazardous materials as raw materials	Dilute treated wastes in air or water	Store wastes in surface impoundments

© 2014 Cengage Learning

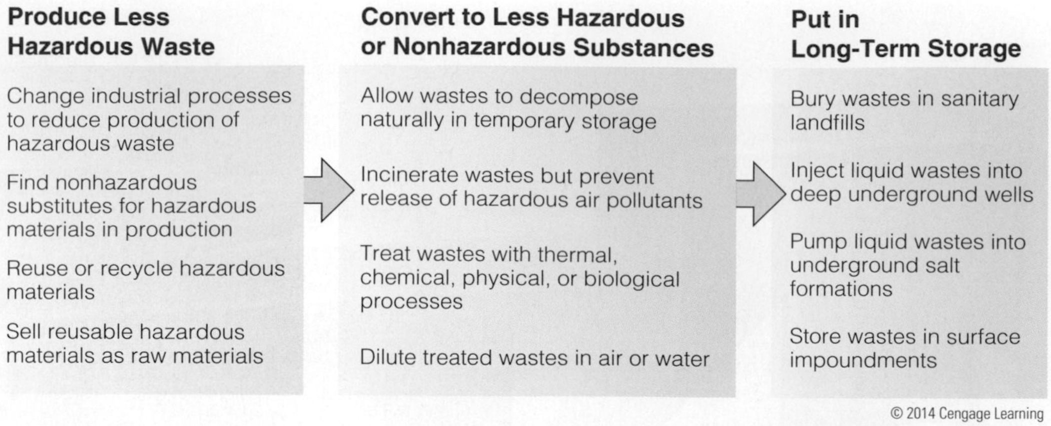

Cutting Production of Hazardous Wastes

Figure 14.23 outlines an integrated management approach for dealing with hazardous wastes, as suggested by the U.S. National Academy of Sciences. In order, the three recommended levels of priority are: produce less hazardous waste; convert as much of it as possible to less hazardous substances; and put the rest in long-term, safe storage. On a national level, only the country of Denmark follows these priorities.

Industries can use several methods to implement the number-one priority, reducing hazardous waste. They can try to find substitutes for hazardous or toxic materials, reuse or recycle those materials within industrial processes, or use them as raw materials to making other products (see the following *For Instance*).

KEY idea The order of priorities recommended by scientists for dealing with hazardous waste is to produce less of it, to reuse or recycle it, to convert it to less hazardous material, and to safely store what is left.

FOR INSTANCE...

Industrial Ecosystems that Copy Nature

For industry, a promising way to reduce wastes and pollution is to redesign manufacturing processes by copying the ways in which nature deals with wastes. Several analysts argue that industries can reuse or recycle most of the minerals and chemicals that they use, instead of burying or burning them or shipping them somewhere else.

One way for industries to do this is to set up *resource exchange networks* in which the wastes of one manufacturer can be sold as raw materials to another. In Europe, this is being done through clearinghouses in which at least one-third of all the continent's industrial hazardous wastes are recycled in exchanges. However, in the United States, only about 10% of the country's hazardous wastes are exchanged in these clearinghouses.

Industrial systems that mimic nature in this and other ways are called **industrial ecosystems**. One such system has been set up in Kalundborg, Denmark (Figure 14.24), where an electric power plant and nearby industries, homes, and farms work together to reduce the amount of wastes they produce and to save money. Kalundborg's industrial ecosystem also reduces the need to dig mineral and energy resources from the earth's crust, and it cuts the outputs of pollution and wastes that would result from processing these raw materials.

These and other industrial forms of *biomimicry* that copy nature's solutions to waste problems can provide businesses with a variety of economic benefits. For example, by encouraging recycling and pollution prevention, resource exchange systems reduce the costs of dealing with solid and hazardous wastes, controlling pollution outputs, and complying with pollution regulations. These systems also reduce the chances of expensive lawsuits related to the outputs of harmful wastes and pollutants by manufacturers. In addition, they can help companies to cut their health insurance costs by reducing their employees' exposure to hazardous and toxic materials during the manufacturing process. Copying nature in this way can also encourage companies to come up with new and less harmful products and processes that they can sell worldwide.

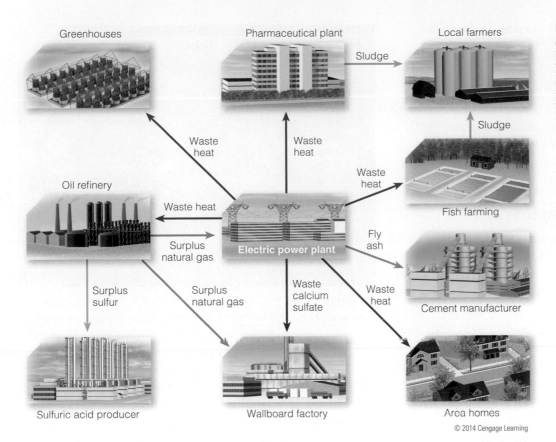

Greenhouses

Pharmaceutical plant

Local farmers

Sludge

Waste heat

Waste heat

Sludge

Oil refinery

Waste heat

Waste heat

Electric power plant

Surplus natural gas

Fish farming

Fly ash

Surplus sulfur

Surplus natural gas

Waste calcium sulfate

Waste heat

Cement manufacturer

Sulfuric acid producer

Wallboard factory

Area homes

FIGURE 14.24 This *industrial ecosystem* in Kalundborg, Denmark, produces less pollution and solid and hazardous wastes than conventional industrial parks do, and it helps its owners to save money through its resource exchange system.

© 2014 Cengage Learning

Dealing with Hazardous E-Waste

The European Union (EU) is the world's leader in reducing e-waste, which contains a number of hazardous and toxic materials such as lead and mercury. The EU has banned e-waste from landfills and incinerators, and it requires manufacturers to take back electronic products at the end of their useful lives for repair or recycling. It covers the costs of this program through a recycling tax on the electronic products sold in the marketplace.

The United States produces about half of the world's e-waste, by weight, but recycles only about 27% of it. However, this percentage is beginning to grow. Thirty-seven states now ban the disposal of television sets and computers in landfills and incinerators, and several cities and states have laws that require manufacturers to take back and recycle most electronic devices. Some U.S. electronics manufacturers have free recycling programs for consumers. Several nonprofit groups also help people to donate, reuse, and recycle their used electronic devices.

Despite such encouraging efforts, it will be difficult for recycling to keep up with the rapid growth of e-waste. Another obstacle to e-waste recycling is the fact that lots of money can be made by illegally shipping e-wastes to other countries. According to Jim Puckett, coordinator of the Basel Action Network, the best long-term solution to the e-waste problem is a *prevention* approach. It would encourage or require manufacturers to make electronic products that do not contain hazardous and toxic materials, and that are designed for easy repair or recycling.

Detoxifying Hazardous Wastes

We can use physical, chemical, and biological methods to detoxify hazardous wastes. The nation of Denmark has a model system, using these methods to detoxify three-fourths of its hazardous wastes. All hazardous wastes produced by industries and households in Denmark are collected and transported to transfer stations located throughout the country. The wastes are then sent to a central detoxification facility and, after detoxification, are buried in a carefully designed and monitored landfill.

We can also incinerate hazardous wastes in order to break them down and convert them to harmless or less harmful chemicals. This approach has the same major pros and cons as the incineration of solid waste has (Figure 14.22). However, the incineration of hazardous wastes can release other air pollutants such as toxic dioxins. It also produces a highly toxic ash that must be stored safely and permanently in landfills or in vaults designed for such storage.

Let's REVIEW

- List in order the three priorities recommended by experts for dealing with hazardous waste.

- Define and give an example of an *industrial ecosystem*.

- What are two ways to deal with hazardous e-waste? Why might these efforts not succeed in dealing with all e-waste? What is the best long-term solution to this problem?

- Summarize the pros and cons of detoxifying hazardous wastes.

Storing Hazardous Wastes

According to experts, burying wastes below or on the land's surface should be used only as the last resort after the first two priorities for dealing with hazardous wastes—waste reduction and detoxification—have been carried out (Figure 14.23).

Deep-well disposal is widely used to deal with liquid hazardous wastes. This involves using high pressure to pump the wastes into various dry and porous rock formations deep underground. These sites are often located beneath *aquifers*, or underground bodies of water that are used to supply irrigation and drinking water. Roughly 64% of the liquid hazardous wastes produced in the United States are injected into such deep wells.

A number of scientists contend that the U.S. government needs to improve the existing regulations for this type of hazardous waste disposal. They are especially concerned about the leakage of hazardous materials into groundwater supplies during and after pumping. Figure 14.25 lists the major pros and cons of deep-well disposal of liquid hazardous wastes.

Ken Sherman/Bruce Coleman USA, Inc.

FIGURE 14.26 This surface impoundment in Niagara Falls, New York, is used to store liquid hazardous wastes.

Liquid hazardous wastes can also be stored in lined *surface impoundments* such as ponds, pits, and lagoons (Figure 14.26). As the water contained in the waste evaporates, the hazardous chemicals settle and become more concentrated. However, volatile harmful chemicals can evaporate into the air and percolate into groundwater lying beneath ponds that have either no liners or leaky liners. This can result in pollution of the air, groundwater, and nearby surface waters. Figure 14.27 lists the major pros and cons of using this method.

According to EPA studies, about 7 of every 10 hazardous waste surface impoundments in the United States have no liners, and up to 90% of those sites could threaten

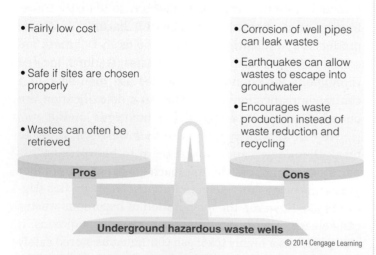

Pros
- Fairly low cost
- Safe if sites are chosen properly
- Wastes can often be retrieved

Cons
- Corrosion of well pipes can leak wastes
- Earthquakes can allow wastes to escape into groundwater
- Encourages waste production instead of waste reduction and recycling

Underground hazardous waste wells

© 2014 Cengage Learning

FIGURE 14.25 *Weighing the pros and cons* of injecting liquid hazardous wastes into deep underground wells.

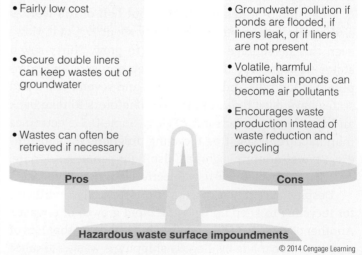

Pros
- Fairly low cost
- Secure double liners can keep wastes out of groundwater
- Wastes can often be retrieved if necessary

Cons
- Groundwater pollution if ponds are flooded, if liners leak, or if liners are not present
- Volatile, harmful chemicals in ponds can become air pollutants
- Encourages waste production instead of waste reduction and recycling

Hazardous waste surface impoundments

© 2014 Cengage Learning

FIGURE 14.27 *Weighing the pros and cons* of storing liquid hazardous wastes in surface impoundments.

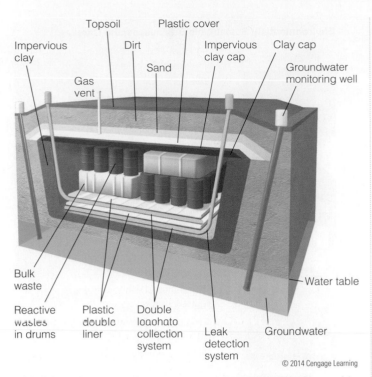

FIGURE 14.28 Hazardous wastes can be isolated and stored in a secure hazardous waste landfill.

groundwater supplies. The EPA also warns that, eventually, all liners in these impoundments are likely to leak, which could then lead to pollution of groundwater.

Some highly toxic materials such as mercury and lead cannot be destroyed, detoxified, or safely buried. The best way to deal with these substances is to stop or sharply reduce our use of them, for example, by finding less hazardous substitute materials, as has been done in many instances. Another option is to store these wastes in metal drums or other secure containers in carefully designed and well-monitored *secure hazardous waste landfills* (Figure 14.28). This costly method is not widely used.

Let's REVIEW

- Summarize the major pros and cons of injecting liquid hazardous wastes into deep underground wells.
- Summarize the major pros and cons of storing liquid hazardous wastes in ponds, pits, and lagoons.
- Describe a secure hazardous waste landfill.

Hazardous Waste Regulation

Only about 5% of the hazardous wastes produced in the United States—those judged to be the most hazardous—are regulated under the Resource Conservation and Recovery Act (RCRA), passed in 1976 and amended in 1984. Under this law, the EPA sets standards for managing several types of hazardous waste and then issues permits, each of which allows a company or other permit holder to produce and dispose of a certain amount of wastes. Permit holders must track their hazardous wastes and prove to the EPA that this tracking record is accurate. The other 95% of the hazardous wastes in the United States are not regulated because of the very high costs of doing so.

NUMB3RS

95%
The percentage of all U.S. hazardous wastes that are not regulated

In 1980, the U.S. Congress passed the *Comprehensive Environmental Response, Compensation, and Liability Act*, commonly known as the CERCLA or *Superfund* program. Its purposes are to identify and clean up sites where hazardous wastes have polluted the environment, often called *Superfund sites*.

Sites that pose immediate and severe threats to human health are put on a *National Priorities List (NPL)* for the earliest possible cleanup. Under the law, the EPA is to find the polluters responsible for these sites and to require them to pay cleanup costs. When responsible parties cannot be found, the EPA is to clean up the sites using funds provided by a tax on oil and chemical companies. In 2011, about 1,300 sites were on the NPL, and more than 350 sites had been cleaned up and removed from the list.

The Superfund law had two important effects. First, it led to a sharp drop in the number of illegal hazardous waste dumpsites, because polluters had to pay for cleaning up any of those sites that they created. Second, the production of hazardous wastes declined and manufacturers began to reuse and recycle more of these wastes, because they feared lawsuits related to the harmful effects caused by the wastes.

However, facing pressure from politically and economically powerful oil and chemical companies, the U.S. Congress in 1995 decided not to renew the tax on those companies that had financed the Superfund. As a result, the Superfund had no funding as of 2010. In addition, U.S. taxpayers, not polluters, are now paying the cleanup costs at Superfund sites whenever the responsible parties cannot be found.

Some international regulations are more strict. Since 1995, a treaty known as the Basel Convention has banned all shipments of hazardous wastes from industrial countries to less-developed countries. By 2011, this agreement had been ratified by 175 countries, but not by the United

States, Afghanistan, or Haiti. This ban has helped to slow international shipping of hazardous wastes but has not eliminated this highly profitable, illegal shipping business.

In 2000, delegates from 122 countries developed a global treaty known as the Stockholm Convention on Persistent Organic Pollutants (POPs). *Persistent pollutants* break down very slowly and thus they last for a long time in the environment. The POPs treaty regulates the use of 21 widely used persistent organic pollutants that can accumulate in the fatty tissues of humans and other animals at levels hundreds of thousands of times higher than those in the general environment. By 2010, the treaty had been signed by 152 countries, but not by the United States or by Russia.

Also in 2000, the Swedish Parliament enacted a law that, by 2020, will ban all persistent organic pollutants in that country. This law also requires industries to show that the chemicals they use are not toxic to humans and other organisms. In other words, chemicals are assumed to be guilty until proven innocent, which is the reverse of the current policy in the United States and in most other countries.

Let's REVIEW

- Describe two laws that are used to regulate hazardous waste production and cleanup in the United States. How much of U.S. hazardous waste is unregulated?
- Describe the Basel Convention and the Stockholm Convention on Persistent Organic Pollutants (POPs).

Including Environmental Costs in Market Prices

A key reason that we have pollution and wastes is that the market prices of the goods and services we use do not include the harmful environmental and health costs of those products. Most ecological economists call for including these costs in market prices—a solution called *full-cost pricing* (a social science principle of sustainability, see Figure 1.28, p. 28, in Module 1). Including all costs in prices would make environmentally harmful products more expensive, thereby making consumers less willing to buy them. This would, in turn, encourage producers to find or invent less environmentally harmful products and methods of production. The end result could be an overall reduction in resource waste and pollution, and improvements in human health.

At first, this change would likely result in a loss of jobs and profits in environmentally harmful businesses, as more consumers switched to less harmful products

Environmentally Sustainable Businesses and Careers

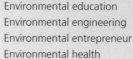

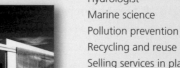

Aquaculture
Biodiversity protection
Biofuels
Climate change research
Conservation biology
Ecotourism management
Energy-efficient product design
Environmental chemistry
Environmental design and architecture
Environmental economics
Environmental education
Environmental engineering
Environmental entrepreneur
Environmental health
Environmental law
Environmental nanotechnology
Fuel cell technology
Geographic information systems (GIS)
Geothermal geologist
Hydrogen energy
Hydrologist
Marine science
Pollution prevention
Recycling and reuse
Selling services in place of products
Solar cell technology
Sustainable agriculture
Sustainable forestry
Urban gardening
Urban planning
Waste reduction
Watershed hydrologist
Water conservation
Wind energy

© 2014 Cengage Learning

FIGURE 14.29 These are some of the businesses and careers that are expected to grow and prosper in coming decades if the countries of the world can shift away from high-consumption, high-waste economies and toward more environmentally sustainable economies.

Photo credits (top to bottom): Rich Gribble/USDA Natural Resources Conservation Service; Ryan McVay/Getty Images; NOAA; Jim Tetro/National Renewable Energy Laboratory; NOAA.

and services. But jobs and profits would grow in environmentally beneficial businesses (Figure 14.29). Phasing in a shift to full-cost pricing over a decade or two would give many environmentally harmful businesses enough time to transform themselves into environmentally beneficial businesses. It would also give consumers enough time to change their buying habits in favor of more environmentally beneficial products and services.

Full-cost pricing has yet to be used widely for two major reasons. First, most producers of environmentally harmful products would have to charge more for their products and services, and they resist doing so. Second, estimating the harmful environmental and health costs of various products is generally very difficult. However, ecological and environmental economists argue that making the best possible estimates of those costs is an important step toward creating more sustainable economies and is far better than simply ignoring these costs.

Let's look at a few economic tools that we can use to phase in full-cost pricing and thereby reduce our ecological footprints.

Taxing Pollution and Wastes

According to 2,500 economists, including eight Nobel Prize winners in economics, most countries have a tax system that is backwards. It *discourages* the growth of what we want—jobs, income, and innovation—and *encourages* the growth of what we don't want—pollution, resource waste, and environmental degradation. A better approach, these economists say, would be to *lower* taxes on labor, income, and wealth, and *raise* taxes on environmentally harmful activities that produce pollution, wastes, and environmental degradation.

Proponents of such a tax shift point out three requirements for implementing it. First, taxes on environmentally harmful manufacturing processes and goods would have to be phased in over 10–20 years so that businesses can plan for the future. Second, income, payroll, or other taxes would have to be reduced to balance the increase in taxes from the first step. Finally, the poor and lower-middle class would have to have a safety net in the form of lower prices on essentials such as food and fuel, until the new pricing system became stabilized. Polls in the United States and Europe have indicated that, once such a tax shift is explained to voters, 70% of them support it.

Rewarding Environmentally Beneficial Businesses

For many years, governments have been giving subsidies and tax breaks to various industries in order to help them grow, to stimulate production, and to strengthen national economies. With the growth of environmental problems, the use of some subsidies has itself become a problem because many of these government payments are supporting environmentally harmful businesses, even though these businesses no longer need large subsidies to make a profit. One way to encourage a shift to full-cost pricing is to phase out such subsidies and tax breaks, and to phase in subsidies and tax breaks for environmentally beneficial businesses (Figure 14.29), technologies (Figure 14.30), and practices such as pollution prevention (see *For Instance*, p. 406).

Understandably, economically and politically powerful interests that receive these subsidies oppose losing

FIGURE 14.30 Many environmental scientists and economists support shifting government subsidies from mature and highly profitable oil, coal, and natural gas industries to the emerging thin-film solar cell industry (a) and the ocean-based wind power industry (b) as a way to help implement the solar energy sustainability principle (Figure 1.24, p. 26).

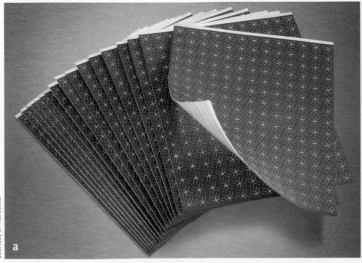

Courtesy of Nanosolar

ssuaphotos/Shutterstock.com

them. They also oppose giving subsidies and tax breaks to more environmentally beneficial businesses that compete with them.

Selling Services Instead of Things

Some economists and other analysts have proposed a new economic model in which, instead of selling certain *goods*, businesses lease or rent out the *services* that the goods provide. Using this approach, a manufacturer places its product in a home or business and then services it. Eventually the manufacturer retrieves the product, recycles or repairs it, and then leases it to another customer. The manufacturer makes more money when the leased product uses the minimum amount of materials, lasts as long as possible, and is easy to maintain, repair, or recycle.

For example, some manufacturers of air conditioners and furnaces are trying out or considering programs focused on providing customers with a comfortable indoor environment, as opposed to selling furnaces and air conditioners. In providing this service, a company could make higher profits by installing energy-efficient heating and air conditioning equipment (Figure 14.31) that is durable and

Christian Delbert/Shutterstock.com

FIGURE 14.31 This is an ultra-high-efficiency heat pump used to heat and cool a home or business.

easy to maintain, recycle, or rebuild. The company would not have to invest as much in raw materials and manufacturing and this could reduce its overall use of resources and its production of wastes.

FOR INSTANCE...

Pollution Prevention, 3M, and Profits

Some companies have had great success in reducing their output of solid and hazardous wastes into the environment. An example is the 3M Company, based in Minnesota, which makes about 60,000 different products.

In 1975, the company began its Pollution Prevention Pays (3P) program. It redesigned its equipment and processes to reduce pollution and wastes, and it focused on using fewer hazardous raw materials for manufacturing. It also identified the toxic chemicals that were generated by its processes and recycled these chemicals or sold them as raw materials to other companies.

Between 1975 and 2010, the company's 3P program prevented more than 3 billion pounds of pollution and wastes from reaching the environment. In the process, the company saved more than $1.4 billion in their costs of raw materials, waste disposal, and compliance with U.S. pollution laws and regulations, making the 3P program a superb example of how pollution prevention pays. Since 1990, a growing number of companies have adopted similar pollution and waste prevention programs.

Let's **REVIEW**

- What is full-cost pricing and why is it so important for reducing our outputs of solid and hazardous wastes?
- Explain how the following strategies can promote full-cost pricing: taxing pollution and wastes instead of profits and wages; rewarding environmentally beneficial businesses; and selling services instead of things.
- How did 3M save money by reducing its output of pollution and wastes into the environment?

Our growing problems with both solid and hazardous wastes represent an environmental issue that is critical now and will be well into the future (see *A Look to the Future*, at right). They also represent opportunities to make changes that could benefit us all in several ways.

A Look to the Future

Shifting from a High-Waste to a Low-Waste Economy

Many environmental scientists and economists as well as business leaders believe that the best long-term solution to our environmental and resource problems is to shift from our dependence on eventually unsustainable *high-consumption, high-waste economies* (see Module 1, Figure 1.22, p. 21) to more sustainable **low-consumption, low-waste economies**—economic systems that reward efficient use of resources (lower consumption), reduction of waste production (low waste), and pollution prevention (less pollution).

How can we make such a shift? Environmental scientists and economists urge us to begin by learning about how nature has sustained an amazing variety of species and habitats for at least 3.5 billion years, despite several major changes in environmental conditions during that time. Then, they argue, we can apply nature's sustainability lessons to help us design and phase in more sustainable economies during this century.

Based on those lessons, we have derived three scientific principles of sustainability (see Module 1, Figure 1.24, p. 26) and three social science principles of sustainability (see Figure 1.28, p. 28). First, the earth's life is sustained by energy from the sun, so we can get much more of our energy directly from the sun as well as from wind and flowing water, which are indirect forms of solar energy.

Second, there is essentially no waste in nature because all of the nutrients on which the earth's living organisms depend are endlessly recycled. According to experts, we could recycle up to 80% of the resources we use and thus greatly reduce our production of wastes and pollution.

Third, the earth's biodiversity provides a variety of ways for species to adjust to new environmental conditions. Even when large portions of the world's species have been wiped out by catastrophic changes, as has happened several times during the earth's history, surviving species have rebuilt the planet's biodiversity within 5 to 10 million years, by reproducing and evolving as they adjust to changing environmental conditions over time.

In addition, competition among the world's diversity of species for limited resources such as food, water, and habitat puts a limit on the population size of any species. In other words, nature has developed ways to control the long-term population growth of all species.

There is considerable evidence that we are not applying these sustainability lessons from nature and that this is leading to severe degradation of the earth's natural capital on which we and all other species depend. We get most of our energy from nonrenewable fossil fuels, and by burning these fuels, we add climate-changing carbon dioxide and other harmful pollutants to the atmosphere. We recycle less than a third of the metal, paper, and other resources that we use when we could be recycling as much as 80%.

Also, we are sharply reducing rather than conserving the earth's life-sustaining biodiversity. Wildlife experts project that during this century, at least a fourth and perhaps half of the world's known land species are likely to become extinct because of human activities that destroy their habitats and otherwise threaten their existence. Examples of such activities include deforestation (Figure 14.32), air pollution (Figure 14.33, p. 408), which is contributing to climate change, and water pollution (Figure 14.34, p. 408). Because we depend on biodiversity for our own health and well-being, this elimination of species will

FIGURE 14.32 A large area of this diverse tropical rain forest in Sabah, Borneo, has been replaced by oil palm plantations. Such deforestation threatens the existence of many species by destroying or degrading their habitats.

FIGURE 14.33 Air pollution can reduce habitats for species by damaging forests and other habitat areas.

help to make our current lifestyles and economies less sustainable.

Not only are we degrading our own life-support system, but we are doing so at a rapid rate. If we cannot make a shift to more-sustainable economies during this century, it could pose a severe threat to the human population,

economies, and lifestyles. Thus, in making the shift, we would be helping to save our own species as well as the one-fourth or more of the earth's species that are projected to go extinct during this century, largely because of harmful human activities.

Environmental scientists and economists urge us to apply the three scientific principles of sustainability to our own lifestyles and economies by *relying on solar energy*, *recycling matter*, and *preserving biodiversity*. They argue that our goal should be to develop more sustainable, low-consumption, low-waste, and low-pollution lifestyles and economies.

Making such a shift could utilize several strategies that involve the three social science principles of sustainability (see Module 1, Figure 1.28, p. 28). We could phase in *full-cost pricing* by including the environmental costs of products and services in their market prices. This would help consumers to avoid making environmentally harmful purchases. It would also help businesses to become more environmentally and economically sustainable while still maintaining their profits.

We could also focus on finding *win-win solutions* in our political processes by encouraging elected officials and ordinary citizens to work together to solve major environmental problems. This would mean finding ways to benefit the environment and human health while growing the economy. For example, economies that put a high priority

FIGURE 14.34 Solid wastes have severely polluted this river along with many of the world's lakes and oceans. Such wastes degrade aquatic habitats and threaten many species that live there.

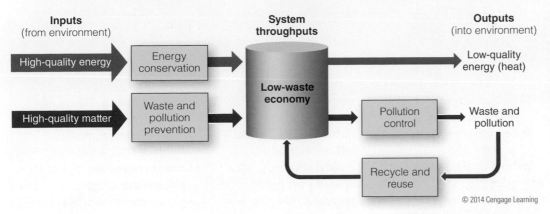

Inputs
(from environment)

High-quality energy → Energy conservation

High-quality matter → Waste and pollution prevention

System throughputs

Low-waste economy

Outputs
(into environment)

Low-quality energy (heat)

Pollution control → Waste and pollution

Recycle and reuse

© 2014 Cengage Learning

FIGURE 14.35 We can use the lessons in sustainability provided by nature to help us make the shift from high-waste economies to more environmentally sustainable *low-waste economies* that are focused more on energy conservation, pollution prevention, and waste reduction.

on recycling and reusing materials would provide jobs in the recycling industry while helping to reduce solid and hazardous waste.

Finally, we could use our educational systems and media to promote an *ethical responsibility* to leave the earth's life-support systems in as good a condition as that which we inherited, or better. Figure 14.35 is a simplified representation of how a low-consumption, low-waste economy works, based on energy conservation, waste reduction, pollution prevention, and resource recycling and reuse.

Let's **REVIEW**

- What is a *low-consumption, low-waste economy*?
- What are three important scientific principles of sustainability and how can we apply these lessons from nature to help us make a shift toward low-consumption, low-waste economies?
- What are three important social science principles of sustainability and how can we can use them to help us make this shift?

What Would You Do?

In this module, we have seen that human societies produce huge amounts of solid and hazardous wastes, largely through the high-waste economies of many of the world's more-developed countries. Several governments and businesses are trying ways to reduce this production of wastes. However, as with other environmental problems, the key to solving this one lies in the choices and habits of individuals. A growing number of people are recognizing this and making changes to reduce waste in their own daily lives.

Here are three key strategies based on the 3 Rs—Reduce, Reuse, and Recycle—that many people are trying.

Reducing unnecessary resource use.
- Buying fewer new products, getting more out of what they already own, and saving money.
- Questioning their motives for each purchase: will it satisfy a real need or can they get along without it?
- Trying to borrow or buy used items to satisfy some needs instead of buying everything new.

Reducing unnecessary resource waste.
- Reusing items and selling or giving away things that are no longer needed.
- Avoiding the use of throwaway items when reusable or refillable versions are available.
- Completing the recycling loop by buying items that are made from recycled materials to help ensure that recycled materials actually get used.

Becoming educated.
- Being aware of and learning about waste production by reading from the growing number of sources on the Internet and in publications.
- Learning about what happens to the trash they produce; finding out how much of it is recycled, composted, buried, or burned, and how these processes are completed.
- Considering full costs; learning about how goods and services they regularly use are produced and what the resulting environmental effects are, and using Internet resources that help consumers to learn about the full environmental costs of various goods and services.

KEYterms

THINKINGcritically

1. For one week, measure and record the amount of solid waste that you produce every day. Enlist one or more friends to do the same, and compare your results with those of your friends. Using this data, consider whether or not you can reduce your output of solid waste and list some ways in which you could do so.

2. Would you oppose having a waste incinerator, a hazardous waste landfill, or a deep-injection well for hazardous waste storage located in or near your community? For each of these possibilities, explain your response. If you are against having these waste-disposal facilities in your community, how do you believe the hazardous wastes generated in your community should be managed?

3. How does your school dispose of its solid and hazardous wastes? Does it have a recycling program? How well does it work? Does it have a hazardous waste collection system? If so, what does it do with these wastes? List three ways to improve your school's waste-management system and three ways in which solid waste production could be reduced at your school.

4. Explain why you would agree or disagree with the argument that industries and businesses should be charged a tax or fee for each unit of solid or hazardous waste they produce. Would you also apply such a tax or fee to solid or hazardous wastes generated by households? Explain.

5. What do you think of the idea of using low-interest loans, tax breaks, and other financial incentives to encourage industries that produce hazardous waste to reduce, reuse, recycle, or decompose this waste? Assume you are a legislator arguing for or against this proposal. Write a summary of your argument.

6. Of the nine strategies that many people are using for reducing waste in their own lives (see *What Would You Do?*), which three do you think would be the most effective in helping us to make a transition to a low-consumption, low-waste economy? Explain. Which of these strategies, if any, are you using? Which, if any, would you like to try?

LEARNINGonline

Access an interactive eBook and module-specific interactive learning tools, including flashcards, quizzes, videos and more in your Environmental Science CourseMate, accessed through **CengageBrain.com**.

15

Environmental Health Hazards

Why Should You Care about Environmental Health Hazards?

For people lucky enough to live in a more-developed country with all the resources they need for living comfortably, environmental health hazards may not seem like a big issue. Most people in these countries have fairly clean air to breathe, access to clean drinking water, and a generally safe food supply.

On the other hand, many people, especially the poor who live in less-developed parts of the world, can be exposed to a number of environmental hazards. They often go without enough food (see module-opening photo) and are subject to polluted air, contaminated water and food, and infectious diseases. For these people, environmental hazards are a daily life-threatening issue. In terms of annual deaths, the largest environmental health hazard by far is poverty, with the associated effects of increased susceptibility to infectious disease, malnutrition, and even starvation.

One reason for those of us who are relatively free of these environmental hazards to care about them is that they threaten billions of our fellow human beings. To make matters worse, certain trends are increasing the overall dangers of some environmental hazards for almost everyone. For example, infectious diseases that are transmitted from person to person can spread faster when people travel easily from one side of the globe to the other, and can thus spread through any community almost anywhere. Smoking, the second biggest health hazard in terms of annual deaths (Figure 15.1a), also exposes nonsmokers to this health hazard through second-hand smoke.

Another example of growing environmental health hazards is our steadily increasing use of fossil fuels, which exposes us to a variety of air pollutants that can cause health problems and deaths. An additional hazard for many people is indoor air pollution caused by the burning of wood and charcoal in open indoor fires (Figure 15.1b). This helps to explain why the third largest killer of humans globally is exposure to indoor and outdoor air pollution.

The burning of fossil fuels is also adding to atmospheric warming. Climate scientists warn that this will very likely change the earth's climate during this century and have potentially harmful health effects on just about everyone unless we reduce our use of fossil fuels dramatically, starting now.

Thus, environmental health hazards are a growing problem for our own generation and for future generations. In this module, we introduce you to the major types of environmental health hazards, and we consider the harm they can cause as well as some possible ways to prevent or reduce our exposure to such hazards.

What Do You Need to Know?

Types of Health Hazards

In this module, we discuss four major types of health hazards:

- **Biological hazards** are threats from organisms that can cause disease in other organisms. These organisms,

FIGURE 15.1 These young people in Russia (a) are taking a cigarette break during a party. They may not know that smoking tobacco and exposure to indoor air pollution from open cooking and heating fires (b) together kill about 7 million people every year—an average of more than 19,000 per day—and cause serious health problems for hundreds of millions of other people.

Kuzma/Shutterstock.com

Joerg Boethling/Peter Arnold, Inc.

FIGURE 15.2 The *Anopheles* mosquito spreads malaria by transmitting a parasite through its bloodstream to the bloodstreams of people it bites.

called **pathogens**, include tiny bacteria, viruses, and parasites that can only be seen with a microscope. Some of these diseases, such as malaria, are spread by insect bites that transmit the pathogens (Figure 15.2).

- **Chemical hazards** come from harmful chemicals in the air, water, and soil, and they can also be found in our food and in manufactured products. Some of these hazards come from natural sources, such as volcanic eruptions, but most come from human sources such as indoor cooking and heating fires (Figure 15.1b), interior

FIGURE 15.3 Air pollution from coal-burning power plants (a) and from motor vehicles (b) is a major source of chemical health hazards.

building materials, factories, coal-burning power plants (Figure 15.3a), motor vehicle exhausts (Figure 15.3b), and various types of wastes (Figure 15.4, p. 414).

- **Cultural hazards** are threats that develop within human societies and they include poverty (see module-opening photo), war and other forms of conflict, motor vehicle accidents, and unsafe working conditions.
- **Lifestyle health hazards** are those that originate in our own choices and habits. Examples include smoking tobacco, using heroin and other addictive drugs, eating unhealthy foods, and not getting enough exercise.

In this module, we focus on environmental hazards that affect humans. However, other animal species also face health and environmental threats from human activities. These threats result from a combination of the growing size of the human population and the rising rates of resource consumption per person. Together, these two factors are rapidly expanding our total and per capita ecological footprints (see Module 1, Figure 1.23, p. 24), which threaten all life on earth.

Now, let's look more closely at some of the major environmental hazards we face.

Infectious Diseases

Pathogens such as bacteria, viruses, and parasites survive by invading the bodies of other organisms and reproducing within their hosts' cells and tissues. An **infectious disease** is any illness that is caused by the invasion of a body by a pathogen. Examples are the flu (influenza), measles, malaria, and tuberculosis (TB).

FIGURE 15.4 This mountain lake has been polluted by various forms of solid waste.

An infectious disease that can be transmitted from one person to another is called a **transmissible disease** (or a *contagious* or *communicable disease*). Examples are influenza and TB, which can be transmitted directly through the air from one person to another. Malaria, on the other hand, is spread only by a certain species of mosquito (Figure 15.2)

FIGURE 15.5 Infectious diseases can be spread along several pathways.

that first bites a person infected with malaria and then bites others, thereby spreading the malaria pathogen from person to person.

Figure 15.5 shows the major pathways for infectious diseases—namely air, water, food, and body fluids such as blood, urine, feces, and mucus droplets sprayed by sneezing and coughing. Another major pathway is biting organisms such as mosquitoes, ticks, and fleas. When an infectious disease spreads widely in any large area, it is called an *epidemic*. When an epidemic spreads around the world, it is called a *pandemic*.

Nontransmissible or **noninfectious diseases** are those illnesses that do not spread from person to person and are not caused by pathogens but are instead caused by genetic, environmental, or lifestyle factors. Examples are heart disease, stroke, cancers, and arthritis.

Let's REVIEW

- Why should we care about environmental health hazards?
- Describe the four major types of health hazards covered in this module and give an example of each.
- Define *pathogen* and give three examples.
- Define and distinguish among *infectious diseases, transmissible diseases,* and *nontransmissible,* or *noninfectious, diseases,* and give an example of each.

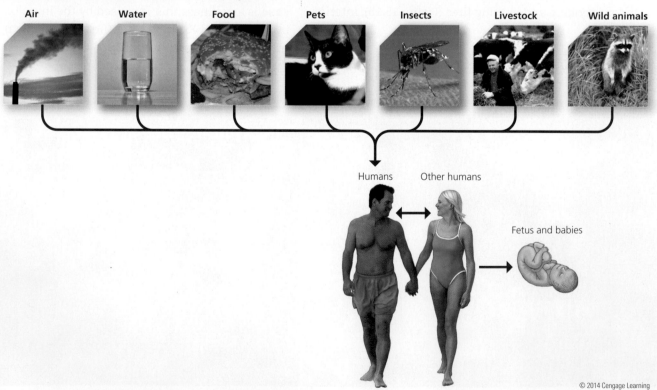

Toxic Chemicals

The presence of hazardous, or toxic, chemicals in the environment is a major environmental issue. A **toxic chemical** is one that can cause temporary or permanent harm or death to humans and other species. According to the U.S. Environmental Protection Agency (EPA), the five most threatening chemicals are arsenic, lead, mercury, vinyl chloride (used to make certain plastics), and polychlorinated biphenyls (PCBs). However, there are thousands of other chemicals that can threaten human health.

We can group toxic chemicals into several major categories according to how they affect human health. In this module, we will look at four of these groups. The first group is **carcinogens**, chemicals that can cause or promote *cancer*—a disease in which the body's cells malfunction and multiply out of control, creating tumors that can cause severe illness and death. After one is first exposed to a carcinogen, it can take 10 to 40 years for the signs of cancer to be detected. Chemicals that have been found to be *carcinogenic* include arsenic, benzene, chloroform, formaldehyde, PCBs, vinyl chloride, and certain chemicals in tobacco smoke.

The second group consists of **mutagens**, chemicals that can cause mutations, or changes, to the DNA molecules in the cells of our reproductive systems. Some mutations can lead to cancers and other serious health problems. Examples of mutagens are benzene, nitrous acid, bromine, and sodium azide (found in many car airbag systems).

The third group of toxic chemicals is **teratogens**, chemicals that can harm an embryo or fetus and cause birth defects. For example, ethyl alcohol found in alcoholic drinks can act as a teratogen. Women who drink alcohol during pregnancy risk having babies with low birth weight and certain physical, developmental, behavioral, and mental problems. (Smoking presents similar risks for pregnant women.) Examples of other teratogens include benzene, cadmium, formaldehyde, lead, mercury, PCBs, phthalates, and vinyl chloride. These chemicals have been known to leak from rusting toxic waste storage drums (Figure 15.6) and to get into groundwater used as sources of drinking water.

The fourth group of toxic chemicals is called **neurotoxins**, chemicals that can harm the nervous systems of humans and other animals. Exposure to these chemicals has been linked to learning disabilities, developmental delays, and attention deficit disorder. Acute responses to neurotoxin exposure can include paralysis and death. Examples of neurotoxins are mercury, arsenic, lead, and certain pesticides.

There are hundreds of ways in which we can be exposed to toxic chemicals. Figure 15.7 (p. 416) shows a few of these potential pathways for such chemicals.

Lisa Rivali/Shutterstock.com

FIGURE 15.6 Rusting toxic waste storage drums can leak hazardous chemicals into water supplies.

FIGURE 15.7 Toxic chemicals move through the environment on several different pathways.

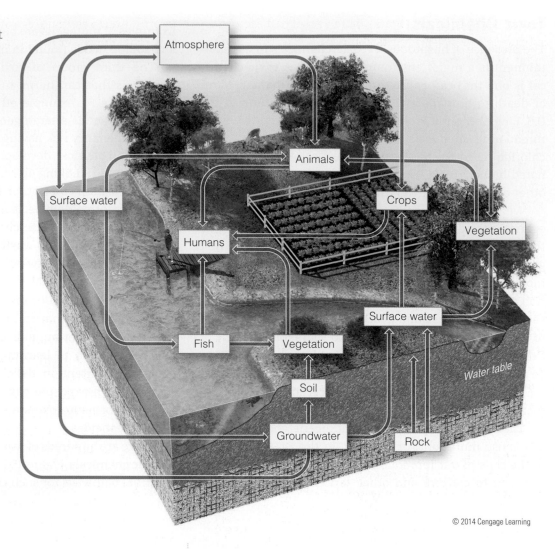

Atmosphere

Animals

Surface water

Crops

Vegetation

Humans

Fish

Vegetation

Surface water

Soil

Water table

Groundwater

Rock

© 2014 Cengage Learning

Let's REVIEW

- What is a *toxic chemical*?
- Define and distinguish among *carcinogens*, *mutagens*, *teratogens*, and *neurotoxins* and give an example of each.

Determining the Toxicity of Chemicals

Finding out what chemicals are toxic is a complex and difficult job. The same is true of determining any chemical's **toxicity**—the measure of how harmful a substance is, based on its tendency to cause injury, illness, or death to a living organism. The field of science in which people study the harmful effects of chemicals on humans and other organisms is called **toxicology**.

One of the basic rules of toxicology is *any chemical can be harmful if taken in a large enough quantity*. Many people tend to believe that all natural chemicals are safe and all synthetic chemicals are harmful. In fact, there are synthetic chemicals in a number of the foods we eat and in

the medicines we take, and they are quite safe if used as intended. On the other hand, many natural chemicals, including mercury and lead, are extremely toxic.

KEY idea

Any chemical can be harmful if ingested, inhaled, or absorbed in a large enough quantity.

Determining the quantities at which chemicals become harmful is the work of *toxicologists*. It is challenging work, partly because there are many variables involved and it is often difficult to establish that a particular harmful effect is caused by a specific chemical and not by some other variable or chemical in the environment.

Another difficulty is that numerous factors can affect the level of harm caused by a chemical. One is its *solubility*, or how quickly and completely the chemical will dissolve in another substance such as water. A second factor is a substance's *persistence*, or its ability to last in the environment without breaking down into simpler chemicals. A third factor is *biological magnification*, in which the concentrations of a toxic chemical increase as it moves through a food chain. For example, if the toxic pesticide DDT is present in tiny organisms that live in a body of water, fish that eat those organisms will ingest DDT. As bigger fish eat smaller fish, the DDT moves up through this food chain and becomes magnified in animals at the higher levels (Figure 15.8).

When determining the toxicity of a chemical, one key measure that toxicologists consider is the **dose**, or the amount, of a harmful chemical that a person has ingested, inhaled, or absorbed through the skin. Another measure, called the **response**, is the health effects caused by exposure to a chemical. There are different types of responses. One is called an *acute effect*, an immediate or rapid harmful reaction to an exposure, ranging from dizziness and nausea to death. Another type of response is called a *chronic effect*, a permanent or long-lasting illness such as lung damage or cancer resulting from exposure to a harmful substance.

Toxicologists use laboratory experiments to determine the toxicity of various chemicals. To test a chemical, these scientists often expose a population of live laboratory animals, usually mice or rats, to measured doses of the chemical under carefully controlled conditions so as to keep all other variables unchanged. They determine the *lethal*, or deadly, dose for the test animals and then use this information to estimate the amount that could harm or kill a human, based on physical differences between the test animal's body and the human body.

It is difficult to estimate the toxicity of just one chemical, and even harder to evaluate how mixtures of two or more potentially toxic substances might interact to increase or decrease their combined toxicity. It requires numerous experiments to isolate various factors and evaluate more than one substance at a time.

Outside of the laboratory, toxicologists also use *case reports*, usually made by physicians, to get information about human responses to potentially toxic chemicals. These reports can involve accidental poisonings, drug overdoses, suicide attempts, and industrial accidents. Another source of information is *epidemiological studies* that compare the health of people who have been exposed to a particular chemical with the health of a similar group of people not exposed to the chemical. For example, scientists often study the health of people whose drinking water source has been contaminated by industrial pollution and compare it to the health of people who use an unpolluted source of drinking water.

Let's REVIEW

- Define *toxicity*. What is *toxicology*? What is a basic rule of toxicology that applies to all chemicals, both natural and synthetic?
- Name and briefly describe three factors that help to determine the level of harm caused by any chemical.
- Define *dose* and *response*. What are two general types of responses?
- How do toxicologists working with lab animals estimate the dose of a chemical that can be harmful to humans?
- Outside of the laboratory, what are two ways in which toxicologists gather information about chemicals?

Cultural and Lifestyle Hazards

Cultural hazards are often closely related to biological and chemical hazards. For example, poverty-related health problems together make up the leading cause of death in

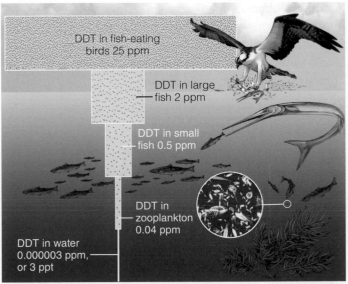

DDT in fish-eating birds 25 ppm

DDT in large fish 2 ppm

DDT in small fish 0.5 ppm

DDT in zooplankton 0.04 ppm

DDT in water 0.000003 ppm, or 3 ppt

© 2014 Cengage Learning

FIGURE 15.8 This diagram represents an actual case of biological magnification of DDT that took place in Long Island Sound off the coast of New York State.

the world, and this is a prime example of a cultural hazard. In some less-developed parts of the world where poverty runs rampant, people are crowded together by the thousands in slums with little in the way of clean water and a safe food supply (Figure 15.9). In these situations, people are especially susceptible to water-borne diseases (a biological hazard) and indoor air pollution from open cooking fires (Figure 15.1b) or leaky stoves burning wood, coal, or animal dung (a chemical hazard).

Other cultural health hazards commonly arise from unsafe working conditions. For example, in China, India, Pakistan, and some African countries, people have set up electronic waste recycling centers where workers, many of them children, dismantle computers, cell phones, and other discarded electronic devices, usually shipped from more-developed countries (Figure 15.10). The workers dismantle, burn, and treat the e-waste with acids to recover valuable metals and reusable parts. These recycling operations present great dangers to the workers' health, but in poverty-stricken countries, people often have to settle for this dangerous work to help support their families. Thus, the overall health hazard from these working conditions could be considered largely a cultural hazard.

KEY idea Poverty is the world's leading cause of numerous health problems and deaths, in part, because it puts people at greater risk from biological and chemical health hazards.

Peter Essick/Aurora Photos/Alamy

FIGURE 15.10 These children in Karachi, Pakistan, work under hazardous conditions to disassemble electronic waste for reuse and recycling.

Lifestyle choices can also be hazardous to human health. An obvious example is smoking, which we examine later in this module. Other such choices are eating foods high in saturated fat, getting a deep tan in the sun or in a tanning booth, having unprotected sex, and texting or talking on a phone while driving.

Risk Assessment

We can evaluate hazards in our lives in terms of **risk**—the chances of suffering injury, disease, death, economic loss, or damage from any particular hazard.

All of us take risks every day. Every time we drive or ride in a car, eat high-fat foods, drink alcoholic or sugary drinks, smoke or spend time in an enclosed space with a smoker, and lie in the sun or use a tanning booth, we are assuming some level of risk. The challenge is to be aware of how serious each of the risks we face is, to decide whether the benefits of risky activities are worth the risks, and to make careful decisions based on this information.

A.S. Zain/Shutterstock.com

FIGURE 15.9 The urban poor suffer from diseases and other health problems because they typically live in crowded slums or shantytowns with little or no access to clean water.

Risk Assessment

Hazard identification
What is the hazard?

Probability of risk
How likely is the event?

Consequences of risk
What is the likely damage?

Risk Management

Comparative risk analysis
How does it compare with other risks?

Need for risk reduction
How much should it be reduced?

Risk reduction strategy
How will the risk be reduced?

Financial commitment
How much money should be spent to reduce risk?

© 2014 Cengage Learning

FIGURE 15.11 Risk assessment is a necessary part of risk management. These two sets of skills can help each of us to live our lives with less risk. They also help environmental scientists to advise governments and corporations when making decisions that affect environmental and human health.

We can think of any particular risk in terms of **probability**—a measure of how likely it is that we will suffer some harm from any particular hazard. For example, a scientist might say that a lifetime smoker faces a probability of 1 in 250 of developing lung cancer from smoking one pack of cigarettes per day. This means that 1 of every 250 people who smoke a pack of cigarettes every day will very likely develop lung cancer over a typical lifetime.

Risks are estimated through a process of **risk assessment**, the scientific process of using statistical methods to estimate how much harm a particular hazard can cause to human health or to the environment. By estimating the probabilities of risks, we can compare them and set priorities for managing or avoiding them. This process of deciding whether and how to reduce or avoid risks, and at what cost, is called **risk management**. Figure 15.11 shows how risk assessment and risk management are related.

Because the news and entertainment media tend to sensationalize some risks more than others, many people worry about the highly unlikely possibilities of some minor risks such as being killed by a shark or a python. Many also ignore the high probabilities of harm from more common major risks such as driving a motorcycle or a car.

Let's **REVIEW**

- Describe two health hazards related to poverty.
- List four lifestyle choices that can be hazardous to human health.
- Define and distinguish between *risk* and *probability*.
- Define and distinguish between *risk assessment* and *risk management*. How are the two related?

What Are the Problems?

The Continuing Threat of Infectious Disease

While many infectious diseases have been controlled by the use of medicines and by improvements in living conditions, they are still major health threats to billions of people, especially in less-developed countries. The World Health Organization (WHO) estimates that the seven most threatening infectious diseases, some of which are caused by bacteria, others by viruses, kill about 26,000 people every day (Figure 15.12).

Bacterial Infections

Bacteria are single-celled microorganisms that live in soil, water, organic matter such as animal waste, and the bodies of plants and animals. Most of these microbes, including those that live in our intestines and help us to digest food, are beneficial. Indeed, without the help of trillions of bacteria, we would not be alive. However, some bacteria can cause illnesses such as tuberculosis (see *For Instance*, p. 421), pneumonia, and staph infections.

Harmful bacteria can enter our bodies in several ways. We can inhale them, ingest them with food or drink, or bring them in through kissing and sexual contact. Our

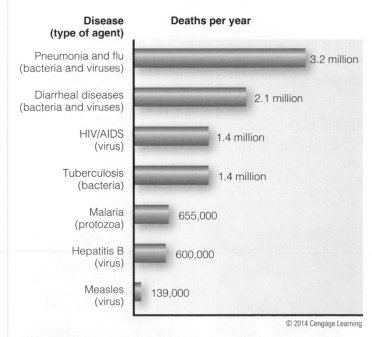

Disease (type of agent)	Deaths per year
Pneumonia and flu (bacteria and viruses)	3.2 million
Diarrheal diseases (bacteria and viruses)	2.1 million
HIV/AIDS (virus)	1.4 million
Tuberculosis (bacteria)	1.4 million
Malaria (protozoa)	655,000
Hepatitis B (virus)	600,000
Measles (virus)	139,000

© 2014 Cengage Learning

FIGURE 15.12 Each year, the seven most threatening infectious diseases, represented in this graph, kill an estimated 9.5 million people—most of them poor people in less-developed countries. (Data from the World Health Organization, 2011)

immune systems help us to fight off invading harmful bacteria by producing white blood cells in the marrow of our bones. Mature white blood cells then travel in the blood to the spleen and lymph nodes, where they are stored and available for fighting harmful bacteria. However, if the bacteria can multiply and spread fast enough in our bodies, they can overcome a weakened immune system or even a healthy one.

We help our bodies fight disease-causing bacteria with medicines called *antibiotics*. However, we face a growing problem resulting partly from the overuse of antibiotics. That is, many disease-carrying bacteria have developed **genetic resistance** to widely used antibiotics, which means that their populations have changed genetically over time to make them able to withstand the antibiotics that were designed to kill them. This is because bacteria can reproduce at an astonishing rate. For some species, a single bacterium can initiate the reproduction of well over 16 million offspring within 24 hours. A percentage of the individuals in such a large population have genes that make them resistant to the effects of various antibiotics (Figure 15.13).

While bacteria can produce a new generation in 20 minutes, it can take 10 years or more for scientists to develop a new antibiotic. When an antibiotic is widely used, bacteria can develop genetic resistance to it within several years. So in the long run, we are highly unlikely to win any battle against rapidly reproducing harmful bacteria. All we can do is hold the bacteria off temporarily with currently available antibiotics while we look for new chemical weapons.

The growing problem of genetic resistance to antibiotics is perhaps best represented by the bacterium known as *methicillin-resistant Staphylococcus aureus*, or MRSA (usually pronounced "MER-sa"). This bacterium has become resistant to most common antibiotics. It can cause severe pneumonia, flesh wounds that grow and do not easily heal, and a quick death if it infects the bloodstream. Every year, MRSA infections play a key role in the deaths of about 19,000 people in the United States.

Other factors help to increase genetic resistance in certain bacteria. One of these factors is international travel and trade, which helps to spread bacteria around the globe. Another is the overuse of antibiotics. Several studies have found that at least half of all antibiotics routinely prescribed by doctors are not necessary for treating the illness for which they are prescribed. In some countries, antibiotics are available without a prescription, which allows people to overuse them further. The widespread use of antibiotics in animal feed used in livestock feedlots has also contributed to bacteria becoming resistant to many of those antibiotics, some of which are also intended for human use. The growing use of antibacterial soaps and cleansers is another factor that can promote genetic resistance in bacteria.

Because of the combination of these factors, every major disease-causing bacterium now has strains that are genetically resistant to at least one of the roughly 200 antibiotics used to treat bacterial infections. This problem is especially troublesome in hospitals. According to the U.S. Centers for Disease Control and Prevention (CDC), bacterial genetic resistance is thought to play a role in the deaths of at least 100,000 of the 1.7 million people per year who become infected with harmful bacteria while staying in U.S. hospitals. The situation is much worse in hospitals in some other countries.

KEY idea Rapidly multiplying infectious bacteria are becoming genetically resistant to a growing number of widely used antibiotics.

FIGURE 15.13 This diagram illustrates how genetic resistance to an antibiotic can develop in a population of bacteria.

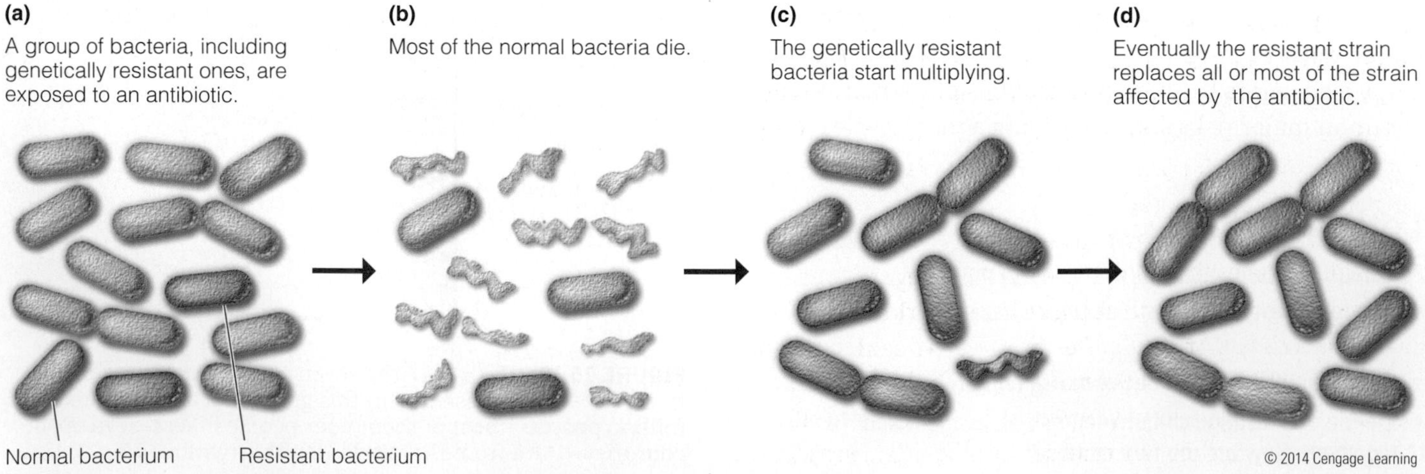

(a) A group of bacteria, including genetically resistant ones, are exposed to an antibiotic.

(b) Most of the normal bacteria die.

(c) The genetically resistant bacteria start multiplying.

(d) Eventually the resistant strain replaces all or most of the strain affected by the antibiotic.

Normal bacterium   Resistant bacterium

© 2014 Cengage Learning

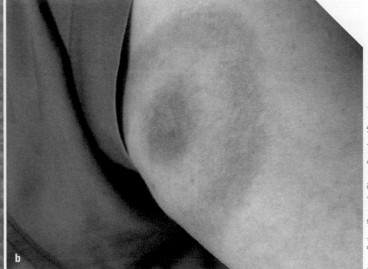

FIGURE 15.14 This deer tick (a) carries the Lyme disease bacterium from mice and from white-tailed deer to humans, which usually results in a growing, circular red "bull's-eye" rash (b) at the site of the tick bite.

Another infectious disease caused by a bacterium is Lyme disease. This pathogen can move from wild deer and field mice through certain species of hard-bodied ticks to humans (Figure 15.14). The disease does not affect the mice or deer, but it can be severely debilitating to humans. People who spend time in wooded areas, including hikers and suburban dwellers, are more likely to come into contact with these ticks and thus are more at risk of getting Lyme disease.

FOR INSTANCE...

Tuberculosis—A Global Threat

Tuberculosis (TB) is an extremely contagious bacterial disease that usually affects the lungs but can affect other parts of the body such as the kidneys, bones, skin, and joints. Its symptoms include low-grade fever, night sweats, fatigue, and a persistent cough. If left untreated, people with TB usually die when the TB bacteria destroy their lung tissue (Figure 15.15).

Tuberculosis was once under control in most of the more-developed countries. Since 1990, however, it has been spreading rapidly again throughout the world. People who are malnourished or who suffer from poor health are especially susceptible to TB. According to the WHO, tuberculosis now infects about 9 million people per year and kills an estimated 1.4 million—or one death every 22 seconds, on average. More than 8 of every 10 of these deaths occur in less-developed countries.

One reason for the rapid spread of this disease is that many TB-infected people do not appear to be sick, and about half of them do not

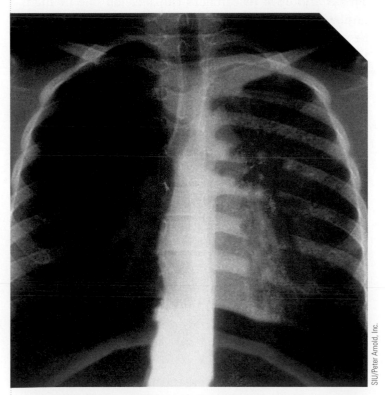

FIGURE 15.15 The red areas in this chest X-ray show lung damage caused by the tuberculosis bacterium.

know they are infected and thus can easily infect other people. Another reason is that most strains of the TB bacterium have developed genetic resistance to most of the antibiotics used to fight it. Another problem affecting TB's spread is that there are too few TB screening and control programs, especially in less-developed countries where the disease is spreading most rapidly. Also, greater population densities in growing cities, along with the rise of global air travel, have greatly increased the person-to-person contacts that spread the disease.

TB can be treated with the use of four inexpensive drugs that cure 90% of the cases of active TB. However, to slow the spread of TB, doctors must identify and treat infected people as early as possible, which is a difficult challenge. Another problem is that the drugs must be taken every day for 6 to 8 months. However, because the symptoms of TB disappear after a few weeks of drug treatment, many infected people feel better and stop taking the drugs before they are completely cured. This allows the disease to recur in a strain that can be more resistant to the drugs used to treat those patients, who then continue spreading it to other people.

We now face a new threat from TB—an apparently incurable form known as *multidrug-resistant TB*. According to the WHO, each year there are about 490,000 new cases of this form of TB resulting in about 120,000 deaths. Unless and until researchers can develop one or more antibiotics to kill this microbe, its victims will have to be isolated from the rest of society for the remainder of their lives. Carriers of this pathogen who have not been diagnosed threaten everyone with whom they come in contact.

Let's **REVIEW**

- List the five most threatening infectious diseases.
- Define *bacteria* and describe how harmful bacteria can enter our bodies.
- Define *genetic resistance* and explain why it is a problem. Explain how it occurs.
- What is tuberculosis, and what are its effects? How can it be treated?

Viral Infections: HIV and Flu

Viruses are simple microorganisms that insert themselves into living cells in order to reproduce. Viral diseases kill millions of people every year. Antibiotics do not affect

viruses, and scientists have developed vaccines that can prevent only a few viral diseases, two of which are polio and smallpox.

One deadly viral infection, that of the *human immunodeficiency virus* (HIV), has become a major global health threat. Every year, this virus infects about 2.7 million people (43,000 in the United States)—half of them 15–24 years old. HIV is transmitted from one person to another through unsafe sex, sharing of needles by drug users, and exposure to infected blood. Also, infected mothers can pass the virus on to their children before or during birth. There is no vaccine to help prevent an HIV infection.

HIV was first identified in 1981, and since then, has spread rapidly throughout the world. In 2010, according to the WHO, about 34 million people worldwide (about 1.1 million in the United States) were living with HIV.

HIV does not directly kill the people it infects, but it cripples the human body's immune system, which makes its victims vulnerable to potentially deadly infections such as TB and certain forms of cancer. This weakened immune system condition is known as *acquired immune deficiency syndrome (AIDS)*. At least half of all HIV-infected people develop AIDS within 10 to 15 years after they are infected. During this long period, HIV victims can spread the virus without even knowing they themselves are infected.

There is no cure for AIDS and because of their weakened immune systems, most AIDS victims eventually die from some infectious disease, many of them from TB. Antiviral drugs can slow the progress of AIDS, but they are expensive. Each year, AIDS kills about 1.4 million people worldwide and about 17,000 in the United States. Worldwide, AIDS is the leading cause of death among persons 25–44 years old.

CONSIDER this

- Between 1981 and 2010, more than 29 million people (617,000 in the United States) died of AIDS-related diseases.
- Every year, AIDS claims about 1.4 million more lives, worldwide—an average of about 160 deaths every hour.

Each year, various strains of the easily transmitted *influenza* or *flu virus*, along with pneumonia often caused by the infection, kill 3.2 million people, on average, according to the WHO. Death tolls vary from year to year. According to the CDC, the annual death toll from flu in the United

States has ranged widely. Between 1976 and 2006, CDC estimates varied from 3,000 to 49,000 deaths per year, most of them from pneumonia caused by flu.

Parasitic Infections

A **parasite** is an organism that lives on or in, and usually does harm to, a living plant or animal, called its host. Some parasites such as tapeworms live inside their hosts. Sea lampreys are parasites that attach themselves to the outsides of their hosts, while ticks and fleas move from one host to another.

The biggest parasitic threat, worldwide, is *malaria*, a deadly disease that has ravaged many tropical countries in the world, especially less-developed African countries (Figure 15.16). In 2010, according to the WHO, there were about 225 million people suffering with malaria and there were an estimated 655,000 deaths from the disease. More than 80% of these deaths occurred in sub-Saharan Africa—most of them among children younger than age 5. Scientists project that, as the earth's atmosphere warms during

FIGURE 15.16 Almost half of the people in the world live in areas where malaria is prevalent, most of which are tropical. (Data from the World Health Organization and U.S. Centers for Disease Control and Prevention)

this century, malaria could spread to areas with temperate climates such as the United States.

CONSIDER this The WHO estimates that malaria kills about 75 people per hour, and that over the course of human history, malaria has killed more people than the number killed in all the wars ever fought.

Malaria is caused by four different species of parasites, called *Plasmodium parasites*, which are carried from person to person by about 60 species of *Anopheles* mosquitoes (Figure 15.2). The mosquitoes pick up the parasite from infected persons. Later, when one of these mosquitoes bites an uninfected person, the parasites can move into that person's bloodstream, where they destroy red blood cells. The parasites eventually move into the victim's liver where they multiply. Malaria can also be transmitted by blood transfusions and through contaminated hypodermic needles.

An infected person is prone to intense fever, chills, sweats, severe abdominal pain, and sometimes vomiting, headaches, and extreme weakness. Victims are also much more susceptible to other diseases. In addition, many of

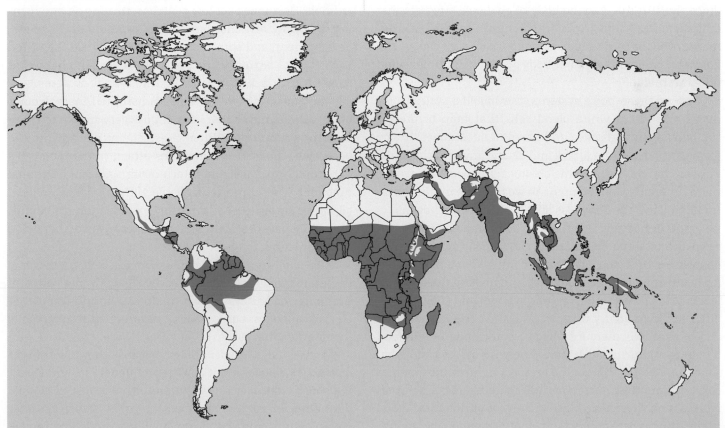

© 2014 Cengage Learning

the children who get malaria and survive suffer brain damage or impaired learning ability.

There was a war on malaria during the 1950s and 1960s, when swamplands and marshes where mosquitoes bred were drained or sprayed with insecticides. Also, health care workers gave victims drugs to kill the parasites in their bloodstreams. This battle was largely successful for awhile. However, since 1970, malaria has come back, largely because a majority of the species of the *Anopheles* mosquito have become genetically resistant to most insecticides. At the same time, the *Plasmodium* parasites have become genetically resistant to the drugs that were originally used to kill them.

Let's REVIEW

- Define *virus* and explain how a virus multiplies.
- Summarize the story of the global AIDS epidemic.
- Describe the health effects of the flu virus.
- Define *parasite* and explain how a parasite can cause malaria.
- Explain why, after malaria was once brought under control, it is on the rise again.

Chemical Hazards

Since the 1970s, a growing body of evidence suggests that long-term exposure to some chemicals in the environment, even in very small, or *trace*, amounts, can wreak havoc on some of the major body systems in humans and other animals.

One of these body systems, the *immune system*, produces and stores white blood cells that help to protect the body against disease and harmful substances. It also forms antibodies that attack and disable invading bacteria, viruses, and protozoa. Some chemicals, especially arsenic, methylmercury, and dioxins, can weaken the immune system and leave the body less able to fight off pathogens.

Another body system that can be harmed by long-term exposure to some chemicals is the *nervous system*, which consists of the brain, the spinal cord, and the nerves that branch throughout the body. Neurotoxins are those chemicals that can harm the nervous system, and they include PCBs, methylmercury, arsenic, lead, and certain pesticides.

For example, mercury (Hg) is a potent neurotoxin (and a teratogen) that can disrupt the nervous system and brain function. About a third of the mercury in the atmosphere is released into the air from natural sources—rocks, soil, volcanoes, and the ocean. Most of the rest of this toxic metal in the atmosphere comes from coal-burning power plants, waste incinerators, and chemical manufacturing plants. In the atmosphere, some mercury is converted to even more toxic compounds that can wind up in aquatic systems. In the air and in water, certain bacteria can convert these compounds to highly toxic methylmercury that, like DDT, can be biologically magnified in food chains and webs (Figure 15.8). As a result, high levels of methylmercury are often found in the tissues of fishes that many people eat, including albacore tuna, king mackerel, walleye, and marlin.

Exposure to even low levels of methylmercury can cause brain damage in fetuses and young children. By various estimates, 30,000 to 60,000 children born each year in the United States are likely to have reduced IQs and other neurological problems because of such exposure. In 2007, the EPA estimated that one of every six women of childbearing age in the United States had enough mercury in her blood to harm a developing fetus. Consequently, the U.S. Food and Drug Administration (FDA) and the EPA have warned nursing mothers, pregnant women, and women who may become pregnant not to eat certain kinds of fish. Methylmercury can also harm the heart, kidneys, and immune systems of children and adults.

A third body system that can be disrupted by chemicals in the environment is the *endocrine system*, a set of glands that release tiny amounts of natural chemicals called *hormones* into the bloodstreams of humans and other animals. Hormones act like switches, turning on and turning off organ systems that control reproduction, growth, development, and some aspects of learning and behavior. Each type of hormone molecule has a certain shape that allows it to attach to cells within these body systems.

The endocrine system can be disrupted by molecules of certain chemicals that have shapes very similar to those of natural hormones. When these chemicals get into the body, their molecules sometimes attach to the molecules of natural hormones and interfere with normal hormonal functions. These interfering molecules are called *hormonally active agents* (HAAs), and they include mercury, aluminum, PCBs, the pesticide DDT, and a widely used herbicide called atrazine.

A special set of HAAs, called *hormone blockers*, disrupts the endocrine system by interfering with the work of hormones called androgens (male sex hormones). Other HAAs, called *estrogen mimics*, interfere with estrogens (female sex hormones). Together, these chemicals are sometimes called *gender benders*, because they have effects on sexual development and reproduction in some animals. For example, higher-than-normal levels of female hormones in males can cause smaller penises, lower sperm counts, and the presence of both male and female sex organs (hermaphroditism). Examples of HAAs include DDT, several herbicides, mercury, PCBs, and bisphenol A, or BPA (see *For Instance* , p. 426).

Another set of chemicals being studied for possible harmful effects are the *phthalates* (pronounced THALL-ates). Chemical manufacturers use them to soften polyvinyl chloride (PVC) plastic that is used to make a variety of products such as soft vinyl toys, teething rings, blood storage bags, and the plastic tubing used in hospitals. Phthalates are also used as solvents in a number of other consumer products, including many perfumes, hair sprays, deodorants, nail polishes, baby powders, and lotions, as well as shampoos for adults and babies.

In some laboratory animals, high doses of various phthalates have caused birth defects, kidney and liver damage, liver cancer, premature breast development, malfunctioning immune systems, and abnormal sexual development. The European Union and at least 14 other countries have banned phthalates. However, U.S. toy makers argue that they have used phthalates for more than 20 years without harmful effects. The U.S. scientific community has not reached agreement on whether or not phthalates pose a threat to human health.

Toxicologists are especially concerned about exposure to toxic chemicals that affect children. Infants and young children are more vulnerable to the effects of toxic substances than adults are, partly because their bodies are smaller and they generally breathe more air, drink more water, and eat more food per unit of body weight than adults do. Also, many children who frequently put their fingers and other objects in their mouths are exposed to toxins on the ground, in soils, and in the grass. In addition, children's immune systems and body cleansing processes are not as fully developed and functioning as are those of adults.

In 2005, the Environmental Working Group did a study involving 10 randomly selected newborns in U.S. hospitals. Analyzing blood from the infants' umbilical cords, the scientists detected 287 chemicals. Of those, 180 are known to cause cancers in humans and other animals, 217 have damaged the brain and nervous systems of test animals, and 208 have caused birth defects or abnormal development in test animals. We don't yet know the effects that most of these chemicals have on humans.

The EPA has proposed that, in order to give children the best possible protection against potentially toxic chemicals, we should assume that children face a risk 10 times higher than that faced by adults for any chemical exposure. Some health scientists argue that we should assume the risk for children is 100 times that of adults.

Almost everyone on the planet has now been exposed to potentially harmful chemicals, and in most people, these chemicals have built up to trace levels in their blood and other parts of their bodies. These chemicals can be found in most modern homes in the world's more-developed countries (Figure 15.17, p. 426). Also, through urine and feces, our bodies release trace amounts of birth control drugs, blood pressure medicines, antidepressants, painkillers, and many other medications. Sewage treatment plants and septic systems do not remove these chemicals from wastewater. As a result, these medications are entering waterways where some of them have been shown to have effects on aquatic animal species, such as the feminizing of some male fish.

We do not yet know how exposure to low levels of these chemicals can affect humans in the long run. However, some scientists are alarmed about these exposures because of their potentially harmful long-term effects on the human immune, nervous, and endocrine systems. Others point out that average life expectancies have been rising in most countries, especially in more-developed countries, and thus they conclude that low-level exposures to these chemicals must not be threatening.

For most of the various synthetic chemicals in our air, water, food, homes, and bodies, scientists simply do not agree on how dangerous trace amounts of these chemicals are. This is primarily because it is very difficult to determine the health effects of long-term exposures to low levels of these chemicals. To account for this uncertainty, and to be on the safe side, scientists and regulators typically set safety levels for exposure to toxic substances at 1/100 or even 1/1,000 of the estimated harmful levels.

CONSIDER this

- The U.S. National Academy of Sciences has estimated that only 10% of the roughly 80,000 registered synthetic chemicals in commercial use have been thoroughly screened for toxicity.

- Just 2% of these chemicals have been adequately tested to determine whether they are carcinogens, teratogens, or mutagens.

- Hardly any of them have been screened for possible damage to the human nervous, endocrine, and immune systems.

- Because of a lack of data and the high costs of regulation, about 99.5% of all chemicals used commercially in the United States are unregulated.

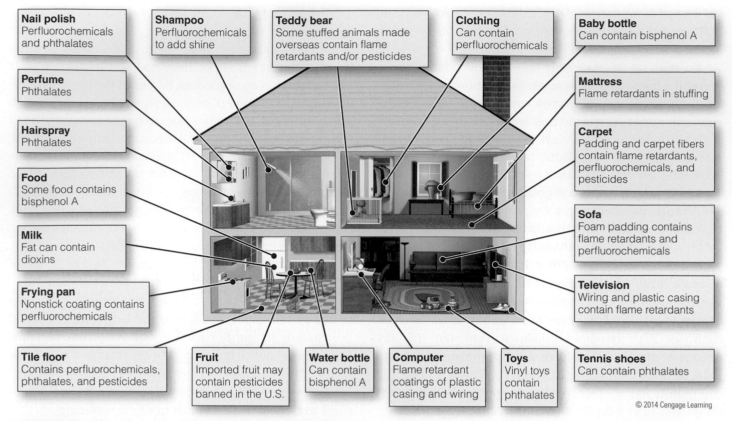

Nail polish
Perfluorochemicals and phthalates

Perfume
Phthalates

Hairspray
Phthalates

Food
Some food contains bisphenol A

Milk
Fat can contain dioxins

Frying pan
Nonstick coating contains perfluorochemicals

Shampoo
Perfluorochemicals to add shine

Teddy bear
Some stuffed animals made overseas contain flame retardants and/or pesticides

Clothing
Can contain perfluorochemicals

Baby bottle
Can contain bisphenol A

Mattress
Flame retardants in stuffing

Carpet
Padding and carpet fibers contain flame retardants, perfluorochemicals, and pesticides

Sofa
Foam padding contains flame retardants and perfluorochemicals

Television
Wiring and plastic casing contain flame retardants

Tile floor
Contains perfluorochemicals, phthalates, and pesticides

Fruit
Imported fruit may contain pesticides banned in the U.S.

Water bottle
Can contain bisphenol A

Computer
Flame retardant coatings of plastic casing and wiring

Toys
Vinyl toys contain phthalates

Tennis shoes
Can contain phthalates

© 2014 Cengage Learning

FIGURE 15.17 This typical home in a more-developed country has contained many potentially harmful chemicals since the day it was built. (Data from U.S. Environmental Protection Agency, Centers for Disease Control and Prevention, and New York State Department of Health)

FOR INSTANCE...

The BPA Controversy

The plastics industry has used bisphenol A (BPA) to create hardened plastics that have been widely used in many different products, including polycarbonate baby bottles (Figure 15.18) and sipping cups, reusable water bottles, sports drink and juice bottles, and food storage containers. BPA has also been used to make the liners of many food and soft drink cans. In 2009, product testers working for *Consumer Reports* magazine found BPA in the contents of many popular brands of canned foods including soups, juices, tuna, and green beans. BPA is also found on nearly all U.S. paper money and in the ink found on about half of cash register receipts printed on thermal paper.

According to lead researcher Karin Michels of Harvard Medical School, in another 2009 study, scientists found that certain drink bottles release BPA into liquids stored in them, even when those liquids are not heated. Yet another 2009 study found that some BPA remains in the body for at least 24 hours after being ingested. An earlier study by the CDC indicated that 93% of Americans over the age of 6 had trace levels of BPA in their urine above the EPA safety level. That study also found that children and adolescents had higher BPA levels than adults had.

However, the question of whether or not exposure to such trace levels of BPA can be a serious threat to human

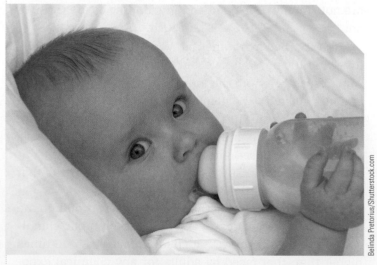

FIGURE 15.18 Some older baby bottles contain BPA. Many manufacturers of these products have switched to BPA-free materials.

Belinda Pretorius/Shutterstock.com

health remains to be answered. More than 100 studies by independent laboratories noted responses in test animals to exposures of very low levels of BPA that have caused some concerns among researchers. These effects include brain damage, early puberty, prostrate disease, breast cancer, reduced sperm count, impaired immune function, type 2 diabetes, hyperactivity, impaired learning, and obesity in unborn test animals.

On the other hand, 12 studies funded by the chemical industry found no evidence or weak evidence of such effects from low-level exposure to BPA in test animals. In addition, the German Society of Toxicology concluded in 2011 that the levels of exposure to BPA from food and beverages are not harmful to humans.

Canada, taking a cautious approach, has declared BPA a toxic substance and banned its use in the manufacturing of baby bottles. In the United States, the FDA has not yet determined that BPA is toxic to humans, but in 2010, the agency said it would look for ways to reduce the use of BPA in food packaging. Meanwhile, some states are taking matters into their own hands. For example, Maine has banned the use of BPA in reusable food and beverage containers, and Oregon is considering a similar ban.

FIGURE 15.19 This is a comparison of the global numbers of deaths per year due to various causes in 2009. (Data from World Health Organization)

Let's REVIEW

- What are three body systems that can be disrupted by toxic chemicals? How is each system disrupted?
- Why is mercury in the environment a threat to human health?
- What are HAAs and why are they a problem? Give three examples of HAAs.
- What are phthalates and why are they a potential environmental problem?
- Give three reasons why scientists are especially concerned about the effects of toxic chemicals on children.
- Summarize the story of the BPA controversy.

Cultural and Lifestyle Hazards

Judging by what you see and hear in the news, you might think that some of the greatest risks to your health are flying in a jet airliner or being killed by a shark. In fact, the greatest risks that most of us face today are rarely covered in the news. When we compare risks as measured by the numbers of premature deaths per year (Figure 15.19) and reduced life spans (Figure 15.20, p. 428), one of the leading risks is poverty. Large numbers of poor people die because of malnutrition, infectious diseases transmitted by unsafe drinking water, and increased vulnerability to normally nonfatal diseases as a result of weakness caused by malnutrition.

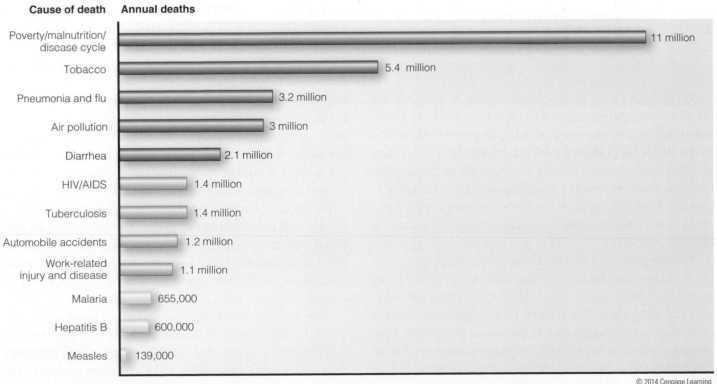

Cause of death	Annual deaths
Poverty/malnutrition/disease cycle	11 million
Tobacco	5.4 million
Pneumonia and flu	3.2 million
Air pollution	3 million
Diarrhea	2.1 million
HIV/AIDS	1.4 million
Tuberculosis	1.4 million
Automobile accidents	1.2 million
Work-related injury and disease	1.1 million
Malaria	655,000
Hepatitis B	600,000
Measles	139,000

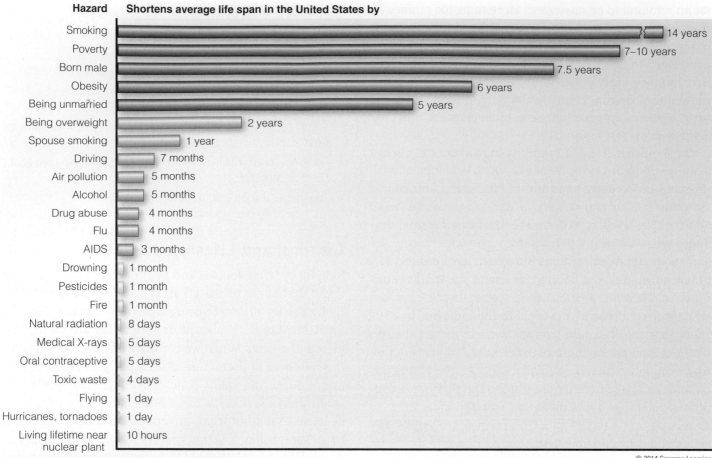

Hazard	Shortens average life span in the United States by
Smoking	14 years
Poverty	7–10 years
Born male	7.5 years
Obesity	6 years
Being unmarried	5 years
Being overweight	2 years
Spouse smoking	1 year
Driving	7 months
Air pollution	5 months
Alcohol	5 months
Drug abuse	4 months
Flu	4 months
AIDS	3 months
Drowning	1 month
Pesticides	1 month
Fire	1 month
Natural radiation	8 days
Medical X-rays	5 days
Oral contraceptive	5 days
Toxic waste	4 days
Flying	1 day
Hurricanes, tornadoes	1 day
Living lifetime near nuclear plant	10 hours

© 2014 Cengage Learning

FIGURE 15.20 This comparison of some of the risks people in the United States face is expressed in terms of shortened average life spans—generalized estimates of how much time is taken off the average person's life span by each of the risks. Individuals respond differently to these risks because of differences in genetic makeup, family medical history, and other factors. (Data from Bernard L. Cohen)

Poverty also puts people in situations where they are far more vulnerable to chemical and biological health hazards than people who are not living in poverty. According to the WHO, more than 1.1 billion people (more than one of every seven people on the earth, most of them poor) live in urban areas where the outdoor air is unhealthy to breathe. They also tend to live in densely populated parts of cities in less-developed countries where clean water is harder to come by.

The biggest environmental threat for most poor people is indoor air pollution coming from open fires or leaky stoves that burn wood, charcoal, coal, or dung for cooking or heating a dwelling (Figure 15.1b). Cigarette smoke is often part of this deadly mix of indoor air pollutants. Also, many poor people have to work in inadequately ventilated and highly polluted spaces. According to the WHO, indoor air pollution kills about 1.6 million people every year—an average of nearly 4,400 deaths per day—and causes health problems for hundreds of millions of other people.

Some of the greatest risks of premature death arise from lifestyle choices that people make, especially choosing to smoke tobacco (see the following *For Instance*) and being overweight from eating too much and getting too little exercise. A 2005 study involving 100 scientists around the world led by Majid Ezzati concluded that one-third of the world's 7 million annual cancer deaths could be prevented if individuals were to follow these guidelines: avoid smoking and exposure to tobacco smoke; lose excess weight and exercise regularly; eat only enough to be satisfied and eat a variety of fruits and vegetables; drink no more than two alcoholic drinks in a single day and avoid alcohol on most days; avoid getting too much sunlight; and practice safe sex.

One problem is that most of us are not good at evaluating and comparing various risks. Many people don't realize, or they simply ignore the fact, that some of the activities they enjoy carry fairly high risks of injury or death. For example, smoking one pack of cigarettes per day creates a 1-in-250 chance of dying by age 70. Driving a car without wearing a seatbelt carries a 1-in-3,300 chance of dying with each trip you take. (The probability goes down to 1 in 6,070 when you wear a seatbelt.) Indeed, driving or riding in a car is the single greatest daily risk that most people take. Compare this to some probabilities related to events that many people greatly fear, such as dying in a commercial airplane crash (1 in 9 million), or being killed by a shark (1 in 281 million).

Factors other than poor information can also warp a person's view of how risky a particular activity, product, or technology can be. One of these factors is fear. Plenty of scientific studies have shown that fear causes people to overestimate risks related to events such as plane crashes, tornadoes, shark attacks, fires, and terrorist attacks. Fear also causes people to distrust new, unknown products and technologies more than they do older, more familiar ones. For example, most people fear nuclear power plants more than they fear the more commonly used coal-fired power plants that are major sources of dangerous air pollution.

Still another factor is the instant gratification that some people get from certain activities that are risky. Smoking, for example, can be highly pleasurable as can eating lots of ice cream or getting a tan. The pleasure from these activities comes instantly, while their potential harm comes much later. This causes many people to overlook the risks.

The Dangers of Smoking

Most young people who choose to smoke, when they light up their first cigarette, are not thinking about the risks it poses. Even if they are aware that tobacco use is highly addictive, they might think they can avoid or overcome any addiction. If they are aware that smoking and other forms of tobacco use can raise a person's chances of dying prematurely, they usually think that they will avoid that risk. Tobacco use provides an immediate and intense pleasure for those who enjoy it, and that reward can outweigh the risks of addiction and early death in most people's minds.

The fact that cannot be avoided, however, is that the harmful effects of tobacco use kill an average of about 14,800 people every day, worldwide, or about one every 6 seconds, according to the WHO. In 2010, this amounted to 5.4 million deaths—about 1.2 million of them in China, where people smoke nearly a third of all the cigarettes smoked in the world. Tobacco use contributes heavily to deaths from illnesses such as heart disease, stroke, lung cancer, and emphysema (Figure 15.21). The WHO projects that by 2030, the annual death toll from these and other smoking-related diseases will reach more than 8 million, with 3.5 million of them in China.

In the United States, according to the CDC, smoking kills about 442,000 people every year. That amounts to an average of more than 1,200 deaths per day, or nearly one every minute (Figure 15.22, p. 430).

FIGURE 15.21 These normal, healthy human lungs (a) came from a nonsmoker, while the diseased lungs (b) came from a person who died of emphysema, the major causes of which are prolonged smoking and exposure to air pollutants.

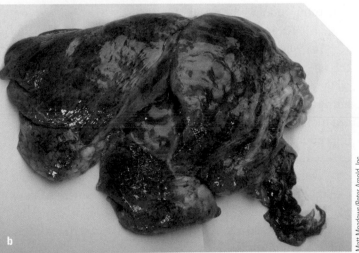

Matt Meadows/Peter Arnold, Inc.

Matt Meadows/Peter Arnold, Inc.

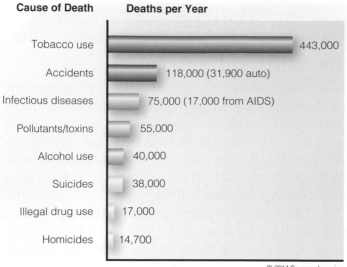

Cause of Death	Deaths per Year
Tobacco use	443,000
Accidents	118,000 (31,900 auto)
Infectious diseases	75,000 (17,000 from AIDS)
Pollutants/toxins	55,000
Alcohol use	40,000
Suicides	38,000
Illegal drug use	17,000
Homicides	14,700

© 2014 Cengage Learning

FIGURE 15.22 This comparison of the causes of death in the United States in 2010 shows that tobacco use results in more deaths every year than the combined total of all the other categories in this figure. (Data from U.S. National Center for Health Statistics, Centers for Disease Control and Prevention, and the U.S. Surgeon General)

Scientific evidence strongly indicates that the drug called nicotine contained in tobacco products is highly addictive. A British government study showed that at least 8 of every 10 adolescents who smoke more than one cigarette become long-term smokers. Only 1 in every 10 people who try to quit smoking actually kicks the habit. This is about the same rate of success as that seen by people who are trying to quit drinking and by people who are trying to kick their addictions to heroin or crack cocaine.

Breathing *secondhand smoke*—the smoke released into the air by smokers—is also a serious health hazard, especially for children and adult relatives and friends of smokers. Studies have shown that children who live with smokers are more likely to develop allergies and asthma. Nonsmoking spouses of smokers have a 30% higher risk of both heart attack and lung cancer than spouses of nonsmokers have.

Several countries have banned smoking in public places. And while tobacco companies still produce trillions of cigarettes per year, the average number of cigarettes smoked per person in the United States and in most other countries is dropping.

NUMB3RS

14

The number of years by which an average lifelong smoker's life is shortened because of tobacco use

CONSIDER this

- The WHO has estimated that tobacco use has contributed to the deaths of 100 million people during the 20th century.
- A 2010 study by British researchers and a 2007 study by the CDC found that every year, about 600,000 people worldwide and 46,000 people in the United States die from exposure to secondhand smoke.

Smokers who want to quit have reason to be encouraged. A 50-year study published in 2004 by Richard Doll and Richard Peto found that cigarette smokers who quit smoking—even at the age of 50—can cut their health risk in half. If they quit by the age of 30, they can avoid nearly all the risk of harmful health effects from smoking.

KEY idea

Tobacco use is the most preventable of all the major environmental health threats.

Let's REVIEW

- What is the leading health risk in terms of deaths per year and shortened life spans?
- Give three reasons why poor people are more at risk from environmental health hazards than are wealthy people. What is the biggest environmental threat for most people living in less-developed countries?
- What are three examples of common risks that we face, and what are the odds of dying from these risks?
- List three factors that can warp a person's assessment of the risks of any particular activity.
- Why is smoking dangerous? What is second-hand smoke and why is it a threat?

What Can Be Done?

Fighting Infectious Diseases

Since 1900, when infectious disease was the leading cause of death in the world, doctors and scientists have learned how to reduce the incidences of bacterial infections by using antibiotics and how to use vaccines to prevent the spread of some infectious viral diseases. The average life expectancy has risen considerably since 1900 in most countries, and the world's leading cause of death is now nontransmissible cardiovascular disease.

Since 1970, improvements have been even more dramatic, according to the WHO, as the global death rate from infectious diseases has dropped by about two-thirds and is projected to continue dropping. Also, since the 1970s, the percentage of children in less-developed countries who have been protected by vaccines from diphtheria, typhoid fever, tetanus, measles, and polio rose from 10% to 84%. This improvement has saved about 10 million children's lives every year since 1971.

CONSIDER this

The cost of protecting a child with a basic set of vaccines is about $30—comparable to the cost of a modest dinner for two in some of the world's more-developed countries.

Infectious disease is still a problem in many less-developed countries, but there are simple, inexpensive ways to prevent deaths from these diseases (see *Making a Difference*, p. 432). For example, many children die from dehydration due to severe diarrhea. But a simple oral rehydration therapy has helped to reduce the annual number of deaths due to diarrhea from 4.6 million in 1980 to 2.1 million in 2009. This therapy is a simple solution of boiled water, salt, and sugar or rice, and it costs only a few cents per person.

Another example of inexpensive preventive measures is the use of bed nets designed to keep mosquitoes off sleeping people in tropical areas (Figure 15.23) in order to reduce the number of malarial infections. The Gates Foundation, founded by Bill and Melinda Gates, has invested in an effort to supply African nations with 70 million bed nets. As a result, in Zambia, the annual number of malaria cases has been cut in half.

Figure 15.24 lists a number of measures promoted by health scientists and public health officials that can help to prevent or reduce occurrences of infectious diseases—especially in less-developed countries. The WHO has estimated that implementing these solutions could save the lives of as many as 4 million children younger than age 5 every year. These preventive solutions are very affordable compared to the costs of treating diseases.

One disease that is not going away is AIDS. The WHO has argued for a global strategy to slow the spread of AIDS, based on six major priorities. First, reduce the annual number of new infections to a level below the annual number

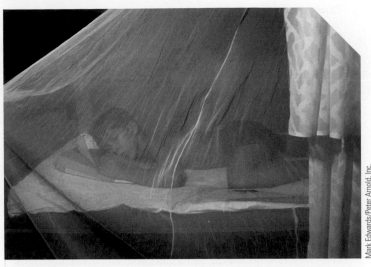

FIGURE 15.23 This boy, who lives in the Amazon rain forest of Brazil, is sleeping under an insecticide-treated bed net to reduce his risk of being bitten by mosquitoes that carry the malaria parasites. These nets cost about $5 each.

Mark Edwards/Peter Arnold, Inc.

of deaths. Second, focus prevention efforts on the societal groups that are most likely to spread the disease, such as sex workers and intravenous drug users. Third, provide free HIV testing to these and other high-risk groups, and be aggressive in promoting such HIV testing. Fourth,

Preventing or Reducing Infectious Diseases

Reduce poverty

Improve basic nutrition

Improve drinking water quality

Provide oral rehydration for diarrhea victims

Immunize children against major viral diseases

Reduce unnecessary use of antibiotics

Educate people to take all of any antibiotic prescription

Sharply reduce antibiotic use to promote livestock growth

Conduct a global campaign to reduce HIV/AIDS

Step up research on tropical diseases and vaccines

© 2014 Cengage Learning

FIGURE 15.24 Here are 10 ways to prevent or reduce occurrences of infectious diseases.

Photo credits (top to bottom): U.S. Navy; U.S. Navy; Ryan McVay/Getty Images.

Tuberculosis (TB) is not necessarily a life-threatening disease if it is diagnosed early enough, but that is a big if, especially in less-developed countries. In the traditional method of diagnosis, a sample of a TB victim's feces must be carefully scanned through a microscope by a highly trained technician. It is a costly, tedious, and time-consuming process.

In poorer areas of the world, these samples have to be sent to labs, sometimes far away from where the victim lives. It can take months to get the lab test results, and during that time, a TB victim can become gravely ill and die while spreading this highly contagious disease to other people. On top of that, according to international health experts, an estimated 40% of TB cases are missed through this traditional method of diagnosis.

In 2008, Daniel Jeck, a biomedical engineering student at North Carolina State University, decided to tackle this problem as part of his senior project. He teamed up with two other NCSU engineering students, Hersh Tapadia and Pavak Shah, and they set a goal of finding a quicker, less expensive way to diagnose TB. With initial help from their advisor, Dr. Howard Shapiro, the three began experimenting with a laptop computer and off-the-shelf electronic components. At one point, they were spending 25 to 30 hours a week on their project.

When they were done, Jeck, Tapadia, and Shah had invented a new device for detecting TB bacteria in samples of saliva from suspected TB victims. As a result, rather than having to peer for hours at a highly magnified microscope slide, a technician can now simply insert the slide into the new device and view it on a computer screen. If the sample glows green, the patient has TB. This process is so simple and quick that a person with minimal training can now diagnose TB in seconds at a cost of about a dollar per trial.

The new device has the potential to save thousands of lives in poor areas of the world where TB is spreading rapidly. Jeck, Tapadia, and Shah have done some work trying to adapt their invention to other purposes, such as diagnosing malaria, AIDS, and other infectious diseases. They plan to make a business out of building and selling their new devices, and through their work, they might eventually save millions of lives.

---

conduct global mass-advertising and education programs, both for adults and for schoolchildren, that focus on prevention. Fifth, provide free or low-cost drugs to slow the progress of the disease. Sixth, increase funding for research on the development of microbiocides, drugs designed to kill microbes; for example, a vaginal gel containing such a drug could help women to protect themselves from getting HIV/AIDS.

Let's REVIEW

- List six ways to prevent or reduce the incidence of infectious diseases, especially in less-developed countries.
- Summarize the six-point global strategy recommended by the WHO for slowing the spread of AIDS.

Dealing with Chemical Hazards

There is still much to learn about the possible health risks that result from exposure to chemicals such as BPA (see *For Instance*, p. 426), phthalates, and many other potentially toxic agents. It is particularly difficult to assess the possible harmful health effects of exposures to very low levels of various chemicals that are widely found in the environment and in products that we use regularly. It will take decades of expanded research to evaluate these chemicals.

For this reason, many scientists argue that we should broadly apply the **precautionary principle** in dealing with potentially hazardous chemicals. According to this principle, when substantial preliminary evidence indicates that an activity can harm human health or the environment, we should take precautionary measures to prevent or reduce this harm even if some of the cause-and-effect relationships have not been fully established scientifically.

In the case of chemicals, this approach suggests that chemical consumers and manufacturers, as well as government regulatory agencies, should assume that a chemical is harmful when preliminary evidence indicates that it is. For example, many scientists believe that as a reasonable precaution, consumers should strictly avoid using chemicals such as BPA that might be hormone disrupters, especially in products used for infants or young children. These scientists also call for manufacturers to search for less harmful substitutes for these chemicals, and for governments to require that manufacturers take these precautionary measures. Manufacturers generally oppose such measures and favor the current approach, which assumes that many chemicals are harmless until very strong evidence shows otherwise.

A practical way to apply the precautionary principle is through **pollution prevention**—a strategy focused on preventing pollutants from entering the environment, as opposed to cleaning them up after they are emitted. This approach is receiving more attention from businesses and governments around the world because they have had to spend valuable resources on pollution cleanup—an after-the-fact approach to dealing with hazardous pollutants in the environment.

A glaring example of this was the disastrous oil leak in the Gulf of Mexico resulting from the explosion of the British Petroleum (BP) *Deepwater Horizon* oil-drilling rig in April of 2010 (Figure 15.25). The accident killed 11 workers on the platform, sank the rig, ruptured the ocean-bottom wellhead, and released massive amounts of crude oil into the gulf—the largest oil spill ever in U.S. waters. This oil wreaked havoc on ocean floor and coastal ecosystems, threatened the health of cleanup workers exposed to toxic chemicals that were used in the cleanup effort, and disrupted the lives of millions of people who depend on the gulf's fishing and tourism businesses.

BP will eventually spend many millions of dollars on the cleanup and compensation costs. The company could probably have prevented these expenses, along with the resulting bad publicity it received, by making a relatively

U.S. Coast Guard

FIGURE 15.25 On April 20, 2010, oil and natural gas escaping from an oil-well drilling operation ignited, and caused an explosion on the BP *Deepwater Horizon* drilling platform in the Gulf of Mexico.

Preventing or Reducing Mercury Pollution

Prevention

Phase out waste incineration

Remove mercury from coal before it is burned

Switch from coal to natural gas and renewable energy resources

Control

Sharply reduce mercury emissions from coal-burning plants and incinerators

Collect and recycle batteries and other products that contain mercury

Label all products that contain mercury

© 2014 Cengage Learning

FIGURE 15.26 This chart lists ways to prevent the release of mercury into the environment from human sources—primarily coal-burning power plants and incinerators—and methods for controlling these inputs.

small investment in better preventive technology for the oil well's drill head and in better management practices. Improved enforcement of existing oil drilling regulations by the U.S. government would also have helped to prevent this accident.

A less dramatic but equally important example of how pollution prevention can help to safeguard public health is the case of mercury pollution. There are many ways to prevent mercury from entering the environment, as summarized in Figure 15.26.

Some companies have made pollution prevention an important part of their businesses. An example is the 3M Company, based in Minnesota, which makes more than 60,000 different products in 100 manufacturing plants. In 1975, it began a pioneering Pollution Prevention Pays (3P) program. To reduce or eliminate its output of air and water pollutants and toxic wastes, 3M began making more nonpolluting products, redesigned its manufacturing processes and equipment, recycled or reused waste materials, and decreased its use of hazardous raw materials as well as its output of toxic chemicals into the environment.

Between 1975 and 2010, 3M's 3P program prevented more than 3 billion pounds of pollutants from reaching the environment—an important contribution to improving human health. This program also saved the company more than $1.4 billion in its costs for raw materials, waste disposal, and compliance with U.S. pollution laws and regulations. This is a superb example of why pollution prevention pays. Since 1990, a growing number of companies have implemented similar pollution and waste prevention programs.

Using pollution prevention is also a way to help people in poverty to reduce health threats to their lives. The worst threat to poor people, especially those living in less-developed countries, is indoor air pollution (Figure 15.1b). There are several new models of inexpensive cooking stoves (Figure 15.27) that can burn the same fuels that

FIGURE 15.27 This energy-efficient cooking and heating stove greatly reduces indoor air pollution and decreases the biggest health threat for families in less-developed countries such as this family in India.

people now burn in leaky stoves and open fires. These devices could be provided as a form of financial aid or made available with small loans to poor people, and this would go a long way toward protecting poor families from indoor air pollution and illness.

Let's REVIEW

- State the *precautionary principle* and define *pollution prevention*.
- How are these two concepts related? Give an example of how pollution prevention can be less expensive than pollution cleanup.
- Summarize the nature and benefits of 3M's Pollution Prevention Pays program.
- How can pollution prevention be used to improve the lives of many poor families?

Implementing Pollution Prevention

Scientists, public officials, and chemical industry representatives have debated about how far we need to go in applying the precautionary principle by using pollution prevention. Some argue that anyone who wants to introduce a new chemical or technology should bear the burden of establishing its safety. This would mean that potentially harmful new chemicals would be assumed to be harmful until scientific studies could show that they were not. It would also mean that potentially hazardous chemicals now in use should be removed from the market until their safety can be established scientifically.

Some countries are moving in this direction. In 2000, a number of countries signed a treaty that banned or phased out the use of 12 of the most notorious *persistent organic pollutants (POPs)*. Animal studies have shown that various POPs act as carcinogens, mutagens, teratogens, and hormonal disrupters. The list, also called the *dirty dozen*, includes PCBs, DDT, and eight other pesticides. Other chemicals have since been added to this list. The POPs treaty went into effect in 2004 but has not been ratified and implemented by the United States.

The European Union has enacted regulations known as REACH (for **R**egistration, **E**valuation, **A**uthorisation and **R**estriction of **Ch**emical substances). It requires the registration of 30,000 untested, unregulated, and potentially harmful chemicals used in EU nations. In the REACH process, the most hazardous substances are not approved for use if safer alternatives exist. When there is no such alternative, producers must present a research plan aimed at finding one.

Under REACH, the burden is on chemical manufacturers to show that chemicals are safe. Historically, the burden of showing that a chemical is dangerous has been on governments, as is still the case under U.S. chemical regulation programs. At congressional hearings in 2009, experts testified that the U.S. system makes it almost impossible for the government to limit or ban the use of toxic chemicals. The hearings found that the EPA had required testing for only 200 of the more than 80,000 chemicals used in industry and, primarily because of budget limitations, had issued regulations to control just five of those chemicals.

Manufacturers and businesses contend that the REACH approach puts too big a burden on the chemical industry. They argue that it could make it too expensive to introduce any new chemicals. Proponents of the REACH approach agree that it is possible to go too far in using the precautionary principle. But they argue that we should push for pollution prevention in working toward the important goal of protecting human health and the earth's life-support system. They also point out that regulations on toxic chemicals can spur scientists, engineers, and businesses to use their creativity and resources to find pollution prevention solutions.

In fact, preventive solutions have been found and are being implemented, as companies such as 3M have found. These companies have realized that pollution prevention helps them to deal efficiently with pollution regulations, and it reduces health risks for their employees and customers. These companies also make money selling safer products and the innovative technologies they have developed in response to regulations. Their pollution prevention programs have also improved their public images as good corporate citizens.

Let's **REVIEW**

- Give two examples of how governments are implementing pollution prevention programs.
- What are the arguments for and against focusing government regulations on pollution prevention?
- What are the three components of REACH? Summarize the arguments for and against using the REACH approach.

Evaluating and Reducing Risk

An important, but often overlooked, early step in the effort to deal with health hazards is to learn how to evaluate and compare risks so that we can put our best efforts toward reducing the worst of those risks.

Risk analysis experts have developed guidelines that governments, organizations, businesses, and individuals alike can use to reduce their risks in the most efficient and effective ways:

- *Determine how much risk is acceptable.* For example, when asked to pick a threshold beyond which they will act to avoid an environmental hazard, most people settle on a chance of around 1 in 100,000 of dying or suffering serious harm from the hazard. However, in establishing safety standards, the EPA generally assumes that a 1 in 1 million chance of dying from an environmental hazard is the acceptable threshold. Determining acceptable risk requires some research and study on the part of any government agency, organization, business, or individual.
- *Determine the actual risk involved in any particular decision.* This is a complicated process for governments and corporations dealing with broad, complex risks. But in evaluating more local or personal risks, we need to realize that news media, for example, usually exaggerate the daily risks we face in order to capture our interest and sell newspapers and magazines, or to gain television viewers. As a result, most people have an inaccurate understanding of how various risks compare.
- *Compare risks and carefully choose how to use resources in the best way to reduce risks.* Most activities involve some level of risk. In evaluating any particular risk, the question is "how risky is this activity (or technology, or product), compared to other risks?" Then, the government, organization, business, or individual comparing risks must decide how much effort to put into reducing or avoiding those risks.

Perhaps the most important benefit we can get from learning how to evaluate and reduce risk is to prevent human health hazards before they become costly and deadly. Some environmental and health scientists are urging businesses and governments to focus a great deal more of their attention and resources on the prevention of health hazards in order to save large amounts of money and untold numbers of human lives. Economic studies have shown that, just as pollution prevention works, so does the broader strategy of working to prevent health hazards before they arise (see *A Look to the Future*, p. 436).

Evaluating Technology in Risk Management

Governments and businesses have to evaluate their systems and processes in terms of risk, which can be a complex undertaking. For example, one approach is to express the overall probability that a technological system will

FIGURE 15.28 This nuclear power plant (a) is located in Florida. Such plants are run from control rooms like this one (b), found in another nuclear power plant.

complete a task without failing (its *reliability*) as a percentage that is the product of two factors:

System reliability (%) = Technology reliability × Human reliability

When it is carefully designed, controlled, and maintained, a highly complex system such as a nuclear power plant or a deep-sea oil drilling rig can achieve a high degree of reliability. But because of the millions of factors affecting human behavior (a much more complex system than any machine we can build), human reliability often is much lower than technological reliability and is almost impossible to predict.

For example, say the technological reliability of a nuclear power plant (Figure 15.28) is 95% (0.95) and human reliability is 75% (0.75). Then, the overall system reliability is the product of the two: 71% (0.95 × 0.75 = 71%). Even if we could make the power plant 100% reliable (1.0), the overall system reliability would still be only 75% (1.0 × 0.75 = 75%). The fact that even the most carefully designed systems depend on largely unpredictable human reliability helps to explain tragedies that were once thought to be highly unlikely, such as the 1986 explosion at the Chernobyl nuclear power plant in Ukraine and the massive release of oil into the Gulf of Mexico in 2010 (Figure 15.25).

One way to make a system more fail-safe is to remove as much of the human element as possible, transferring as much of the system's functioning and control as possible to more reliable technology. But automatic control systems can be damaged by chance occurrences such as tornadoes, fires, and—as was demonstrated by Japan's Fukushima Daiichi nuclear power plant accident in 2011—earthquakes and tsunamis. In addition, the components of any automated control system are manufactured, assembled, tested, and maintained by human beings, as are the computer software programs used to monitor and control the systems. Thus, no technological system can be 100% risk-free.

Let's REVIEW

- Why is it important to learn how to evaluate and compare risks?
- What are three steps that individuals, businesses, and governments can take in evaluating risks?
- Why is it that no technological system can be 100% risk-free?

A Look to the Future

The Prevention Approach to Health Hazards

Human health care is not just about curing people of diseases and injuries. It is about preventing them from getting sick or hurt in the first place. As crucial as this sounds, it is not recognized as widely as you might guess. In Great Britain, for example, according to research done by one of this book's authors, Norman Myers, the National Health Service is seen by some to behave as if it were primarily a national disease service. It spends much more time and money on curing people of disease than on helping them to avoid it in the first place. As much as 95% of health-care spending in Great Britain is for curing illness, leaving just 5% for preventive measures. The situation in the United States is similar.

There are many examples of how it is better to avoid having to treat a disease that has spread through a population by eliminating it before it spreads. During the 20th century, 300 million people died from smallpox, but in 1978, the disease was finally eradicated through a vaccination program. Since that time, the savings gained by giving vaccinations (a preventive approach) have amounted to more than $1 billion per year, or three times what it cost to eliminate the disease.

Another example is malaria, which kills about 655,000 people every year, mostly children in Africa where the disease costs a total of at least $12 billion a year. There is an ambitious campaign underway to cut the death toll from malaria in half. This has included efforts to develop new and more effective drugs and to raise the number of protective bed nets in use by ten-fold (Figure 15.23). As a result, annual numbers of malaria deaths have been reduced by 44% in Kenya and by 50% or more in Eritrea and Rwanda.

Measles, one of the world's most prevalent infectious diseases, could also be eliminated fairly readily. It costs less than $1 to vaccinate a child against measles, yet this disease claims the lives of 380 children a day, most of them poor. In 1999, the United Nations launched a vaccination drive with the goal of cutting the measles death rate by 90% (Figure 15.29). By 2007, Africa had achieved an 89% reduction.

In 2003, it looked as though polio—a crippling disease—would become the second disease (after smallpox) to finally be eradicated after it had been isolated to just six countries. When an intensive polio eradication program began in 1988, the disease existed in 125 countries and was paralyzing 350,000 people every year. By 2010, the annual number of new cases of polio had been cut to 1,290 through the use of polio vaccines.

Along with these vaccination programs, there are many other examples of preventive efforts that could eventually save large numbers of lives and lots of money. Most notable are various efforts to get people to quit smoking, to prevent motor vehicle accidents, and to reduce the occurrence of obesity. Public health officials are finding that the money saved through these prevention programs is typically three or more times the cost of the programs. The savings can be even higher for programs that prevent cardiovascular disease and diabetes—diseases that are often related to preventable unhealthy lifestyle choices.

Another prominent health risk for which prevention could make a huge difference is that of cancer. National economies can be significantly impacted by cancer cases because of lowered work productivity, the economic costs of medical treatments for the disease, and premature deaths. In the United States, these three factors cost the economy about $200 billion a year. Yet less than a nickel of every health-care dollar is spent on preventive research

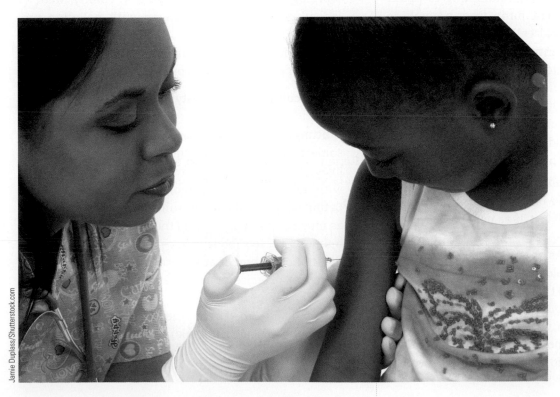

FIGURE 15.29 This nurse is vaccinating a child in Africa against measles.

of all kinds, and cancer research gets much less than what would be necessary to make solid gains against the disease. In the United States, just the medical treatments for cancer cost about $60 billion per year, or 33 times the roughly $1.8 billion spent by the National Cancer Institute each year on research. As was stated by a spokesperson for Lasker Charitable Trust, a foundation that funds medical research, "if you think research is expensive, try disease."

Overall, the potential for savings through a preventive health-care approach is stunning. Studies show that with an annual investment of $66 billion, we could save 8 million lives a year and generate economic benefits worth at least six times the investment. Thus, by citing cost as a reason for not supplying funds to promote preventive measures, governments are turning away potentially huge health and economic gains.

The prevention approach will only grow more important in the future because as the human population continues to expand and age, more diseases and other health hazards will likely arise. Since 1970, for example, some 32 new diseases previously unreported in humans have been identified, including hepatitis C, Legionnaires' disease, Ebola virus infections, severe acute respiratory syndrome (SARS), and avian flu. It is likely that, as the human population continues to grow and expand into tropical forests and other natural environments, more pathogens will attack humans and more new diseases will arise.

A few scientists are now trying to apply the prevention approach to their research by attempting to detect unknown pathogens that might jump from animals to humans—a relatively new field called *ecological medicine*. This and other sorts of preventive efforts could be—and many scientists argue, should be—the new frontier for health-care professionals.

What Would You Do?

Most of us can take responsibility for avoiding many health hazards. For example, we can practice good hygiene and steer clear of dangerous habits such as using tobacco and overeating.

We can also make choices that might help to lessen health hazards for other people. For example, if we want to dispose of a computer, we can try to make sure it is recycled responsibly instead of shipped overseas to a place where poor people risk their health working in dangerous electronics recycling shops. If enough people took this step, it might lessen the demand for such hazardous recycling.

We can also work on the political front to get laws passed to outlaw these hazardous recycling processes.

There are many other ways in which people are working to reduce their own health hazards and improve the environment for those around them.

Avoiding infectious diseases.
- Washing hands thoroughly and frequently using warm water and a mild soap, while staying away from people who have bacterial or viral infections.
- Avoiding the use of antibacterial soaps, liquids, and sprays. Several studies have shown that these antibacterial products are no more effective in preventing infections than regular soap and warm water.
- Practicing good hygiene when using health clubs and other public places: using available cleansers to clean off exercise equipment; making sure that restaurant utensils and dishes are clean; and avoiding areas in restaurants that appear to be unclean.

Working to reduce chemical hazards in their own lives and in the environment.
- Trying to find and use nature-based substitutes for chemicals used in cooking, hygiene, cosmetics, house cleaning, and lawn care.
- Doing business with companies that are working to reduce or prevent their contributions to air and water pollution; avoiding products made by companies that are known for polluting and letting them know why.
- Working politically to try to get city, county, state, or federal governments to sign on with pollution prevention and control laws and treaties; supporting and working for political candidates who are in favor of pollution prevention and control.

Avoiding cultural and lifestyle hazards as much as possible.
- Quitting or avoiding tobacco products altogether is something many people are doing to lengthen their lives, and smokers can try to avoid endangering others with their second-hand smoke.
- Eating whole foods (fresh fruit, vegetables, and grains) as much as possible to avoid the synthetic chemicals in many processed foods; not overeating; drinking alcohol in moderation, if at all; practicing safe sex; getting plenty of exercise; and avoiding heavy tanning from the sun or in tanning booths.
- Becoming informed about the risks in their lives, comparing those risks, and making careful choices in what they do, what they eat, and how they live.

KEYterms

bacteria, p. 419
biological hazards, p. 412
carcinogens, p. 415
chemical hazards, p. 413
cultural hazards, p. 413
dose, p. 417
genetic resistance, p. 420
infectious disease,
   p. 413

lifestyle health hazards,
   p. 413
mutagens, p. 415
neurotoxins, p. 415
noninfectious diseases,
   p. 414
nontransmissible
   diseases, p. 414
parasite, p. 423

pathogens, p. 413
pollution prevention,
   p. 433
precautionary principle,
   p. 432
probability, p. 419
response, p. 417
risk, p. 418
risk assessment, p. 419

risk management, p. 419
teratogens, p. 415
toxic chemical, p. 415
toxicity, p. 416
toxicology, p. 416
transmissible disease,
   p. 414
viruses, p. 422

THINKINGcritically

1. In dealing with malaria, some scientists have suggested that the banned insecticide DDT would help people in malaria-prone areas of the world to fight off mosquitoes carrying the malaria parasites, and that the risk of harm from DDT is worth the benefit to be gained in protecting people from malaria. Others argue that this is simply trading in one major biological hazard for a major chemical hazard—the danger of damage to animal and human health from DDT. Which of those arguments do you agree with and why?

2. Assume you are a public health official working for a community. Write a brief plan of action for reducing the threat to your community from both **(a)** tuberculosis and **(b)** HIV/AIDS.

3. Representatives of the chemical industry argue that regulations that put the burden of proof on their industry to show that chemicals are safe before they can be widely sold would cost too much and would thus harm their business and in turn the larger economy. Write a brief argument in favor of this position and a brief argument opposed to it.

4. Do you believe that smoking indoors should be banned in all public buildings and private businesses? Explain. If you do not believe in this idea, how would you regulate smoking, if at all? Explain.

5. Think of two major cultural or lifestyle risks that you have taken at least once in your life. Which of these risks did you take voluntarily and which did you have to take without being given much of a choice? How could you have reduced these risks, if at all?

6. Think of the health hazard that frightens you most and do some research to find out what the real risk of this hazard is. That is, find out what your odds are of suffering harm or death from this risk and compare it with two other risks that frighten you. Write a brief report on your findings.

LEARNINGonline

Access an interactive eBook and module-specific interactive learning tools, including flashcards, quizzes, videos and more in your Environmental Science CourseMate, accessed through **CengageBrain.com.**

Glossary

A

acid deposition Acids and acid-forming compounds (droplets or particles) falling from the atmosphere to the earth's surface; commonly known as *acid rain*, a term that refers to the deposition of acidic droplets.

active solar heating system System that uses solar collectors to capture energy from the sun and store it as heat for space heating and water heating. Liquid or air pumped through the collectors transfers the captured heat to a storage system such as an insulated water tank or rock bed. Pumps or fans then distribute the stored heat or hot water throughout a building as needed. Compare *passive solar heating system*.

age structure Percentage of the population (or number of people of each gender) at each age level in a population.

agrobiodiversity The genetic variety of animal and plant species used to provide food.

air pollution One or more chemicals in high enough concentrations in the air to harm humans, other animals, vegetation, or materials. Such chemicals or physical conditions are called air pollutants. See *primary air pollutant*, *secondary air pollutant*.

aquaculture Growing and harvesting of fish and shellfish for human use in freshwater ponds and lakes, or in cages or fenced-in areas of coastal lagoons and estuaries or in the open ocean.

aquatic life zones Saltwater and freshwater portions of the biosphere. Examples include freshwater life zones (such as lakes and streams) and ocean or marine life zones (such as estuaries, coastlines, and the open ocean).

aquifer Porous, water-saturated layers of sand, gravel, or bedrock that can yield an economically significant amount of water.

area strip mining Type of surface mining used where the terrain is flat and the mineral resource lies close to the surface. An earthmover strips away the overburden and a power shovel digs a trench to remove the mineral deposit. The trench is then filled with overburden, and a new trench is made parallel to the previous one; the process is repeated over the entire site. Compare *mountaintop removal mining, open-pit mining, subsurface mining*.

artificial selection Process by which humans select one or more desirable genetic traits in a population of a plant or animal species and then use selective breeding to produce populations containing many individuals with the desired traits. Compare *genetic engineering*.

atmosphere The whole mass of air surrounding the earth. See *stratosphere, troposphere*. Compare *biosphere*.

atom The smallest unit of an element that can exist and still have the unique characteristics of that element; it is the basic building block of all chemical elements and thus all matter. Compare *ion, molecule*.

B

background extinction rate Historic extinction rate that existed before humans dominated the planet; biologists estimate that it was about 0.0001% per year, or 1 species for every 1 million species on the planet, on average. Compare *mass extinction*.

bacteria One-celled organisms, some of which transmit diseases. Most act as decomposers and obtain the nutrients they need by breaking down complex organic compounds in the tissues of living or dead organisms into simpler inorganic nutrient compounds.

biodegradable wastes Materials that can be broken down into simpler substances (elements and compounds) by bacteria or other decomposers in a period of time ranging from days to decades. Compare *nondegradable wastes*.

biodiversity　The variety of different species, the genetic variability among individuals within each species, and the variety of ecosystems, as well as the functions such as energy flow and matter cycling needed for the survival of species and biological communities.

biodiversity hotspot　An area especially rich in plant species that are found nowhere else and that are in great danger of extinction. All such areas have seen serious ecological disruption, primarily because of rapid human population growth and the resulting pressure on natural resources.

biofuel　Gas (such as methane) or liquid fuel (such as ethyl alcohol or biodiesel) made from plant materials (biomass).

biological hazards　Threats from organisms that can cause disease in other organisms. These organisms include bacteria, viruses, and parasites that can be seen only with a microscope. See *pathogens*.

biome　A terrestrial region characterized by the predominance of a certain set of species of plants and animals. Examples include various types of deserts, grasslands, and forests.

biomimicry　Process of observing certain changes in nature, studying how natural systems have responded to those changing conditions over many millions of years, and applying what is learned from these observations to meeting some environmental challenge.

biomining　The use of natural or genetically engineered microorganisms to remove desired metals from ores found in underground deposits; taking place underground, this biological approach does not disturb the surrounding environment and avoids much of the air and water pollution associated with other forms of mining and mineral processing.

biosphere　Zone of the earth where life is found, consisting of parts of the atmosphere, hydrosphere (mostly surface water and groundwater), and crust (mostly soil and surface rocks and sediments on the bottoms of oceans and other bodies of water). Compare *atmosphere*.

C

carbon capture and storage (CCS)　Process of removing carbon dioxide gas from the exhausts of coal-burning power and industrial plants, and storing it somewhere (usually underground or under the ocean floor). To be effective, it must be stored essentially forever, so that it cannot be released into the atmosphere.

carbon cycle　Cyclic movement of carbon in different chemical forms within the environment and the earth's life-support systems.

carcinogen　Chemical, ionizing radiation, or virus that causes or promotes the development of cancer. Compare *mutagen, teratogen*.

carrying capacity　Maximum population of a particular species that a given habitat can support over a given period of time.

chemical change　Interaction between chemicals in which the chemical composition of the elements or compounds involved changes. Compare *physical change*.

chemical cycling　The circulation of chemicals from the environment (mostly from soil and water) through organisms and back to the environment.

chemical element　A fundamental type of matter that has a unique set of properties and cannot be broken down into simpler substances by chemical means. Compare *compound*.

chemical formula　Shorthand way to show the number of atoms (or ions) in the basic structural unit of a compound. Examples include H_2O, $NaCl$, and $C_6H_{12}O_6$. See *compound*.

chemical hazards　Harmful chemicals in the air, water, and soil that can also be found in our food and in manufactured products; some come from natural sources, such as volcanic eruptions, but most come from human sources such as coal-burning power plants, factories, motor vehicles, indoor cooking and heating fires, interior building materials, and wastes.

chemical reaction　See *chemical change*.

city　See *urban area*.

climate　General pattern of atmospheric conditions in a given area over periods ranging from at least 30 years to thousands of years; based primarily on the area's average annual temperature and average annual amount of precipitation. Compare *weather*.

climate change Long-term changes in the earth's average temperature, precipitation, and other environmental factors.

coal Solid, combustible mixture of organic compounds, formed in several stages out of the remains of plants subjected to heat and pressure over millions of years; with 30–98% carbon by weight, mixed with various amounts of water and small amounts of sulfur and nitrogen compounds.

coastal wetland Land along a coastline or inland from an estuary, that is covered with saltwater all or part of the year. Examples include coastal marshes, tidal flats, and mangrove swamps. Compare *inland wetland*.

coastal zone Warm, nutrient-rich, shallow part of the ocean that extends offshore from the high-tide mark on land to the edge of a shelflike extension of continental land masses known as the continental shelf. Compare *open sea*.

cogeneration Production of two useful forms of energy, such as high temperature heat or steam and electricity, from the same fuel source.

combined heat and power (CHP) See *cogeneration*.

commensalism An interaction between organisms of different species in which one type of organism benefits and the other type is neither helped nor harmed to any great degree. Compare *mutualism*.

community The populations of all species living and interacting in a given area at a particular time.

composting A process that involves mimicking nature's recycling of nutrients by allowing decomposer bacteria to break down organic wastes to produce compost, a material that can be added to soil to supply it with plant nutrients.

compound Combination of two or more elements held together in a fixed proportion. Examples are NaCl, CO_2, and $C_6H_{12}O_6$. Compare *chemical element*.

consumer Organism that gets its nutrients by feeding on the tissues of producers or of other consumers. In economics, one who uses economic goods. Compare *producer*.

contour strip mining Form of surface mining used on hilly or mountainous terrain. A power shovel cuts a series of terraces into the side of a hill. An earthmover removes the overburden, and a power shovel extracts the coal. The overburden from each new terrace is dumped onto the one below. Compare *area strip mining, mountaintop removal mining, open-pit surface mining, subsurface mining*.

coral reef Formation found in the coastal zones of warm tropical and subtropical oceans, produced by colonies of tiny coral animals, called polyps, that secrete a stony substance (calcium carbonate) around themselves for protection. When the corals die, their empty outer skeletons form layers; coral reefs grow with the accumulation of such layers.

core Extremely hot innermost zone of the earth, consisting of a solid inner core and an outer core of hot, fluid rock. Compare *crust, mantle*.

crossbreeding A process in which farmers and scientists select one or more desirable characteristics in a population of a plant or animal species and breed certain individuals of that population to generate new populations that contain large numbers of individuals with the desired characteristics.

crude oil Gooey liquid consisting primarily of hydrocarbon compounds and small amounts of compounds containing oxygen, sulfur, and nitrogen. Extracted from underground accumulations, it is sent to oil refineries, where it is converted to heating oil, diesel fuel, gasoline, tar, and other materials.

crust Solid outermost and thinnest zone of the earth. It consists of oceanic crust and continental crust. Compare *core, mantle*.

cultural eutrophication Overnourishment of aquatic ecosystems with plant nutrients (mostly nitrates and phosphates) that result from human activities such as agriculture, urbanization, and discharges from sewage treatment plants. See *eutrophication*.

cultural hazards Threats that develop within human societies that include poverty, war and other forms of conflict, motor vehicles accidents, and unsafe working conditions.

D

dam   A structure built across a river to control the river's flow or to create a reservoir. See *reservoir*.

decomposers   Organisms, mostly bacteria or fungi, that digest parts of dead organisms, as well as cast-off fragments and wastes of living organisms, by breaking down the complex organic molecules in those materials into simpler inorganic compounds and then absorbing some of those nutrients. Compare *consumer*, *producer*.

deforestation   Removal of trees from a forested area.

demographic transition   Hypothesis that countries, as they become industrialized, have declines in death rates followed by declines in birth rates.

depletion time   The time it takes to use a certain fraction (usually 80%) of the known or estimated supply of a nonrenewable resource at an assumed rate of use. Finding and extracting the remaining 20% usually costs more than it is worth.

desalination   Purification of saltwater or brackish (slightly salty) water by removal of dissolved salts.

desert   Biome in which evaporation exceeds precipitation and the average amount of precipitation is less than 10 inches per year. Deserts have little vegetation or have widely spaced, mostly low-growing vegetation.

desertification   Conversion of rangeland, rain-fed cropland, or irrigated cropland to desertlike land. Can be caused by a combination of overgrazing, soil erosion, prolonged drought, and climate change.

dose   Amount of a potentially harmful substance an individual ingests, inhales, or absorbs through the skin. Compare *response*.

drainage basin   See *watershed*.

drought   Condition in which an area does not get enough water because of lower-than-normal precipitation or higher-than-normal temperatures that increase evaporation.

E

ecocity (green city)   A city that puts its highest priority on minimizing its ecological footprint and improving the quality of life for its inhabitants.

ecological footprint   Amount of biologically productive land and water needed to supply a population with the renewable resources it uses and to absorb or dispose of the wastes from such resource use. It is a measure of the average environmental impact of populations in different countries and areas. See *per capita ecological footprint*.

ecological niche   A species' role, or total way of life, in an ecosystem. It includes all physical, chemical, and biological conditions that a species needs in order to live and reproduce in an ecosystem.

ecological restoration   Deliberate alteration of a degraded habitat or ecosystem to restore as much of its ecological structure and function as possible.

ecological succession   Process in which communities of plant and animal species in a particular area are replaced over time by a series of different and often more complex communities. See *secondary ecological succession*.

ecological tipping point   Point at which an environmental problem reaches a threshold level, which causes an often irreversible shift in the behavior of a natural system.

ecology   Biological science that studies the structure and functions of nature in terms of the relationships between living organisms and their environment.

economic depletion   Exhaustion of 80% of the estimated supply of a nonrenewable resource. Finding, extracting, and processing the remaining 20% usually costs more than it is worth. May also apply to the depletion of a renewable resource, such as a fish or tree species.

economic growth   Increase in the capacity of an economic system to provide goods and services designed for a human population. See *economic system*.

economic system   The system of production, distribution, and consumption of goods and services to satisfy people's needs and wants.

ecosystem   One or more communities of different species interacting with one another and with the chemical and physical factors making up their nonliving environment.

ecosystem services   Natural services that support life on the earth and are essential to the quality of human life and the functioning of the world's economies. Examples are the chemical cycling, natural pest controls, and natural purification of air and water. See *natural resources*.

electromagnetic radiation   Forms of kinetic energy traveling as electromagnetic waves. Examples include radio waves, TV waves, microwaves, infrared radiation, visible light, ultraviolet radiation, X-rays, and gamma rays.

electronic waste (e-waste)   The fastest-growing category of municipal solid waste, it includes products such as TV sets, computers, cell phones, and other devices that contain electronic components.

endangered species   Wild species with so few individual survivors that the species could soon become extinct in all or most of its natural range. Compare *threatened species*.

energy   Capacity to do work or to transfer heat; can involve mechanical, physical, chemical, or electrical tasks, or heat transfers between objects at different temperatures.

energy conservation   Any reduction in the use or waste of energy.

energy efficiency   Measure of how much work we can get from each unit of energy we use. See *energy quality*, *net energy*.

energy quality   Measure of the ability of a form of energy to do useful work. High-temperature heat and the chemical energy in fossil fuels are examples of concentrated high-quality energy. Low-quality energy such as low-temperature heat is dispersed or diluted and cannot do much useful work. See *high-quality energy*, *low-quality energy*.

environment   All external conditions, factors, matter, and energy, living and nonliving, that affect any living organism or other specified system.

environmental degradation   Depletion or destruction of a potentially renewable resource such as soil, grassland, forest, or wildlife caused by our using it faster than it is naturally replenished. See also *natural capital degradation*.

environmentalism   Social movement dedicated to protecting the earth's life-support systems for the benefit of all species and ecosystems.

environmental refugees   People who have been forced to leave their homes because of environmental degradation that has left them without the resources, such as soil, food, and water, that they need in order to survive and live well.

environmental science   Interdisciplinary study that uses information and ideas from the physical sciences (such as biology, chemistry, and geology) as well as those from the social sciences and humanities (such as economics, political science, and ethics) to learn how nature works, how we interact with the environment, and how we can deal with environmental problems.

estuary   Partially enclosed coastal area at the mouth of a river from which freshwater, carrying fertile silt and runoff from the land, mixes with salty seawater.

eutrophication   Physical, chemical, and biological changes that take place after a lake, estuary, or slow-flowing stream receives inputs of plant nutrients—mostly nitrates and phosphates—from natural erosion and runoff from the surrounding land basin. See *cultural eutrophication*.

evaporation   Conversion of a liquid into a gas; in the water cycle, the conversion of liquid water in oceans, rivers, lakes, wetlands, and soils into water vapor in the atmosphere.

evolution by natural selection   Scientific theory that accounts for individuals in a population of any species having differing *traits*, or characteristics defined by their genes. Some have traits that give them a better chance of surviving and reproducing under a particular set of environmental conditions than other individuals that do not have these traits. As a result, individuals with the beneficial traits are more likely to produce more offspring than are individuals without such traits.

e-waste   See *electronic waste*.

exponential growth   Growth in which some quantity, such as population size , increases at a constant rate per unit of time. An example is the growth sequence 2, 4, 8, 16, 32, 64, and so on, which increases by 100% at each interval. When the increase in quantity over time is plotted, this type of growth yields a curve shaped like the letter J.

extinction   The process by which a species ceases to exist on the earth. See also *endangered species*, *mass extinction*, *threatened species*.

F

family planning   Providing information, clinical services, and contraceptives to help people choose the number and spacing of children they want to have.

first law of thermodynamics (law of conservation of energy) Whenever energy is converted from one form to another in a physical or chemical change, no energy is created or destroyed. This law does not apply to nuclear changes, in which large amounts of energy can be produced from small amounts of matter. See *second law of thermodynamics*.

fish farming See *aquaculture*.

fishery Population of wild fish or shellfish that is regularly harvested by commercial fishing operations.

food chain A sequence of organisms in which energy is transferred from producers to various consumers. Compare *food web*.

food insecurity Condition under which people live with chronic hunger and malnutrition that threatens their ability to lead healthy and active lives. Compare *food security*.

food security Condition under which every person in a given area has daily access to enough nutritious food to have an active and healthy life. Compare *food insecurity*.

food web Complex network of interconnected food chains and feeding relationships. Compare *food chain*.

forest systems Land areas that are dominated by trees. Compare *old-growth forest*, *second-growth forest*, and *tree plantation*.

fossil fuel Product of partial or complete decomposition of plants and animals as a result of exposure to heat and pressure in the earth's crust over millions of years; occurs as crude oil, coal, natural gas, or heavy oil. See *coal*, *natural gas*.

fracking See *hydraulic fracturing*.

freshwater aquatic systems Areas containing water with little or no salt content, including rivers, streams, inland wetlands, and most lakes, along with the aquatic species that live in them.

full-cost pricing Setting market prices of goods and services to include the hidden, harmful environmental and health costs of producing and using them.

G

generalist species Species with a broad ecological niche. They can live in many different places, eat a variety of foods, and tolerate a wide range of environmental conditions. Examples include flies, cockroaches, mice, rats, and humans. Compare *specialist species*.

genetic engineering Insertion of an alien gene into an organism to give it a beneficial genetic trait. Compare *artificial selection*.

genetic resistance The acquired ability of populations of disease-causing organisms and pest populations such as weeds that have changed genetically over time to withstand the antibiotics or pesticides that were designed to kill them. See *pesticide*.

geology Study of the earth's dynamic history. Geologists study and analyze rocks and the features and processes of the earth's interior and surface.

geothermal energy Heat transferred from the earth's underground concentrations of dry steam (steam with no water droplets), wet steam (a mixture of steam and water droplets), or hot water trapped in fractured or porous rock.

grassland Biome found in regions with enough annual average precipitation to support the growth of grass and small plants but not enough to support large stands of trees. Compare *desert*, *forest*.

green city See *ecocity*.

greenhouse effect The ability of the lower atmosphere to temporarily store some of the energy received from the sun as heat. This warming occurs primarily because of the presence of four naturally occurring greenhouse gases in the lower atmosphere: water vapor (H_2O), carbon dioxide (CO_2), methane (CH_4), and nitrous oxide (N_2O). As some of the sun's energy is reflected from the planet's surface back toward space, greenhouse gases interact with this energy and release heat into the earth's lower atmosphere. See greenhouse gases.

greenhouse gases Gases, such as carbon dioxide (CO_2), methane (CH_4), water vapor (H_2O), nitrous oxide (N_2O), chlorofluorocarbons, and ozone (O_3), in the earth's lower atmosphere that interact with solar energy to have a warming effect on the earth's atmosphere. See *greenhouse effect*.

groundwater   Water that sinks into the soil and is stored in slowly flowing and slowly renewed underground reservoirs. See *aquifers*. Compare *surface water*.

H

habitat   Place or type of place where an organism or population of organisms lives. Compare *ecological niche*.

hazardous waste (toxic waste)   Any waste that threatens human health or the environment because it is poisonous, dangerously chemically reactive, flammable, or corrosive. Examples include industrial solvents, car and household batteries (containing acids as well as toxic lead and mercury), household pesticides, hospital medical waste, and toxic ash from incinerators and coal-burning power plants. Electronic waste is also a fast-growing source of hazardous wastes. See *electronic waste*.

high-consumption, high-waste economy   An economic system that attempts to stimulate economic growth by using more matter and energy resources to produce more goods and services for more people. Such economies convert much of the high-quality matter and energy resources they use into waste, pollution, and low-quality heat that flows into the environment.

high-grade ore   Ore containing a large amount of a desired mineral. Compare *low-grade ore*.

high-input agriculture   See *industrialized agriculture*.

high-quality energy   Energy that is concentrated and has great ability to perform useful work. Examples include high-temperature heat and the energy in electricity, coal, oil, gasoline, sunlight, and nuclei of uranium-235. Compare *low-quality energy*.

HIPPCO   Acronym used by conservation biologists for the six most important secondary causes of extinction hastened by human activities: **H**abitat destruction, degradation, and fragmentation; **I**nvasive (nonnative) species; **P**opulation growth (too many people consuming too many resources); **P**ollution; **C**limate change; and **O**verexploitation.

hydraulic fracturing   Process of drilling horizontally through shale rock layers and then pumping large volumes of a mixture of water, sand, and chemicals between the layers under high pressure to fracture the rock and release oil or natural gas.

hydrologic cycle   See *water cycle*.

hydropower   Electrical energy produced by falling or flowing water.

I

igneous rock   Rock formed when molten rock material (magma) wells up from the earth's interior, cools, and solidifies into rock masses within the earth's crust. Compare *metamorphic rock*, *sedimentary rock*. See *rock cycle*.

indicator species   Species whose decline serves as early warnings that a community or ecosystem is being degraded. Compare *keystone species*, *native species*, *nonnative species*.

industrial ecosystem   Manufacturing process that mimics the way nature deals with wastes by reusing or recycling most of the minerals and chemicals that it uses, instead of burying or burning them, or shipping them somewhere else.

industrialized agriculture   Production of large quantities of crops or livestock involving the use of large inputs of energy from fossil fuels (especially oil and natural gas), water, fertilizer, and pesticides.

industrial smog   Type of air pollution consisting mostly of a mixture of sulfur dioxide, suspended droplets of sulfuric acid formed from some of the sulfur dioxide, and suspended solid particles. Compare *photochemical smog*.

industrial solid waste   Solid waste produced by mines, factories, refineries, food growers, and businesses that supply people with goods and services. Compare *municipal solid waste*.

infant mortality rate   Number of babies out of every 1,000 born each year who die before their first birthday.

infectious disease   Disease caused when a pathogen such as a bacterium, virus, or parasite invades the body and multiplies in its cells and tissues. Examples are flu, HIV, malaria, tuberculosis, and measles. See *transmissible disease*. Compare *nontransmissible disease*.

inland wetland   Land away from the coast, such as a swamp, marsh, or bog, that is covered all or part of the time with freshwater. Compare *coastal wetland*.

integrated coastal management   A community-based effort to develop and use coastal resources more sustainably, and to prevent pollutants from entering coastal aquatic systems.

integrated pest management (IPM)   Combined use of biological, chemical, and cultivation methods in proper sequence and timing to keep the size of a pest population below the level that causes an economically unacceptable loss of a crop or a livestock animal.

integrated waste management   A variety of strategies for both waste reduction and waste management designed to deal with the solid wastes we produce. See *waste management, waste reduction*.

interspecific competition   Attempts by members of two or more species to use the same limited resources in an ecosystem.

ion   Atom or group of atoms with one or more positive (+) or negative (−) electrical charges. Examples are Na^+ and Cl^-. Compare *atom, molecule*.

IPAT model   A simple model that brings together three major factors that impact the environment: population size (P), resource consumption per person, or affluence (A), and the harmful and beneficial environmental effects of technologies (T). The IPAT model shows that the environmental impact (I) of human activities depends primarily on how these three factors interact, as represented by this simple equation: Impact (I) = Population (P) × Affluence (A) × Technology (T).

K

keystone species   Species that play roles affecting many other organisms in an ecosystem. Compare *indicator species, native species, nonnative species*.

kinetic energy   Energy that matter has because of its mass and speed, or velocity. Compare *potential energy*.

L

land degradation   Decrease in the ability of land to support crops, livestock, or wild species in the future as a result of natural or human-induced processes.

law of conservation of matter   In any physical or chemical change, matter is neither created nor destroyed but merely changed from one form to another; in physical and chemical changes, existing atoms are rearranged into different spatial patterns (physical changes) or different combinations (chemical changes).

less-developed country   Country that has low to moderate industrialization and low to moderate average income and rate of resource use per person. Most are located in Africa, Asia, and Latin America. Compare *more-developed country*.

life expectancy   Average number of years a newborn infant can be expected to live in a given country and time period.

lifestyle health hazards   Those hazards that originate in our own choices and habits. Examples include smoking tobacco, using heroin and other addictive drugs, eating unhealthy foods, and not getting enough exercise.

limiting factor   Single factor that limits the growth, abundance, or distribution of the population of a species in an ecosystem.

liquefied natural gas (LNG)   Natural gas converted to liquid form by cooling it to a very low temperature.

liquefied petroleum gas (LPG)   Mixture of liquefied propane (C_3H_8) and butane (C_4H_{10}) gas removed from natural gas and used as a fuel.

low-consumption, low-waste economy   Economy based on working with nature by recycling and reusing discarded matter; preventing pollution; conserving matter and energy resources by reducing unnecessary waste and use; and building things that are easy to recycle, reuse, and repair. Compare *high-consumption, high-waste economy*.

low-grade ore   Ore containing a small amount of a desired mineral. Compare *high-grade ore*.

low-quality energy   Energy that is dispersed and has little ability to do useful work. An example is low-temperature heat. Compare *high-quality energy*.

M

mantle   Zone of the earth's interior between its core and its crust, composed mostly of fluid rock, making up about 84% of the earth's volume. Compare *core, crust*.

marine aquatic systems   Aquatic life zones associated with oceans and their accompanying bays, estuaries, coastal wetlands, shorelines, coral reefs, and mangrove forests, as well as with some inland seas and lakes with salty waters.

mass extinction Catastrophic, widespread, often global event in which major groups of species go extinct over a short period of time. Compare *background extinction rate*.

matter Anything that has mass (the amount of material in an object) and takes up space.

metamorphic rock Rock produced when preexisting rock is subjected to high temperatures (which may cause it to melt partially), high pressures, chemically active fluids, or a combination of these agents. Compare *igneous rock*, *sedimentary rock*. See *rock cycle*.

migration Movement of humans and other species into or out of a specific geographic area.

mineral Any naturally occurring inorganic substance found in the earth's crust as a solid with a regularly repeating internal structure (a crystalline solid). See *mineral resource*.

mineral resource Concentration of mineral materials in or on the earth's crust in a form and amount such that their extraction and conversion into useful materials or items are currently or potentially profitable. Mineral resources are classified as *metallic* (such as iron and tin ores) or *nonmetallic* (such as sand).

model Approximate representation or simulation of a system being studied.

molecule Combination of two or more atoms of the same chemical element (such as O_2) or different chemical elements (such as H_2O) held together by chemical bonds. Compare *atom*, *ion*.

monoculture Form of crop farming in which one plant species is planted in a large area. Compare *polyculture*.

more-developed country Country that is highly industrialized and has a high average income and rate of resource use per person. Compare *less-developed country*.

mountaintop removal mining Type of surface mining that uses explosives, massive power shovels, and large machines called draglines to remove the top of a mountain and expose seams of coal underneath. Compare *area strip mining*, *contour strip mining*.

MSW See *municipal solid waste*.

municipal solid waste (MSW) Solid materials discarded by homes and businesses in or near urban areas. See *solid waste*. Compare *industrial solid waste*.

mutagen Agent such as a chemical or a form of radiation that increases the frequency of mutations, or random changes, in the DNA molecules found in cells. See *carcinogen*, *teratogen*.

mutualism Type of species interaction in which both participating species generally benefit. Compare *commensalism*.

N

nanotechnology Use of science and engineering to manipulate and create materials out of atoms and molecules at the ultra-small scale of less than 100 nanometers.

native species Those species that normally live and thrive in a particular ecosystem. Compare *indicator species*, *keystone species*, *nonnative species*.

natural capital Natural resources and natural services that keep us and other species alive and support our economies. See *natural resources*, *natural services*.

natural capital degradation The waste, depletion, or destruction of any of the earth's natural capital. See *environmental degradation*.

natural gas Mixture of gases found underground, consisting mostly of methane gas (CH_4) and widely used as a fuel for heating, cooking, and other uses.

natural resources Resources such as air, water, and soil, and various forms of energy in nature that are essential or useful to humans. See *natural capital*, *resource*.

natural services Processes of nature, such as pest control and the purification of air and water, that support life and human economies. See *ecosystem services*, *natural capital*.

net energy yield Total amount of useful energy available from an energy resource or energy system over its lifetime, minus the amount of energy used, automatically wasted because of the second law of thermodynamics, and unnecessarily wasted in finding, processing, and transporting it to users.

neurotoxins A hazardous chemical in the environment that can harm the nervous systems of humans and other animals.

niche   See *ecological niche*.

nitrogen cycle   Cyclic movement of nitrogen in different chemical forms within the environment and the earth's life-support systems.

nondegradable waste   Material that is not broken down by natural processes. Examples include the toxic elements arsenic, lead, and mercury. Compare *biodegradable waste*.

noninfectious disease   See *nontransmissible disease*.

nonnative species   Species that migrate into an ecosystem or are deliberately or accidentally introduced into an ecosystem by humans. Compare *native species*.

nonpoint source   Broad and diffuse area, rather than a specific point, from which pollutants enter the air or bodies of surface water. Examples include burning forests, runoff of chemicals and sediments from cropland, logged forests, parking lots, lawns, and golf courses. Compare *point source*.

nonrenewable energy resource   A resource such as oil, coal, or natural gas that cannot be replenished by natural processes on a human time scale. See *renewable energy resource*.

nonrenewable resource   Resource that exists in a fixed amount (stock) in the earth's crust and has the potential for renewal by geological, physical, and chemical processes taking place over hundreds of millions to billions of years. Examples include copper, aluminum, coal, and oil. We classify these resources as *exhaustible* because we are extracting and using them at a much faster rate than they are formed. Compare *renewable resource*.

nontransmissible, noninfectious disease   Disease that is not caused by living organisms and does not spread from one person to another. Examples include most cancers, diabetes, cardiovascular disease, and malnutrition. Compare *transmissible disease*.

nuclear fission   Nuclear change in which the nuclei of certain isotopes with large mass numbers (such as uranium-235 and plutonium-239) are split apart into lighter nuclei when struck by a neutron. This process releases more neutrons and a large amount of energy. Compare *nuclear fusion*.

nuclear fuel cycle   Includes the mining of uranium, processing and enriching the uranium to make fuel, using it in a reactor, safely storing the resulting highly radioactive wastes for thousands of years until their radioactivity falls to safe levels, and retiring the nuclear power plant by taking it apart and storing its high- and moderate-level radioactive parts safely for thousands of years.

nuclear fusion   Nuclear change in which two nuclei of an isotope of an element with a low mass number (such as hydrogen-2 and hydrogen-3) are forced together at extremely high temperatures until they fuse to form a heavier nucleus (such as helium-4). This process releases a large amount of energy. Compare *nuclear fission*.

nutrient   Any chemical that an organism must ingest in order to live, grow, or reproduce.

O

ocean acidification   A process that begins with ocean waters absorbing CO_2 as part of the natural carbon cycle; the CO_2 reacts with the water to form a weak acid called carbonic acid, which increases the acidity of ocean water.

ocean bottom   The deepest zone in the open ocean and the third major marine life zone; it contains many species that get their food from the large numbers of dead and decaying organisms that drift down from the upper zones.

oil shale   A type of rock that contains a mixture of hydrocarbons called *kerogen*, which is composed of organic chemical compounds that form a portion of the organic matter in sedimentary rock. Compare *shale oil*.

old-growth forest   A forest that has not been seriously disturbed by human activities or natural disasters for 200 years or more and contains trees that are often hundreds—sometimes thousands—of years old.

open dump   Field or large pit where garbage is deposited and sometimes covered with soil. They are rare in developed countries, but are widely used in many developing countries, especially to handle wastes from cities. Compare *sanitary landfill*.

open-pit surface mining   The removal of minerals such as sand and metal ores by digging them out of the earth's surface and leaving an open pit behind. Compare *area strip mining, contour strip mining, mountaintop removal mining, subsurface mining*.

open sea   Part of any ocean that lies beyond the continental shelf. Compare *coastal zone*.

ore   Part of a metal-yielding material that can be economically extracted from a mineral deposit. See *high-grade ore, low-grade ore*.

organic agriculture   Growing crops with no use of synthetic pesticides, synthetic fertilizers, or genetically modified crops; raising livestock without the use of synthetic growth regulators and feed additives; and using only organic fertilizer (manure, legumes, compost) and natural pest controls (bugs that eat harmful insects, plants that repel harmful insects, and environmental controls such as crop rotation).

organic fertilizer   Organic material such as animal manure, green manure, and compost applied to cropland as a source of plant nutrients.

organism   Any form of life.

overburden   Layer of soil and rock overlying a mineral deposit, to be removed by surface mining.

overconsumption impacts   In more-developed countries, the environmental degradation that typically results when affluence is the biggest factor in a country's total environmental impact. In these countries, the high rate of resource use per person leads to high levels of waste, pollution, and resource depletion and degradation.

overgrazing   The destruction of vegetation caused by too many grazing animals that feed for too long on a specific area of pasture or rangeland and exceed the carrying capacity of that rangeland or pasture area.

overnutrition   Diet so high in calories, saturated (animal) fats, salt, sugar, and processed foods, and so low in vegetables and fruits that the consumer runs a high risk of developing diabetes, hypertension, heart disease, obesity, and other health hazards.

overpopulation impacts   The results of population size being the biggest factor in a country's total environmental impact, common in less-developed countries where environmental degradation, such as deforestation and depletion of topsoil, often results from a growing number of poor people trying to survive by using these resources; while average resource use per person in these countries is low, the total resource use is high because of the large and growing population.

ozone depletion   Decrease in concentration of ozone (O_3) in the stratosphere. See *ozone layer*.

ozone layer   Layer of gaseous ozone (O_3) in the stratosphere that protects life on earth by filtering out most of the sun's harmful ultraviolet radiation.

P

parasite   Consumer organism that lives on or in, and feeds on, a living plant or animal, known as the host, over an extended period. The parasite draws nourishment from and gradually weakens its host; it may or may not kill the host. See *parasitism*.

parasitism   Interaction between species in which one organism, called the parasite, preys on another organism, called the host, by living on or in the host.

passive solar heating system   System that, without the use of mechanical devices, captures sunlight directly within a structure and converts it into low-temperature heat for space heating or for heating water for domestic use. Compare *active solar heating system*.

pathogen   Living organism that can cause disease in another organism. Examples include bacteria, viruses, and parasites.

peer review   The process of scientists reporting details of the methods and models they used to test a hypothesis, the results of their experiments, and the reasoning behind their hypothesis for other scientists working in the same field (their peers) to examine and criticize.

per capita ecological footprint   Amount of biologically productive land and water needed to supply each person in a population with the renewable resources he or she uses and to absorb or dispose of the wastes from such resource use. It measures the average environmental impact of individuals in populations in different countries and regions. Compare *ecological footprint*.

per capita water footprint   A measure of how much water it takes to meet the needs of an average person in a country for a year. In addition to the water used for domestic purposes, this measure also includes virtual water, the water used to produce goods and services.

permafrost   Perennially frozen layer of the soil that forms when the water in it freezes and remains frozen for long periods of time; found in Arctic tundra.

pesticide   Any chemical designed to kill or inhibit the growth of an organism that people consider undesirable.

petrochemicals   Chemicals obtained by refining crude oil; used as raw materials in manufacturing most industrial chemicals, fertilizers, pesticides, plastics, synthetic fibers, and paints, as well as many other products.

petroleum   See *crude oil*.

phosphorus cycle   Cyclic movement of phosphorus in different chemical forms within the environment and the earth's life-support systems.

photochemical smog   Complex mixture of air pollutants produced in the lower atmosphere by the reaction of hydrocarbons and nitrogen oxides under the influence of sunlight. Especially harmful components include ozone, peroxyacyl nitrates (PANs), and various aldehydes. Compare *industrial smog*.

photosynthesis   Chemical process used by green plants to make their own nutrients, or the chemicals necessary for their survival, by converting solar energy to chemical energy that plants store in their tissues.

photovoltaic (PV) cell   Device that converts radiant (solar) energy directly into electrical energy. Also called a solar cell.

physical change   Process that alters one or more physical properties of an element or a compound without changing its chemical composition. Examples include changing the size and shape of a sample of matter (crushing ice or cutting aluminum foil) and changing a sample of matter from one physical state to another (boiling or freezing water). Compare *chemical change*.

point source   Single identifiable source that discharges pollutants into the environment. Examples include the smokestack of a power plant or an industrial plant, the drainpipe of a meatpacking plant, the chimney of a house, or the exhaust pipe of an automobile. Compare *nonpoint source*.

pollutant   Substance, chemical, or form of energy that can adversely affect the health, survival, or activities of humans or other living organisms. See *pollution*.

pollution cleanup   Device or process that removes or reduces the level of a pollutant after it has been produced or has entered the environment. Examples include automobile emission control devices and sewage treatment plants. Compare *pollution prevention*.

pollution prevention   Device, process, or strategy used to prevent a potential pollutant from forming or entering the environment or to sharply reduce the amount entering the environment. Compare *pollution cleanup*.

polyculture   Form of crop farming in which various species of plants maturing at different times are planted together. Compare *monoculture*.

population   Group of individual organisms of the same species living in a particular area.

population density   Number of organisms in a particular population per unit of area or volume in which they are found.

potential energy   Energy stored in an object because of its position, the position of its parts, or its chemical content. Compare *kinetic energy*.

poverty   Condition in which people are unable to meet their basic needs for food, clothing, and shelter.

precautionary principle   When there is significant scientific uncertainty about potentially serious harm from chemicals or technologies, decision makers should act to prevent harm to humans and to the environment. See *pollution prevention*.

precipitation   Water in the form of rain, sleet, hail, and snow that falls from the atmosphere onto land and bodies of water.

predation   Interaction in which an organism of one species (the predator) captures and feeds on an organism of another species (the prey).

primary air pollutant   Chemical that has been added directly to the air by natural events or human activities and occurs in a harmful concentration. Compare *secondary air pollutant*.

principles of sustainability   See *scientific principles of sustainability* and *social science principles of sustainability*.

probability   Mathematical statement about how likely it is that something will happen.

producer   Organism that uses solar energy (green plants) or chemical energy (some bacteria) to manufacture the organic compounds it needs as nutrients from simple inorganic compounds obtained from its environment. Compare *consumer, decomposer*.

proven oil reserves   Identified deposits from which conventional crude oil can be extracted profitably at current prices with current technology.

R

rangeland Land that supplies forage or vegetation (grasses, grasslike plants, and shrubs) for grazing and browsing animals.

rate of extinction The estimated number of species or the percentage of known species that go extinct during a certain time period, typically a year. See *background extinction rate*.

recycle To collect and reprocess discarded materials so that they can be made into new products; one of the three Rs of resource use. An example is collecting aluminum cans, melting them down, and using that aluminum to make new cans or other aluminum products. Compare *reduce and reuse*.

reduce To consume less and live a simpler lifestyle; one of the three Rs of resource use. Compare *recycle and reuse*.

reliable surface runoff Surface runoff of water that generally can be counted on as a stable source of water from year to year.

renewable energy resource A source of energy that natural processes can replenish within seconds or decades. Examples are energy from the sun, wind, and flowing water.

renewable resources Resource that can be replenished rapidly (in hours to several decades) through natural processes as long as it is not used up faster than it is replaced. Examples include trees in forests, grasses in grasslands, wild animals, fresh surface water in lakes and streams, most groundwater, fresh air, and fertile soil. If such a resource is used faster than it is replenished, it can be depleted. Compare *nonrenewable resource*. See also *environmental degradation*.

reserves Resource deposits that have been identified and from which a usable mineral or fossil fuel can be extracted profitably at present prices with current mining or extraction technology.

reservoir Artificial lake created when a stream is dammed. See *dam*.

resource Anything obtained from the environment to meet human needs and wants. It can also be applied to other species. See natural *resource*, *nonrenewable resource*, and *renewable resource*.

response Reaction to exposure to a certain dose of a harmful substance or form of radiation; can include damage to one's health. See *dose*.

reuse To use a product over and over again in the same form. An example is collecting, washing, and refilling glass beverage bottles. One of the 3 Rs of resource use. Compare *reduce and recycle*.

risk Probability that something undesirable will result from deliberate or accidental exposure to a hazard. See *risk assessment*, *risk management*.

risk assessment Process of gathering data and making assumptions to estimate short- and long-term harmful effects on human health or the environment from exposure to hazards associated with the use of a particular product or technology.

risk management Use of risk assessment and other information to determine options for, and make decisions about, reducing or eliminating risks. See *risk assessment*.

rock Any solid material that makes up a natural, contiguous part of the earth's crust. See *mineral*.

rock cycle Largest and slowest of the earth's cycles, consisting of geologic, physical, and chemical processes that, over millions of years, form and modify rocks and soil in the earth's crust.

rural area Any populated area that is not classified as an urban area. Compare *urban area*.

S

salinization See *soil salinization*.

sanitary landfill Waste disposal site in which waste is spread in thin layers, compacted, and regularly covered with a fresh layer of clay or plastic foam. Compare *open dump*.

science Methodical effort to discover order in nature and use that knowledge to make projections about what is likely to happen in nature. See *scientific data*, *scientific hypothesis*, *scientific law (law of nature)*, *scientific theory*.

scientific data Factual information, including observations and measurements that scientists gather during the process of conducting an experiment.

scientific hypothesis An educated guess that attempts to explain a set of scientific observations. Compare *scientific law (law of nature)*, *scientific theory*.

scientific law (law of nature) Description of what scientists find happening in nature repeatedly in the same way, without known exception. See *first law of thermodynamics (law of conservation of energy)*, *law of conservation of matter*, *second law of thermodynamics*. Compare *scientific hypothesis*, *scientific theory*.

scientific model See *model*.

scientific principles of sustainability Principles by which life on the earth has sustained itself for billions of years through its reliance on solar energy, biodiversity, and nutrient cycling.

scientific theory A well-tested and widely accepted scientific hypothesis. Compare *scientific hypothesis*, *scientific law (law of nature)*.

secondary air pollutant The harmful chemicals that form in the air by reacting with primary air pollutants or with chemicals naturally found in the air. Compare *primary air pollutant*.

secondary ecological succession Ecological succession in an area in which natural vegetation has been removed or destroyed but the soil or bottom sediment has not been destroyed. See *ecological succession*.

second-growth forest A stand of trees resulting from secondary ecological succession; these forests develop after the trees in an area have been removed by human activities, such as clear-cutting for timber or cropland, or by natural forces such as fire, hurricanes, or volcanic eruptions.

second law of thermodynamics Whenever energy is converted from one form to another in a physical or chemical change, we end up with lower-quality or less usable energy than we started with. In any conversion of heat energy to useful work, some of the initial energy input is always degraded to lower-quality, more dispersed, less useful energy—usually low-temperature heat that flows into the environment. See *first law of thermodynamics (law of conservation of energy)*.

sedimentary rock Rock that forms from the accumulated products of erosion and in some cases from the compacted shells, skeletons, and other remains of dead organisms. Compare *igneous rock*, *metamorphic rock*. See *rock cycle*.

shale oil A heavy oil that must be extensively processed to be useful. Compare *oil shale*.

site and service Program in which a city government supplies a couple or family with a piece of land—the *site*—big enough for a small house, and the household members pay a low rent to the government in exchange for basic *services* such as roads, running water, streetlights, latrines, sewers, and other community essentials; used to alleviate poverty in city slums and shantytowns, mostly in less-developed countries.

smart growth Form of urban planning that recognizes that urban growth will occur but uses zoning laws and other tools to prevent sprawl, direct growth to certain areas, protect ecologically sensitive and important lands and waterways, and develop urban areas that are more environmentally sustainable and more enjoyable places to live.

smelting Process in which a desired metal is separated from the other elements in a mineral ore.

social science principles of sustainability Three principles that could help us to make a transition to more sustainable economies and societies: *full-cost pricing* (from economics), in which we find ways to include the harmful environmental and health costs of producing and using goods and services in their market prices; *win-win solutions* (from political science), in which we learn to work together to deal with environmental problems and find solutions that will benefit the largest possible number of people, as well as the earth itself; and *a responsibility to future generations* (from ethics) in which we accept our responsibility to pass on to future generations the earth's life-support systems in as healthy a condition as that which we have enjoyed.

soil Complex mixture of inorganic minerals (clay, silt, pebbles, and sand), decaying organic matter, water, air, and living organisms.

soil conservation Methods used to reduce soil erosion, prevent depletion of soil nutrients, and restore nutrients previously lost by erosion, overgrazing, and excessive crop harvesting.

soil degradation A condition that occurs when topsoil loses some of its ability to support plant growth.

soil erosion   Movement of soil components, especially topsoil, from one place to another, usually by wind, flowing water, or both. This natural process can be greatly accelerated by human activities that remove vegetation from soil. Compare *soil conservation*.

soil salinization   A form of soil degradation in which salts build up in the upper layers of soil as a result of repeated irrigation; can make soil unable to support plant growth.

solar cell   See *photovoltaic cell*.

solar thermal system   A type of active solar heating system that collects and concentrates direct solar energy to a temperature high enough to boil water and produce steam for generating electricity.

solid biomass   Wood, crop residues, and animal manure that provide about 95% of the energy used for heating and cooking in the world's poorest countries.

solid waste   Any unwanted or discarded material that is not a liquid or a gas. See *industrial solid waste*, *municipal solid waste (MSW)*.

specialist species   Species with a narrow ecological niche. They may be able to live in only one type of habitat, tolerate only a narrow range of climatic and other environmental conditions, or use only one type or a few types of food. Compare *generalist species*.

speciation   Formation, usually over thousands of years, of two species from one species when two or more different populations of a species remain separated for a very long time and evolve independently because they are exposed to different environmental conditions. Eventually, the genetic makeup of individuals from the separated populations of a sexually reproducing species may become so different that they cannot produce live, fertile offspring if they happen to come into contact with, and attempt to mate with, one another. Compare *extinction*.

species   Group of organisms sharing key distinguishing characteristics and, for sexually reproducing organisms, a set of individuals that can mate and produce fertile offspring. Every organism is a member of a certain species.

spoils   Unwanted rock and other waste materials produced when a material is removed from the earth's surface or subsurface by mining, dredging, quarrying, or excavation.

stratosphere   Second layer of the atmosphere, extending about 11–30 miles above the earth's surface. It contains a layer of gaseous ozone (O_3), which filters out about 95% of the incoming harmful ultraviolet radiation emitted by the sun. Compare *troposphere*. See *ozone layer*.

subsidies   Government payments or other forms of economic support designed to help an industry.

subsurface mining   Extraction of a metal ore or fuel resource such as coal from a deep underground deposit with the use of tunnels and shafts. Compare *surface mining*.

sulfur cycle   Cyclic movement of sulfur in various chemical forms within the environment and the earth's life-support systems.

surface mining   Removing large areas of soil, subsoil, and other layers of earth and then extracting a mineral deposit found fairly close to the earth's surface. See *area strip-mining*, *contour strip-mining*, *mountaintop removal mining*, *open-pit surface mining*. Compare *subsurface mining*.

surface runoff   Water flowing off the land into bodies of surface water. See *reliable surface runoff*.

surface water   Precipitation that does not infiltrate the ground or return to the atmosphere by evaporation or transpiration. Found in streams, rivers, lakes, and wetlands. Compare *groundwater*.

sustainability   Ability of earth's various systems, including human cultural systems and economies, to survive and adapt to changing environmental conditions indefinitely.

synthetic inorganic fertilizer   A fertilizer made of various combinations of nitrogen, phosphorus, potassium, and trace amounts of various minerals.

T

tar sands   Deposit of a mixture of clay, sand, water, and varying amounts of a tarlike heavy oil known as *bitumen*, which can be extracted from tar sand by heating, and then purified and upgraded to synthetic crude oil.

tectonic plates   Various-sized pieces of the earth's crust that move slowly around atop the mantle. Most earthquakes and volcanoes occur around the boundaries of these plates.

teratogen   Chemical, ionizing agent, or virus that causes birth defects. Compare *carcinogens, mutagen*.

terrestrial area   The land portions of the earth that are characterized by the types of soil in each area, the varieties of plants and animals that live in it, and the area's climate.

threatened species   A wild species that is still abundant in its natural range but is likely to become endangered because of a decline in numbers. Compare *endangered species*.

topsoil   A thin layer of rich organic matter forming the top layer of a body of soil.

total fertility rate (TFR)   Estimate of the average number of children that women in a given population will have during their childbearing years.

toxic chemical   Chemical that can cause harm to an organism. See *carcinogen, mutagen, teratogen*.

toxicity   Measure of the harmfulness of a substance.

toxicology   Study of the adverse effects of chemicals on health.

toxic waste   See *hazardous waste*.

traditional intensive agriculture   Production of enough food for a farm family's survival with a surplus that can be sold. This type of agriculture uses higher inputs of labor, fertilizer, and water than traditional subsistence agriculture. See *traditional subsistence agriculture*. Compare *industrialized agriculture (high-input agriculture)*.

traditional subsistence agriculture   Production of enough crops or livestock for a farm family's survival and, in good years, a surplus to sell or put aside for hard times. Compare *industrialized agriculture (high-input agriculture), traditional intensive agriculture*.

transmissible disease   Disease that is caused by living organisms (such as bacteria, viruses, and parasitic worms) and can spread from one person to another through air, water, food, or body fluids, or in some cases through insects or other organisms. Compare *nontransmissible, noninfectious disease*.

transpiration   Process in which water is absorbed by the root systems of plants, moves up through the plants, passes through pores in their leaves or other parts, and evaporates into the atmosphere as water vapor.

tree plantation (tree farm)   Site planted with one or only a few tree species. When the trees mature, they are usually harvested by clear-cutting and then replanted. Tree farms are normally used to raise rapidly growing tree species for firewood, timber, or pulpwood. Compare *old-growth forest, second-growth forest*.

troposphere   Innermost layer of the atmosphere. It contains about 75% of the mass of earth's air and extends about 11 miles above sea level. Compare *stratosphere*.

U

urban area   Geographic area containing a community with a population of 2,500 or more. The minimum number of people used in this definition varies among countries, from 2,500 to 50,000.

urban growth   Growth of an urban population.

urbanization   Creation or growth of urban areas, or cities, and their surrounding developed land. See *urban area*.

urban sprawl   Growth of low-density development on the edges of cities and towns. See *smart growth*.

V

virtual water   Water that is not directly consumed but is used to produce food and other products.

virus   Infectious agent that is smaller than a bacterium and works by invading a cell and taking over its genetic machinery to copy itself; it then multiplies and spreads throughout the body, causing a viral disease such as flu or AIDS.

W

waste management   Managing wastes to reduce their environmental harm without seriously trying to reduce the amount of waste produced. See *integrated waste management*. Compare *waste reduction*.

waste reduction   Reducing the amount of waste produced; wastes that are produced are viewed as potential resources that can be reused, recycled, or composted. See *integrated waste management*. Compare *waste management*.

waste-to-energy incinerator Incinerator that burns municipal solid waste (MSW) and uses the resulting heat to boil water and make steam to heat water or interior spaces, or to produce electricity.

water cycle Biogeochemical cycle that purifies and distributes the earth's fixed supply of water within the environment and the earth's life-support systems.

water footprint A rough measure of the volume of water used directly and indirectly to keep a person or group of people alive and to support their lifestyles.

waterlogging Saturation of soil with irrigation water or excessive precipitation so that the water table rises close to the surface.

water pollution Any physical or chemical change in surface water or groundwater that can harm living organisms or make water unfit for certain uses.

watershed (drainage basin) Land area that delivers water, sediment, and dissolved substances via small streams to a larger stream, river, or lake.

weather Short-term changes in the temperature, humidity, precipitation, cloud cover, wind direction and speed, and other conditions in the troposphere in a given place and time period. Compare *climate*.

wind A moving mass of air created by differences in solar heating between the earth's equator and its poles, and by the earth's rotation.

Note: Page numbers in **boldface** type indicate key terms. Page numbers followed by italicized *f* or *t* indicate figures and tables.

Hazard(s). *See* Environmental health hazards; Hazardous waste; Human health

Hazardous waste, **390**–91
cutting production of, 400
detoxifying, 401
disposal sites for, as pollution source, 310
electronic waste as, 390, 391, 418*f*
harmful chemicals in homes and, 329*f*, 390*f*, 426*f*
integrated management of, 400*f*
international agreements on, 391, 403–4
liquid, 402
radioactive, 164, 167, 170, 172, 390, 391*f*
regulation of, 403–4
storing, 402–3
in U.S., 389*f*, 401, 402, 403

Health. *See* Human health

Heat as kinetic energy, 10

Heating
of buildings, 120, 121*f*, 125*f*, 133–34, 406
geothermal, 124, 125*f*
high-temperature industrial, 119*f*, 147*f*
solar, 120, 121*f*
space, 119*f*, 147*f*
water, 134

Heat pump, ultra high-efficiency, 406*f*

Heavy oils, 150–51
advantages/disadvantages of, 168
problems caused by, 158–59

Herbicides, 90. *See also* Pesticides

Herbivorous species, 85

Hidden costs, 126, 385

Hidden water, 266. *See also* Virtual water

High-consumption, high-waste economies, 21, 384, 407

Higher-level consumers in ecosystems, 16*f*

High-grade ore, **182**–83

High-input agriculture, **87**. *See also* Industrialized agriculture

High-level radioactive waste, 164, 167, 170, 172

High-quality energy, **11**

High-speed rail, 70, 71*f*, 133*f*

High temperature, gas-cooled (HTGC) nuclear reactors, 171

High temperature industrial heat, net energy ratios for energy systems producing, 119*f*, 147*f*

High-waste economies, 21

HIPPCO, 212 (Habitat loss; Habitat degradation; Habitat fragmentation; Invasive species, Population growth; Pollution; Climate change; Overexploitation), **212**–17, 222

HIV (human immunodeficiency virus) and AIDS, 419*f*, 422, 431–32

Holdren, John, 42

Homes
air pollution in, 328, 329*f*, 340, 341*f*, 347
efficient air conditioning and heating of, 406*f*
environmental impact of energy use in, 29
harmful chemicals in, 329*f*, 390*f*, 425, 426*f*
reducing energy waste in, 134*f*
smart, 134
solar-heated, 120, 121*f*
space heating of, 121*f*, 124, 125*f*, 434*f*
water conservation in, 283–84

Honeybees, 98

Hoover Dam, 267*f*

Horizons, soil, 86*f*, 227, 228, 229*f*

Hormonally-active agents (HAAs), 424–25

Hormone(s)
human, 424
insect, 106–7

Hormone blockers, 424

Hormone mimics, 424

Host of parasites, 208

Houses. *See* Homes

Human(s)
environment and impact of (*see* Human environmental impact)
health of (*see* Human health)
population (*see* Human population; Human population growth)

Human capital, 383, 384*f*

Human environmental impact, 22*f*–23*f*
air pollution and, 328–40
on chemical cycles, 300–301
climate change and, 321, 322, 323*f*, 324*f*, 358–61 (*see also* Atmospheric warming; Climate change)
consumption, waste, and, 21
Earth's carrying capacity and, 44
economic growth, threats to natural capital, and waste production, and, 384–91
on ecosystems, 238–39
environmental degradation as, 22–23 (*see also* Environmental degradation)

flooding and, 74, 284–85
food production and, 29, 94–103
freshwater, depletion of, 268–78
home energy use and, 29
human ecological footprint and, 21, 24*f*, 32, 40–42 (*see also* Ecological footprint, human)
human population growth and, 40–44
individual action to reduce (*see* Individual action)
IPAT model of, 42–43*f*
land degradation as, 239–49
from mineral resource mining and use, 186–98
on natural cycles, 300–301
of nonrenewable energy sources, 156–65
in renewable energy, 126–30
resource use and, 29
species extinction and, 209–17
steps to reduce, 29
of transportation, 29
in urban areas, 59–69
water pollution as, 301–10
water resource depletion and degradation, 268–78

Human health
acid deposition effects on, 334
air pollution effects on, 64, 156–57, 161, 320, 335–36*f*
assessment of risks to, 418–19
asthma, 335, 336*f*
biological hazards to, 412, 413, 414, 419–24, 430–32 (*see also* Infectious disease)
chemical hazards to, 413, 415–17, 424–27, 432–34
climate change as threat to, 367–68
cultural hazards to, 413, 417–18, 427–29
food security, nutrition, and, 94–95
individual action on protection of, 438
lifestyle health hazards, 413, 417–18, 427–30
mercury and (*see* Mercury)
mining and threats to, 160
nanotechnology as potential hazard to, 198–99
ozone depletion, ultraviolet radiation, and, 339, 344*f*
pesticides and, 99, 415, 417*f*, 425
precautionary principle and pollution prevention applied to, 432–35